Developments in Agricultural Engineering 8

Mechanics of Agricultural Materials

OTHER TITLES IN THIS SERIES

1. Controlled Atmosphere Storage of Grains
by J. Shejbal (Editor)
1980 viii+608 pp.

2. Land and Stream Salinity
by J. W. Holmes and T. Talsma (Editors)
1981 iv+392 pp.

3. Vehicle Traction Mechanics
by R.N. Yong, E.A. Fattah and N. Skiadas
1084 xi+307 pp.

4. Grain Handling and Storage
by G. Boumans
1984 xiii+436 pp.

5. Controlled Atmosphere and Fumigation in Grain Storages
by B.E. Ripp et al. (Editors)
1984 xiv+798 pp.

6. Housing of Animals
by A. Maton, J. Daelemans and J. Lambrecht
1985 xii+458 pp.

7. Soil Cutting and Tillage
by E. McKyes
1985 vii+215 pp.

Developments in Agricultural Engineering 8

Mechanics of Agricultural Materials

GYÖRGY SITKEI

Department of Woodworking Machines, University of Forestry and Wood Science
Sopron, Hungary

ELSEVIER

Amsterdam · Oxford · New York · Tokyo 1986

This is the revised English version of
„*A mezőgazdasági anyagok mechanikája*"
published by Akadémiai Kiadó, Budapest

English translation by
S. Bars

The distribution of this book is being handled by the following publishers

for the USA and Canada
Elsevier Science Publishing Co., Inc.
52 Vanderbilt Avenue
New York, New York 10017, USA

for the East European countries, Democratic People's Republic of Korea,
People's Republic of Mongolia, Republic of Cuba, and Socialist Republic of Vietnam
Kultura Hungarian Foreign Trading Company
P. O. Box 149, H-1389 Budapest 62, Hungary

for all remaining areas
Elsevier Science Publishers
25 Sara Burgerhartstraat
P. O. Box 211, 1000 AE Amsterdam, The Netherlands

Library of Congress Cataloging-in-Publication Data

Sitkei, György.

Mechanics of agricultural materials.
(Developments in agricultural engineering; 8)
Translation of: A mezőgazdasági anyagok mechanikája.
Bibliography: p.
Includes index.
1. Agricultural engineering. 2. Materials.
3. Agricultural physics. I. Title. II. Series.
S727.S6513 1986 631.3 85-29292
ISBN 0-444-99523-4

ISBN 0-444-99523-4 (Vol. 8)
ISBN 0-444-41940-3 (Series)

Joint edition published by Elsevier Science Publishers, Amsterdam, The Netherlands and Akadémiai Kiadó, Budapest, Hungary

Printed in Hungary

CONTENTS

PREFACE

The importance of economical production of agricultural materials, especially crops and animal products serving as base materials for foodstuffs, and of their technological processing (mechanical operations, storage, handling, etc.) is ever-increasing. During technological processes agricultural materials may be exposed to various mechanical, thermal, electrical, optical and acoustical (e.g. ultrasonic) effects. To ensure optimal design of such processes, the interactions between biological materials and the physical effects, acting on them, as well as the general laws governing the same, must be known.

The behavior of most agricultural materials deviates essentially from that of the generally known elastic materials. The flow properties of cereals, of granular materials and of those materials which fall into the category of non-Newtonian liquids, also deviate essentially from the corresponding ideal liquid behavior. The work of agricultural engineers has been aggravated by these circumstances, and this is the main reason why agricultural engineering has relied for a long time on empirical data.

The mechanics of agricultural materials, as a scientific discipline, is still being developed at present, and in many cases has no exact methods as yet. However, the methods developed so far can already be utilized successfully for designing and optimizing machines and technological processes.

The present work is the first attempt to summarize the calculation methods developed in the main fields of agricultural mechanics, and to indicate the material laws involved, on the basis of a unified approach, with all relevant physicomechanical properties taken into account.

The author expresses sincere gratitude to his co-workers at the Agricultural College of Körmend, who have contributed through important results of their own research work to the general development of the subject. Special thanks are due to Professor J. Janik, Director, Dr. I. Bajsz, Professor of Agricultural Mechanics, and Dr. J. Fehér, Senior Assistant Lecturer.

I express my thanks also to Professor J. Galambos, who assisted by giving valuable counsel during review of the manuscript, and concerning the final construction of the book.

G. Sitkei

1. THE DEVELOPMENT AND IMPORTANCE OF AGRICULTURAL MECHANICS

Certain elements in the mechanics of agricultural products may be observed as far back as the turn of the century, and then in the period between the two World Wars when Jannsen [146], and later Gutjar [147], accounted for the mechanical behavior of bulk materials in their silo theory, whereby reliable rating of large silo storage structures became possible. The study of the laws of compaction of fibrous forage materials for the rating of baling presses, as well as the examination and analysis of cutting processes, also began in this period with the works of Goriachkin among others.

The range of these studies was extended in the period after 1945, and intensive research began all over the world. Professor Zheligovski initiated publication of the series "Zemledelcheskaia Mekhanika" ("Agricultural Mechanics"), whose volumes contained papers presented at annual or biannual conferences. In the early 1960s, Professor Macepuro in Minsk edited numerous volumes of the series "Voprosi Selskokhoziaistvennoi Mekhaniki" ("Problems of Agricultural Mechanics"), publishing detailed elaborations in specialized fields. Also, the study of agricultural products and the determination of material laws, and their application, gained ever-increasing importance at the annual conferences of the ASAE in the United States.

The subject of the mechanics of agricultural products has been extended continually since its beginning, not only to the material laws related to mechanical loading and its methods of application, but also to the study of other characteristic properties relevant to agricultural engineering. Thus, thermal, optical and electrical properties, and their application, now appear regularly in the thematic range of agricultural mechanics. Moisture content affects the mechanical and other (thermal, optical and electrical) properties of most agricultural materials decisively, and so investigations of the adsorption and desorption of water by these materials must also be included among the tasks of the research worker.

Agricultural engineering has relied for a long time on empirical data and knowledge. The multiplicity of agricultural products, their complex biological structure and the continuous variation of their properties and of the laws governing their behavior, involving very complicated interactions and relationships, prevented the elaboration of reliable theories and dimensioning methods for a long time. A certain amount of progress became possible when more systematic experimental results became available, of which thorough analysis and correlation permitted certain generalizations. This situation has not changed greatly even now: although a large amount of experimental data is already available, the majority either has not yet been evaluated suitably to permit any generalization, or this evaluation is still in progress.

One of the most important tasks in the further study of the mechanics of agricultural materials is the elaboration of calculation methods permitting the utilization of known material laws and physical properties in the design and simulation of machines and technological processes. Since the material laws involved are complex (they depend on numerous factors, and are nonlinear), the elaboration and application of calculation methods also presents a difficult task.

In most branches of agricultural production, mechanization has become general. However mechanization has unwelcome implications in several fields: losses at harvest may be high, and products may be damaged during mechanical operations, whereby their quality is reduced or they eventually become valueless. These losses are especially high during gathering and handling of fruits and vegetables. Mechanization is also significantly limited by the need to preserve the germination capacity in the case of seeds.

Reduction of losses due to damage, and preservation of the quality of products, are possible only by taking the relevant material laws and properties into account intentionally. Knowledge of the properties of agricultural materials permits the design of more modern machines and technological processes with improved work quality characteristics, involving lower losses and more efficient operation. In this respect the mechanics of agricultural materials is of great importance, and this explains the fast development of this young discipline. It may be foreseen with certainty that the mechanics of agricultural materials will become the basic subject of agricultural engineering.

2. PHYSICAL PROPERTIES OF AGRICULTURAL MATERIALS

2.1 Shape and size

The functioning of many types of agricultural machines (e.g., sifters, sowing machines, pneumatic transport systems, etc.) is influenced decisively by the shape and size of the objects participating, and so in order to study a given process they must be described accurately.

In certain cases a process may depend not only on the shape (*a*) and size (*b*), but for example, also, on the orientation (*c*), the extent of compaction (*d*), etc. of the product. In this case, the process relationship may be written in the general form

$$F = f(a, b, c, d, \dots) \tag{1}$$

As an example, the case of determining how many fruits of a given size can be packed in a container may be mentioned. If the variables appearing in eqn. (1) are denoted by $x_1, x_2, \dots, x_n$, it may be written in polynomial regression form as

$$F = b_1 x_1 + b_2 x_2 + b_3 x_3 + \dots + b_n x_n \tag{2}$$

To evaluate eqn. (2), the effect of each variable x on F must be determined, by measuring specimens sampled from the actual material, by means of multiple-correlation and variance analysis [1].

The shapes of the various seeds, fruits and plants are generally irregular, and so a very great number of measurement data would be needed to describe them accurately. However, practical measurements show that the various shapes may generally be characterized well by specifying purposely selected orthogonal axes: for example, seeds are usually characterized by their length, width and thickness. In certain cases, even a single linear dimension (e.g., chaff length) suffices to characterize a product.

The dimensions of agricultural products are not uniform, but scatter around a mean value. Therefore, it is necessary to determine the distribution of individual sizes and the mean size on the basis of this distribution. The quality of processing (e.g., in chopping and milling) may be characterized by a product's mean size and mean standard deviation σ, or these data may be used to organize

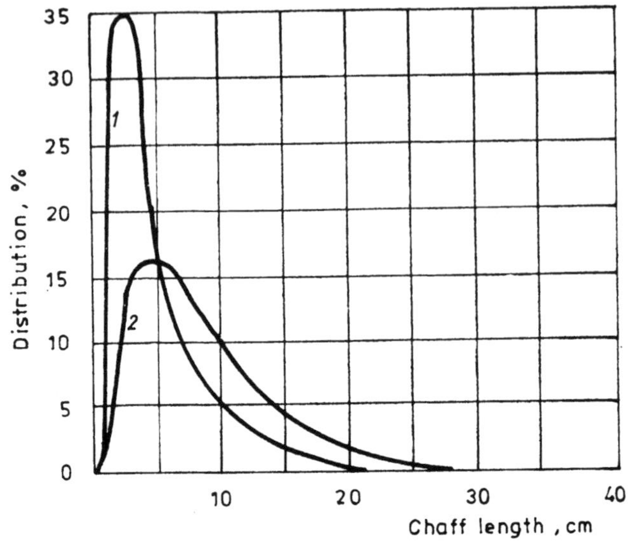

Fig. 1. Size distribution of chopped forage for two different cutter heads. *(1)* Cylinder; *(2)* flywheel

a technological process or in designing certain structural elements (e.g., mesh dimensions of sifters for the calibration of seeds, or dimensions of screen holes).

The distribution of individual dimensions is generally presented in the form of a distribution diagram as used in earlier studies (Fig. 1), or using the recently introduced probability scale (Fig. 2). The advantage of the latter method is that in the case of a normal distribution a straight line is obtained, and the mean value, as well as the standard deviation σ, may be read off simply.

Generally, as yet there exists no appropriate method for describing exactly the shapes of agricultural products. The shape of certain plants and products may be compared, on the basis of their longitudinal dimensions and cross-sections, to charted standard forms. Such standard shapes have been established for apples, peaches, potatoes, etc. [1].

Visual comparison of the shape of a given product with a standard shape is very simple but is not exempt from error, owing to its reliance on the subjective judgment of the observer. Therefore, in cases where a technological process is influenced significantly by shape, it is advisable to use objective measuring indexes.

The shape of a product affects its packing coefficient in a container. Figure 3 shows the possible ways of accommodating single items relative to one another. The packing coefficient is defined by the ratio of the volume V_{fr} of material packed to the total volume V_0, or

$$\lambda = V_{fr}/V_0$$

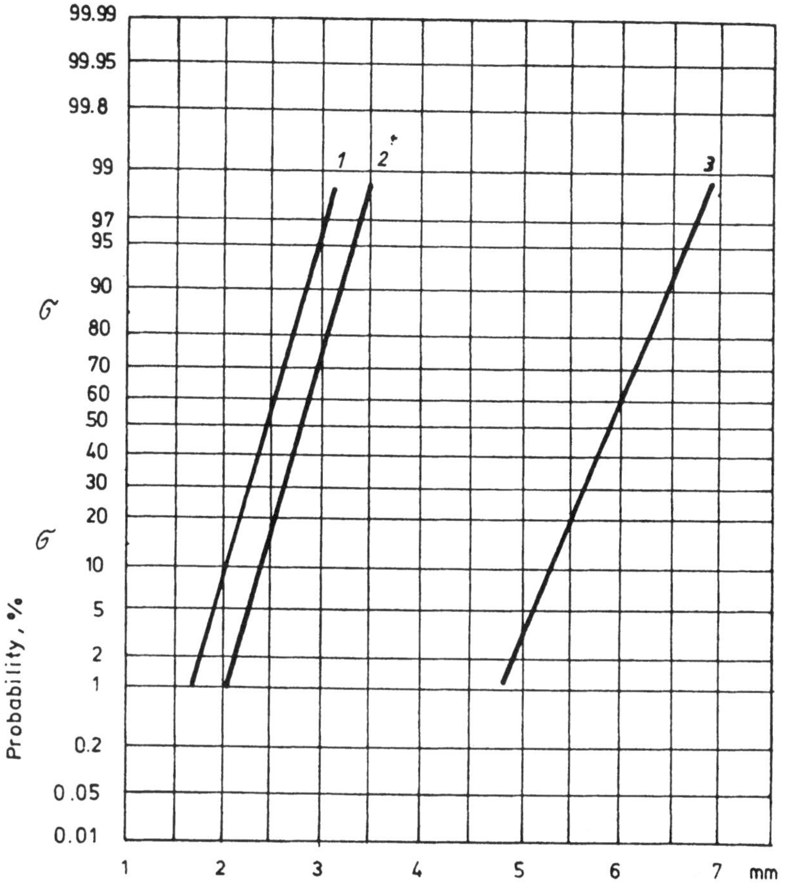

Fig. 2. Size distribution of wheat grains on a probability scale. *(1)* Thickness; *(2)* width; *(3)* length

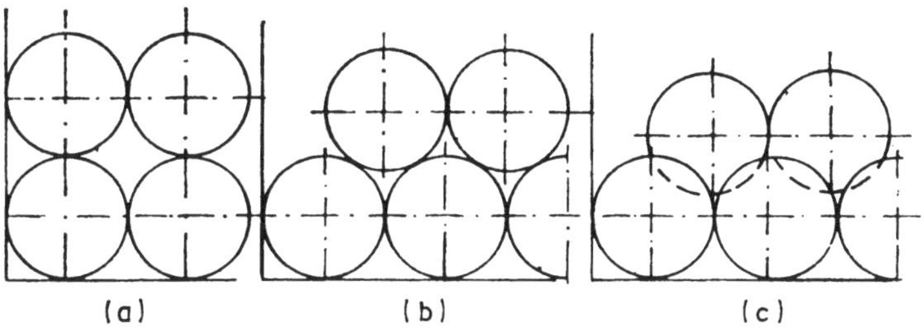

Fig. 3. Packing of fruits in a container

The theoretical value of the packing coefficient may be calculated easily if a spherical shape is assumed. For the three cases appearing in Fig. 3, the following values are obtained: $\lambda_1=\pi/6=0.5236$; $\lambda_2=\pi/3\sqrt{3}=0.6046$ and $\lambda_3=\pi\sqrt{2}/6=$ $=0.7405$. In practical cases, the packing coefficient is influenced by the deviation of the shapes of single items from spherical, by deformation and by less than ideal packing at lateral walls.

The objective measuring indexes used to characterize a given shape may differ, depending on the nature of the task. Such indexes may be the roundness, the roundness ratio, the sphericity, the axial ratio, the degree of inequality of projected areas, etc.

Roundness may be defined in several ways (Fig. 4). According to one definition,

$$\text{roundness} = F_m/F_c$$

where F_m is the largest projected area of the object, and F_c the area of the circumscribing circle. According to another definition,

$$\text{mean roundness} = \sum r/nR$$

where r is the radius of curvature, R the radius of the inscribed circle, and n the number of corners.

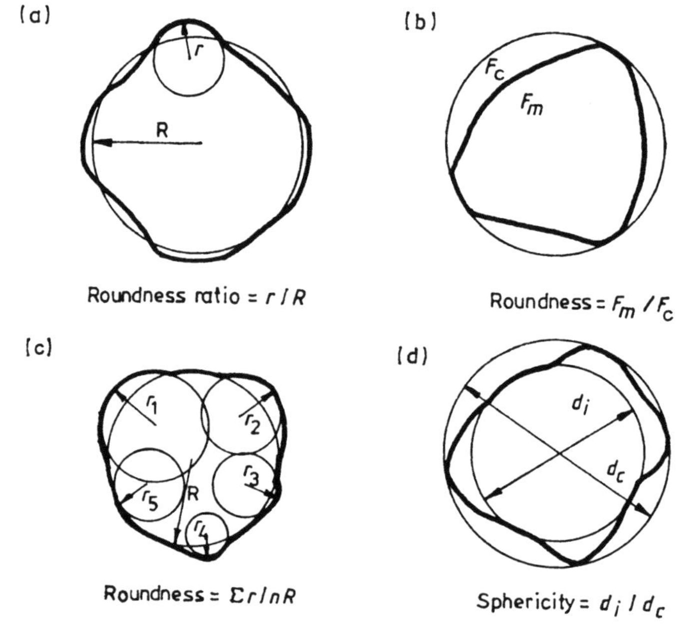

Fig. 4. Determination of roundness and sphericity

The roundness ratio, also illustrated in Fig. 4, is the ratio of the radius of the smallest corner, to the mean radius of the object. The axial ratio, from the point of view of technological processes is the ratio of the shorter axis to the longer.

Sphericity may be defined by the equation

$$\text{sphericity} = d_i/d_c$$

where d_i is the diameter of a sphere whose volume is identical to that of the object, and d_c the diameter of the circumscribing sphere. If it is assumed that the volume of the object equals that of the corresponding triaxial ellipsoid, and that the diameter of the circumscribing sphere equals the longest axis of the ellipsoid, the following expression is obtained:

$$\text{sphericity} = [(\pi/6 \cdot abc)/\pi/6 \cdot c^3]^{1/3} = (abc)^{1/3}/c$$

where a, b, and c are the lengths of the axes of the ellipsoid. Accordingly, the sphericity is the ratio of the mean geometrical diameter to the longest diameter. According to another definition,

$$\text{sphericity} = d_i/d_c$$

where d_i is the diameter of the greatest inscribed circle, and d_c that of the smallest circumscribed circle (Fig. 4).

The equivalent diameter of irregularly shaped bodies is given by the diameter of a sphere of identical volume:

$$d_e = \sqrt[3]{6G/\gamma\pi}$$

where G is the mass (weight) and γ the density (volumetric weight) of the body. In designing a sizing machine, the mean projected cross-sectional area of the product to be sized determined for any arbitrary position of the product according to Fig. 5, will be of importance [2]. The mean projected area obtained in this way is related to the volume of the body according to the equation

$$F_m = KV^{2/3} \tag{3}$$

where $K = 1.21$ for a sphere and is greater for other convex bodies. Figure 6 shows the relationships between F_m and V for carrots, potatoes and lemons [2]. Note that sphericity may also be characterized by the value of K. The closer K is to 1.21, the more spherical the shape of the body.

The theory of distributions. The sizes of individual particles in bulk products (seeds, ground materials, chaff) met with in agriculture may be conceived as random quantities, and their distribution may be described by the equations of mathematical statistics, or by empirical relationships.

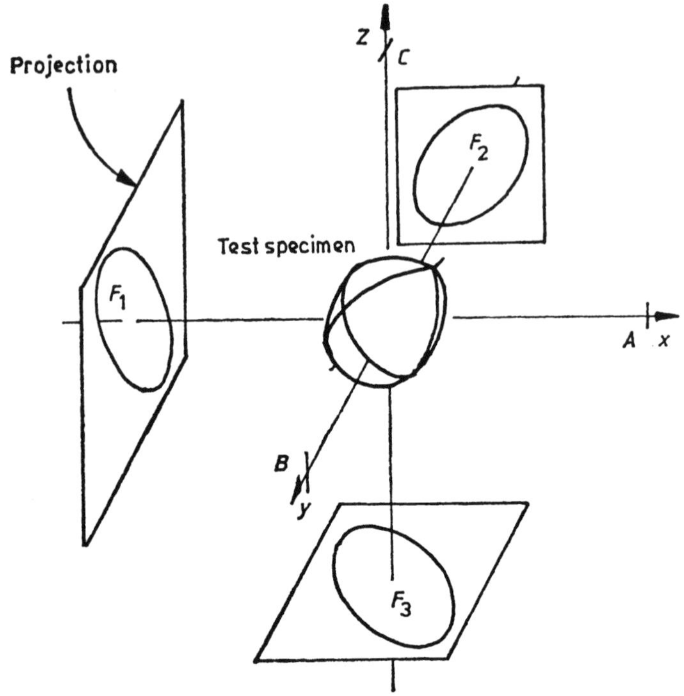

Fig. 5. Determination of mean projected cross-sectional area of agricultural products

The size distribution of some bulk products met with in practice corresponds to the Gaussian normal distribution (Fig. 7). The normal distribution function can be treated mathematically very easily, and so it plays an important role in theoretical studies. The curve of the normal distribution is symmetrical, and may be described by the equation

$$y = y_0 e^{-(x-\bar{x})^2/2\sigma^2}$$

where $\bar{x}$ is the mean value, and σ the standard deviation. The area under the curve is

$$\int_{-\infty}^{+\infty} y_0 e^{-(x-\bar{x})^2/2\sigma^2} dx = y_0 \sigma \sqrt{2\pi}$$

Therefore, the equation of the curve with unit area under it is

$$y = (1/\sigma\sqrt{2\pi}) e^{-(x-\bar{x})^2/2\sigma^2} \tag{4}$$

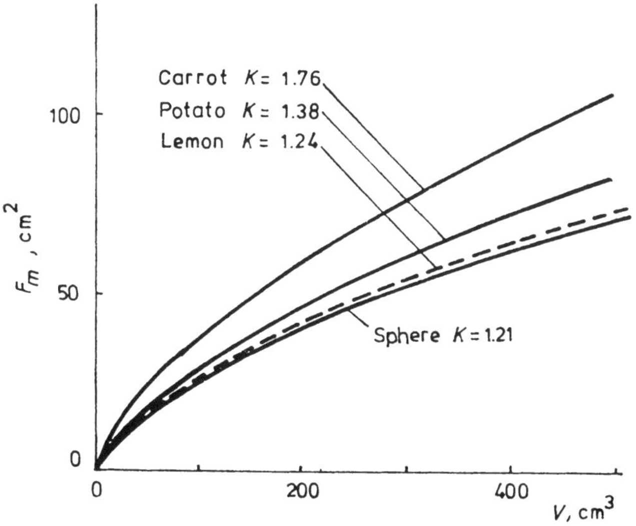

Fig. 6. Relationship between mean projected cross-sectional area and volume

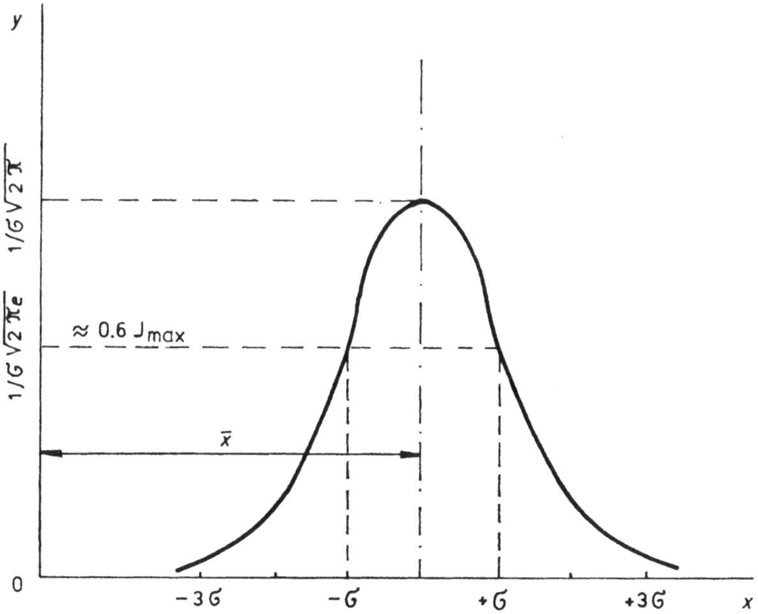

Fig. 7. Gaussian normal distribution

The standard deviation σ characterizing the distribution may be calculated using the equation

$$\sigma^2(x) = \int_{-\infty}^{+\infty} x^2(1/\sigma\sqrt{2\pi})e^{-(x-\bar{x})^2/2\sigma^2}\,dx - \bar{x}^2$$

The curve is symmetrical about the arithmetic mean, the probabilities of a value deviating from the value $\bar{x}$ downwards or upwards are identical. The ordinate of the curve is greatest at the point $\bar{x}$ and its value is

$$y_{\max} = 1/\sigma\sqrt{2\pi}$$

The curve has two inflection points, whose abscissa values are

$$x_{1,2} = \bar{x} \pm \sigma(x)$$

If the center of the system of coordinates is chosen to coincide with the arithmetic mean, then the equation of the curve becomes simpler:

$$y = (1/\sigma\sqrt{2\pi})e^{-x^2/2\sigma^2}$$

In this latter case, the probability of a random quantity x_i falling in the interval $a-b$ is

$$P(a < x_i < b) = \int_a^b (1/\sigma\sqrt{2\pi})e^{-x^2/2\sigma^2}\,dx \tag{5}$$

The integral of the term $e^{-(x^2/2\sigma^2)}$ cannot be expressed in terms of elementary functions; therefore tables have been compiled for calculating values of integrals of the type

$$\Phi(z) = (1/\sqrt{2\pi})\int_0^z e^{-z^2/2}dz$$

By introducing the substitutions $x/\sigma = z$ and $dx/\sigma = dz$, it becomes possible to write eqn. (5) in the form

$$P(a < x < b) = \int_{a/\sigma}^{b/\sigma} (1/\sqrt{2\pi})e^{-z^2/2}dz = \Phi(b/\sigma) - \Phi(a/\sigma) \tag{5a}$$

Practice proves that the set of random quantities determined by natural processes shows a normal distribution in most cases. For example, the size, tensile and compressile strength, shear resistance, etc., show such a distribution for many products (seeds, root bulbs, fruits).

At the same time, the set of random quantities created by artificial action can be described in most cases only by log–normal or other empirical equations. Such sets are encountered, for example, for ground products, chopped forages, etc.

The log-normal relationship describes an asymmetrical distribution, giving the normal distribution if a logarithmic dimensional coordinate (x-coordinate) is applied. Accordingly, the size distribution in a bulk product may be written as

$$N(x) = (N/\sqrt{2\pi}\,\sigma x) \int_0^x e^{-[\ln(x/\bar{x})/\sigma]^2/2}\,dx$$

where N is the number of particles in the aggregate considered. Frequently, it is inconvenient in practice to determine the number of particles in individual fractions, whereas their weight can be obtained easily. This is the case for screen analysis. Thus in such cases it is advisable to transform the above equation so that it expresses the distribution by weight. The weight of an assumedly cubic particle characterized by a given dimension x is

$$s(x) = x^3 \gamma n(x)$$

and the total weight is

$$S = \int_0^\infty s(x)\,dx$$

or

$$S = (\gamma N/\sqrt{2\pi}\,\sigma) \int_0^\infty x^2 e^{-[\ln(x/\bar{x})/\sigma]^2/2}\,dx$$

whose integration gives

$$S = N\gamma\bar{x}^3 e^{9\sigma^2/2} \tag{6}$$

The proportion by weight of particles falling into the interval between x and $x+\Delta x$ may be expressed using the density distribution function by weight, in the form $q(x)\Delta x$, where the term $q(x)$ is

$$q(x) = s(x)/S = (x^2/\sqrt{2\pi}\sigma\bar{x}^3 \cdot e^{9\sigma^2/2}) \int_0^\infty x^2 e^{-[\ln(x/\bar{x})/\sigma]^2/2}\,dx$$

Integration of this equation gives the distribution function by weight:

$$Q(x) = (1/\sqrt{2\pi}\sigma\bar{x}^3 e^{9\sigma^2/2}) \int_x^\infty x^2 e^{-[\ln(x/\bar{x})/\sigma]^2/2}\,dx \tag{7}$$

On introducing the notation

$$z = \ln(x/\bar{x})/\sigma - 3\sigma$$

eqn. (7) reduces to the simpler form

$$Q(x) = (1/\sqrt{2\pi}) \int_z^\infty e^{-z^2/2}\,dz \tag{7a}$$

whose values may be found in integral tables. The parameters $\bar{x}$ and σ characterizing the distribution may be determined by the method of moments. The expression for the first order moments is

$$M_1 = \int_0^\infty xq(x)\,dx = \sum_{i=1}^{k} x_i p_i$$

and that for the second-order moments is

$$M_2 = \int_0^\infty x^2 q(x)\,dx = \sum_{i=1}^{k} x_i^2 p_i$$

where p_i is the relative quantity of material retained on the ith screen.

Accordingly,

$$\bar{x} = M_1^8/(\sqrt{M_2})^7$$

and

$$\sigma = \sqrt{\ln (M_2/M_1^2)}$$

Plots of experimental data on Gaussian paper have shown in numerous cases that the distribution curves are not straight, but are inflected at the ends. The explanation is that the theoretical distribution is valid between size limits of zero and infinity while in reality grains bigger than the mesh size of the screen, or the original grain size, do not occur. The resulting distortion of the log–normal distribution may be corrected by a renorming procedure. This means that the upper size limit of the log–normal density function is not infinity, but the greatest grain size occurring (x_{max}). Thus the distribution in a real pile of grain may be derived from a theoretical log–normal granular pile which has fallen through a certain screen mesh size.

The weight distribution function $Q^*(x)$ for a granular pile cut off at x_{max} is related to the ideal function $Q(x)$ as follows:

$$Q^*(x) = Q(x)/Q(x_{max}) = (1/\lambda)Q(x) \tag{8}$$

i.e., the weight distribution in a cut-off granular pile is obtained by multiplying the ideal weight distribution by $1/\lambda$. This means that the distribution in a real pile has a constant $\times$ log–normal form. The value of the neglected weight-percentage is $100(1-\lambda)$. On Gaussian scale the function $Q^*(x)$ will be not a straight line but a monotonically increasing curve which is convex on the under-side, approximating the straight line $x=x_{max}$ asymptotically. In practice, λ is determined by multiplying the given measured data [$Q^*(x_1)$, $Q^*(x_2)$, $Q^*(x_3)$, which may be screen-residue or screen-pass data] by various values $0<\lambda<1$ in such a way that the points $\lambda Q^*(x)$ fall on a straight line in the plot.

The size distribution of ground agricultural products is frequently described by the Rosin–Rammler–Bennet law, as expressed by the equation

$$D(x) = 1 - e^{-(x/x_0)^n} \tag{9}$$

where $D(x)$ is the cumulative weight of fines beneath size x, x_0 the characteristic grain size, at which the sieve residue is 36.8% ($1/e$), and n an exponent characterizing the uniformity of the pile, with a value varying between 0.5 and 1.3.

One of the simplest distribution functions is the equation due to Schumann

$$D(x) = 100(x/K)^n \tag{10}$$

where K is the size modulus (the characteristic size, relative to which all grains are theoretically smaller), n is the distribution modulus.

2.2 Surface area

The surface area of certain parts of agricultural materials plays an important role in various technological processes. The surface area of a leaf is characteristic of its capacity for photosynthesis and of its growth rate, and plays a role in determining plant–soil–water relations, in the application of plant-protecting chemicals, and, in the case of certain plants (e.g., the tobacco) characterizes the quantity and quality of the crop. The surface area of crops and fruits is also of importance in certain tests, for example, in the respiration measurements, in determinations of reflectance and color, in heat-transfer phenomena, etc.

Various methods are used for measuring the surface area of leaves. An image of a leaf may be printed by the contact method onto light-sensitive paper, and the surface obtained in this way then planimetered. An essentially identical method is to trace the leaf contour onto millimeter paper and planimeter the resulting surface. A faster method is to measure the length and width of the leaf and read its surface area from a plot giving the area as a function of the product of the two dimensions. Such an empirical relationship for tobacco leaves is shown in Fig. 8. The method obviously requires similarity of shape of leaves of differing sizes, and its accuracy thus depends on the extent of similarity.

The above method is less accurate than those based on direct measurements. The surface area of leaves may be measured rapidly and accurately using an air-flow planimeter (Fig. 9) [7]. The essential components of the instrument are a constant-speed rotary pump and two perforated plates (100 holes per cm^2) mounted on the top of a suction funnel. One of the perforated plates may be closed and opened arbitrarily by a plane slide. The measurement is performed in

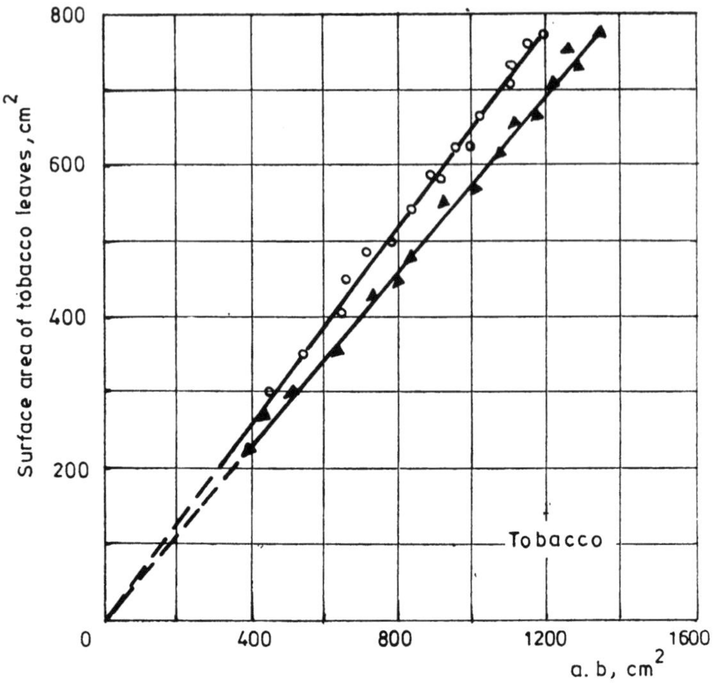

Fig. 8. Surface area of tobacco leaves as a function of the product of length and width, for two different varieties

the following way. First, the measuring grid is closed completely and the pressure drop measured. Then the leaf is placed on the other perforated plate and the closed plate is opened until the pressure drop is identical to the preceding value. The surface area of the leaf equals the open surface area of the measuring grid, which can be read directly on the closing slide.

The surface areas of fruits may be determined by direct measurement, by calculation, or from plots based on an easily measurable linear dimension (some diameter), on a quadratic dimension (some cross-sectional area), or on weight. Direct measurements may be performed by cutting a fruit into narrow strips and summing the surface areas of the individual strips. This method is slow, and is recommended only for finding relationships. If the shape of a fruit is sufficiently similar to a rotational ellipsoid (e.g., as for certain kinds of plum), then the surface area is given approximately by

$$f = ac\pi$$

where a and c are the lengths of the orthogonal minor and major axes.

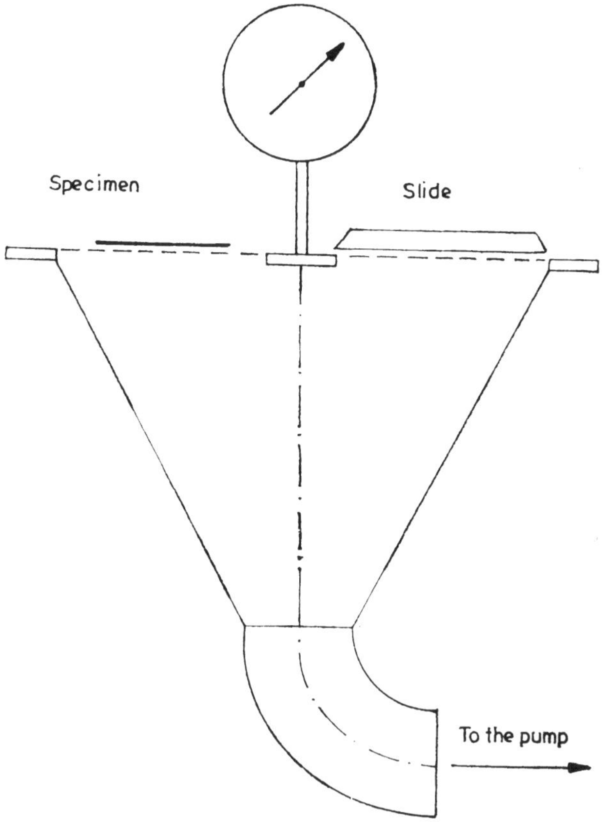

Fig. 9. Schematic diagram of air-flow planimeter

The surface areas of fruits are determined most frequently on the basis of their measured diameter or weight. Knowing the diameter or weight of a fruit, its surface area may be calculated using empirical equations, or read from an appropriate plot (Fig. 10) [8].

In the case of spherical bodies, the relationship between surface area and weight is

$$f = \left(4.836/\gamma^{2/3}\right) G^{2/3}$$

where γ is the volumetric weight.

The surface area of a fresh egg may be determined using the equation

$$f = kG^{2/3} \ (\mathrm{cm}^2)$$

where the value of k varies between 4.6 and 5.0 (for units of cm). For calculating the surface area of potatoes, $k = 4.76$ may be used.

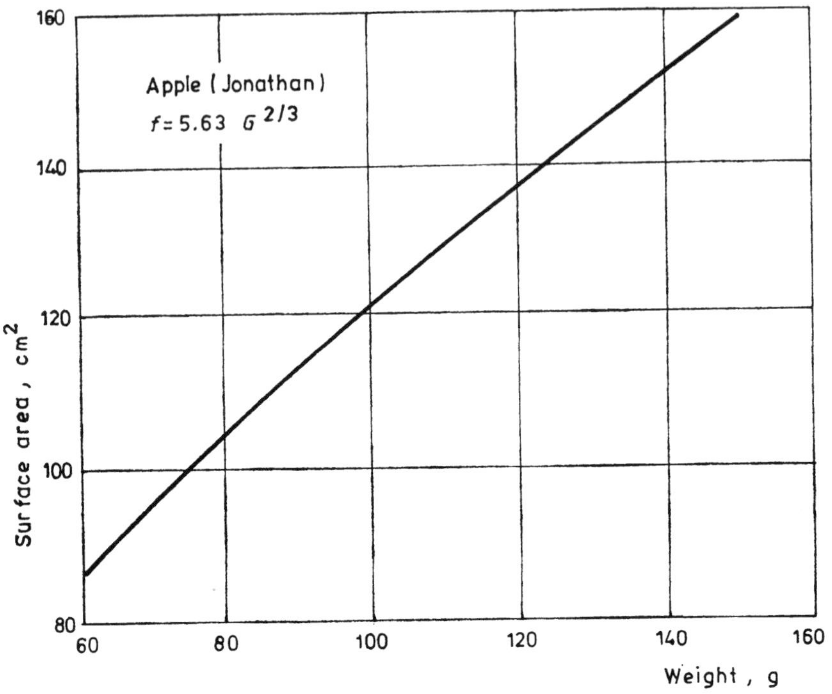

Fig. 10. Relationship between surface area and weight of a product

2.3 Volume and density

The volume and density of various agricultural products play an important role in numerous technological processes and in the evaluation of product quality. Examples which may be mentioned are the drying and storage of hays, the design of silos and other storage structures, the stability of feed pellets, the sorting of impurities (e.g., stones), establishing the ripeness of fruits, and characterizing the quality of crops (e.g., wheat), in all of which the density or specific weight appears as a decisive factor in some form.

The volume and density of larger objects such as fruits are measured by the water displacement method (Fig. 11). The fruit is first weighed in air, then submerged in water, and the weight of water displaced is weighed. The volume of the fruit is

$$V = G_w/\gamma_w$$

where G_w is the weight of water displaced. The specific weight of the fruit is

$$\gamma_{fr} = G_{fr}\gamma_w/G_w$$

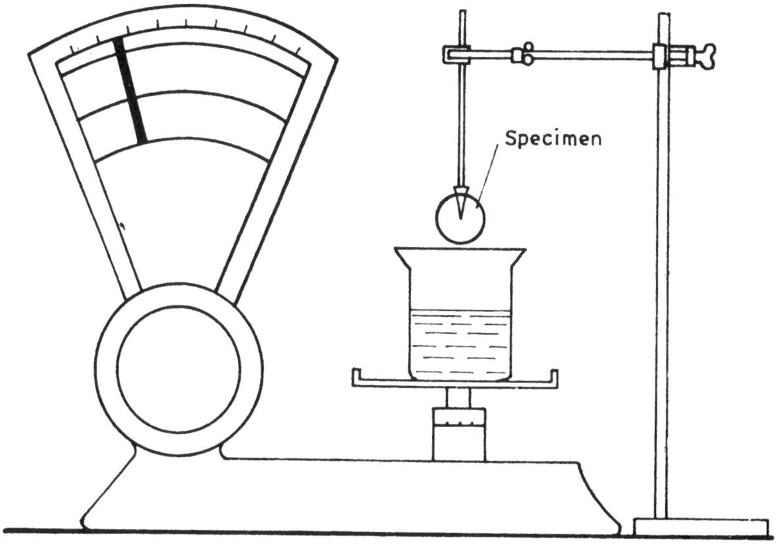

Fig. 11. Volume measurement

where G_{fr} is the weight of the fruit in air. The volume and specific weight of smaller seeds are measured by the generally known pycnometer method. However, it is advisable to use toluene ($C_6H_5CH_3$) instead of water, since it is absorbed by seeds to a lesser extent, and its surface tension is low, so that it fills even shallow dips in a seed; furthermore, its dissolution power is low.

Recently, commercially available air-comparison pycnometers have been used widely (e.g., that of Beckman Instruments, Inc.). These instruments consist essentially of two chambers, two pistons, a valve connecting the two chambers, and a differential manometer. The volume of a chamber is reduced by placing a product into it, and so the pressure of a given volume of air introduced into the chamber will be higher. The volume of a product placed into the instrument chamber can be read off directly. The instrument indicates the real volume of the sample: if the external (apparent) volume of a porous material is to be determined, then the test specimen is covered by a thin wax layer, to prevent the penetration of air into the pores.

An important characteristic of porous and thus of granular bulk materials is their porosity, which is the ratio of the volume of the cavities found within a pile to its total volume. Porosity plays an important role in drying and ventilation processes, since the air resistance of a bulk layer and the corresponding movement of air depend greatly on the porosity. Porosity may also be measured using an air-comparison pycnometer (Fig. 12). The operation of the pycnometer as developed by Mester (see in ref. [4]) is as follows. The sample volume (usually 300 cm^3)

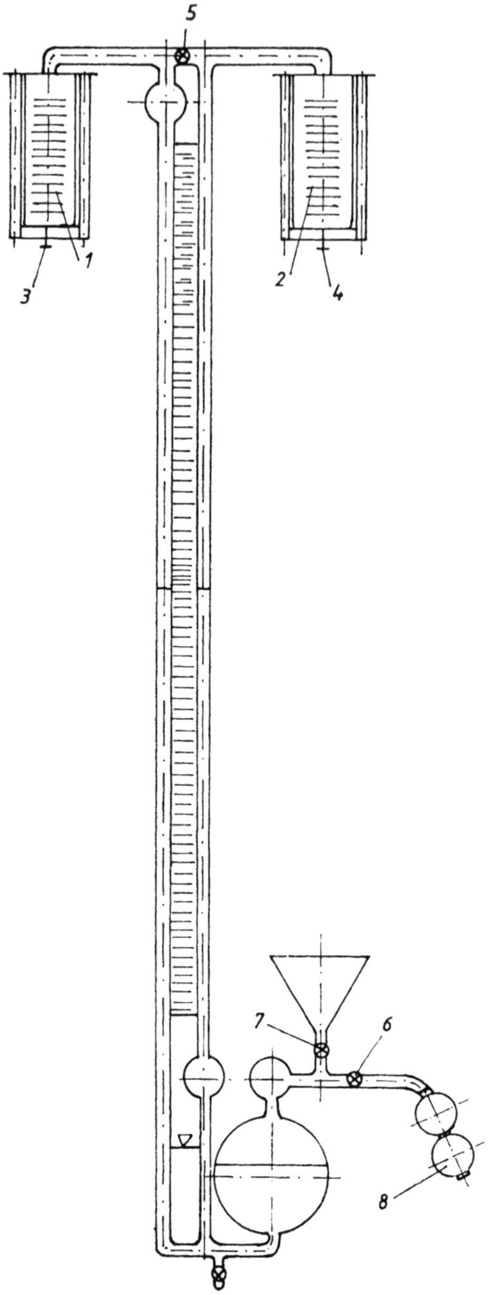

Fig. 12. Principles of construction of an air-comparison pycnometer. *(1, 2)* Scales; *(3, 4)* chambers; *(5, 6, 7)* valves; *(8)* rubber ball

of material to be measured is placed into the left-hand measuring chamber, and the water level adjusted to a given mark with both valves *5* and *6* open. Then valve *5* is closed and water is pumped into the measuring tubes until the two rising liquid levels attain the same height. Then valve *6* is closed and the value is read off the scale, calibrated in units of cm^3. The instrument compares the compressibility of the air volume reduced by the volume of the grains, with the compressibility of the air volume without the grains. Each of the measuring tubes contains an auxiliary chamber (a sphere), one at the top, the other at the base. When the water is pumped upwards, the lower auxiliary chamber is filled with water, pushing the air into the tube above it and then into the upper expansion space (*2*), while the auxiliary chamber on the side of the measuring cylinder is filled with air. The sensitivity of the instrument is optimal when the true volume to be measured approximately equals the volume of the auxiliary chamber. The accuracy of the reading is ± 0.1 cm^3 in this case.

Porosity may be expressed in terms of the specific weight γ and volumetric weight γ_{vw} of a material, as

$$\varepsilon = 1 - (\gamma_{vw}/\gamma)$$

The volumetric weight of granular materials depends on the shape of

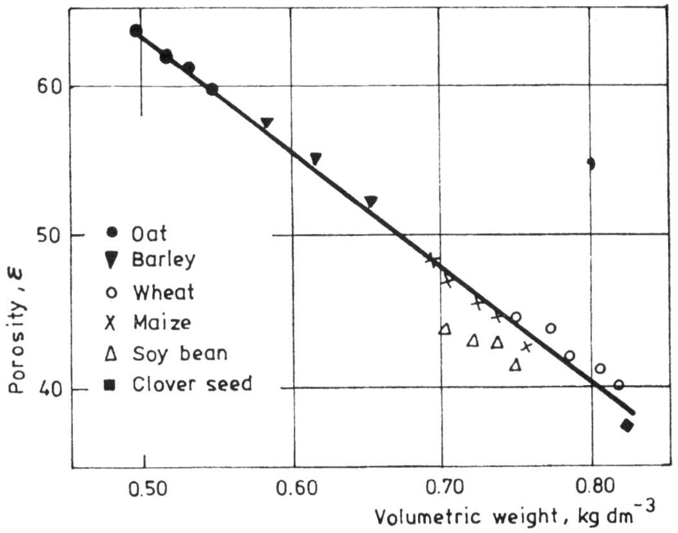

Fig. 13. Porosity of bulk products as a function of volumetric weight

the grains. Maize grains may have very different shapes, and so the volumetric weight and porosity of individual fractions may differ: large, flat maize grains have the maximum, and small, round ones the minimum volumetric weight [38]. The porosity generally varies linearly as a function of volumetric weight. The position of the straight line is influenced by the specific weight of the grains. The differences in specific weight among various agricultural products are not too great; thus the porosity–volumetric weight relationship may be expressed, to a certain approximation, by the same curve for different crops. This relationship is shown in Fig. 13 [3].

2.4 Thermal and hygroscopic expansion

Agricultural products vary in size, as do metallic materials, as a result of temperature variations: with a rise in temperature they expand, and as the temperature drops they contract. Agricultural products always contain water, the quantity of which may vary owing to external circumstances (e.g., the relative humidity of the ambient air) or as the result of artificial treatments (drying). With change of moisture content the volume of many products varies, i.e., they swell or shrink.

The importance of thermal and hygroscopic expansion phenomena is due mainly to the development of heat and moisture stresses consequent upon the formation of temperature and moisture gradients in a material. These stresses may cause a product to crack (stress cracks), whereby its quality and price are reduced.

If the expansion coefficient differs in the main directions in an anisotropic material, stress will appear during variation of temperature and moisture content, even when no temperature or moisture gradients develop (e.g., during slow drying of wood).

For characterizing expansion properties, both linear and volumetric expansion coefficients are used. The relative length variation of a material caused by unit temperature variation is

$$\Delta L = \alpha L_0$$

and the relative variation of its volume is

$$\Delta V = \alpha_v V_0$$

where α and α_v are the linear and volumetric heat expansion coefficients, respectively, and L_0 and V_0 the initial length and volume, respectively. The variation of length and volume caused by moisture variation may be expressed similarly:

$$\Delta L = \beta L_0$$

$$\Delta V = \beta_v V_0$$

where β and β_v are the linear and volumetric coefficients, respectively, of hygroscopic expansion. If the linear expansion coefficients are identical in the three main orthogonal directions, then $\alpha_v \cong 3\alpha$ and $\beta_v \cong 3\beta$ approximately, and in the contrary case, $\alpha_v \cong \alpha_1 + \alpha_2 + \alpha_3$ and $\beta_v \cong \beta_1 + \beta_2 + \beta_3$. Relatively few data concerning thermal and hygroscopic expansion can be found in the literature. The existing data for wheat, maize and rice grains are summarized below.

The coefficient of volumetric thermal expansion for maize kernels can be obtained from the empirical expressions [41]

$$\alpha_v = (19.76 + 0.254U) \times 10^{-5}$$

for the temperature range 18–43 °C and

$$\alpha_v = (22.52 + 0.515U) \times 10^{-5}$$

for the temperature range 43–74 °C, where U is the percentage moisture content (wet basis). Prasad [88] studied the thermal expansion coefficient of brown rice in the temperature range 30–70 °C, and found the relationship:

$$\alpha_v = (9.36 + 1.097U^2 + 0.0329U^3) \times 10^{-6}\ °\mathrm{C}^{-1}$$

For the linear thermal expansion coefficient of the maize, a value of $\alpha = 3.4 \times 10^{-5}$

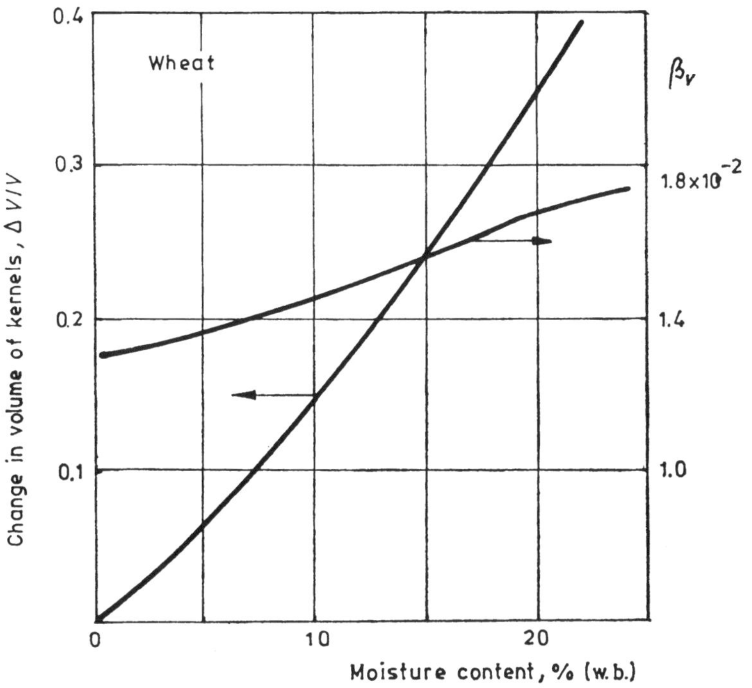

Fig. 14. Change in volume of wheat kernels as a function of moisture content, and coefficient of volumetric moisture expansion

has been found [41]. The coefficient of volumetric moisture expansion for brown rice was also investigated by Prasad [88] and the following relationship was obtained for the temperature range 0–70 °C:

$$\beta_v = 0.0106 + 5.9 \times 10^{-5}\,\vartheta, \quad (U(\%))^{-1}$$

where ϑ is temperature (°C).

Figure 14 shows the relative volume change of wheat grains as a function of moisture content [23]. From the curve it is possible to determine the volumetric expansion coefficients for moisture: the values are somewhat higher than those obtained for rice.

Comparison of the thermal and hygroscopic expansion coefficients shows that the latter is about 100 times as high as the former. Consequently, cracks formed in grains during drying will be caused primarily by moisture gradients, and stresses caused by temperature gradients may in most cases be neglected.

3. MECHANICAL PROPERTIES

During almost all technological processes (sowing, planting, harvesting, etc.) agricultural products are exposed to mechanical effects (forces). A force is always accompanied by deformation, which must be either sufficiently great (for cutting, tearing, pressing, etc.), or sufficiently slight in order to avoid damage (harvesting of vegetables and fruits, threshing of crop products, etc.). Consequently, the mechanical (compressile, tensile and shear) strength of products here plays an important role.

Most agricultural products are viscoelastic, i.e., they behave differently under static tensile, or compressile forces, and under dynamic or repeated dynamic loads (vibration). With knowledge of a product's behavior, it is possible to decide whether, for example, it may be comminuted best by shearing or by impact. The power requirements for compressing a material also vary depending on whether static or dynamic forces are applied.

For viscoelastic materials there exists no uniquely defined relationship between stress and deformation: deformation also depends on time (as in the phenomenon of creep). The modulus of elasticity is not constant, but decreases with deformation and increases during compaction.

In the case of silage and other bulk materials the compressibility, the expansion characteristics (e.g., recovery), the angle of internal friction and the cohesion are all of great importance, as they govern the internal stress state of the product, for example, during loading into a silo, during storage and during unloading.

Cutting of materials is generally the result of combined deformation (by shearing and bending). In practice, it is advisable to determine the cutting resistance as a mechanical property, thus permitting the power requirements for cutting to be determined directly. In numerous design tasks, knowledge of the static and dynamic friction coefficients is also necessary. The dynamic state of a material (e.g., on a sieve, on a rotary spreading disc, in chutes, on oscillating surfaces, etc.) and its stress state (e.g., of silage or of bulk granular materials in a silo) both depend decisively in numerous cases on the value of the friction coefficient.

Aerodynamic and hydrodynamic properties play an important role in pneumatic or hydraulic transport of goods, in sorting foreign bodies out of the same, and in the aeration of granular piles. The aerodynamic properties are also related to the size, shape and density of a material.

An important characteristic of agricultural products is their sensitivity to injury and damage, which depends on their strength characteristics and also on their biological properties. The occurrence of damage must be counted upon primarily during harvesting, handling, sizing and transport of crops. Fruits are sensitive to repeated loads (e.g., vibration during transport), as their texture becomes softer under the effect of repeated loads, and their load-carrying capacity decreases.

Certain agricultural materials are liquid, and thus viscous, with rheological properties deviating from those of Newtonian liquids. We refer here to materials, such as liquid manure, liquid fodder mixtures, fruit juices, mashes, etc., whose transport by pipeline involves many problems and can be dimensioned safely only with a knowledge of the relevant flow properties (e.g., the viscosity) of the material.

The factors characterizing the deformation and the flow of such agricultural materials are termed rheological properties. The subject of rheology is thus the deformation and flow of materials, with time effects also taken into account. The behavior of a material is determined here by three variables: stress, deformation and time.

The mechanical characteristics described above are discussed in detail in the following Chapters.

4. THERMAL PROPERTIES

Certain agricultural materials, of both plant and animal origin, are subjected to heat treatment. The main processes are heating, cooling, drying and freezing. The aim of heat treatment in most cases is preservation or degermination.

Heating and cooling of products may be realized by convection, conduction or radiation. To calculate these processes, knowledge of the thermal characteristics, namely, the specific heat, heat-conduction coefficient, diffusion coefficient and absorption (emission) coefficient, is indispensable.

In the heating and drying of agricultural products it is necessary to know what temperature can be applied and for how long without damage. For example, the germination capacity of seeds decreases rapidly when a certain temperature is exceeded, while the quality of other materials may deteriorate.

4.1 Specific heat

The quantity of heat required to raise the temperature of 1 kg of material by 1 °C is termed its specific heat. Knowledge of the specific heat is indispensable for calculating any heating or cooling process. The specific heat of agricultural products depends essentially on their moisture content and to a lesser extent on temperature.

At ambient temperature, the specific heat of a water-containing material may be calculated from the specific heat values for the dry material and for water:

$$c = c_d(1-U_1)+c_w U_1 \tag{11}$$

where c_d is the specific heat of the dry material, c_w the specific heat of water, and U_1 the moisture content (calculated on a wet basis). The specific heat of dry starch is 1.54 kJ kg^{-1} °C^{-1} (0.37 kcal kg^{-1} °C^{-1}); since that of completely dry cereals (wheat, maize) agrees well with this value, the specific heat of any cereal with an arbitrary moisture content may be determined using the above equation. The specific heat of the dry material of vegetables and fruits is

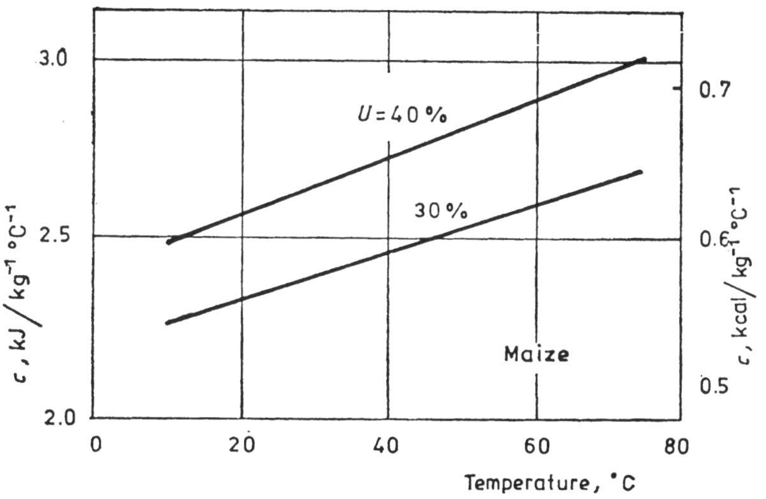

Fig. 15. Specific heat of maize as a function of temperature

0.8–0.9 kJ kg^{-1} °C^{-1}, and that of fatty dried milk is 1.8–1.9 kJ kg^{-1} °C^{-1}, showing that the specific heats of the dry contents of variously structured materials may be very different.

The specific heat of frozen vegetables is about one-half the value measured at ambient temperature. Generally, values between 1.8 and 2.0 kJ kg^{-1} °C^{-1} (0.43–0.48 kcal kg^{-1} °C^{-1}) are found. The lower values relate to materials having a lower moisture content (e.g., green peas).

Generally, with a rise in temperature the specific heat increases slightly, but in some cases the reverse trend is found [6, 42]. Figure 15 shows the specific heat of maize as a function of temperature [31]. The variation may be regarded as linear over the temperature range given.

4.2 Heat-conduction coefficient

The heating and cooling rates of materials, i.e., the temperature profiles developing in them, depend greatly on the heat-conduction coefficient, and so knowledge of this parameter is required unconditionally in order to perform calculations. The heat-conduction coefficient depends, as does the specific heat, on moisture content and temperature, and for porous bulk materials (i.e. granular bulk materials) it also depends on porosity. In the case of materials of fibrous structure, the direction of heat flow, along the fibers or normal to them, is also a factor.

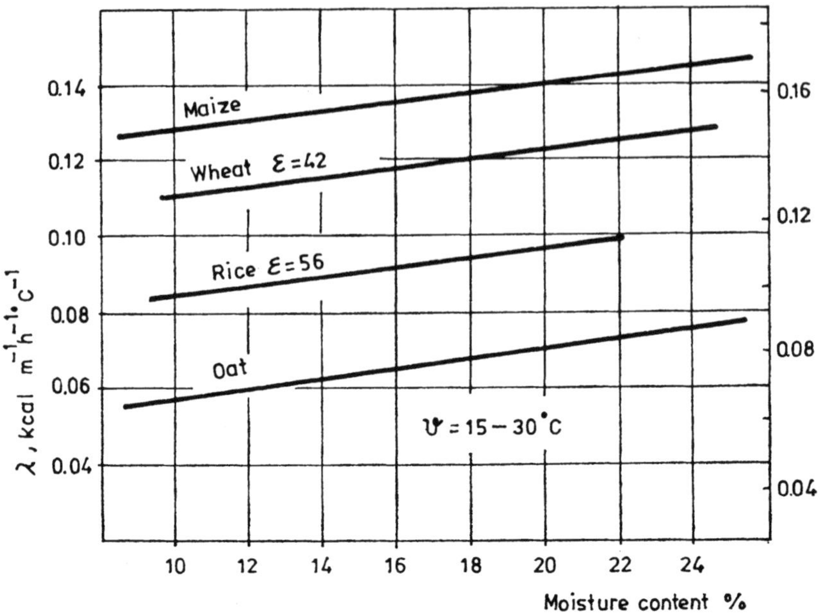

Fig. 16. Heat-conduction coefficients for agricultural products as functions of moisture content

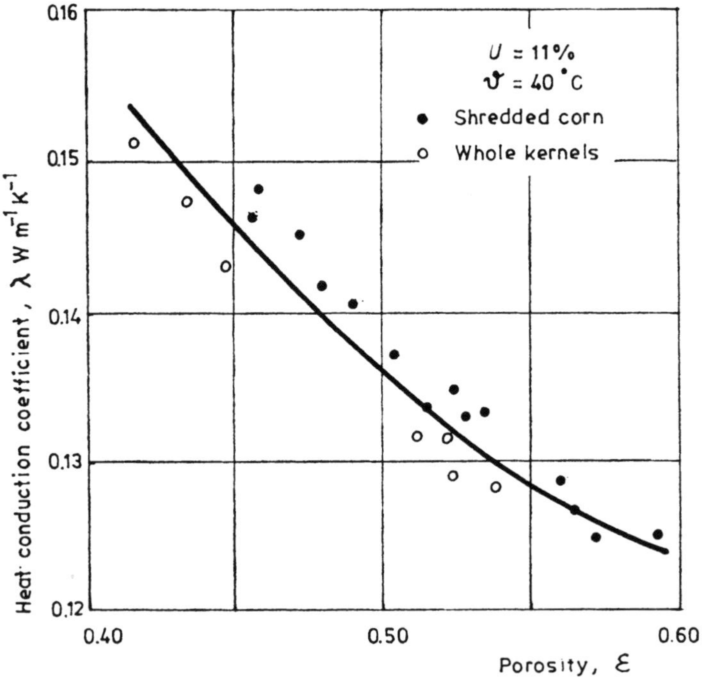

Fig. 17. Dependence of heat-conduction coefficient on porosity

Figure 16 shows the heat-conduction coefficients of various agricultural products as functions of their moisture content [42]. The higher heat-conduction coefficient of wheat is due mainly to its lower porosity [5]. The effect of porosity on the heat-conduction coefficient is shown in Fig. 17. The higher the porosity, the

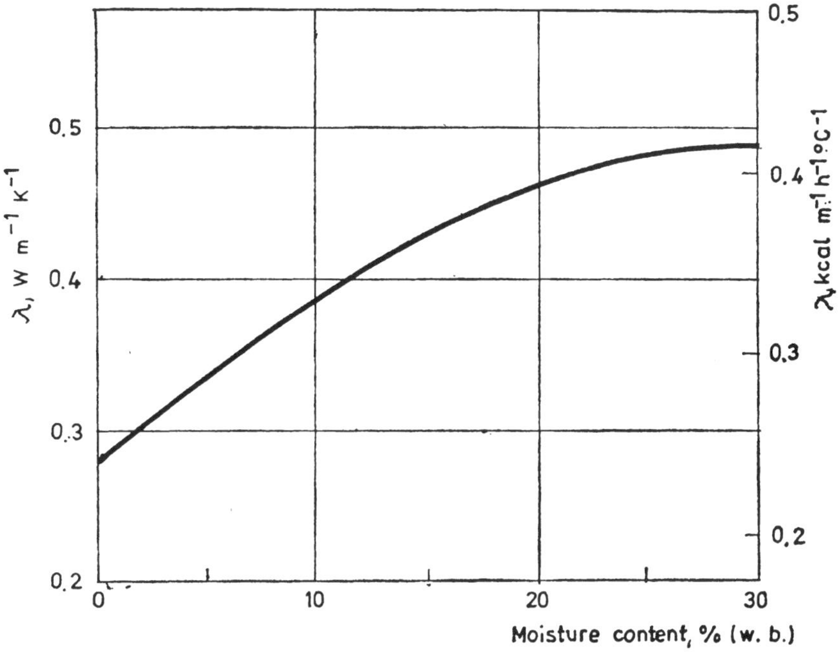

Fig. 18. Heat-conduction coefficient of wheat grains as a function of moisture content

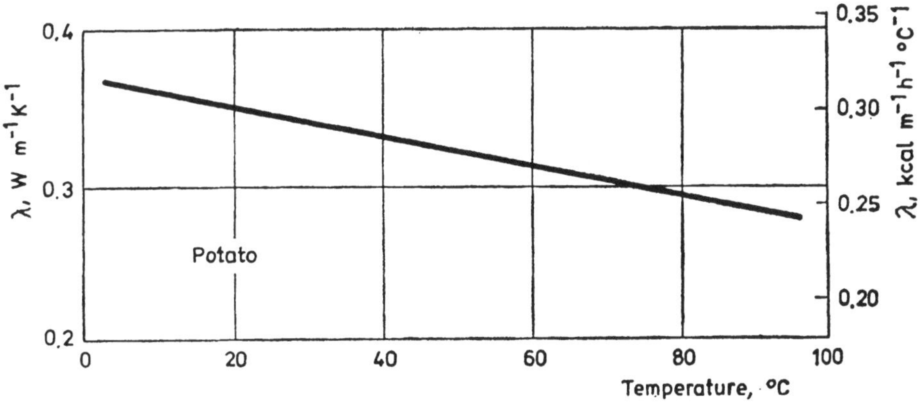

Fig. 19. Heat-conduction coefficient of potatoes as a function of temperature

looser the individual particles relative to each other and the lower the heat-conduction coefficient [193]. The heat-conduction coefficient of the individual grains in a pile deviates considerably from that for the bulk material. Figure 18 presents the heat-conduction coefficient of wheat kernels as a function of their moisture content [42].

The effect of temperature on the heat-conduction coefficient may vary. The heat-conduction coefficient of water increases slightly with temperature, and

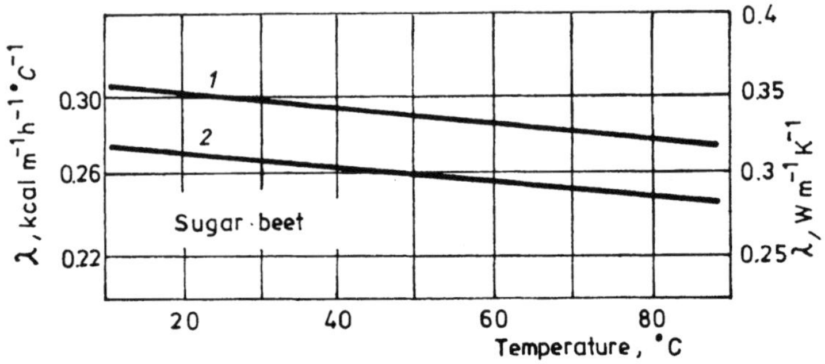

Fig. 20. Heat-conduction coefficient of sugar beet. *(1)* Parallel and *(2)* normal to the direction of fibres

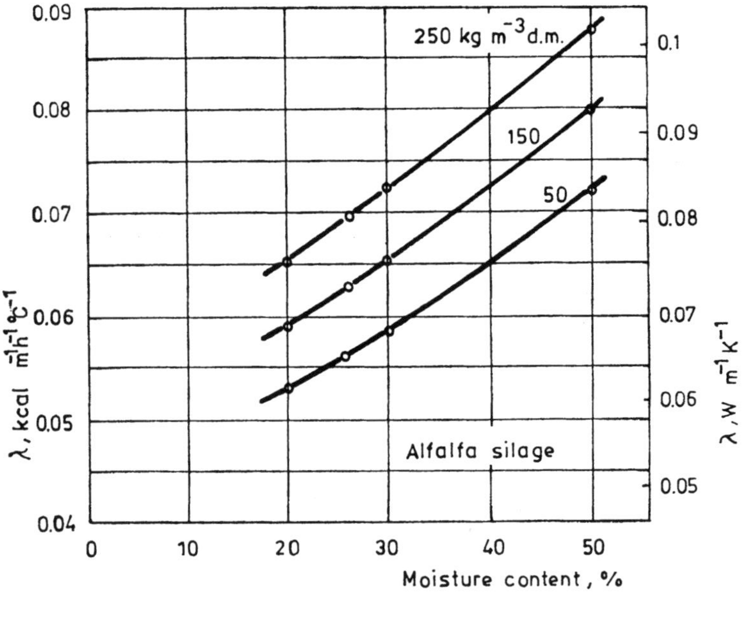

Fig. 21. Heat-conduction coefficient of alfalfa silage as a function of moisture content

accordingly, similar behavior could be expected of materials having a high moisture content, however, measurements show either a contrary trend or an insignificant temperature effect. Figure 19 shows the heat-conduction coefficient of potatoes as a function of temperature [6]. As may be seen, the value of λ decreases slightly with rising temperature.

Figure 20 presents the heat-conduction coefficients of sugar beet, parallel and normal to the direction of the fibers [6]. The value is always higher in the direction of the fibers.

The heat-conduction coefficient of alfalfa silage as a function of moisture content is shown in Fig. 21 for various volumetric weights (relative to the dry content).

As the data reveal, the heat-conduction coefficients of both granular piles and silage are low. Therefore, biological heat generated during the storage of great volumes is transferred very slowly to the environment.

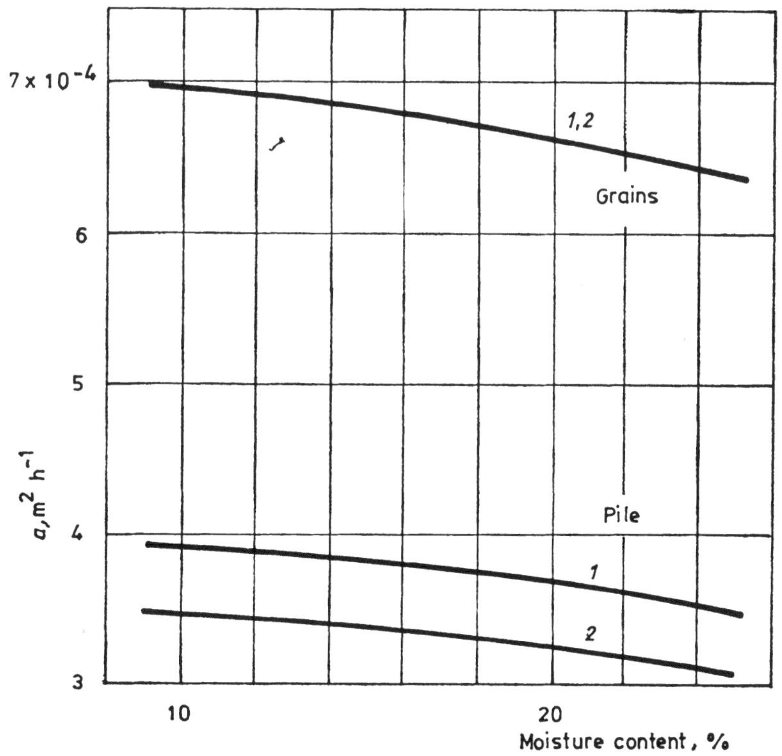

Fig. 22. Temperature conductivities of wheat and maize, as a function of moisture content. *(1)* Maize; *(2)* wheat

4.3 Temperature conductivity

The expression occurring frequently in various heat calculations,

$$a = \lambda/c\gamma$$

is termed the temperature conductivity, or thermal diffusivity. Its value may be calculated easily with a knowledge of the heat-conduction coefficient and the specific heat.

Figure 22 shows the temperature conductivities of wheat and maize as functions of moisture content [42]. It is seen that a decreases slightly as the moisture content increases. Generally, the temperature conductivity increases with temperature (Fig. 23) [6].

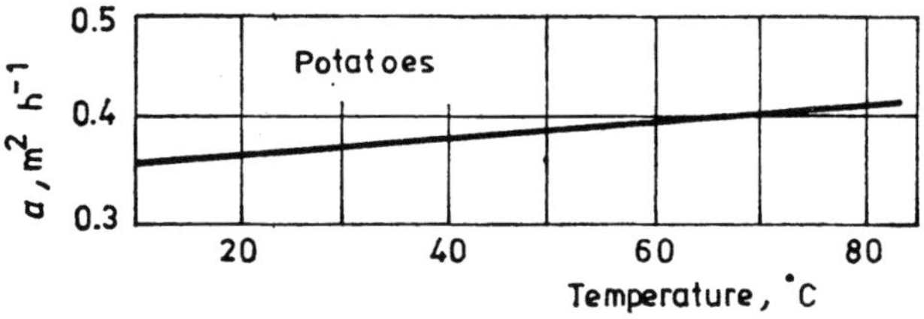

Fig. 23. Temperature conductivity of potatoes as a function of temperature

5. ELECTRICAL PROPERTIES

Among the electrical properties of agricultural products, their dielectric properties, their capacitance and their electrical conductivity play important roles during various technological processes. The moisture content of bulk materials is widely measured by instruments utilizing the electrical conductivity or the capacitance of grains. Frequently, when classifying small seeds, use is made of the electrical properties by virtue of which electrostatic charge is held. The holding of surface charge by seeds is determined mainly by their conductivity.

Figure 24 shows the principles of construction of an electrostatic sorting device. Seeds are loaded by a feed hopper onto a positively charged moving conveyor belt. A negative electrode creates an electrostatic field which attracts the positively charged seeds, according to their charge. Thus the seeds follow various

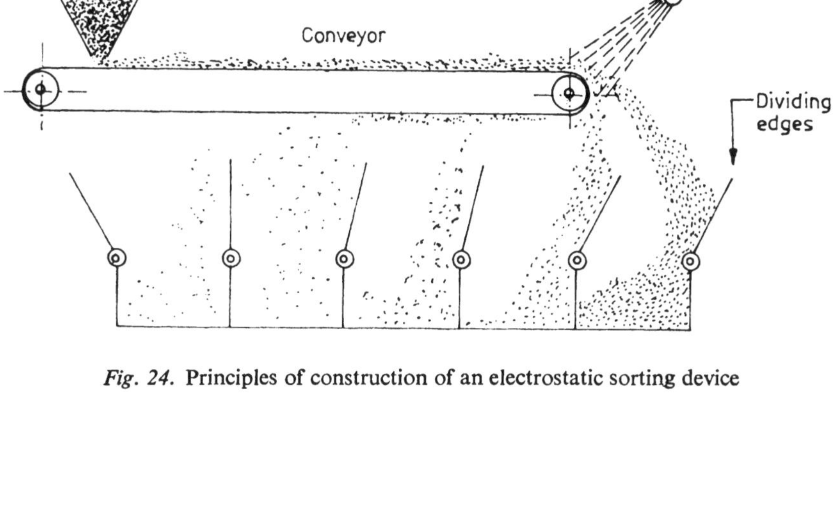

Fig. 24. Principles of construction of an electrostatic sorting device

paths. The negative electrode may also be designed and installed so as to create high-voltage discharges, whereby the seeds acquire a negative charge. Seeds of low conductivity take up and hold the charge more easily, and so these seeds do not drop immediately, owing to the attraction of the belt. This method is suitable not only for the separation of seeds, but also for sorting sterile (ungerminative) from apparently fertile seeds [10], since it has been found that the moisture content and conductivity are higher for less germinative seeds.

In high-frequency drying, the dielectric constant and loss factor of the material to be dried play an important part. The energy released in the form of heat by a material is proportional to its dielectric constant and the tangent of the dielectric loss angle ($\tan \delta$). High-frequency drying permits a product to be dried quickly without causing an excessive temperature rise in the material and so is well suited for drying high quality goods.

The extent of damage to agricultural products and plant tissues is frequently determined by the impedance technique. The electrical impedance, and particularly the capacitance of destroyed tissue is much lower than that of living, sound tissue. Therefore the ratio of the low- and high-frequency impedances is characteristic of the extent of damage to the tissues of a product.

Various types of electromagnetic radiation (radiofrequency, infrared, ultraviolet, X-ray and gamma-ray) are often used for treating agricultural materials. The effect of irradiation may be a simple heat effect (for greater wavelengths and lower energy levels) or a chemical effect (for shorter wavelengths and higher energy levels). Radiofrequency and ultraviolet irradiation are used for sterilization. Often the sprouting of hard seeds may be promoted by suitable irradiation. For all these processes it is necessary to know how the material considered behaves under electromagnetic radiation.

5.1 Dielectric constant and dielectric loss

When studying the dielectric properties of a material, it must be placed in an electrical force field. The field of electrical forces is characterized by a vector of electrical field strengths (E) and a vector of dielectric displacements (D). The following relationship exists between the two characteristics [9, 12]:

$$D = \varepsilon E$$

where ε is the dielectric constant of the material through which the field acts. The value of the dielectric constant of a material is given by

$$\varepsilon = \varepsilon_0 \varepsilon_r$$

where ε_0 is the dielectric constant of a vacuum, and ε_r the relative dielectric constant, which is characteristic of the material. In the case of an alternating voltage, the field strength varies periodically:

$$E = E_0 e^{i\omega t}$$

and so the relationship between the dielectric displacement and the field strength may be written as

$$D = \varepsilon E_0 e^{i(\omega t - \delta)} = (E_0 \varepsilon e^{-i\delta}) e^{i\omega t} = E_0 \varepsilon^* e^{i\omega t}$$

where $\varepsilon^* = \varepsilon e^{-i\delta} = \varepsilon(\cos\delta - i\sin\delta)$, i.e., $\varepsilon^* = \varepsilon' - i\varepsilon''$. The complex dielectric constant ε^* consists of two parts: the real part is termed the dielectric constant, or permittivity, while the imaginary part is termed the loss factor. The quotient of the two parts gives the loss tangent, $\tan\delta = \varepsilon''/\varepsilon'$. The dielectric constant may be measured by means of a Q-meter. The capacitor in the instrument is designed according to the properties of the material to be measured: a concentric capacitor is used in measurement for cereals, and a capacitor consisting of parallel plates for fibrous materials.

The dielectric constant of cereals depends on their moisture content, on temperature and on frequency. Figure 25 shows the dielectric constant of wheat as

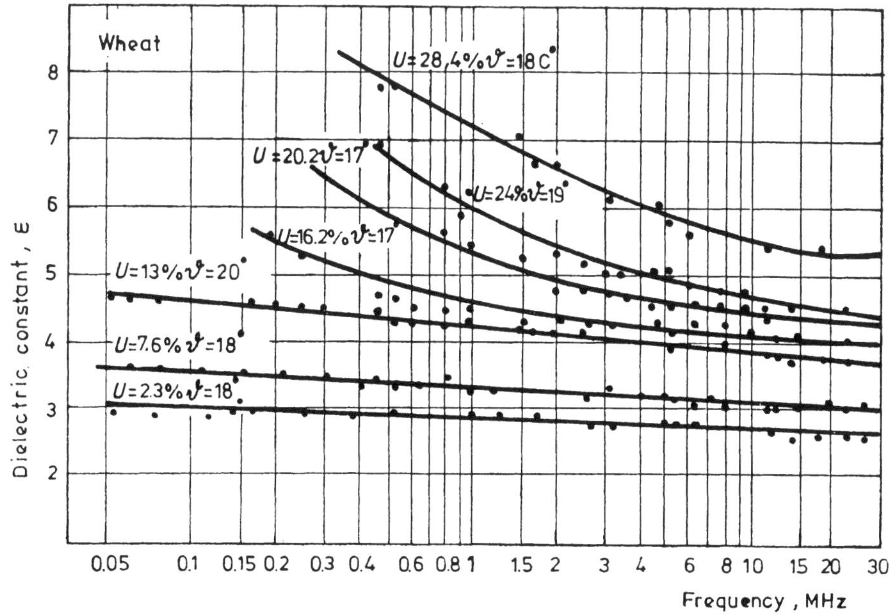

Fig. 25. Dielectric constant of wheat as a function of frequency

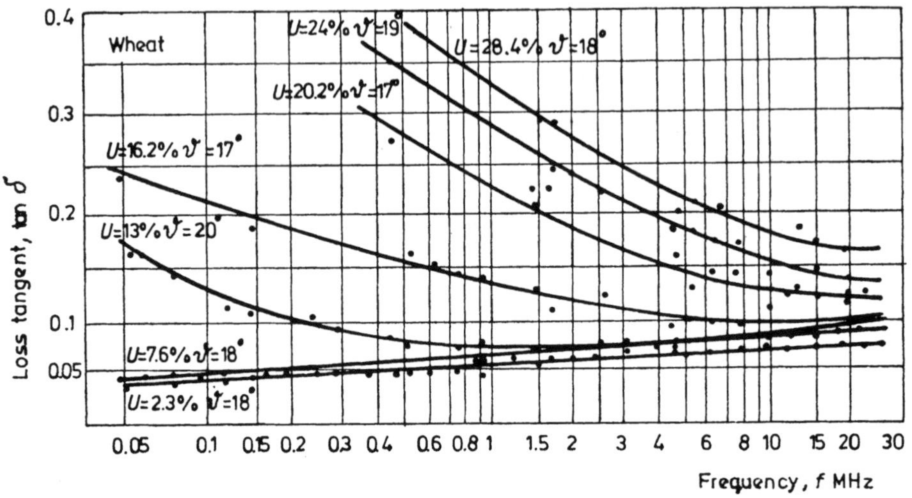

Fig. 26. Loss tangent for wheat as a function of frequency

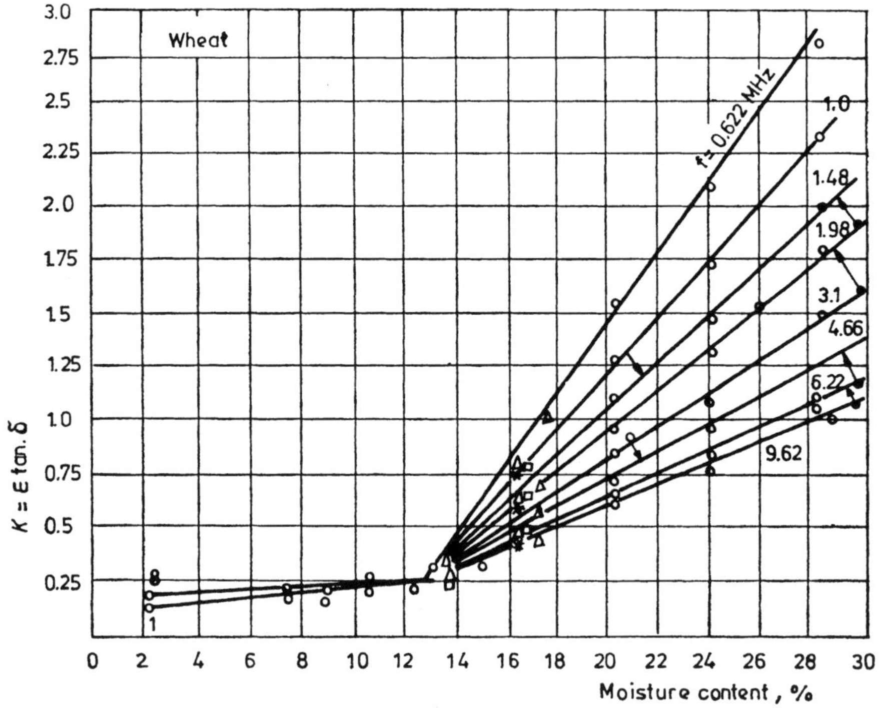

Fig. 27. Loss factor for wheat as a function of moisture content

a function of frequency at various moisture contents [11]. The effect of moisture content is very important, especially at lower frequencies. The effect of frequency increases as the moisture content increases. Figure 26 shows the variation of the loss tangent for wheat as a function of frequency. Here, the effect of the moisture content is much more important than that of frequency [11]. The effect of temperature appears mainly at higher moisture contents, above 13–14%. In this case, the dielectric constant increases with temperature.

In high-frequency drying the loss factor $k = \varepsilon \tan \delta$ plays a major role, and so it is of interest to study the variation of k as a function of moisture content. Figure 27 illustrates the relationship for wheat [11], at various frequencies. As

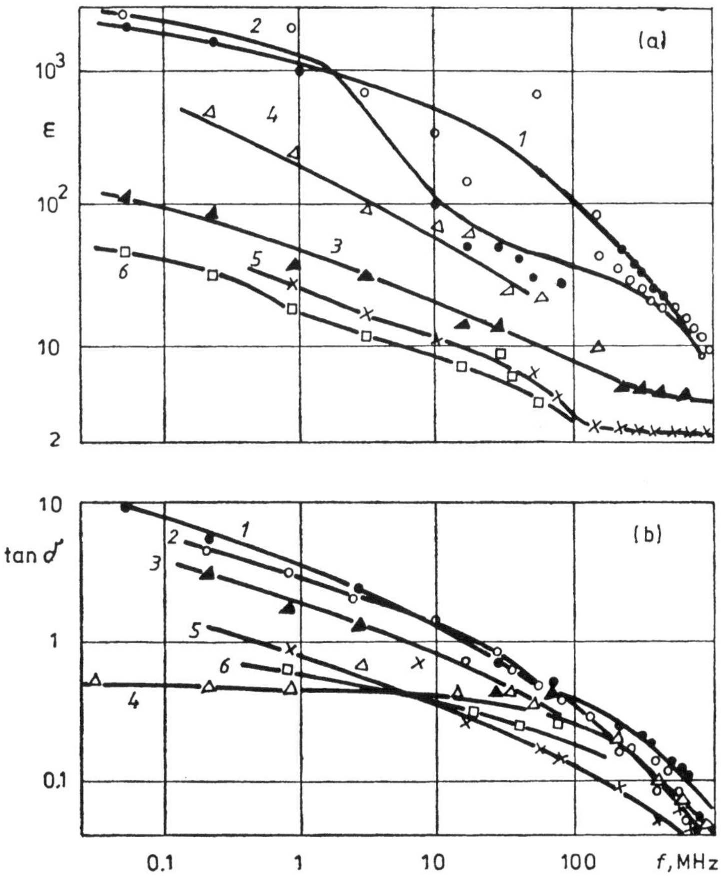

Fig. 28. Dielectric constans and loss tangents for various materials as functions of frequency. *(1)* Potatoes; *(2)* sugar beet; *(3)* soil; $X=16\%$; *(4)* soil; $X=5\%$; *(5, 6)* stones, $X=6–8\%$

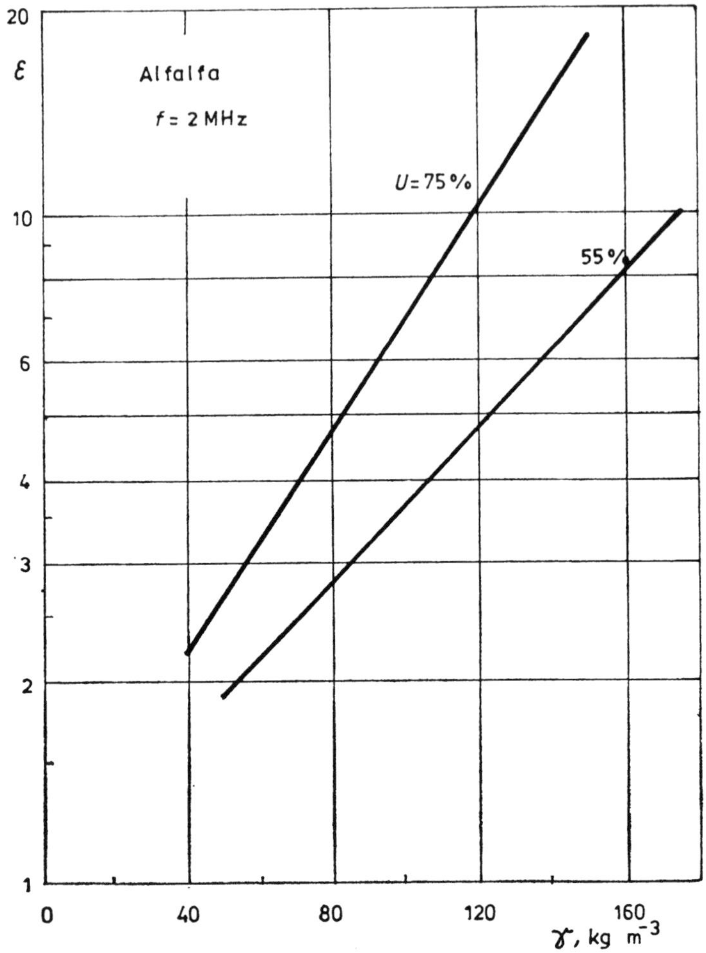

Fig. 29. Dielectric constant of alfalfa as a function of volumetric weight

may be seen, k is nearly constant up to 13% moisture content, independent of frequency. At higher moisture contents the value of k shows an approximately linear increase, which is the steepest in the region of low applied frequency. The reason for this is that for moisture content up to 13% water is present predominantly as adsorbed moisture, while above 13% it is present mainly in the form of capillary moisture. The dielectric loss is governed primarily by the capillary moisture.

The dielectric constants of root bulb plants (potatoes, carrots, etc.), of soil and of stones are very different, a fact which may be utilized for sorting earth clods and stones, using suitable sensing elements. Values of ε and $\tan \delta$ as functions of frequency are presented in Fig. 28 for the above materials. The optimal sorting frequency, at which the difference between the dielectric constants of the materials to be separated is maximum, may be selected on the basis of the figure.

For fibrous materials the dielectric constant depends on the volume weight, as well as on the moisture content. The greater the volume weight, the greater the proportion of space filled by a fibrous material, and the higher the value of the dielectric constant. Figure 29 shows the dielectric constant of alfalfa as a function of volumetric weight at various moisture contents. The dielectric constant decreases considerably with decreasing temperature [12].

6. OPTICAL PROPERTIES

One of the most characteristic properties of the fruits is their color, both external and internal, which determines in most cases unequivocally both their ripeness and quality. Thus, classification of fruits and certain vegetables by their color has recently come increasingly into the foreground. Besides color, other optical properties, such as transmittance and reflectance, are also of importance in quantitative evaluations of various properties. With a change of color, the transmittance and reflectance of a product also change.

As is known, the various colors are electromagnetic radiations of various wavelengths. A body will absorb or reflect light rays of various wavelengths differently, depending on its color. Therefore measurements are performed using monochromatic light of various wavelengths (spectrophotometry).

The spectrum of visible light is usually divided into eight intervals, corresponding to the most characteristic colors. The wavelength bands corresponding to these intervals are shown in Table 1.

Table 1

Wavelength (mμ)	Color	Wavelength (mμ)	Color
380–450	Violet	550–575	Yellow-green
450–480	Light blue	575–585	Yellow
480–510	Blue	585–620	Orange
510–550	Green	620–760	Red

The various colors may be constituted from the three primary colors, red, yellow and blue. Accordingly it is sufficient to know the relative proportions of the primary colors in order to characterize a given color:

$$\text{color} = x'X + y'Y + z'Z$$

where X, Y and Z are the primary colors, and x', y' and z' the color coordinates expressing the proportions of the individual primary colors.

6.1 Reflectance properties

One of the main requirements in treating vegetables and fruits is that they retain their color even after a long storage period, without browning. The extent of browning may be determined from the reflectance of a product as a function of wavelength, since the reflectance of browned material decreases markedly in the orange range (600–650 mμ). Figure 30 shows as an example reflectance values for carrots treated by various methods, after three years' storage [14].

In numerous cases it is advisable to select two characteristic reflectance values R_1 and R_2 corresponding to two chosen wavelengths in order to form an index characterizing the variation of reflectance:

$$\bar{I} = (R_1 - R_2)/R_2$$

which varies more sensitively as a function of the variable studied (storage or ripening period, etc.). For example, $R_1 = 720$ mμ and $R_2 = 678$ mμ have been selected for studying the ripening of lemons, on the basis of the following considerations. The surface chlorophyll of green fruit shows strong absorbance in the wavelength region of 678 mμ, so that the reflectance is relatively lower here. With ripening of the fruit the green color turns to yellow, whereby the level of chlorophyll and the corresponding absorbance are lowered and the reflectance increases. The reflectance at 720 mμ does not vary significantly. The wave-

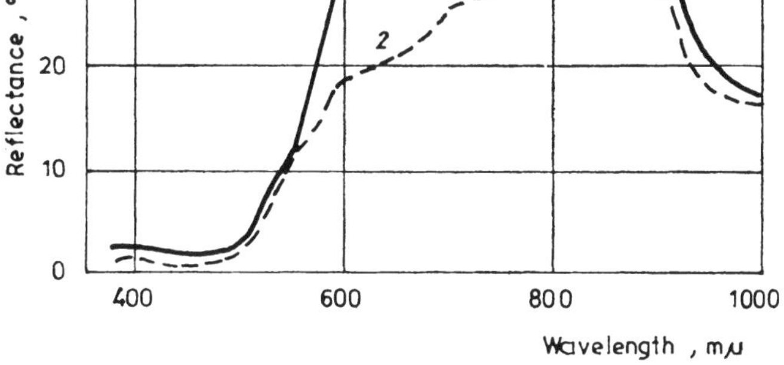

Fig. 30. Reflectance properties of carrots after various treatments. *(1)* High temperature sterilization; *(2)* normal sterilization

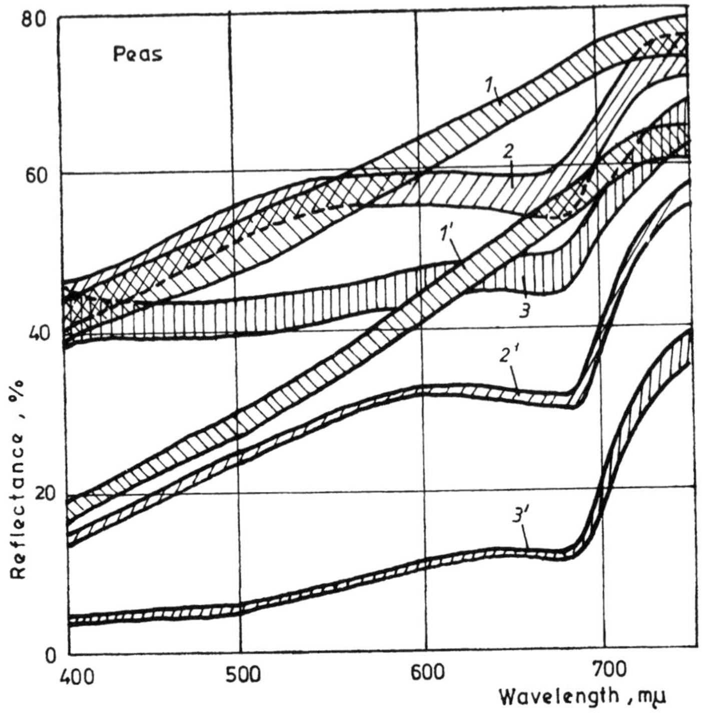

Fig. 31. Reflectance characteristics of peas: *(1)* Green; *(2)* greenish yellow; *(3)* yellowish brown; *(1, 2, 3)* individual peas; *(1, 2, 3)* bulk layers

lengths R_1 and R_2 are selected accordingly, so that R_2 shows the major variation.

Figure 31 shows the reflectances of individual peas and of bulk layers of unripe and ripe peas [13]. If the reflectance index I is calculated for curves (*1*) and (*2*) for the wavelength given above, it is found that the value of I increases by a factor of at least four during ripening. The same figure also shows that the reflectance of brown peas is lower. The differing reflection capacities of various materials may also be utilized to sort foreign bodies from crops (e.g., earth clods from potatoes) [15].

6.2 Transmittance properties

The main deficiency of studies based on reflectance measurements is that they do not permit a determination of the internal properties of a material, but yield information concerning only its external color and appearance. Measurements of transmittance as a function of wavelength, however, do permit determinations of

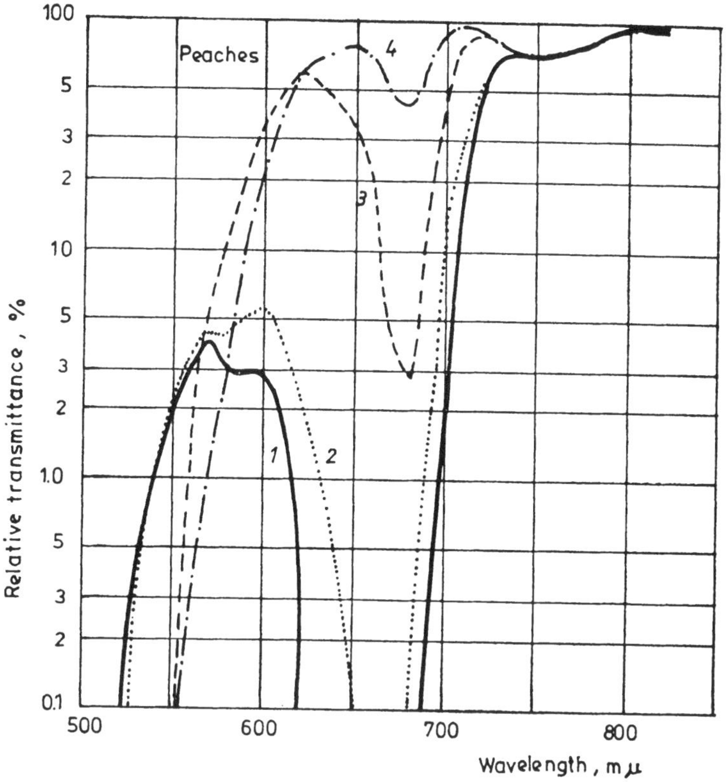

Fig. 32. Transmittance curves for peaches. *(1)* Green; *(2)* greenish; *(3)* nearly ripe; *(4)* ripe

internal color and thus of internal changes (e.g., the presence of water cores in apples, of blood in eggs, internal damage, etc.).

Figure 32 shows transmittance curves for peaches at various stages of ripeness, as a function of wavelength [16]. The transmittance varies greatly with wavelength, and so the use of a logarithmic scale is advisable. It can be seen from the figure that during ripening, the peak found at the shorter wavelengths shifts from 570 to 650 mμ. The greatest change is apparent in the absorption zone of chlorophyll (680 mμ): with the advance of ripening, the transmittance increases abruptly. The ratio of the transmittance values at two suitably selected wavelengths may thus serve to characterize the stage of ripening. In Fig. 33 the logarithm of the ratio of the transmittances at 700 and 740 mμ is plotted as a function of the number of days of ripening. Full ripening was attained between the 25th and 28th days. The results show that the stage of ripening can be determined by measuring the transmittance; moreover, the time of ripening may be predicted on the basis of the ripening curve [16].

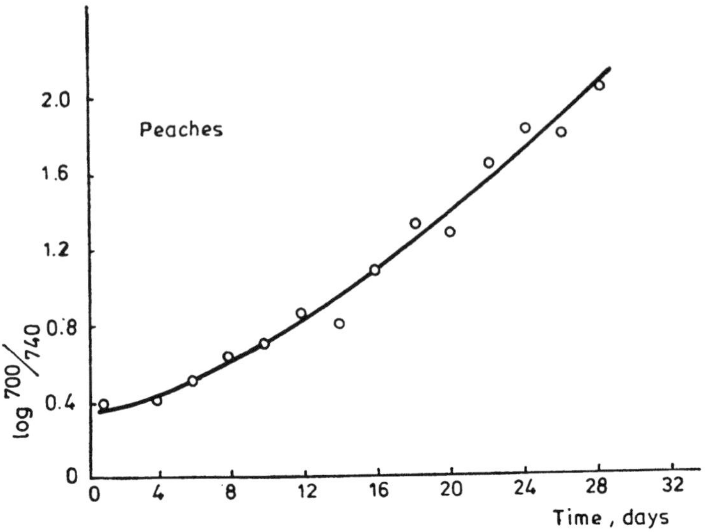

Fig. 33. Variation of transmittance ratio during ripening of peaches

Transmittance values for variously ripe tomatoes are shown as functions of wavelength in Fig. 34 [14]. With the advance of ripening the quantity of chlorophyll decreases, whereby the transmittance increases greatly in the interval 670–680 mμ. The transmittance at 550 mμ changes in the opposite direction. Various possibilities are available for characterizing the stage of ripening. The changeover from a yellow-red to a red color may be characterized well by the ratio of the transmittances at 620 and 670 mμ, which varies by a factor of about 30:1 during ripening. The initial ripening of a completely green tomato may be characterized by the ratio of the transmittances at 520 and 545 mμ, since the first change observed in the transmittance curve at the start of ripening appears in the 520 mμ region. Complete ripening may be characterized by combination of the two ratios, in the form

$$R = (T_{670} - T_{520})/(T_{620} - T_{545})$$

Figure 35 shows their characteristics as a function of ripening period. At the start of the measurements the tomato was still completely green. On the following day the fruit began to color, and on the fourth day its external color was red [14].

As mentioned before, the peak transmittance found at the start of ripening, in the range 500–650 mμ, shifts to the right as a function of ripening. If the wavelength corresponding to the peak is plotted as a function of ripening period (Fig. 36), a curve characterizing the initial phase of ripening is obtained. How-

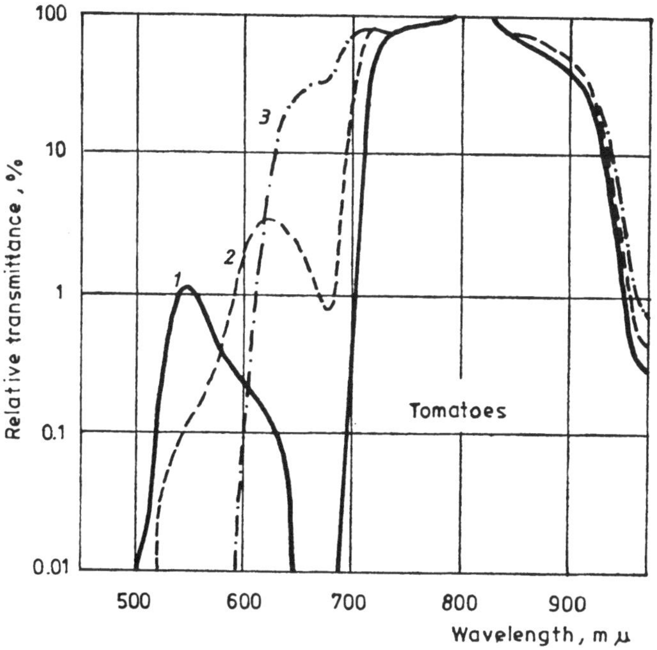

Fig. 34. Transmittance curves for tomatoes. *(1)* Green; *(2)* greenish red; *(3)* red.

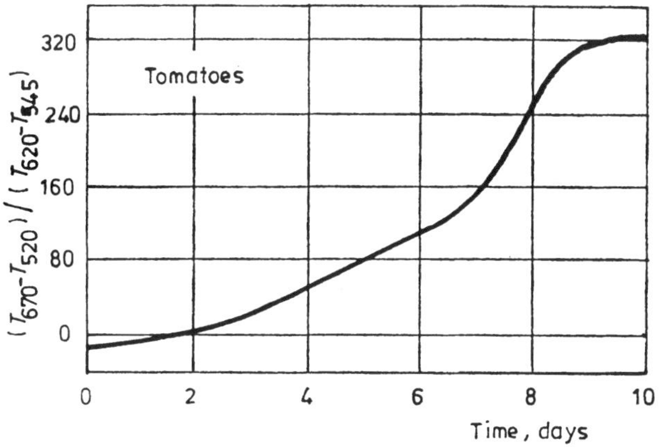

Fig. 35. Characterization of the ripening of tomatoes by the transmittance ratio.

ever, this presentation fails to show the development of internal color, contrarily to Fig. 35, where the steep sloping section of the curve indicates the full development of internal color on the seventh or eighth day [14].

The examples listed reveal that the complete transmittance curve is needed only during development of the measurement method. Generally, measurements of the transmittances at two wavelengths suffice to determine the stage of ripening

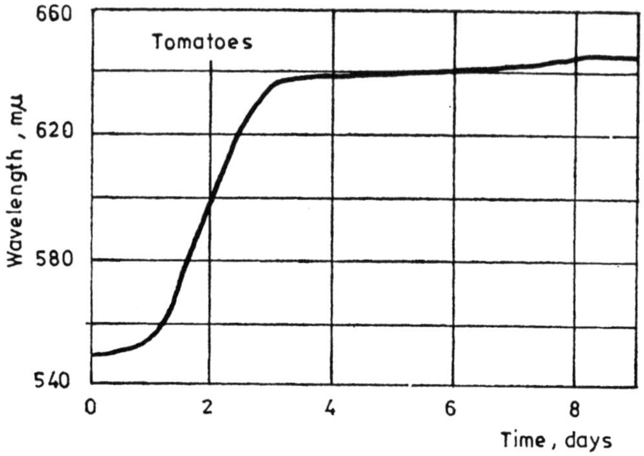

Fig. 36. Shift of transmittance peak during ripening of tomatoes

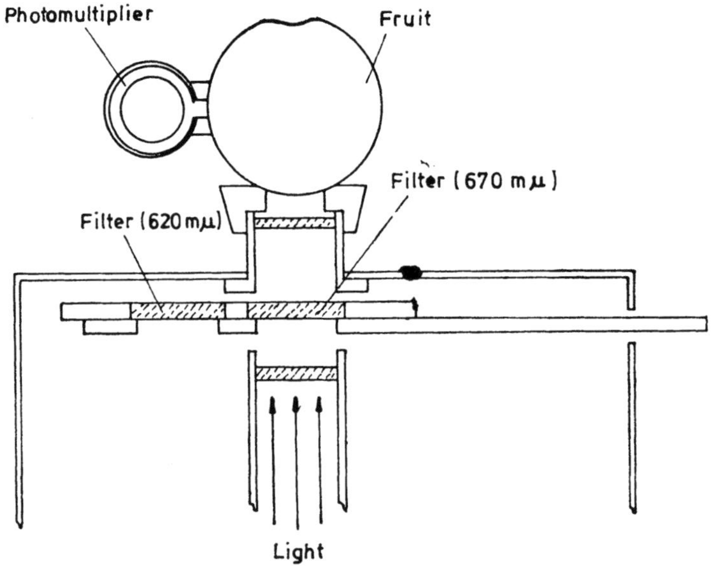

Fig. 37. Apparatus for measuring transmittance

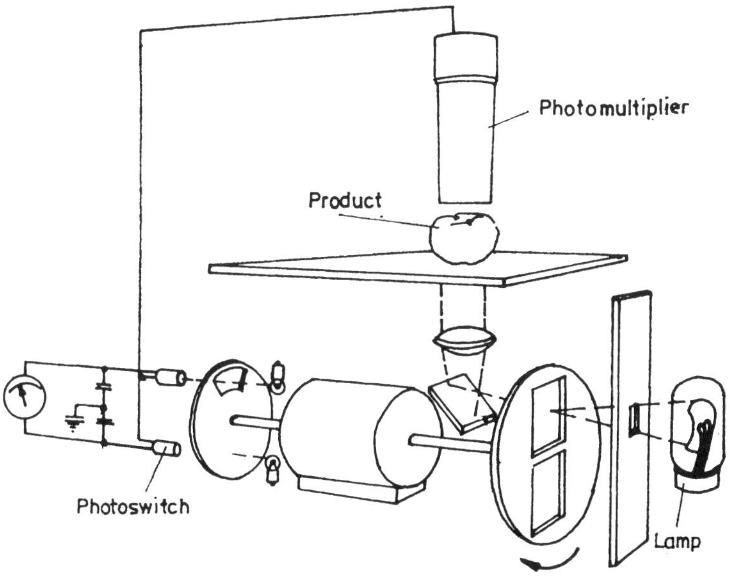

Fig. 38. Apparatus for measuring the transmittance ratio

Such measurements may be performed in a relatively simple way using the equipment shown in Fig. 37 [14]. The fruit to be tested is illuminated from below through interference filters, and the monochromatic light passing through the fruit is measured by a photomultiplier, adjusted to 90°. The two filters, mounted beside each other may be shifted over simply.

Figure 38 shows a schematic drawing of a more complex difference meter which can be used as a quick measuring instrument [43]. A disc containing two filters is rotated by a synchronous motor at 1800 rpm whereby the fruit to be tested is illuminated by monochromatic light at two wavelengths alternately. The transmitted light is sensed by a photomultiplier arranged over the fruit, with corresponding electrical signals applied alternately to two sides of a voltmeter. The latter is controlled by two photoswitches and indicates the differential intensity between the two wavelengths.

Variation of transmittance may be used to characterize not only the ripening process, but also variations appearing in the internal texture. To demonstrate such variations, one of the wavelengths used is adjusted to the absorption band of the material causing the variation: for example, the water cores which sometimes appear in apples may be revealed by adjusting one wavelength to the absorption band of water (760 mμ). Blood has significant absorption properties in the regions of 540 and 575 mμ.

7. WATER STORAGE IN AGRICULTURAL MATERIALS

7.1 Physics of water storage

One of the most important characteristics of biological materials is their moisture content, which affects their physicomechanical properties decisively. The storability of certain products is also greatly influenced by their moisture content, and therefore in these cases superfluous moisture is removed by natural or artificial drying.

Moisture content is usually expressed in terms of percentage values relative to a wet or a dry basis, namely,

$$U = G_w/(G_{dm}+G_w)$$

or

$$X = G_w/G_{dm}$$

where G_w is the weight of moisture contained in the product, and G_{dm} the weight of dry material.

Water is stored by biological materials in two ways: by molecular and by capillary adsorption. There is an essential difference between the two storage modes. Starting from completely dry material, water is bound first by molecular adsorption, and then by capillary adsorption only after a certain moisture content has been attained.

During molecular adsorption, molecules of water are adsorbed very close to the cell walls (at a distance of order of magnitude 10^{-7} cm, comparable to the diameter of the water molecule), and adhere by attractive forces to the surface molecules. The attractive force acting on the first layer of water molecules is the greatest, and the force decreases gradually for subsequent layers. The field of molecular forces holds the water molecules under a pressure which is proportional to the attractive force, whereby the density of the water increases and pressure (compression) develops in the system. Thus the volume of the product increases (i.e., the product swells) on adsorption of water, but by an amount somewhat less than the volume of this water owing to the pressure caused by the attractive forces.

Molecular adsorption is accompanied by heat generation, whose value is 840–1680 kJ kg^{-1} water (200–400 kcal kg^{-1}). Certain properties of water bound by molecular adsorption deviate from those of free water. One of these deviations is that molecularly adsorbed water has practically no power of solution. Its specific weight also exceeds that of free water; for black soil with 1.64% moisture content, $\gamma = 1.74$ g cm^{-3}; for a moisture content of 13.8%, $\gamma = 1.13$ g cm^{-3} [23]. Water bound by molecular adsorption also freezes at a much lower temperature, significantly below 0 °C. Its electrical conductivity is very low, and in a high-frequency force field it behaves differently. This latter property is shown in Fig. 27.

During capillary adsorption, water is retained by means of surface-tension forces in the capillary-sized cavities found in cellular systems. The surface tension of water is 76 dyn cm^{-1} at 0 °C, and decreases linearly with temperature according to the relationship

$$\sigma = 75(1-0.0020\vartheta)$$

where ϑ is temperature (°C).

The molecular pressure above a liquid surface is a function of the surface curvature. The surface of wetting liquids found in capillaries is concave, and for small capillaries the radius of the half-spherical surface is identical to that of the capillary. In this case the capillary pressure is negative, and its value is

$$p = 2\sigma/r \tag{12}$$

where r is the radius of the capillary.

Liquid rises in a capillary tube, owing to wetting and to capillary pressure, to a certain height h, whose value is

$$h = 2\sigma/rg\varrho_w$$

where ϱ_w is the density of the liquid. From this relationship it follows that the capillary rise is inversely proportional to the capillary radius: the smaller the radius, the higher the capillary rise.

A consequence of capillary pressure is that the saturated vapor pressure over a concave syrface is lower than over a plane surface. The relative vapor pressure may be calculated from a formula due to Thompson:

$$\varphi = p_g/p_s = \exp -(2\sigma\varrho_g/p_s\varrho_w r)$$

where p_g and ϱ_g are respectively the pressure and density of the water vapor in the capillary, and p_s is the pressure of the saturated vapor over a free water surface. This relationship may be applied to capillaries whose radius is in the interval 0.5×10^{-7}–10^{-5} cm. The variation of φ as a function of capillary radius at 20 °C is shown in Table 2 [29]:

Table 2

φ	r ($\times10^{-7}$ cm)	φ	r ($\times10^{-7}$ cm)	φ	r ($\times10^{-7}$ cm)
0.05	0.36	0.45	1.56	0.90	21.9
0.10	0.46	0.50	1.80	0.95	26.3
0.15	0.57	0.55	2.11	0.96	35.3
0.20	0.67	0.60	2.50	0.97	53.3
0.25	0.78	0.65	3.01	0.98	106.6
0.30	0.89	0.70	3.73	0.99	1 077.0
0.35	1.17	0.75	6.51	0.999	10 770.0
0.40	1.34	0.85	10.23	1.0	∞

If the radius of the capillary exceeds 10^{-5} cm, then the saturated vapor pressure over the meniscus agrees practically with the vapor pressure over a free water surface. On the basis of this fact, capillaries having a radius greater than 10^{-5} cm are usually termed macrocapillaries, and those of smaller radius, microcapillaries.

A cavity in a body may be termed capillary until the surface of the liquid found in it is determined mainly by surface tension and the effect of the weight force is negligible. This limit is at a radius of about 10^{-3} cm.

On removal of moisture, energy must be imparted to a system. In the case of free water, this energy is the (latent) heat of evaporation. Its value is 2385 kJ kg^{-1}. To remove the water bound in biological materials, additional energy is required, and so the heat of evaporation is higher than in the case of free water. The work required to remove 1 mole of water (assuming an isothermal, reversible process) equals the variation of the free energy, i.e.,

$$A = -\Delta F = -RT \ln (p_g/p_s)$$

According to Gibbs' law the enthalpy of an isothermal process (here the heat of dissociation of bound water) is

$$dH = dF - T(\partial \Delta F/\partial T)$$

which gives, after differentiation with respect to T,

$$Q = RT^2[(\partial \ln p_g/\partial T) - (\partial \ln p_s/\partial T)] = Q_1(U) - Q_0$$

where $Q_1(U)$ is the heat of evaporation of the water found in the material at a given moisture content U, and Q_0 the heat of evaporation of free water. It has been seen above that p_g is always lower than p_s (with the exception of large capillaries), and so $Q_1(U) > Q_0$. From this it also follows that

$$\partial \ln p_g/\partial T > \partial \ln p_s/\partial T$$

i.e., the curve $\ln p_g = f(T)$ is steeper than the curve $\ln p_s = f(T)$ (Fig. 39). As the directional tangent of the $\ln p_g$ curve approximates that of the $\ln p_s$ curve, the difference $Q_1(U) - Q_0$ decreases, and when the two directional tangents are equal, $Q_1(U) = Q_0$.

From this, it follows that the value of $Q_1(U)$ may be determined from the ratio of the directional tangents of the two curves.

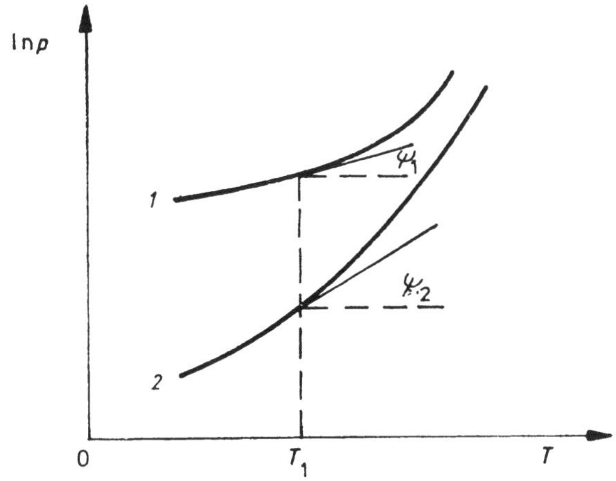

Fig. 39. Water vapor pressures in a material and over a free water surface as a function of temperature

The essence of the procedure may be followed in Fig. 40, which shows the vapor pressure (p_g) above a layer of wheat as a function of temperature, and the corresponding variation of the saturated vapor pressure of free water. A plot as a function of p_s is also possible, because a uniquely defined relationship exists between p_s and T. An advantage of this procedure is that straight lines are obtained in the given system of coordinates. Making use of these curves the following equation is valid:

$$Q_1(U)/Q_0 = (\ln p_{g2} - \ln p_{g1})/(\ln p_{s2} - \ln p_{s1})$$

where the subscripts 1 and 2 refer to two arbitrary points. From Fig. 40 it may be seen that the directional tangents of the curves vary as a function of moisture content. The value of $Q_1(U)$ also varies accordingly, increasing in the direction of lower moisture content. This is illustrated clearly in Fig. 41, obtained on the basis of the data of Fig. 40 [28].

Concerning the method of determination of p_g, values are obtained on the basis of sorption isotherms plotted at various temperatures (see the following section) by multiplying the value of p_s by the actual moisture content. From

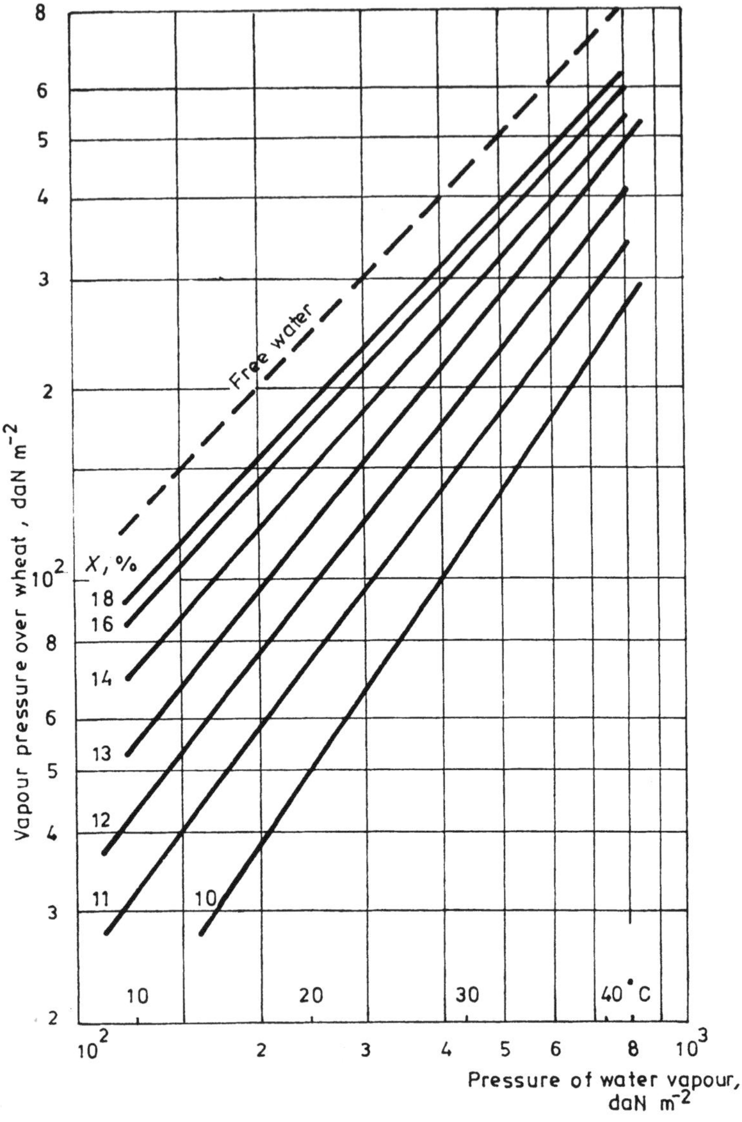

Fig. 40. Relationship between vapor pressure over wheat and over a free water surface

Fig. 41 it may be stated that the heat of evaporation of water bound in wheat is higher than that of free water, and increases with decreasing moisture content.

The most recent development for characterizing the water status of plant tissue concerns the concept of water potential [20]. The water potential is defined

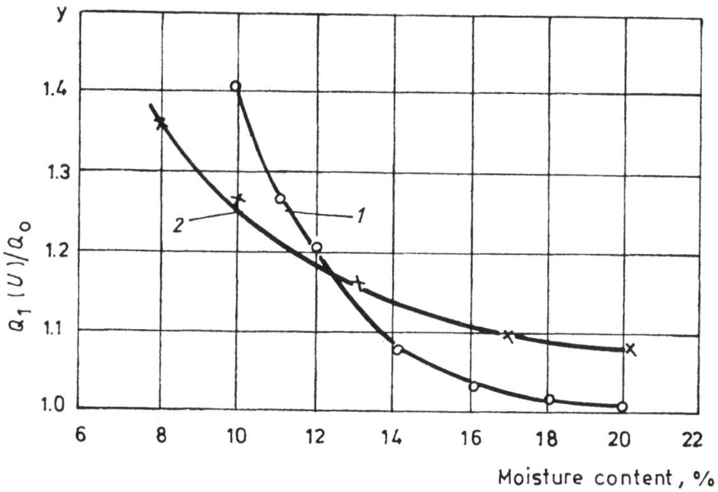

Fig. 41. Heat of evaporation of water bound in crops as a function of moisture content. *(1)* Wheat; *(2)* maize

as the deviation from a reference value of the chemical potential of water divided by the molal volume of water. This quantity has dimensions of energy per unit volume of water.

The water potential of air containing water vapor is given by the expression

$$\psi_{air} = (RT/V_w) \ln \varphi$$

where V_w is the molal volume of liquid water (18 $cm^3\,mol^{-1}$), and φ the relative humidity of the air (decimal). The water potential of an aqueous solution, such as cell sap, can be determined as

$$\psi_{sol} = (RT/V_w) \ln \varphi_w,$$

where φ_w is the mole fraction of water in the solution. The basic relation adopted in practical water-potential measurements for vegetative tissues is that for a system consisting of an aqueous solution and an air–water-vapor atmosphere in chemical equilibrium, i.e., for which $\psi_{air} = \psi_{sol}$.

The mole fraction φ_w in solutions is always less than unity, while the water surrounding a cell externally has $\varphi_w = 1.0$. This means that an inward potential gradient exists, producing water flow into the cell. In response to the water flow into the cell the pressure in it will rise (the so-called turgor pressure), straining the cell wall and giving a higher modulus of elasticity to the whole body.

7.2 Adsorption and desorption of water

It was seen earlier that the bonding of the first layer of water molecules held by molecular adsorption is the strongest. It is thus understandable that removal of this layer is the most difficult. From it follows that both desorption and adsorption of water by products having a low moisture content frequently occur with greater difficulty and at a lower rate than in the case of higher moisture contents. Figure 42 shows the adsorption rate for rice as a function of time at various relative air humidities [18]. The tests were performed by placing grains of equilibrium moisture content at a given (initial) relative humidity into a space with a higher relative humidity. The difference between the two relative humidities may be characterized by the difference in vapor pressure. The higher moisture content corresponds to the higher initial relative humidity. It may be seen from the figure that more moisture is adsorbed in unit time by wetter grains than by those of lower moisture content.

The rate of adsorption by a product depends on its biological construction, on the temperature and on the difference in relative humidity bringing about the adsorption. Figure 43 illustrates the adsorption curve for dried peas and the

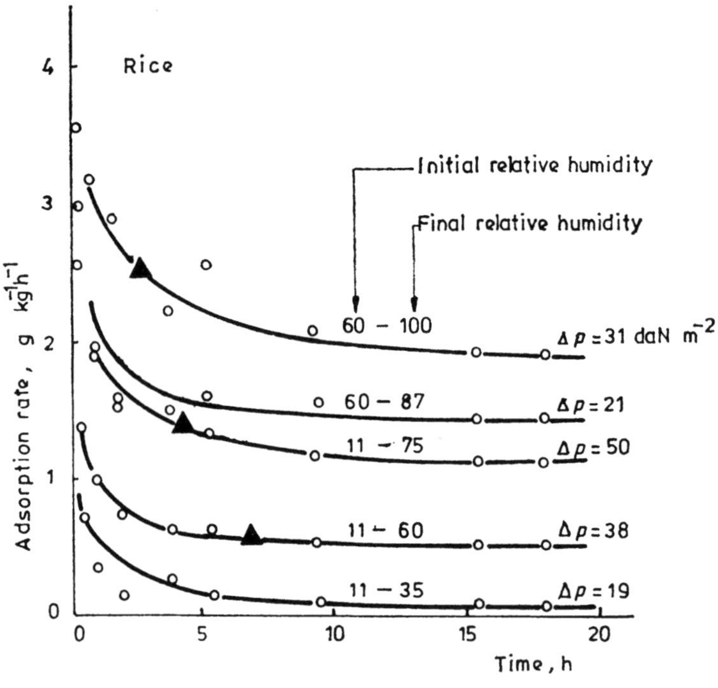

Fig. 42. Rate of water adsorption by rice as a function of time

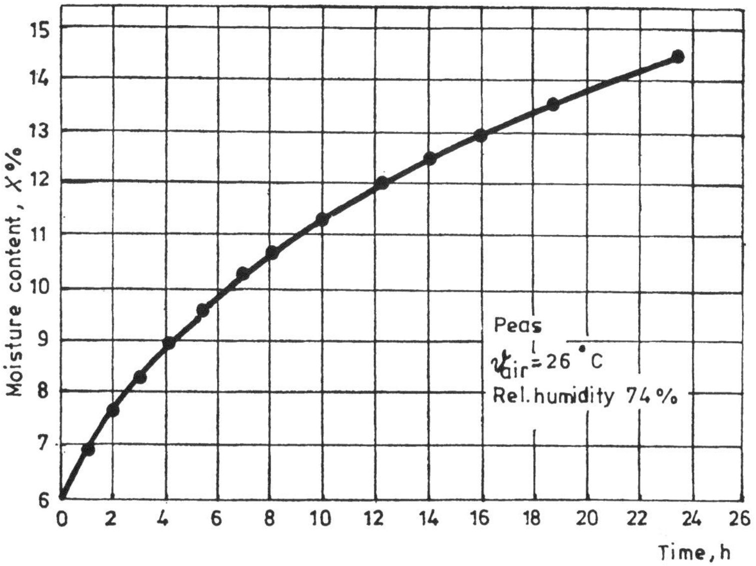

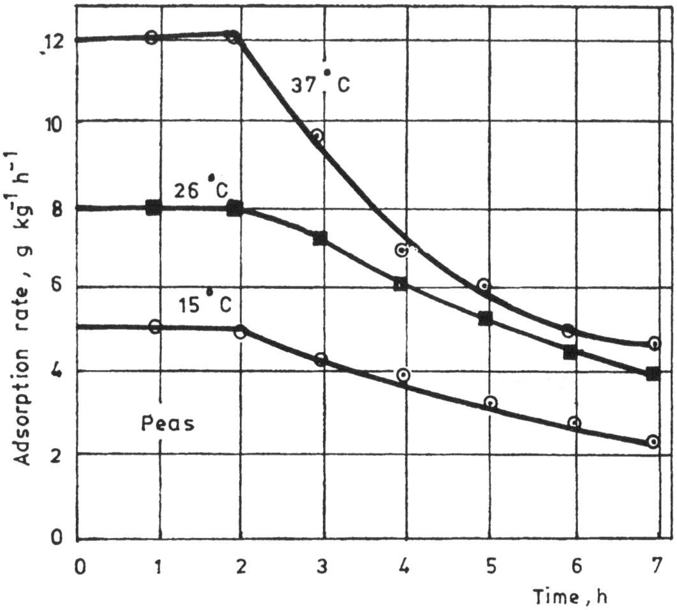

Fig. 43. Water adsorption curve for dried peas

adsorption rate in a medium of 74% relative humidity at 6.2% initial moisture content [22]. As may be seen, an initial phase exists during which the adsorption rate is constant and highest. When the moisture content attains a critical value (in the present case, 7.2%), the adsorption rate decreases. Temperature also influences the adsorption rate significantly [22].

7.3 Equilibrium moisture content

It has been seen in the preceding chapters that agricultural materials will generally adsorb or dissipate moisture when they are placed in media of various relative humidities. At every relative humidity there is one moisture content such that a material neither adsorbs nor dissipates moisture. This is the equilibrium moisture content [17, 36, 37].

If the equilibrium moisture content (relative to a dry basis) at a given temperature is plotted as a function of relative humidity, one of the most important characteristics of a material is obtained, namely, the so-called sorption isotherm (Fig. 44). As may be observed from the figure, the equilibrium moisture con-

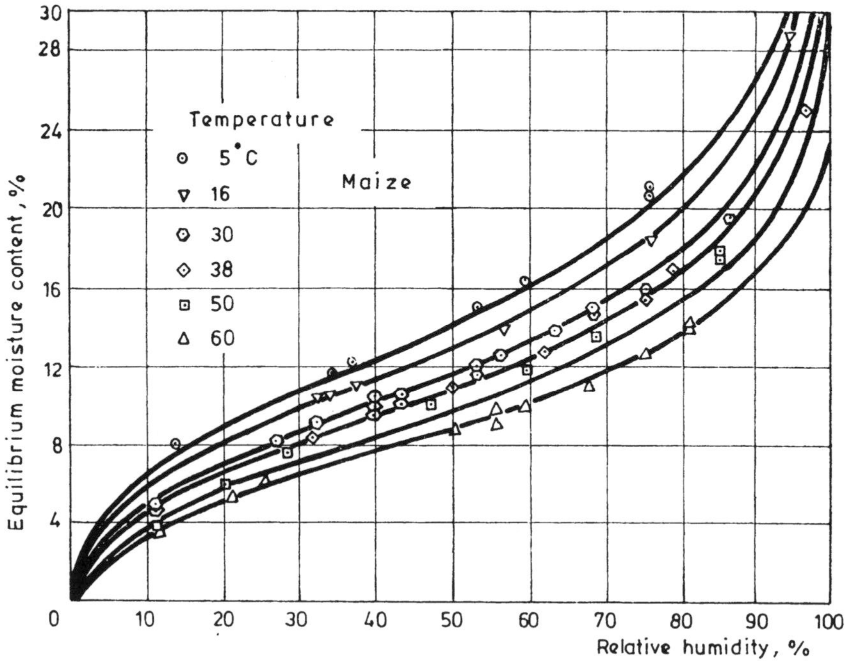

Fig. 44. Water sorption isotherms for maize

tent decreases with increasing temperature. The curves may be expressed by the empirical equation [17]

$$1-\varphi = e^{-kTX_e^n} \tag{13}$$

where φ is the relative humidity, k a material-dependent constant, X_e the equilibrium moisture content, and n an exponent which varies with the material.

Agricultural products consist of biological materials of various constructions (starch, gluten, cellulose, etc.), whose sorption isotherms may deviate significantly from each other. Figure 45 shows sorption isotherms for the individual components of maize kernels [19].

Experiments have shown that on removing moisture (desorption) the equilibrium moisture content is always slightly higher than in the case of adsorption. Thus it is necessary to distinguish between the sorption and desorption isotherms. Figure 46 shows the sorption and desorption isotherms for maize at 22 °C. With

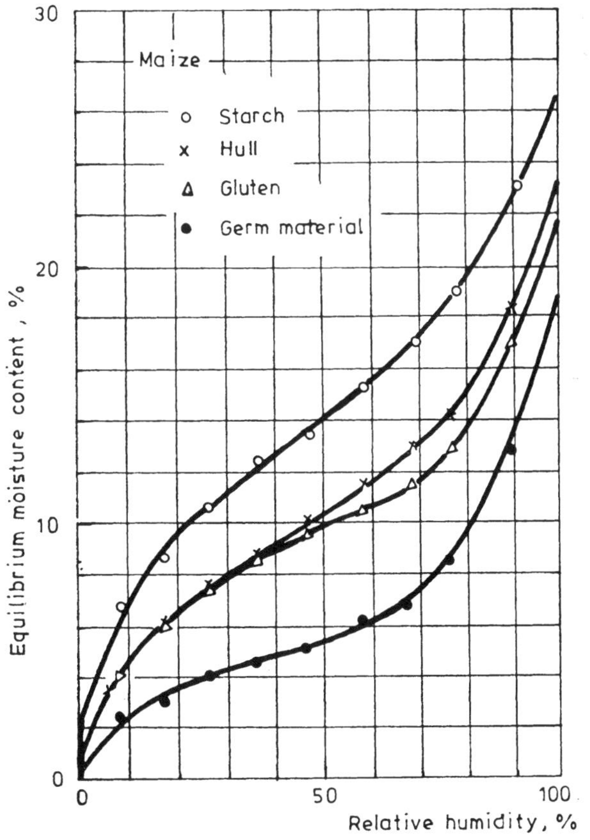

Fig. 45. Water sorption isotherms for individual components of maize

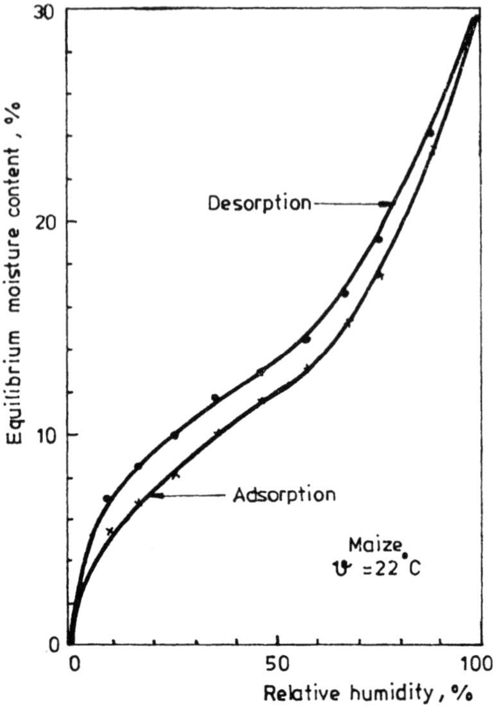

Fig. 46. Water sorption and desorption isotherms for maize

increasing temperature the two curves approach each other more closely, i.e., the hysteresis decreases [19].

Cereals assume their equilibrium moisture content after the passage of a relatively long time. Figure 47 presents the time required to attain equilibrium moisture content for completely dry wheat, as a function of relative humidity [21]. Here again it can be seen that the rate of adsorption is lower for low moisture contents, and the effect of temperature is also obvious.

With variation of grain size the ratio of surface area to volume varies, and so the moisture content of larger grains decreases or increases at a lower rate, owing to the relative reduction in surface area.

Cereal grains cannot be regarded as isotropic, owing to their biological construction. Although water is adsorbed or desorbed through the whole surface of the grains, the rates of these processes differ significantly on the various parts of the surface. The water sorption and desorption rates are the highest for the germ part, and the moisture content of the germ in the equilibrium state is also higher than the mean moisture content of the entire grain, while water adsorption and desorption by the endosperm are considerably slower.

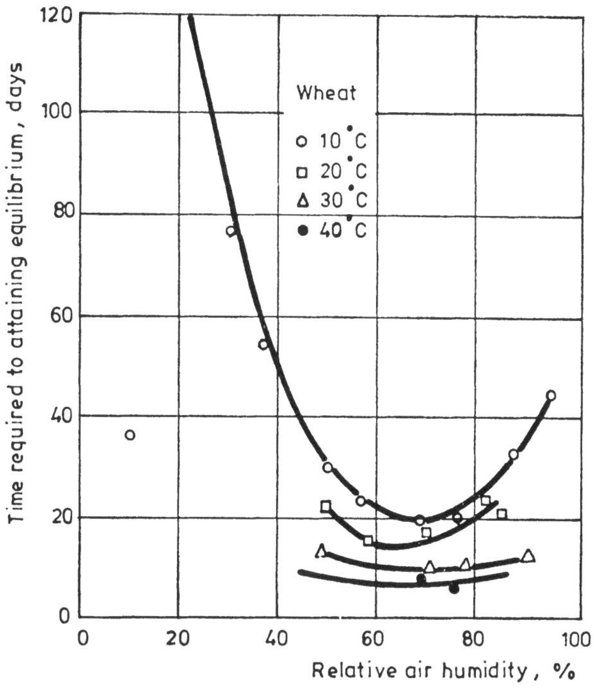

Fig. 47. Time required to attain equilibrium moisture content as a function of relative air humidity

7.4 Moisture adsorption by hygroscopic materials

Certain agricultural materials (e.g., fertilizers, farinaceous materials) adsorb moisture from the air to such an extent that their physicomechanical properties change decisively. These materials are termed hygroscopic.

The bonding of water vapor on the surface of solid bodies is an adsorption process, which may be physical or chemical. We speak of physical adsorption when the water molecules impinging on the surface adhere to the latter under the effect of cohesive forces acting there, and then penetrate into the material by conduction. In the case of chemical adsorption, the vapor molecules reaching the surface are bound by a chemical transformation. Frequently, physical and chemical adsorption occur simultaneously.

Adsorption properties are generally investigated by plotting adsorption isotherms for the equilibrium state. However, hygroscopic materials in most cases attain their equilibrium state after a long time, by which they have adsorbed so much water that they lose their practical value (e.g., the granular fertilizers become liquid). Furthermore, equilibrium measurements give no information whatever concerning the dynamics of water adsorption.

On the basis of the above, hygroscopicity is better understood in terms of the rate of water adsorption from the air, and it is advisable to determine kinetic curves for water adsorption by hygroscopic materials.

In the case of molecular adsorption, the amount of water bound on the surface is proportional to the number of molecules impinging against unit surface area and to the time elapsed. The number of impinging molecules is directly proportional to the partial pressure of the water vapor; it is therefore advisable to express the moisture adsorbed by hygroscopic materials as a function of the partial vapor pressure and of time, or

$$\Delta X = f(p_g, t)$$

where ΔX is the amount of water absorbed relative to a dry basis, expressed as a weight percentage.

Mester [4] studied the properties of hygroscopic fertilizers, and found that the extent of water adsorption is proportional to the square roots of the partial vapor pressure and of time, i.e., $\Delta X = c\sqrt{p_g t}$.

Figure 48 shows experimental results obtained for various fertilizers. The moisture adsorption curves for fertilizers of various hygroscopicities differ from each other only by the constant c, and so the hygroscopicity of these materials may be characterized by the values of the constant. As the value of c is less than unity

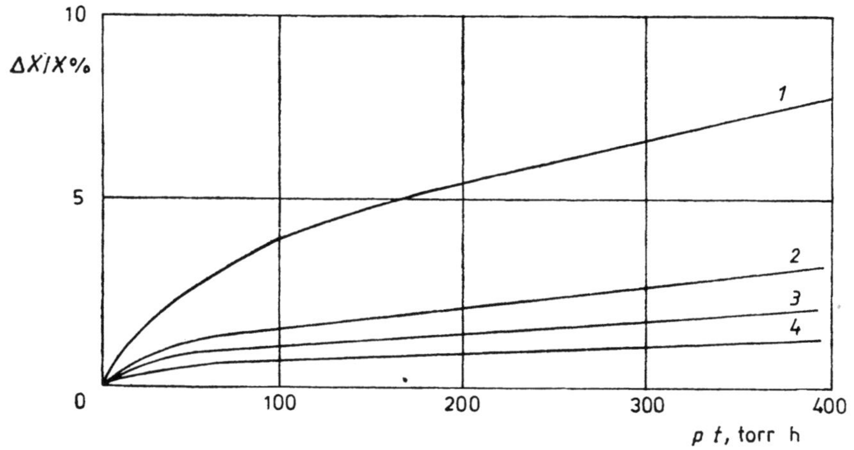

Fig. 48. Hygroscopicity curves for fertilizers. *(1)* Superphosphate; *(2)* saltpeter (nitrogen); *(3)* potassium salt; *(4)* carbamide

for the measuring units selected (torr, h), it is advisable to use the value $c \times 100$ and to denote it by *hy*: then

$$\Delta X = hy(\sqrt{p_g t}/100) \quad (14)$$

Table 3 shows values of *hy* for a number of fertilizers.

Table 3

Fertilizer	*hy*
Carbamide	7.72
Potassium salt	12.02
Peretrix	16.81
Saltpeter (nitrogen)	17.45
NPK	19.55
Superphosphate	41.01

7.5 Internal moisture movement

Most agricultural products may be regarded as colloidal, capillary-porous materials, in which moisture may migrate in the liquid or vapor phase. In the case of isothermal mass transfer, moisture migrates under the effects of partial vapor pressure, osmotic pressure, moisture gradients and, in certain cases, pressure developing in the material. The basic equation for isothermal mass transfer may be written in the form

$$q = -\lambda_m \operatorname{grad} \Theta_m \quad (15)$$

where λ_m is the mass-conduction coefficient, and grad Θ_m the gradient of the mass-transfer potential. The potential Θ_m is to be understood in the general case as the total thermodynamic potential, which is the sum for the individual phases and components of the system and may be expressed in the case of hygroscopic materials by the chemical potential (μ).

Under isothermal conditions the mass-transfer potential may be expressed in terms of the moisture content as

$$\Delta\Theta_m = \Delta u/c_m$$

where c_m is the specific mass, giving the variation of moisture content for unit potential variation, i.e.,

$$c_m = \Delta u/\Delta\Theta_m$$

or

$$c_m = \Delta u/\Delta\mu$$

Equation (15) may be written on the basis of the above as

$$q = (-\lambda_m/c_m\varrho_0)\,\varrho_0 \text{ grad } u = -D\varrho_0 \text{ grad } u \tag{15a}$$

or

$$q = -D\varrho_0 c_m \text{ grad } \mu \tag{15b}$$

where D is the diffusion coefficient ($m^2\ s^{-1}$), μ the chemical potential, and ϱ_0 the density of the dry material.

Under nonisothermal conditions (e.g., on drying), moisture migrates as a result not only of the moisture gradient, but also of the temperature gradient. Under the effect of a temperature gradient, the increase of molecular diffusion, the increase of pressure due to heating of enclosed air, and the variation of capillary suction with temperature all cause moisture movement, always in the direction of heat flow. This means that a temperature gradient brings about moisture movement towards the center of a body during heating and towards the surface during cooling. The specific moisture flow may be written with consideration of the temperature gradient as

$$q = -D\varrho_0 \text{ grad } u - D\varrho_0\delta \text{ grad } \Theta \tag{15c}$$

where δ is the thermogradient coefficient ($kg\ kg^{-1}\ K^{-1}$) and grad θ the temperature gradient. The mass-transfer potential for hygroscopic materials may be expressed (with $\Theta_m = |\mu| = E$) by the thermodynamic equation

$$\mu = RT \ln \varphi \tag{16}$$

where R is the universal gas constant, whose value is $1.979\ cal\ mol^{-1}\ K^{-1}$ or $8.312\ J\ mol^{-1}\ K^{-1}$, and φ is the relative air humidity.

7.6 Mass-transfer at the surface

Moisture reaching the surface from internal parts of a material is generally removed by convection. If there is no artificial air movement and natural convection is also negligible, then the water gets into the environment by diffusion from the surface.

The relationship between the migration of moisture within a material and its removal from the surface may vary. For example, during the aeration of a potato pile the internal migration of moisture is sufficiently rapid to prevent the development of practically any moisture gradient within the tubers, and so the amount of moisture removed from the surface is determined by the permeability of the peel and by the rate of convection. In other cases (e.g., that of drying cereals), the internal migration of moisture cannot ensure a saturated state on the surface or directly under the husk, and so the amount of water which may be removed from the surface is limited by the internal moisture flow.

In the case of a free water surface, the mass-transfer rate may be described by the equation

$$Q = (M\beta F/RT)(p_s - p)$$

where M is the molecular weight, F the area of the evaporating surface, and β the evaporation coefficient.

In most cases agricultural products have a peel of a certain thickness, which behaves as a porous membrane and whose construction differs essentially from that of the internal part (e.g., the peels of fruits or of root bulb products). In these cases the moisture is transferred by diffusion through the peel to the surface, and thence by convection to the environment. These two resistance components are connected serially to each other, and in this case the mass-transfer equation is

$$Q = (p_s - p)/(r\delta RT/MDF + RT/M\beta F)$$

where $r\delta$ is the peel parameter (the product of the specific resistance and the thickness of the peel), and D the molecular diffusion coefficient.

In practical cases, only a part of the total geometrical surface area can be taken into account. In such cases γF is substituted for the surface area F, where $\gamma < 1$ and its value is determined experimentally. For example, the values found for potatoes during aeration are $r\delta = 3.474 \times 10^{-3}$ m and $\gamma = 0.0089$. The value of $r\delta$ may also depend on the partial-pressure difference $(p_s - p)$ as has been found in the case of potatoes [27].

The evaporation coefficient β appearing in the above equations may be determined from the general similarity equation of the form

$$N'u = A\,Re^n$$

where

$$N'u = \beta d/D$$

and

$$Re = u_0 d/6(1-\varepsilon)\nu$$

d being the equivalent diameter of the body, ε the porosity of the pile, u_0 the flow rate across the whole surface area, and ν the kinematic viscosity of the medium.

The following equation has been found [27] to apply during the aeration of potato piles:

$$N'u = 0.344\,Re^{0.539}$$

For liquid drops, or for particles covered by liquid films the similarity equation due to Frössling may be used:

$$N'u = 2 + 0.552\,Re^{1/2}Sc^{1/3}$$

where $Sc = \nu/D$ is the Schmidt number. The molecular diffusion coefficient for water vapor in air at atmospheric pressure may be calculated as a function of temperature from the equation

$$D = 2.15 \times 10^{-6} T^{1.85} (\mathrm{m^2\,h^{-1}})$$

7.7 Mass-transfer coefficients

It may be seen from eqn. (15b) that for concrete calculation it is necessary to know the mass-transfer coefficients c_m and D for the given type of material and conditions.

To determine the value of the mass-transfer coefficient c_m, it is necessary to establish the variation of the chemical potential μ as a function of the moisture content of the material, i.e.,

$$c_m = \partial u / \partial \mu$$

Figure 49 shows the functions $\mu = f(u)$ for wheat flour, maize and rice at 20 °C.

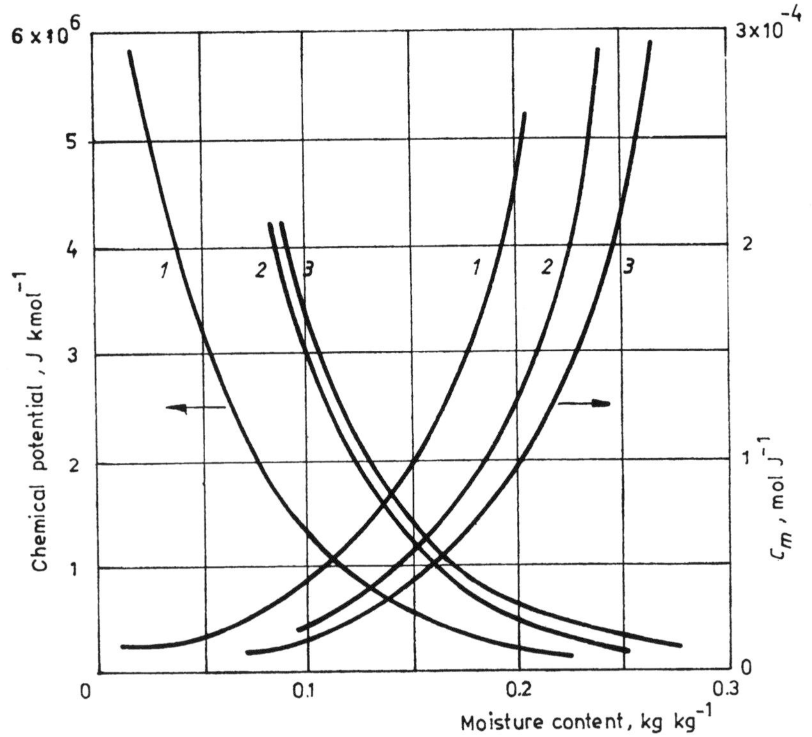

Fig. 49. Dependence of chemical potential on moisture content for various agricultural products *(1)* Wheat flour; *(2)* maize; *(3)* rice

The horizontal axis gives the equilibrium moisture contents to which certain determined relative humidity values correspond (see, for example, Fig. 44). By graphical differentiation of the curves it is possible to obtain the curves $c_m = f(u)$, which increase progressively with increasing moisture content [23].

As is known, the curves representing the equilibrium moisture content (i.e., the sorption and desorption isotherms) show hysteresis. Thus the curves in Fig. 49 are defined uniquely only for one of the curves for equilibrium moisture content (either for increasing or decreasing moisture content). Consequently, there may be cases where the moisture gradient existing at a given relative humidity will not cause any migration of moisture.

The equilibrium moisture content of various materials depends on temperature, and so the chemical potential μ and the mass-transfer coefficient c_m are also temperature-dependent. Figure 50 shows the corresponding curves for wheat, for various moisture contents. At a given moisture content, c_m is seen to increase with temperature [23].

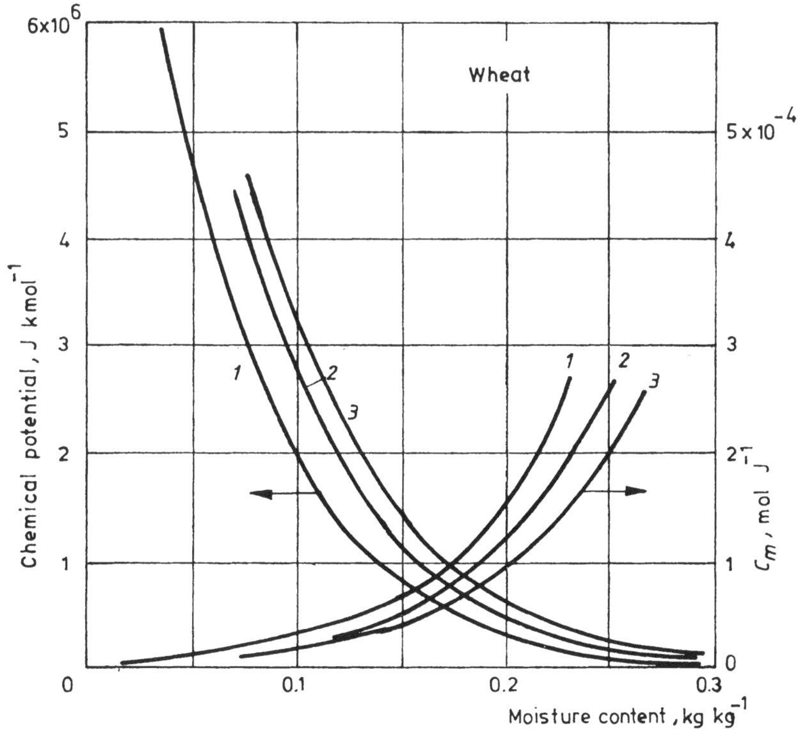

Fig. 50. Effects of temperature on chemical potential and mass-transfer coefficient c_m. *(1)* 50 °C; *(2)* 20 °C; *(3)* 0 °C

The diffusion coefficient D is a basic parameter characterizing moisture migration, determining decisively the velocity of migration. Determination of the diffusion coefficient is rather difficult in the case of cereal grains, owing to their small dimensions. This is why so few experimental data are available and why even these are contradictory in many cases [23]. Figure 51 depicts the diffusion coefficient for wheat as a function of moisture content at various temperatures; D increases with increasing moisture content in agreement with the previous statement that water may be both adsorbed and desorbed more easily by wetter material.

The diffusion coefficient increases greatly with increasing temperature [23]. This dependence is usually expressed by Arrhenius' law in the form

$$D = D_0 e^{-E/RT}$$

where E is the activation energy, whose value is 40–50 kJ mol^{-1} for wheat and about 30 kJ mol^{-1} for rice. The temperature dependence is expressed more conveniently by the simple exponential function

$$D = D_0 (T/273)^n$$

where T is the absolute temperature (K), and n an exponent whose value varies

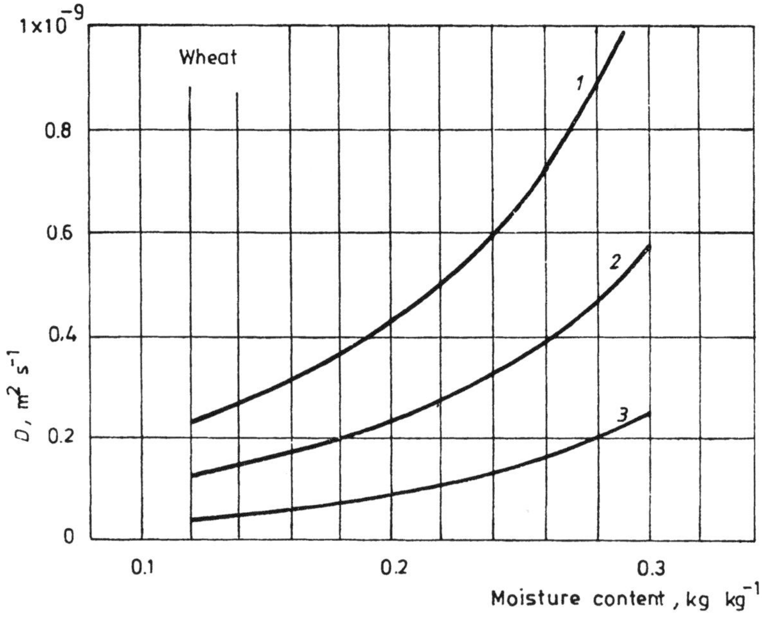

Fig. 51. Diffusion coefficient of wheat as a function of moisture content. *(1)* 60 °C; *(2)* 40 °C; *(3)* 20 °C

between 10 and 12 for wheat and maize. The exponent n is also influenced by the moisture content: with decreasing moisture content, the value of n decreases slightly.

For more rigorous calculations the effect of the density change and the concentration gradient should be taken into account. A composite model describing the effects of shrinkage and concentration on the mass diffusivity may be used, in the form

$$D = D_0 e^{-E/RT} (\varrho_0/\varrho)^{n_1} e^{n_2(C-C_0)}$$

where ϱ_0 and ϱ are the initial and instantaneous dry-mass densities (kg m^{-3}), C_0 and C the initial and instantaneous liquid concentrations (kg m^{-3}), and n_1 and n_2 are constants.

Knowing the material characteristics mentioned above, the approximate rate of moisture flow may be determined from eqn. (15b), with the given conditions taken into account. For small-sized materials of nearly spherical shape (e.g., cereals), eqn. (15b) may be written in the form

$$q = -D\varrho_0 c_m (\mu_1 - \mu_2)/r_m$$

where μ_1 and μ_2 are the chemical potentials respectively on the surface and in the center, and r_m is the mean radius of the material.

As an example, let us determine the moisture flow for drying wheat, when the initial moisture content is $u=0.25$ kg kg^{-1}, the external temperature of the grains is 40 °C, $D=0.3\times10^{-9}$ m^2 s^{-1}, and $c_m=2.4\times10^{-7}$ kmol J^{-1}. For drying air at a relative humidity of $\varphi \simeq 0.1$, the chemical potential on the surface is

$$\mu_1 = RT \ln \varphi = 8.32\times10^3 \cdot 313 \ln 0.1 = -60\times10^5 \text{ J kmol}^{-1}$$

In the center of the grain the initial moisture content is in equilibrium with a relative humidity of $\varphi \simeq 0.9$ and calculations for an internal temperature of 30 °C give

$$\mu_2 = RT \ln \varphi = 8.32\times10^3 \cdot 303 \ln 0.9 = -2.65\times10^5 \text{ J kmol}^{-1}$$

With the above values,

$$q = -0.3\times10^{-9} \cdot 1.2\times10^3 \cdot 2.4\times10^{-7}[-(60-2.65)\times10^5/1.5\times10^{-3}] =$$

$$= 3.30\times10^{-4} \text{ kg m}^{-2} \text{ s}^{-1}$$

Again, few data are available for the thermogradient coefficient δ. The values measured by Likov [29] for wheat vary between 0.00025 and 0.0005 kg kg^{-1} K^{-1}. The lower values were obtained for the lower moisture ranges (5–8%).

7.8 Moisture gradients

According to eqn. (15c), the specific moisture flow may be written in the form

$$q = D\varrho_0(\text{grad } u+\delta \text{ grad } \Theta)$$

from which the expression for the moisture gradient is

$$\text{grad } u = -(q/D\varrho_0)-\delta \text{ grad } \Theta$$

As may be seen, the moisture gradient is determined by the rate of moisture removal (q), by the diffusion coefficient, by the temperature gradient, and by the coefficient δ.

In the case of cereal grains, the temperature difference between the surface and the center does not generally exceed 10 °C. The mean radius of the grains is 2.0 mm, so that the temperature gradient in the case of a linear temperature distribution is 5000 K m^{-1}, while in the case of a parabolic distribution it is 10 000 K m^{-1}. Calculations with $\delta=0.0005$ give a value of δ grad Θ amounting to 5.0 kg kg^{-1} m^{-1} at the most.

The first term on the right-hand side of the equation for grad u is greater by several orders of magnitude, and so the moisture gradient caused by the temperature distribution is negligible to a first approximation. The specific moisture flow relative to a surface area F may thus be expressed in terms of the drying rate in the form

$$q = Wm_{dm}/F$$

where W is the drying rate (kg kg^{-1} s^{-1}), and m_{dm} the dry weight of the material. The surface area F of a grain may be expressed in terms of its equivalent diameter d_e as

$$F = 6m/d_e\varrho$$

where m is the mass of the grain, and ϱ its mass density. Utilizing the above expressions, then,

$$q = Wm_{dm}d_e\varrho/6m = (Wd_e/6)\,\varrho/(1+u_0)$$

where u_0 is the initial moisture content. Now the gradient of moisture content on the surface is

$$(\partial u/\partial x)_F = -(Wd_e/6D)\varrho/\varrho_0(1+u_0)$$

According to calculations performed on the basis of drying test results, the moisture gradient is high at the beginning of the process, and then decreases rapidly with increasing temperature. As the moisture content decreases in the vicinity of the surface, the reduction of the moisture gradient is diminished by

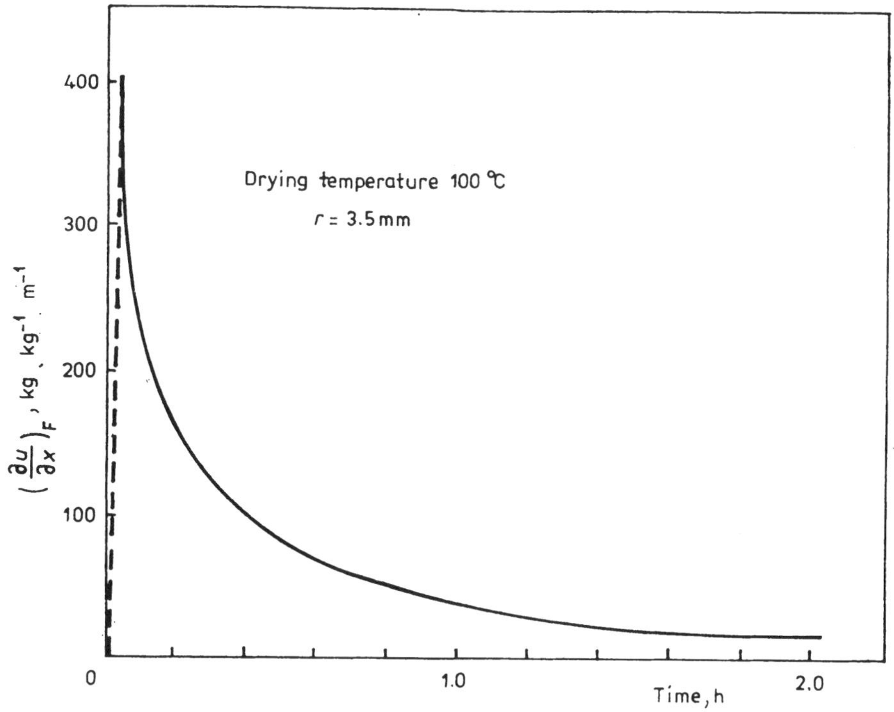

Fig. 52. Variation of the moisture gradient

the decrease of the diffusion coefficient (Fig. 52). The lower the final moisture content attained in drying, the lower the diffusion coefficient at the end of the process and the longer the drying period [23].

The distribution of moisture in a grain can as yet be determined only by approximate methods. Experimental control cannot be implemented because of the small sizes involved. The distribution of moisture at various times may be determined by calculation, using values of the diffusion coefficient and drying rate (see Fig. 58).

Radial moisture gradients will cause shrinkage in the various layers of a body, leading to development of stress. These shrinkage stresses may cause stress-crack phenomena during the drying process, decreasing the quality of the product. Therefore, the development of moisture gradients should be controlled in all cases, if the quality of the product is in any way of importance.

7.9 Contact moisture exchange

During various technological processes (e.g., drying, storage, treatment of seeds), parts having differing moisture contents come into mutual contact, resulting in equalization of their moisture contents. Products of higher moisture content dissipate water, while those of lower moisture content adsorb it. The exchange of moisture takes place partly in the liquid phase, by conduction, and partly in the vapor phase, by desorption and adsorption.

The two types of moisture migration can hardly be separated, either theoretically or practically: i.e., calculations of moisture exchange in this manner are not possible. Information concerning the course of the process must thus be obtained primarily from the results of experiments performed under various conditions.

Figure 53 shows moisture-content curves for a mixture of maize kernels with 10.4 and 28% moisture contents mixed in 1:1 proportions, at 20 °C, as functions of time [23]. The initial sections of the curves show that during the initial phase of moisture equalization the water content of the material with the higher moisture content decreases more rapidly than that of the material with the lower moisture content increases. Consequently, a part of the water desorbed is present in the vapor phase among the kernels, and is adsorbed more slowly by the drier material.

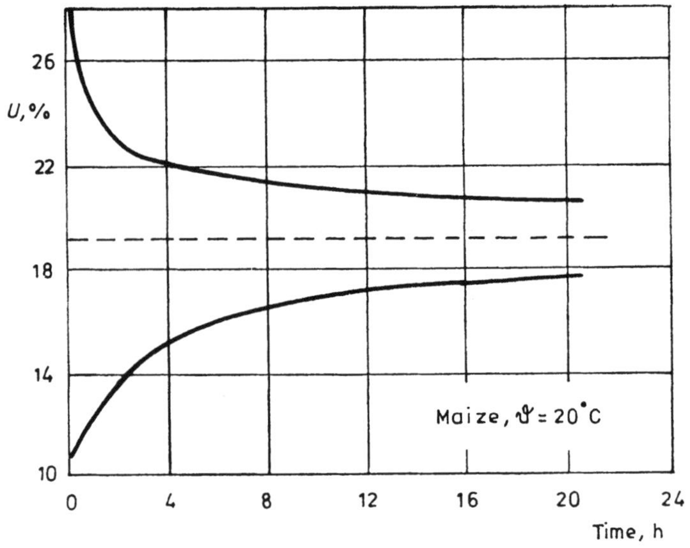

Fig. 53. Equalization curves for maize with various moisture contents

The second important conclusion is that the exchange of moisture does not result in complete equalization of the moisture contents, even after a long time. The reason is the hysteresis between the sorption and desorption isotherms. The extent of hysteresis depends on the properties of the material, on temperature and on the eventual treatment (e.g., drying). After drying at particularly high temperatures the hysteresis increases and then the equalization of moisture is effected to an even lesser extent. Figure 54 presents equalization curves for maize with 8.2, and 24.4% moisture contents, at 38 °C [25]. After 70 h the moisture difference has hardly diminished, even after a prolonged period of contact.

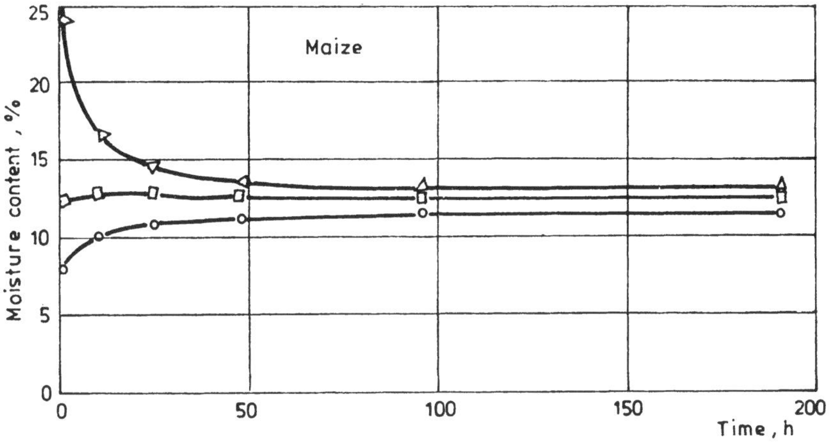

Fig. 54. Equalization of moisture content for maize at 38 °C

The time required for such equalization as does occur depends mainly on temperature. The higher the temperature, the faster the equalization. The effect of temperature on the equalization in the case of maize is shown in Fig. 55 [25]. At 4 °C, about 200, while at 38 °C only 70 h are required for equalization. The residual moisture difference remaining permanently after equalization is again a function of temperature. With increasing temperature the moisture difference decreases.

The temperatures of the materials in contact may also differ and then temperature equalization occurs after mixing. According to measurements, this is much more rapid than moisture exchange. Figure 56 shows the temperature curves for wheat at temperatures of 10 and 52 °C, as functions of time [23]. Equalization occurs in 2 min, which is a negligibly short time compared to the 40–70 h required for moisture exchange.

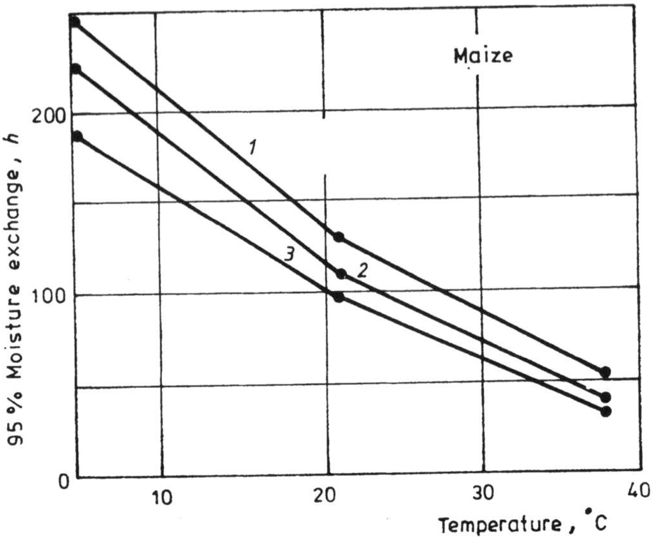

Fig. 55. Effect of temperature on equalization rate for maize.
(1) $U=8\%$; *(2)* $U=20\%$; *(3)* $U=25\%$

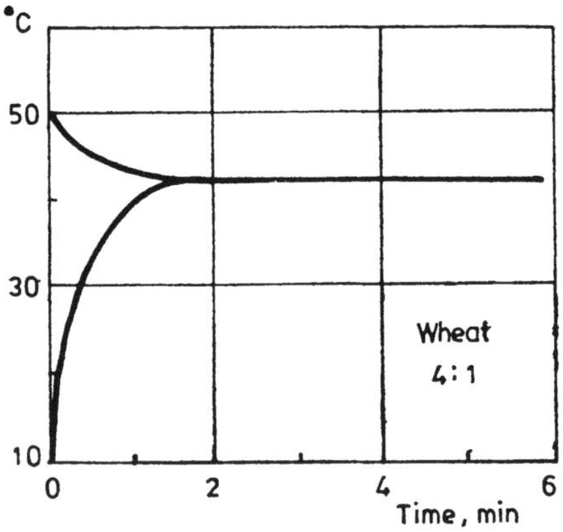

Fig. 56. Equalization curves for wheat at various temperatures (mass ratio: 4 : 1)

7.10 The theory of drying

The task of drying is the reduction of moisture content to a given final value, at which the material being dried can be stored (in the case of cereals, this value is generally 14%). In order to remove or evaporate moisture, heat must be introduced into the material. The energy may be derived from a reduction in the temperature of the material and the water contained in it, or from heat conducted from the surface. Water reaches the surface from the inside of the material by conduction, and is usually removed from the surface by convection. Thus the process of moisture removal may be reduced to the problem of simultaneous heat and mass transfer.

The drying process for bulk agricultural products is in practice a very complex process, owing to various disturbing phenomena. In bulk material lying in several layers a complex two- or three-dimensional flow develops. The shapes of individual grains deviate from spherical, and the material cannot be regarded as homogeneous as concerns moisture conduction. Heat and moisture exchange may also develop between individual grains in contact reliable calculation of which is subject to difficulties as mentioned earlier.

In a spherical body, water vapor entering a differential element of width Δr at radius r increases the vapor concentration and moisture content of the element. The partial differential equation describing the process may be written [39] as

$$(1/r^2)(\partial/\partial r)\left(Dr^2(\partial C/\partial r)\right) = f(\partial C/\partial t)+(1-f)\gamma(\partial X/\partial t)$$

where D is the diffusion coefficient, f the void fraction of the body, and γ the specific weight of the solid material. The quantity of heat entering the element by conduction increases the heat content of the material and of the water contained in it, which is further increased by the latent heat of the moisture adsorbed and by any generation of internal heat. The corresponding differential equation for heat conduction may be written in the form

$$(1/r^2)(\partial/\partial r)\left(r^2\lambda(\partial T/\partial r)\right) = c_k(1-f)\gamma(\partial T/\partial t)+c_w(1-f)\gamma X(\partial T/\partial t)-$$
$$-h(1-f)\gamma(\partial X/\partial t)-Q$$

where λ is the heat-conduction coefficient, c_k and c_w are the specific heats respectively of the solid material and of water, h is the latent heat of vaporization of water, and Q the heat released in unit volume per hour.

The moisture content of a material is determined by the relative humidity of the medium and by the material's temperature, according to the sorption curve for equilibrium moisture content. The analytical expression for the exact relationship is rather complicated (see eqn. (13)), it is therefore advisable in approximate investigations to use the simpler linear relationship

$$X = \alpha+\beta C-\delta T$$

where C is the vapor concentration, (kg m^{-3}), T is the temperature (K), and α, β and δ are constants. The diffusion coefficient as a function of moisture content and temperature is used in the simplified form

$$D = D_1+D_2X+D_3T$$

while the heat-conduction coefficient, with the term for moisture content also taken into account, is

$$\lambda = \lambda_1 + \lambda_2 X$$

For further processing of the differential equations it is convenient to introduce the dimensionless variables $\psi = D_1 t/a^2$, $\Theta = C/C_0$, $\Phi = T/T_0$ and $R = r/a$ where a is the radius of the sphere or grain, C_0 the initial vapor concentration, and T_0 the initial temperature of the body. The above transformations taking into account the differential equations assume the form

$$\partial\Theta/\partial\psi - [(1-f)\delta\gamma T_0/(A_1 C_0)](\partial\Phi/\partial\psi) = [D/(A_1 D_1)](\partial^2\Theta/\partial R^2) + \\ +[2D/(A_1 RD_1)](\partial\Theta/\partial R) + [(D_2\beta C_0)/(A_1 D_1)](\partial\Theta/\partial R)^2 + \\ +[((D_3 - D_2\delta)T_0))/(A_1 D_1)](\partial\Phi/\partial R)(\partial\Theta/\partial R) \tag{17}$$

$$\partial\Theta/\partial\psi - [((c_k + c_v X + h\delta)T_0)/\beta h C_0](\partial\Phi/\partial\psi) = -(\lambda T_0/A_2)(\partial^2\Phi/\partial R^2) - (2\lambda T_0/A_2 R)\times \\ \times(\partial\Phi/\partial R) - (\beta\lambda_2 T_0 C_0/A_2((\partial\Theta/\partial R)(\partial\Phi/\partial R) + (\lambda_2\delta T_0^2)/A_2)(\partial\Phi/\partial R)^2 - (Qa^2/A_2) \tag{18}$$

where

$$A_1 = f + (1-f)\gamma\beta$$

and

$$A_2 = \beta h(1-f)\gamma D_1 C_0$$

Equations (17) and (18) may be solved by the finite-difference method. The partial derivatives with respect to time and place are substituted by finite increments. For example, the partial derivatives with respect to time may be approximated as

$$\partial\Theta/\partial\psi = [\Theta(R, \psi + \Delta\psi) - \Theta(R, \psi)]/\Delta\psi$$

$$\partial\Phi/\partial\psi = [\Phi(R, \psi + \Delta\psi) - \Phi(R, \psi)]/\Delta\psi$$

By substituting the partial differentials the values of $\Theta(R, \psi + \Delta\psi)$ and $\Phi(R, \psi + \Delta\psi)$ may be expressed. If the initial state of the body is known, the new conditions after the elapse of a time increment $\Delta\psi$ may be calculated. The heat transfer and the diffusion of the water vapor may be determined for the whole drying process by iteration.

The radius is divided into 10 equal parts and the variables are determined for each of these parts as the given time increment. The average values for the complete sphere are calculated from the data for the individual parts.

By means of the above-derived equations the effects of various parameters on the drying process may be investigated: primarily, those of the diffusion and heat-conduction coefficients.

The rate of water removal is determined basically by the diffusion coefficient. Figure 57 shows the effect of the diffusion coefficient on the drying period, with

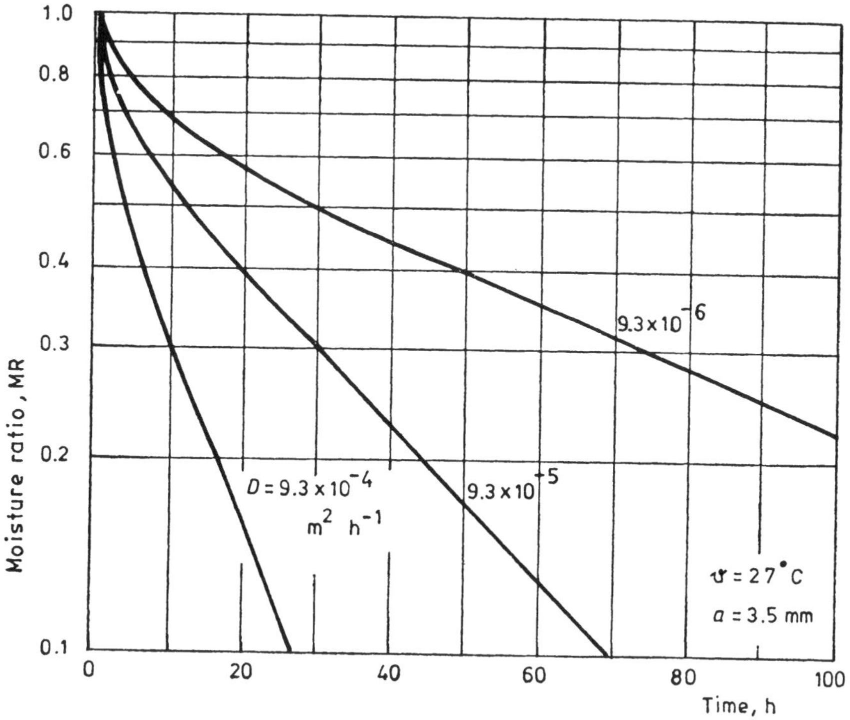

Fig. 57. Effect of diffusion coefficient on drying rate

the assumption that the body is always in thermal equilibrium with its environment, which corresponds to the case of $\lambda=\infty$ [39]. Values of the moisture ratio $MR=(X-X_e)/(X_0-X_e)$ are plotted on the vertical axis. The heat-conduction coefficient λ reduces the velocity of water removal only if it is very low. In practical cases the heat-conduction coefficient is such as to ensure introduction of the heat required and so its effect may be left unconsidered.

The ratio of the thermal and mass diffusivities is characterized by the Lewis number

$$Le = a/D = \lambda/c\gamma D$$

where a is the temperature conductivity (thermal diffusivity), and c the specific heat of the material. If the Lewis number exceeds 0.1, moisture removal is not limited by heat conduction and is determined by the diffusion coefficient.

From eqns. (17) and (18), a modified Lewis number may be derived neglecting the interaction between the vapor concentration and temperature change, in the form [39]

$$Le' = \lambda[f+(1-f)\gamma\beta]/D(1-f)\gamma(c_k+c_v X+h\delta) \simeq \lambda\beta/D(c_k+c_v X+h\delta)$$

According to calculations, the critical value of the modified Lewis number is 60 In the case of lower Lewis values the removal of moisture is limited by heat conduction, and in the case of Le' values exceeding the critical value the limiting factor is the diffusion coefficient.

For drying of cereals the value of the Le is 300–500, that of Le' is $1.0–1.5 \times 10^5$. These values are far above critical, and so the rate of water removal is limited by the diffusion coefficient. To a first approximation the drying period may be assumed to be inversely proportional to the square root of the diffusion coefficient.

Temperature is a decisive factor in drying processes. Both the diffusion coefficient and the concentration gradient increase with temperature, whereby the amount of water removed is also increased. Above 100 °C the partial pressure of water vapor increases in the material such that it may exceed the external pressure. In this case the outward motion of moisture, i.e., the removal of water, is also promoted by the pressure gradient.

The drying time is a complex function of temperature (Fig. 64), since the latter's effect differs in the various phases of drying. Generally, the drying period may be assumed to be inversely proportional to temperature, to the power of 2.0–2.5.

Figure 58 shows the calculated moisture profile developing during the drying

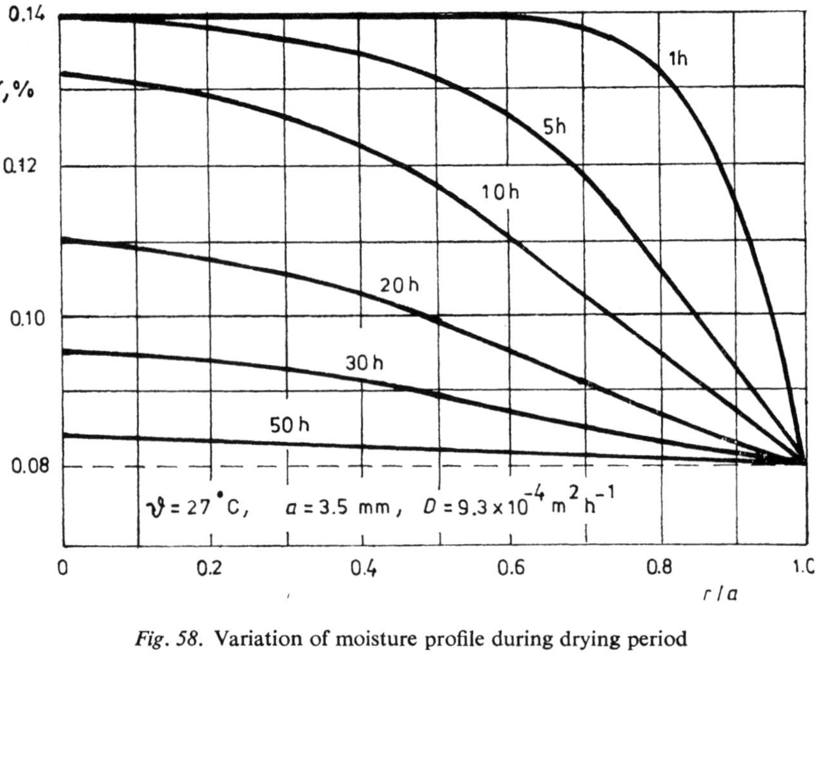

Fig. 58. Variation of moisture profile during drying period

period [39]. The surface is dried in a short time to the equilibrium moisture content, and then the removal of water starts from the inner part. The maximum moisture gradient is obtained under the surface in the initial phase of drying, the gradient then decreases gradually as drying advances.

7.11 General relationships in the drying process

The task of drying is the reduction of moisture content in order to preserve a product, to reduce its weight or its volume, etc. The majority of agricultural products which are dried may be regarded as solid, porous, coarse material in a loose bulk state (in a layer or pile). During drying, the pile is blown through by preheated air, by means of which the energy required for evaporation is provided to the material. The water evaporating from the surface of the material is removed by the air.

The heat and mass transfers occurring in the course of drying constitute a complex process, whose mathematical description is as yet possible only with severe approximations. Concrete application of the approximate solutions is further aggravated by deficient knowledge of the relevant quantities (i.e., the diffusion coefficient) for many materials. Therefore, it is general practice to apply semiempirical methods based on experimental data.

Moisture is removed as a result of the difference in vapor pressure between the surface and the ambient. Moisture migrates to the surface under the effect of the moisture gradient forming between the inner parts and the surface. The drying process lasts until equilibrium is attained between the inner parts and the surface, and between the surface and the ambient.

A detailed study of the drying process has shown that it may be divided into three characteristic stages (zones), in which the drying rate varies differently (Fig. 59).

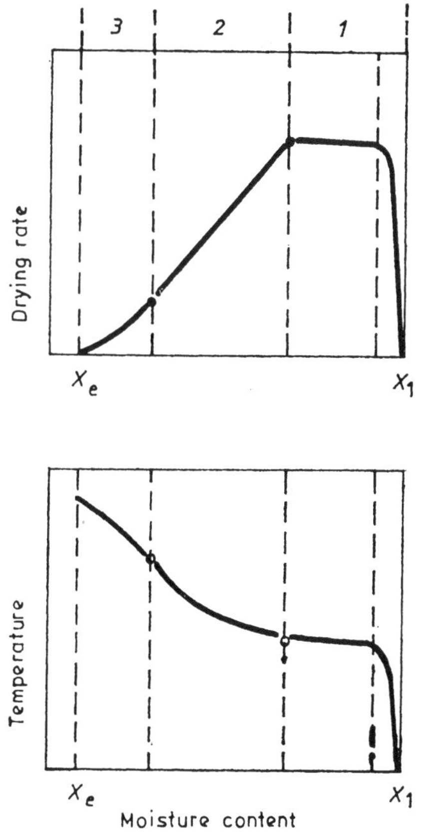

Fig. 59. Stages of drying

In the first stage, starting with a short heating-up period, the drying rate is maximum and constant. In this drying stage the moisture content exceeds the maximum hygroscopic moisture content (see Fig. 44) everywhere in the material. Moisture moves under the effect of capillary and osmotic forces from the inside to the surface of the material, and saturated vapor prevails over the surface. In this case, the drying rate for a given material depends on the characteristics (temperature, relative humidity, velocity) of the drying air. Figure 60 shows the maximum drying rate for maize as a function of the initial moisture content (relative to a dry basis) [31]. The maximum drying rate of cereals may be expressed, on the basis of experimental data, by an empirical function of the form

$$q_{\max} = X_1 v_a^{0.5}(A\vartheta_a + B)$$

where v_a is the velocity of the drying medium, ϑ_a its temperature, and A and B are constants (in the case of maize, $A = 0.0069$, $B = -0.0404$). The end of the first stage is indicated by a break point in the drying rate curve, whose position depends on the initial moisture content and on the air temperature, as may be seen from Fig. 61 [31].

In the second drying stage the moisture content at certain points in the material drops to below the maximum hygroscopic moisture content. The surface of the material dries to the equilibrium moisture content corresponding to the drying air, and the vapor pressure decreases to below the saturated vapor pressure.

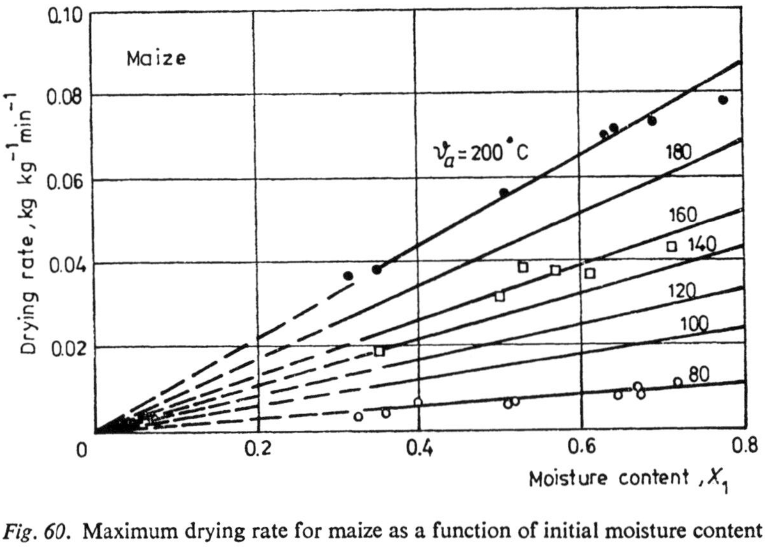

Fig. 60. Maximum drying rate for maize as a function of initial moisture content

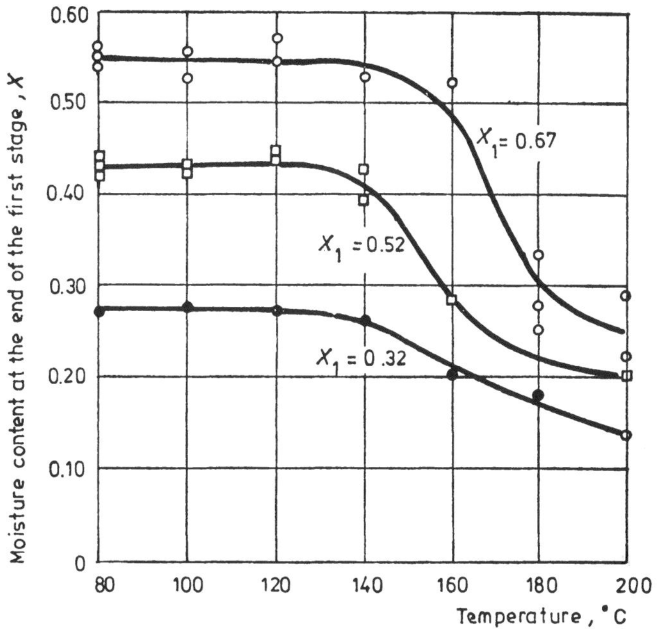

Fig. 61. Moisture content at the end of the first stage of drying, as a function of drying temperature

The evaporation zone advances in the material towards its center; evaporated moisture passes by diffusing through the dry material to the surface. As the moisture content decreases vapor diffusion assumes an ever increasing role in the migration of moisture. The drying rate decreases gradually owing to the higher resistance, and the temperature of the material increases continuously. Figure 62 shows the decrease in the drying rate for maize at various air temperatures [31]. As is seen, increasing the temperature increases the drying rate decisively in this stage also. The drying rate is also affected by the initial moisture content: the higher the initial moisture content, the faster the drying.

The third stage starts when the moisture content of the material is everywhere less than the maximum hygroscopic content (i.e., also in the center). In this case the drying rate decreases further and tends asymptotically to zero. In this stage the average moisture content of the grain is below the storage value (about 14%), and therefore this stage is of no interest practically.

Kinetic curves for drying. During kinetic experiments the decrease of moisture content is plotted as a function of time, for a constant temperature of the drying medium. The curve obtained in this way is termed the kinetic curve for drying. The directional tangent of the curve gives the instantaneous drying rate at the chosen point.

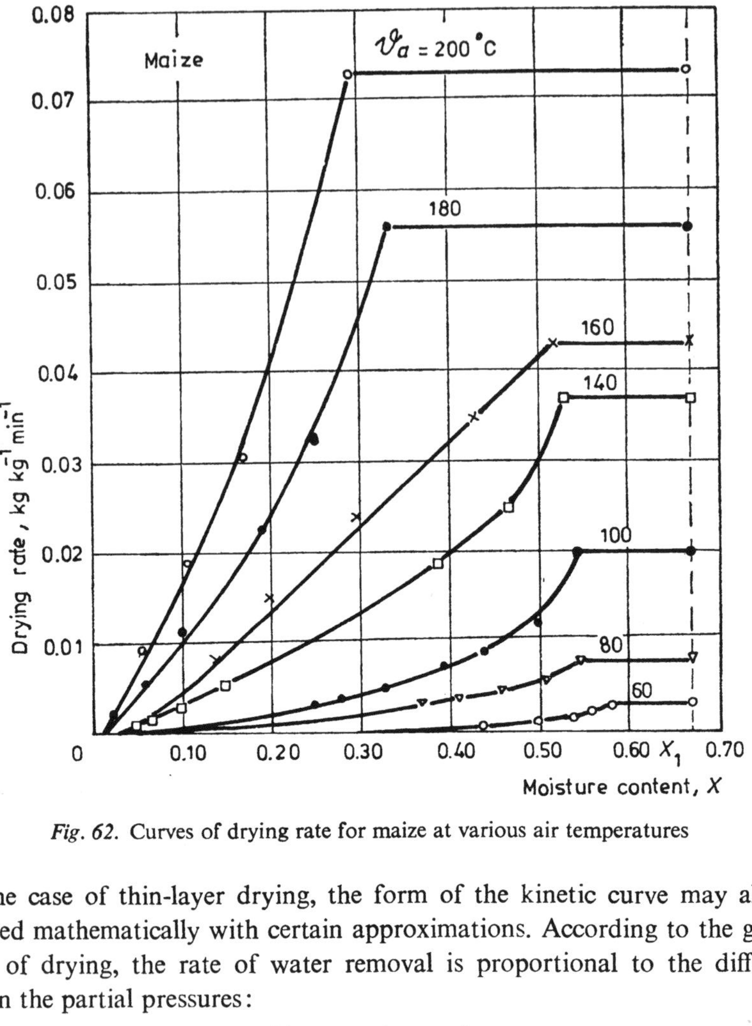

Fig. 62. Curves of drying rate for maize at various air temperatures

In the case of thin-layer drying, the form of the kinetic curve may also be described mathematically with certain approximations. According to the general theory of drying, the rate of water removal is proportional to the difference between the partial pressures:

$$dX/dt = -A(p_F - p_k)$$

where p_F is the partial vapor pressure on the surface of the material, and p_k that in the flowing medium. The partial vapor pressures may be expressed in terms of relative humidities, and so the above equation may be written as

$$dX/dt = -Ap_s(\varphi_F - \varphi_k)$$

where p_s is the saturated vapor pressure at the given air temperature.

Between certain limiting values of the moisture content the sorption isotherm varies approximately linearly and with this assumption it is possible to write

$$dX/dt = -k(X - X_e) \qquad (19)$$

where X_e is the equilibrium moisture content at the given temperature, and k is a drying constant. Integration of eqn. (19) yields the relation

$$(X-X_e)/(X_1-X_e) = e^{-kt} \tag{20}$$

The left-hand side of the equation is termed the moisture ratio, which characterizes the drying process as a decimal number until the equilibrium moisture content has been attained.

The half-period $t_{1/2}$ is understood as the time required for removal of one-half the evaporable quantity of water (X_1-X_e), and its value is

$$t_{1/2} = 0.693/k$$

Figure 63 shows drying half-periods $t_{1/2}$ as functions of the temperature of the

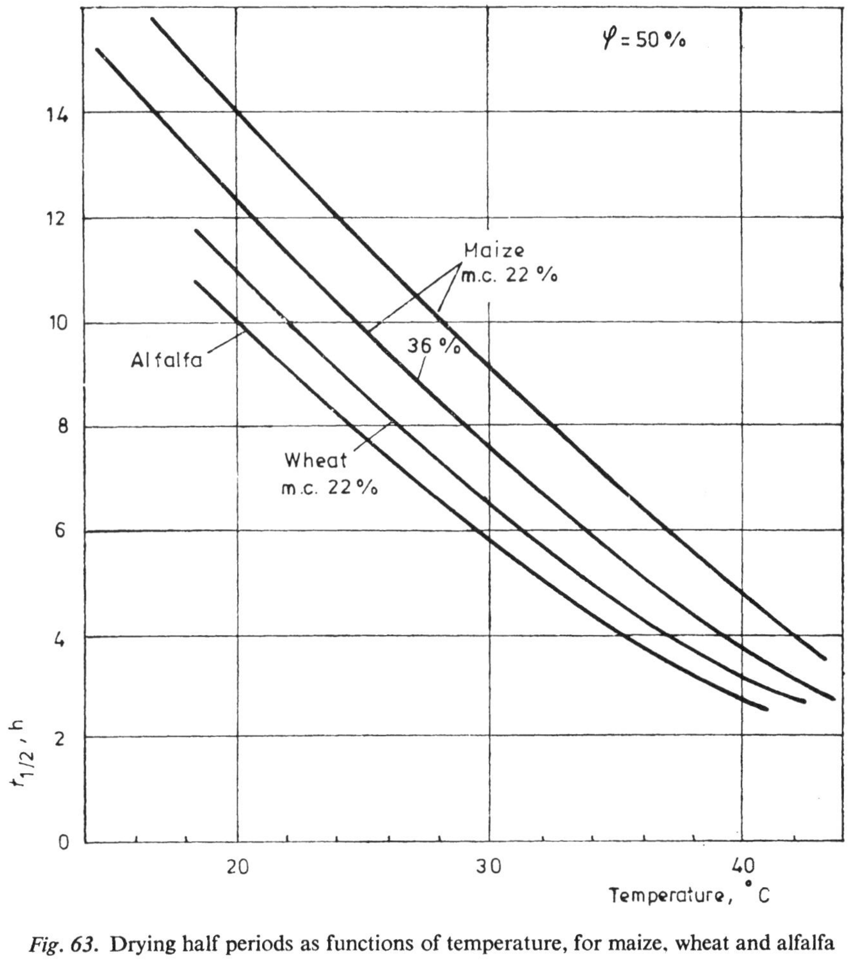

Fig. 63. Drying half periods as functions of temperature, for maize, wheat and alfalfa

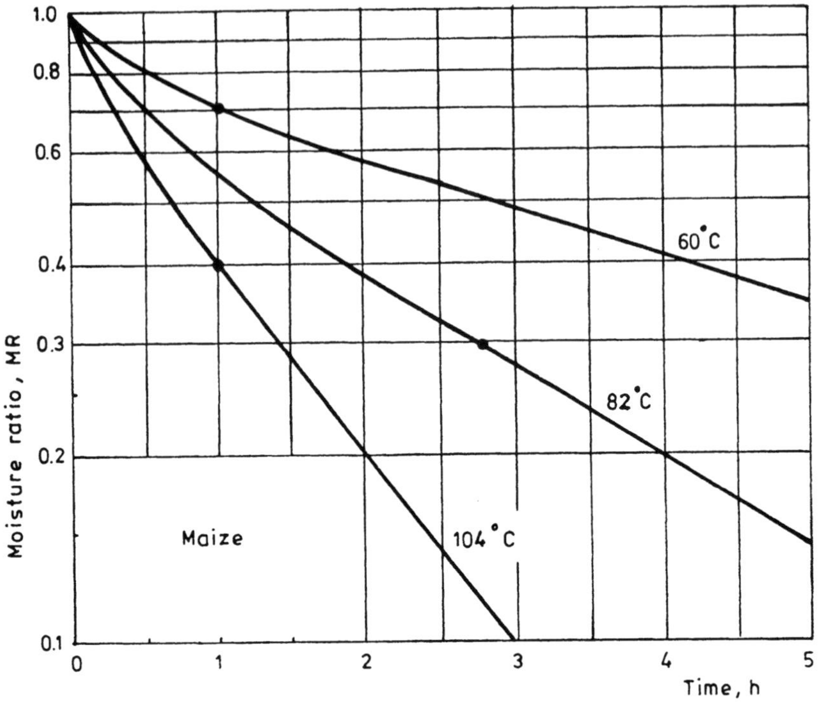

Fig. 64. Effect of drying temperature on the drying process for maize

drying medium for maize, wheat and alfalfa [30]. Figure 64 shows drying curves for maize at various drying temperatures. Evaluation according to eqn. (20) shows that the factor k is highest at the start of drying, and then decreases slowly. The effect of the reduction in k is the greater, the longer the drying period, i.e., the lower the drying temperature. Figure 65 shows the factor k as a function of temperature after various drying periods.

Drying of injured and broken kernels occurs significantly more rapidly than that of a pile consisting of sound grains. This may be explained by the fact that the specific surface area of the grains is greater in the former case, and water is removed more easily from cracks and freshly broken surfaces than through intact husks. Figure 65 also shows the variation of k for half kernels of maize, showing that a considerably higher rate of water removal is obtained [31]. During the harvesting of maize with shelling, about 10–12% of the kernels are injured. This means that these injured kernels are dried to a moisture content lower than average during drying.

At lower temperatures the drying rate is also influenced by the humidity of the ambient air. In this case it is advisable to plot the factor k as a function of the

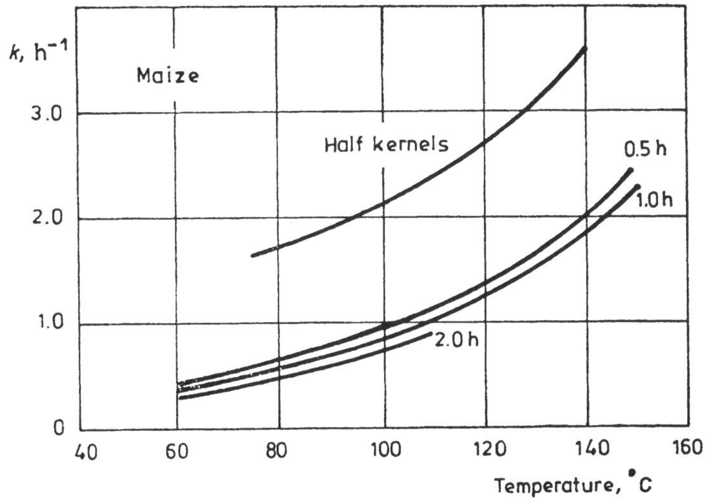

Fig. 65. Drying-rate coefficient k for maize as a function of temperature

partial-pressure difference, rather than of temperature: the partial-pressure difference is a combined function of temperature and relative humidity. For example, in the drying of rye-grass seeds the following relationship has been found to apply for the temperature interval 13–63 °C [34]:

$$k = 0.0287 + 1.11 \times 10^{-5}[p_s(1-\varphi)]^{1.526}$$

where p_s is to be substituted in units of daN m^{-2}.

Heating of material during drying. Material is heated to a certain temperature during drying. The temperature of the material is affected mainly by the temperature of the drying air, but is also influenced by the initial and final moisture contents.

The temperature of a material, such as grain, during drying may be determined on the basis of the heat balance. The quantity of heat transferred to the grain surface increases the temperature of the grain on the one hand, and is spent in evaporating water on the other. Accordingly,

$$-\alpha F(\vartheta - \vartheta_k) = cV\gamma(d\vartheta/dt) - V\gamma_0 Q_0(dX/dt)$$

where ϑ_k is the temperature of the drying air, and α the heat-transfer coefficient. This differential equation may be solved by expressing the rate of moisture removal using eqn. (20), as

$$dX/dt = -(X_1 - X_e)ke^{-kt}$$

Introducing the variable $T = \vartheta - \vartheta_k$ the differential equation may be brought to the general form

$$(dT/dt) + AT = Be^{-kt}$$

where

$$A = \alpha F/c\gamma V \quad \text{and} \quad B = -(\gamma_0 Q_0 k/c\gamma)(X_1 - X_e)$$

Integration of the preceding differential equation gives

$$\vartheta(t) = \vartheta_k + [B/(A-k)](e^{-kt} - e^{At}) + (\vartheta_0 - \vartheta_k)e^{-At}$$

where ϑ_0 is the initial grain temperature. According to this equation the temperature of the grain at first increases rapidly and then approximates asymptotically the temperature of the drying medium. According to measurement results, the equation describes the effective conditions well and may be used for approximate calculations.

If there is no moisture removal, i.e., $dX/dt = 0$, then the initial differential equation simplifies to

$$d\vartheta/dt = -A(\vartheta - \vartheta_k)$$

integration of which gives

$$(\vartheta - \vartheta_0)/(\vartheta_k - \vartheta_0) = 1 - e^{-At}$$

Figure 66 shows characteristic heating curves for maize kernels during drying [31]. The measured data refer to thin-layer drying, for which the temperature of the air flowing through the layer of kernels being dried may be assumed to be constant. It may be seen from the curves that the temperature of the material remains lower by 5–10 °C than that of the drying air, and attains its maximum value within 10–20 min depending on the temperature.

During initial heating a temperature difference develops between the surface and the center of the material, attaining its maximum in the case of cereals nearly at the beginning of the process (i.e., about 30 s after commencement of the instantaneous temperature rise of the medium). Its value may amount to 15–20% of the initial temperature difference between the medium and material [191]. During the progress of heating, the temperature difference within the grains decreases rapidly and its value then does not exceed 1–2 °C. On this basis a homogeneous temperature distribution may be assumed in the calculations. On drying materials of greater size, the temperature difference may be considerably greater during initial heating, and may be calculated by the method due to Heissler (see Ref. [29]).

Relationship between temperature and chemical changes in inner regions. During high-temperature drying a product may undergo browning, implying reduction of

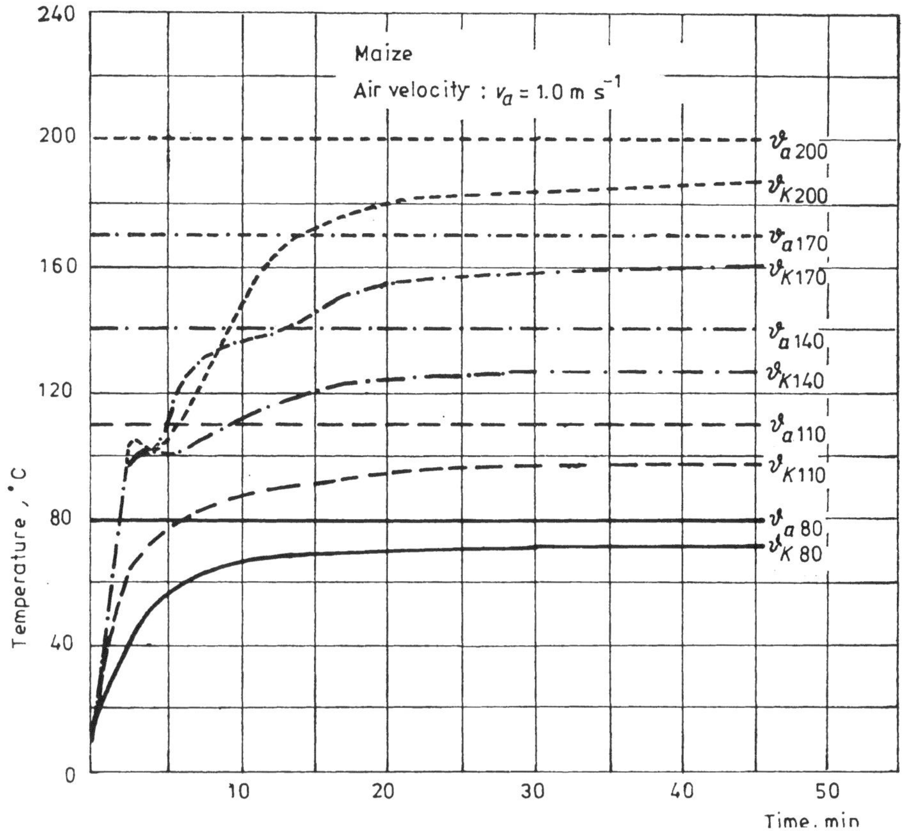

Fig. 66. Heating curves for maize kernels during thin-layer drying.

its feed value. Browning is the result of chemical reactions depending on temperature, on duration of exposure and on the structure of the material. According to experience, browning reactions are related to the protein and lipid contents: the higher the protein content, the more extensive the course of reaction.

In the drying of maize, first the germ part and then the horny endosperm undergoes browning. The farinaceous endosperm keeps its white color practically always, owing to its low protein content. Figure 67 shows the time required for initial browning of maize kernels as a function of drying temperature. At 100 °C browning starts after about 4 h, at 200 °C, after 4 min [31].

The loss of nutritive material may be determined on the basis of the decrease of essential amino acids (lysine, cystine and methionine): close correlation between the lysine content of maize and the gain in weight of animals fed on it is known from feeding experiments. Figure 68 shows the percentage reduction of the lysine

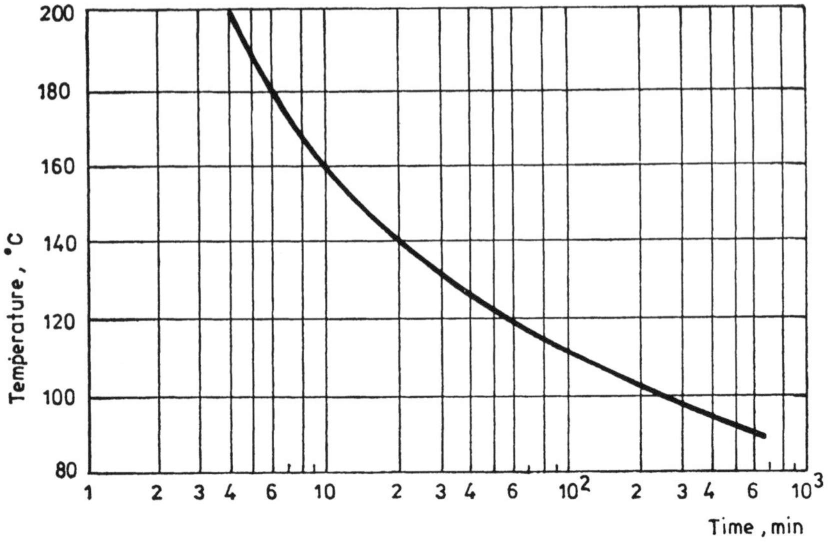

Fig. 67. Time required for browning of maize kernels as a function of drying temperature

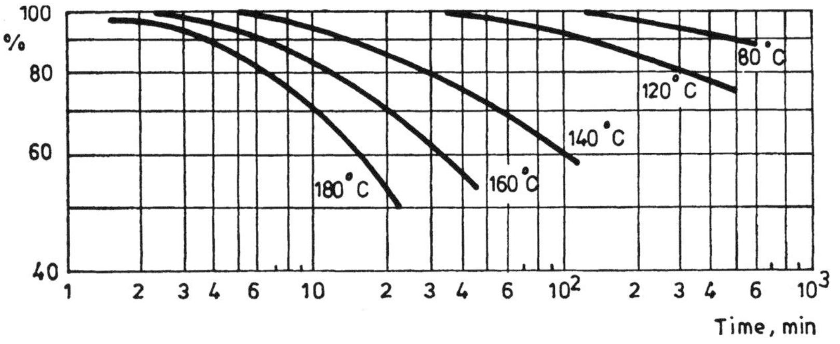

Fig. 68. Decrease of lysine content as a function of time of exposure to drying

content in maize as a function of duration of drying at various temperatures [31]. It is seen from the figure that decomposition of lysine is greatly accelerated by an increase in the drying temperature, especially above 130–140 °C.

Drying of thick layers. In practice, it occurs frequently that the thickness of the evaporation (water-removal) zone is less than that of the product layer. In this case a drying front, whose thickness depends primarily on the velocity and temperature of the air and on the drying constant k, passes through the layer of product (Fig. 69). Where the drying front has passed, the material is dried to the equilibrium moisture content, and so is in equilibrium with the fresh air flowing through it. Before the drying front, the material is in its original state, and so no drying takes place. The semiempirical calculation method due to Hukill (see in Ref. [30]) permits determination of the thickness of the drying zone and of the time required for drying the whole thickness of product. On the basis of thermal equilibrium, the following equation may be written:

$$m_a c_p \Delta\vartheta = (X_1 - X_e)(dm_{dm}/dt)r$$

where m_a is the quantity of air flow per hour, $\Delta\vartheta$ the cooling of the air in the evaporation zone, and r the heat of evaporation. From this equation, the drying rate relative to a dry-material basis is

$$dm_{dm}/dt = m_a c_p \Delta\vartheta/(X_1 - X_e)r \quad (\text{kg h}^{-1}) \tag{21}$$

This equation, with the assumption of total heat utilization, permits determination of the quantity of material drying to the equilibrium moisture content in each hour.

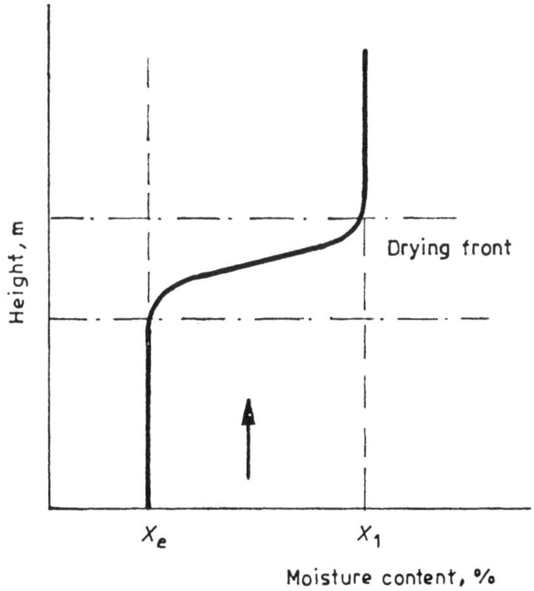

Fig. 69. Formation of a drying front in a thick layer.

To calculate the moisture content, the height of the evaporation zone and the drying period, we introduce a dimensionless quantity. The moisture content can be expressed by the moisture ratio MR appearing on the left-hand side of eqn. (20). The thickness of the evaporation zone is expressed as a multiple of the unit layer thickness. The latter is determined by the quantity m^*_{dm} of dry material drying to the equilibrium moisture content during the half-period $t_{1/2}$. The value of m^*_{dm} may be calculated from the equation

$$m^*_{dm} = (dm_{dm}/dt)\,t_{1/2}$$

and the number of unit layer thicknesses from

$$n = m_{dm}/m^*_{dm}$$

where m_{dm} is the total quantity of dry material. The whole drying period is expressed as a multiple of the half-period:

$$t = \varepsilon t_{1/2}$$

The following relationship exists among the above factors:

$$\mathrm{MR} = 2^n/(2^n+2^{\varepsilon}-1)$$

and is illustrated graphically in Fig. 70 [30].

The use of this calculation method is illustrated by the following example. Suppose that a maize layer 5 m thick is dried by aeration, with the following initial data: $U_1=22\%$, $X_1=28.2\%$ (dry basis), $U_2=14\%$, $X_2=16.3\%$, base area of maize pile $F=1.0\ \mathrm{m}^2$, dry-material weight of maize pile $m_{dm}=2960$ kg, air quantity $m_a=420\ \mathrm{kg\,h^{-1}}$, temperature of air $\vartheta_k=25\ ^\circ\mathrm{C}$, relative air humidity

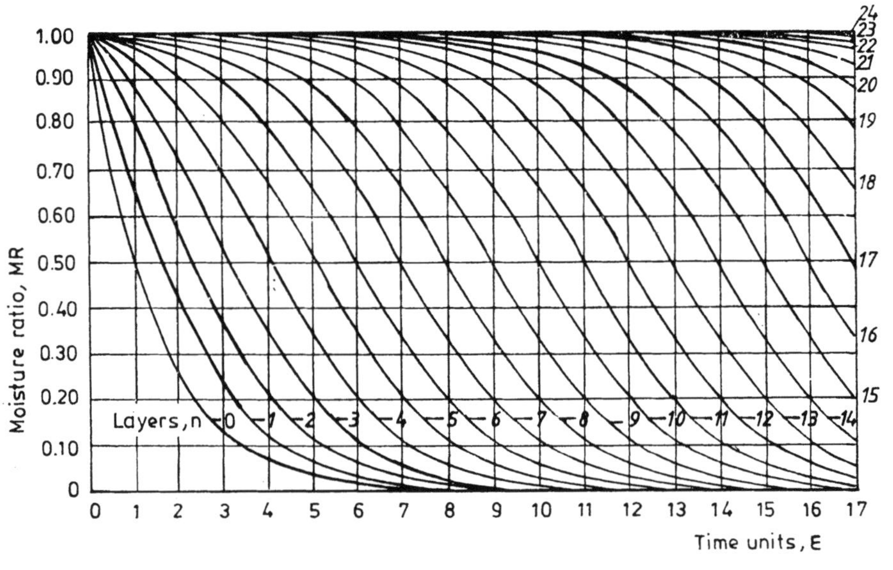

Fig. 70. Relationship between moisture ratio, number of unit layer thicknesses and drying period

$\varphi = 60\%$, wet-bulb temperature 20 °C, equilibrium moisture content of maize $U_e = 14\%$, $X_e = 16.3\%$ and $t_{1/2} = 11.5$ h. For these data,

$$dm_{dm}/dt = 420 \cdot 0.24 \cdot 5/(0.282-0.163)600 = 7.06 \text{ kg h}^{-1}$$

$$m^*_{dm} = 7.06 \cdot 11.5 = 81.2 \text{ kg/layer}$$

$$n = 2960/81.2 = 36.4$$

On the basis of Fig. 70, $\varepsilon \cong 40$, so the whole drying period is

$$t = 40 \cdot 11.5 = 460 \text{ h} = 19.17 \text{ days}$$

Drying in thick layers involves two basic problems: one is the risk of overdrying, the other is wetting, by condensation on the parts still not dried. As seen from the preceding example, if the upper layer is dried to the required moisture content, most of the layers lying under it will have dried to the equilibrium moisture content. At higher temperatures the equilibrium moisture content is considerably less than that necessary for storage, and so a superfluous energy input occurs. On this basis the advised drying temperature for cereals and fibrous materials should be not more than 25–30°C.

Drying at higher temperatures may result in condensation in layers found before the evaporation zone, since in this case the wet-bulb temperature of the drying air may easily exceed the initial temperature of the material. The air passing through the colder parts cools, and a part of its humidity precipitates. If a long time is required for the drying front to reach parts wetted in this way, self-heating and degradation may occur. This phenomenon can be observed especially on drying various hays.

7.12 Heating and cooling of deep piles

During the storage of agricultural products (cereals, potatoes, onions, etc.), it is frequently necessary to remove heat generated by biological reactions, i.e., to recool a product. In certain cases it is advisable to heat products to prevent infections and deterioration (e.g., the *Bothrytis* infection in onions). Heating and cooling of deep piles takes a relatively long time, and so knowledge of the required heating or cooling period is important in such operations.

The whole volume of a pile is not heated or cooled uniformly during these operations; rather, a heating or cooling front passes through the pile. The velocity and thickness of the front are functions of the air velocity, the evaporation rate, the temperature difference and the sizes of the components in the pile. Among the factors enumerated the air velocity is generally of decisive importance.

Mathematical equations describing the heating or cooling front are derived below with certain simplifying assumptions. The most important assumptions are as follows [35]:

(a) The material is isotropic in terms of both thermal and mass diffusivity;

(b) Moisture moves in the material only by diffusion;

(c) The temperature, moisture content and porosity of the pile are each identical everywhere at the beginning of the process, and the porosity does not change during the heating or cooling process;

(d) The temperature of the moving air is not reduced observably by the water evaporated;

(e) The limiting walls are isolated;

(f) Periodical fluctuations of the external temperature are negligible.

Taking the x coordinate axis in the direction of the layer thickness, the following differential equation is valid:

$$\gamma_c(\partial T_i/\partial t+v_x\partial T_i/\partial x) = \lambda(\partial^2 T_i/\partial x^2)-\alpha[6(1-\varepsilon)/d](T_i-T_j) \tag{22}$$

where c is the specific heat, λ the heat-conduction and α the heat-transfer coefficient, ε the porosity of the pile, and d the equivalent diameter of the components forming the pile. The effective velocity v_x may be obtained from the velocity v_0, calculated for the whole cross-sectional area, divided by the porosity

$$v_x = v_0/\varepsilon$$

Making use of eqn. (22) the following equations may be written for an elementary volume of cross-sectional area S and thickness dx, separately for the air and the material forming the pile:

$$\varepsilon Sdx\gamma c_p(\partial T/\partial t+(v_0/\varepsilon)(\partial T/\partial x)) = \varepsilon Sdx\lambda(\partial^2 T/\partial x^2)-\alpha[6(1-\varepsilon)/d](T-T_a)Sdx$$

$$(1-\varepsilon)Sdx\gamma_a c_a(\partial T_a/\partial t) = (1-\varepsilon)Sdx\lambda_a(\partial^2 T_a/\partial x^2)-\alpha[6(1-\varepsilon)/d](T_a-T)Sdx$$

After simplifying and rearranging the above equations, the following differential equations of general validity are obtained:

$$\partial T/\partial t+(v_0/\varepsilon)(\partial T/\partial x) = a(\partial^2 T/\partial x^2)-[6\alpha(1-\varepsilon)/d\gamma c_p\varepsilon](T-T_m) \tag{23}$$

$$\partial T_m/\partial t = a_m(\partial^2 T_m/\partial x^2)-(6\alpha/d\gamma_m c_m)(T_m-T) \tag{24}$$

where $a=\lambda/\gamma c_p$ is the thermal diffusion coefficient. The subscript m appearing in the above equations refers to the material. The equations must be solved with the given initial and boundary conditions taken into account in common. Since the thermal characteristics are constant, i.e., independent of time and place, the equations are linear.

Equations (23) and (24) can be solved only by numerical methods. In the interests of general applicability, the calculation results are presented here as functions of dimensionless quantities [35]. On introducing the abbreviations

$$A' = 6\alpha(1-\varepsilon)/d\gamma c_p v_0$$

and

$$B' = 6\alpha/d\gamma_m c_m v_0$$

the dimensionless quantities applied are the following:

dimensionless coordinate: $\xi = A'x$.

dimensionless time: $\tau = B'[(v_0/\varepsilon)t - x]$,

dimensionless temperatures of the medium and the material:

$$X = (T - T_{\mathrm{mo}})/(T_{\mathrm{in}} - T_{\mathrm{mo}})$$

$$Y = (T_{\mathrm{m}} - T_{\mathrm{mo}})/(T_{\mathrm{in}} - T_{\mathrm{mo}})$$

where T_{mo} is the initial temperature of the material, and T_{in} the temperature of the medium as it enters.

Figure 71 shows the solution of eqns. (23) and (24) [35]. It may be seen from the figure that a heating front passes through the material. The velocity and width of the front are determined by the heat-transfer coefficient, the air velocity, and the size and thermal characteristics of the material.

In numerous cases, conduction is negligible in the air and the material during the heating or cooling stage, and the above differential equations then reduce to the simplified forms

$$\partial T/\partial t + (v_0/\varepsilon)(\partial T/\partial x) = A'(v_0/\varepsilon)(T_m - T) \tag{25}$$

$$\partial T_m/\partial t = B'v_0(T - T_m) \tag{26}$$

On applying the above dimensionless quantities, the differential equations assume the simpler forms

$$\partial X/\partial \xi = Y - X \tag{27}$$

$$\partial Y/\partial \tau = X - Y \tag{28}$$

The initial and boundary conditions may be written as follows: at time $\tau = 0$, $X = Y$ for all values of ξ; at the location $\xi = 0$, $X = 1$ for times $\tau > 0$; $\partial Y/\partial \xi = 0$; on the surface of the bed, $\partial X/\partial \xi = 0$ and $\partial Y/\partial \xi = 0$ for all values of τ.

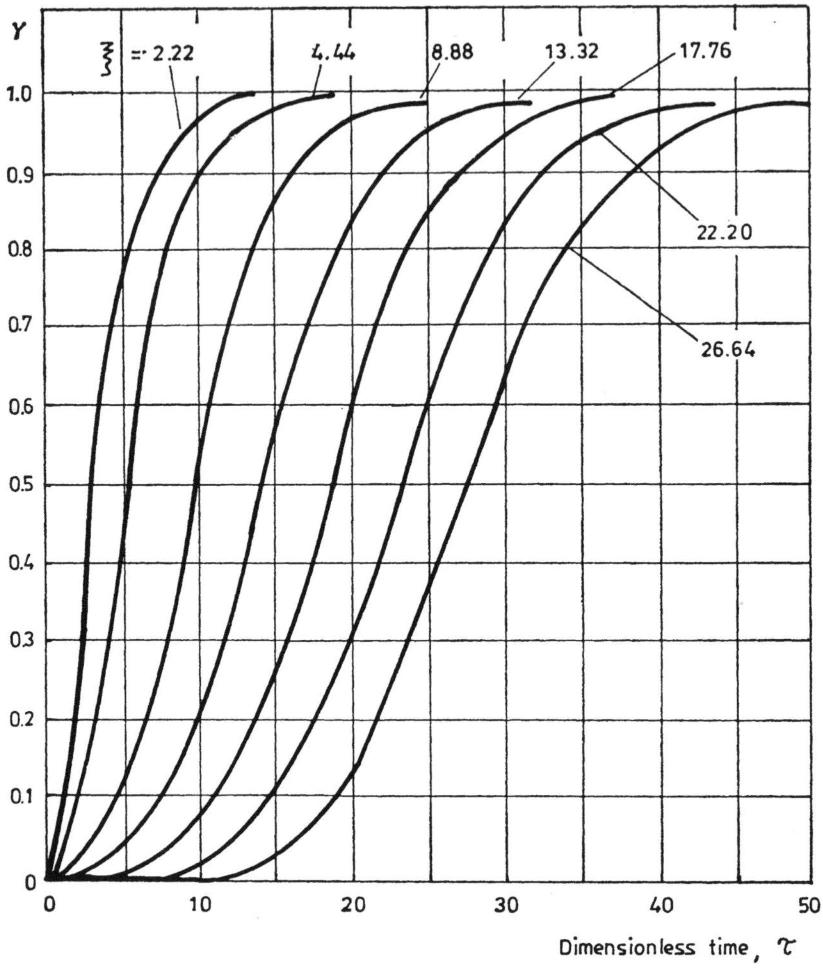

Fig. 71. Solution of eqns (23) and (24)

The above differential equations may be solved by means of Laplace transformation [35]. The dimensionless material temperature may be calculated from the equation

$$Y(\xi, \tau) = \int_0^\tau e^{-(\xi+\tau)} I_0(2\sqrt{\xi\tau})\, d\tau \tag{29}$$

while the temperature difference between the material and the medium is

$$X(\xi, \tau) - Y(\xi, \tau) = e^{-(\xi-\tau)} I_0(2\sqrt{\xi\tau}) \tag{30}$$

where I_0 is the modified Bessel function of the first kind of zero order, with argument $2\sqrt{\xi\tau}$.

The propagation of the heating or cooling front is illustrated better by plotting in a Y–ξ system of coordinates, where the quantity τ appears as a parameter (Fig. 72). The temperature of the material corresponding to any arbitrary distance x at any chosen time may be read from the set of curves.

The applicability of the above calculation methods has been tested for the heating of a deep-bed-packed onion pile, among other cases [35]. The calculated and measured values are shown in the system of dimensionless coordinates in Fig. 73. The basic data were the following: $T_{m0}=8.5\ °C$, $T_{in}=32\ °C$, $v_0=175\ m\ h^{-1}$, $\gamma_m=0.95\ kg\ dm^{-3}$, $c_m=0.93\ kcal\ kg^{-1}\ °C^{-1}$, $d=6.14\ cm$, $\varepsilon=0.4$, $H=3.6\ m$, $\lambda_m=0.52\ kcal\ m^{-1}\ h^{-1}\ °C^{-1}$ and $\alpha=6.39\ kcal\ m^{-2}\ h^{-1}\ °C^{-1}$. On comparing the curves it may be seen that the calculated and measured values agree well up to $Y=0.8$ in the first two-thirds and up to $Y=0.7$ in the upper part of the bed. When the temperature of the material reaches about 90% of the air temperature, noticeable evaporation begins from the surface of the material, and

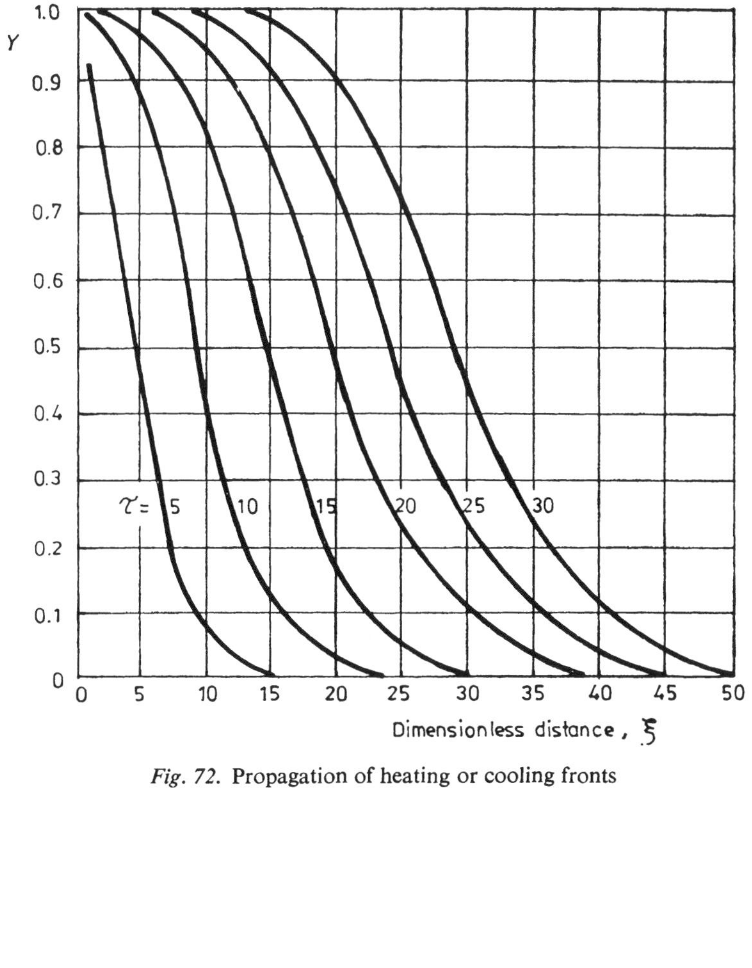

Fig. 72. Propagation of heating or cooling fronts

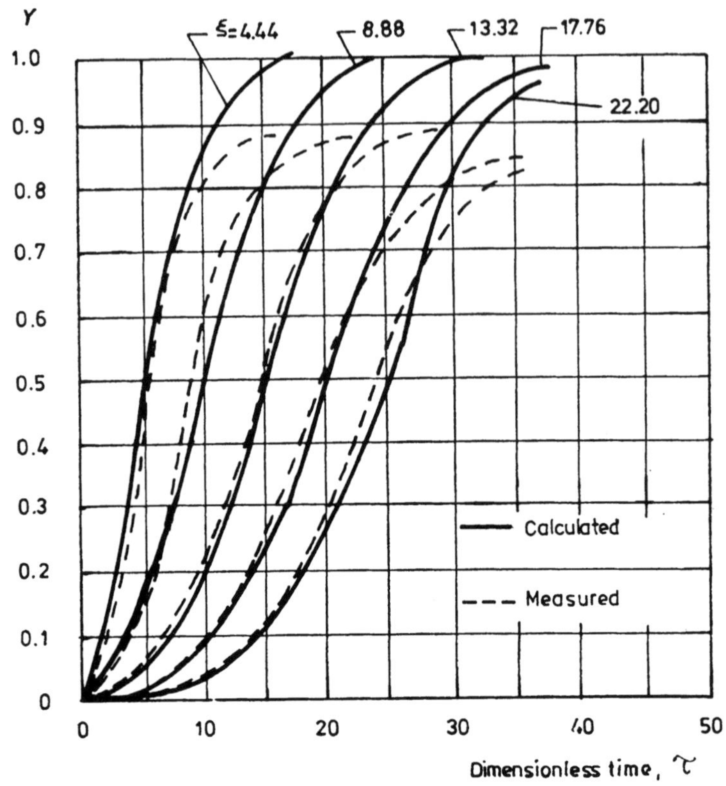

Fig. 73. Heating of a deep-bed-packed onion pile

this implies heat removal, i.e., simultaneous heat and mass transfers are already involved.

The heat-transfer coefficient α appearing in the above equations is determined from the general similarity equation of the form

$$Nu = ARe^n$$

Unfortunaely, such similarity equations are generally not available for bulk agricultural tmaterials. In the range of low Re values (Re=50–1000), the following equation may be used for piles consisting of nearly spherical particles:

$$Nu = 2+0.16Re^{0.67}$$

For $Re>1000$, the equation

$$Nu = 0.3Re^{0.6}$$

may be used. For bulk cylindrical pellets obtained from ground dry forage (of

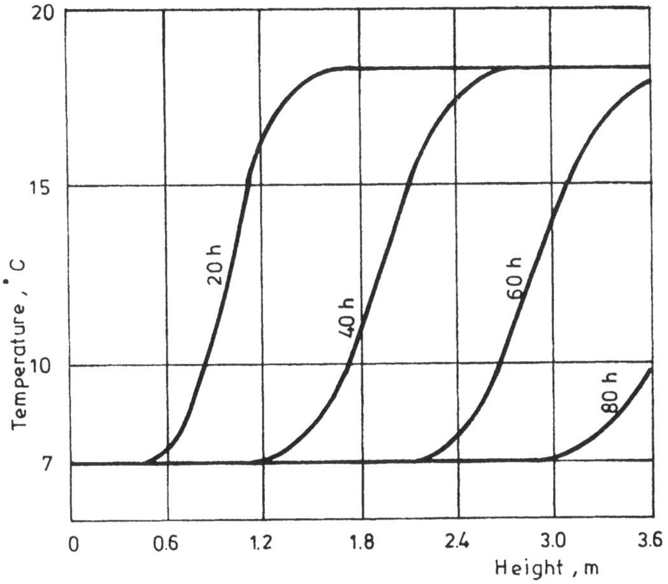

Fig. 74. Recooling of a potato pile after various time periods

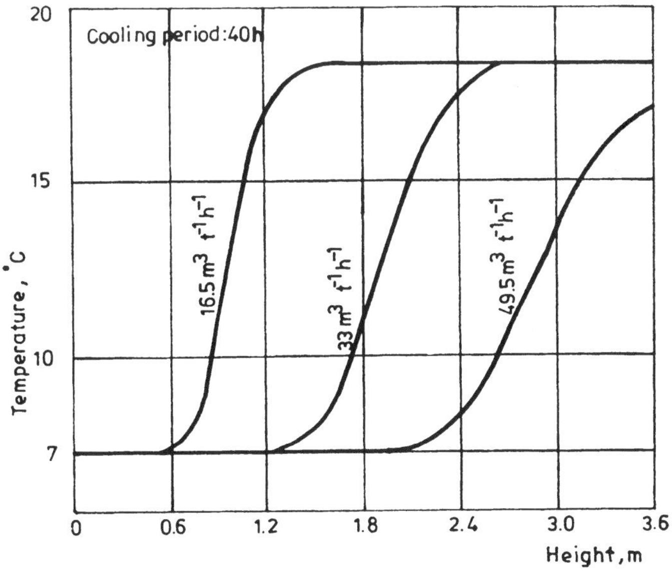

Fig. 75. Effect of quantity of cooling air on position of cooling front

diameter 10–16 mm, length 15–25 mm, and with a density of 1300–1500 kg m^{-3} and a volumetric weight of 600–700 kg m^{-3} in the pile), the following similarity equation has been found:

$$Nu = 0.172 Re^{0.74}$$

The cylindrical shape of the pellets increases the turbulence of the flowing air as compared to a spherical shape, and in all probability this explains the higher exponent of *Re*.

Figure 74 shows data for the recooling of a 3.6 m high potato pile after various time periods. The pile is cooled from 18 °C to 7 °C, the rate of air flow is 33.2 $m^3 h^{-1} t^{-1}$ [26]. The position and width of the cooling zone, as well as the time required for cooling (about 80 h) may be read from the figure. Figure 75 presents cooling curves for the same pile after 40 h, for various rates of flow of cooling air [26]. It may be seen that the rate of propagation of the cooling zone increases in almost direct proportion to the rate of flow of air, while the width of the cooling zone increases simultaneously.

7.13 Heat production in biological materials during storage

Agricultural products, as biological materials, continue to live and respire during storage. The rates of these biological processes are functions of the moisture content and temperature, in addition to the structure of the material. Internal biological processes are paired in most cases with the life phenomena of microorganisms, which are always present, these life phenomena are also functions of the moisture content and temperature.

Both biological processes and the life functions of microorganisms imply oxidation of material according to the general scheme (for oxidation of carbohydrates)

$$C_6H_{12}O_6 + 6O_2 \rightarrow 6CO_2 + 6H_2O + 2834 \text{ kJ}$$

During oxidation of 1% of 1 kg dry substance, 14.7 g of carbon dioxide and 6.0 g of water vapor are formed, and 157 kJ of heat is released. The more extensíve the oxidation, the more carbon dioxide and water vapor formed and the more heat released.

The water vapor and heat released both favor an increase in the rate of oxidation, and so these processes are generally self-accelerating.

As has been seen, carbon dioxide forms during oxidation. The extent of carbon dioxide formation may be used to measure the extent of oxidation. Figure 76 shows the rates of carbon dioxide production from oil seeds and cereals as functions of moisture content [131]. It is seen that the oxidation rate for oil seeds

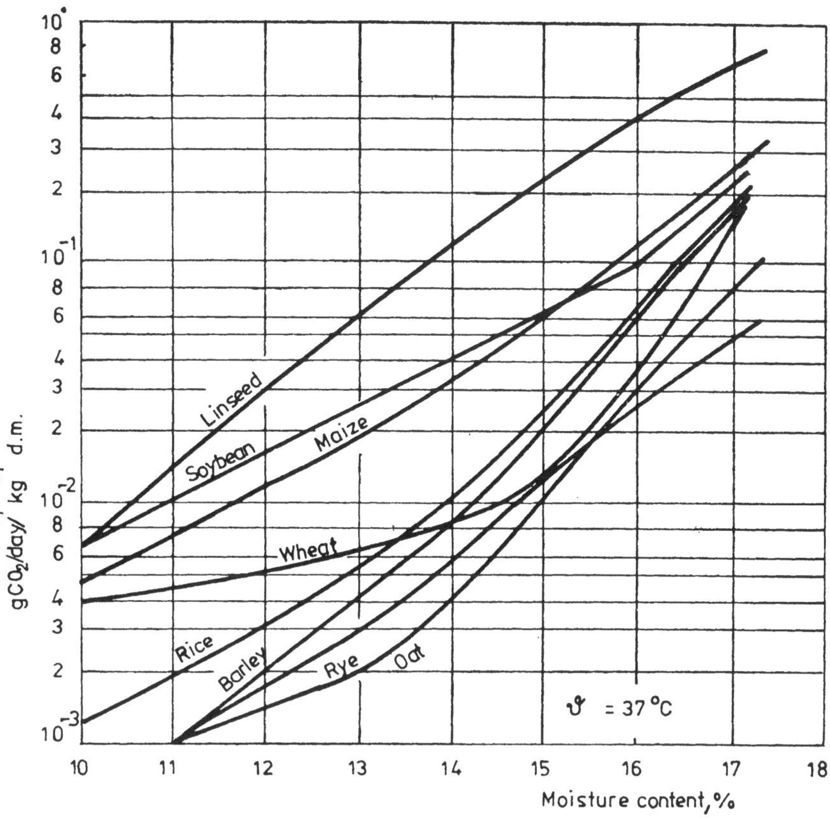

Fig. 76. Oxidation of cereals as a function of moisture content

exceeds that for cereals, and this explains why the moisture content required for safe storage of linseed and sunflower seed is 8–9%, while the corresponding figure is 13–14% in the case of cereals.

The rate of heat generation is of decisive importance from the point of view of storage. Figure 77 presents heat generation curves for maize as functions of kernel temperature, for various moisture contents [32]. The data refer to a pile harvested by a combine, and so also containing damaged kernels. Both oxygen and microorganisms can penetrate more easily into injured kernels: thus the rate of oxidation is higher in this case even under identical external conditions. For example, in a maize pile containing up to 30% damaged kernels the rate of oxidation is practically twice as high as in the case of entirely sound kernels. The rate of heat generation increases in proportion to temperature to the power of 2.5 in the case of maize. The release of heat is considerably increased for moisture

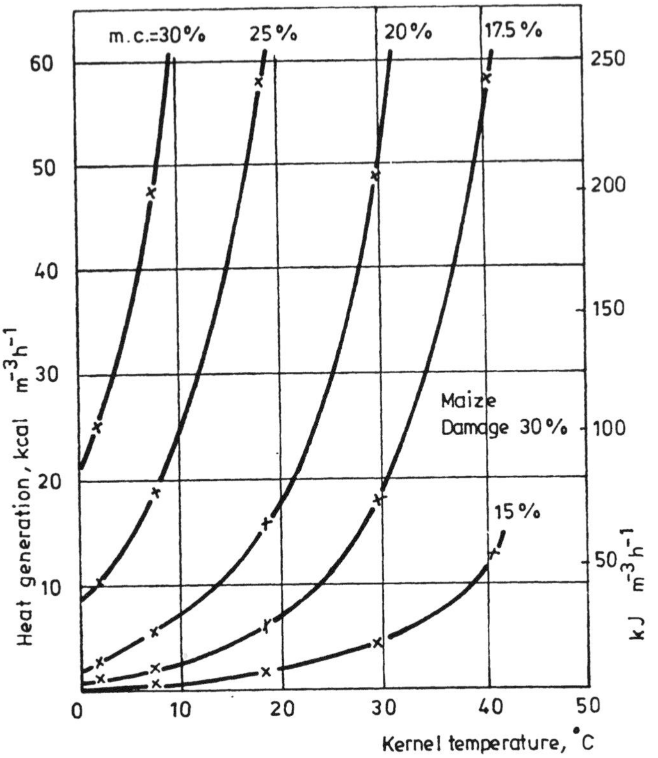

Fig. 77. Heat-generation curves for maize as functions of kernel temperature

contents exceeding 14–15%, and so heat must be removed periodically (by ventilation) in such cases.

Biological heat generation in fruits and vegetables may be calculated from the relationship

$$q = q_0 e^{b\vartheta}$$

where ϑ is the temperature, and the constants q_0 and b are determined experimentally. Table 4 contains q_0 and b data for some important fruits and vegetables [24]. The data in the table are intended only as a guide, as deviations are experienced within each of the species. In addition, the amount of heat generated depend on when (i.e., in which stage of the ripening process) a product is harvested and put into storage. It has also been observed that the rate of heat generation is higher in the first few days than during the subsequent period.

The loss of dry substance as a function of the quantity of heat released is

$$G_m = 2.66 \times 10^{-4} q_0 e^{b\vartheta} \text{ kg kg}^{-1} \text{ h}^{-1}$$

Table 4

Product	q_0		b (°C^{-1})
	kJ kg^{-1} h^{-1}	kcal kg^{-1} h^{-1}	
Green peas	0.4	0.097	0.085
Strawberries	0.16	0.038	0.094
Peaches	0.083	0.020	0.114
Cucumbers	0.071	0.017	0.119
Carrots	0.048	0.0116	0.132
Apples	0.043	0.0104	0.093
Potatoes	0.036	0.0086	0.062
Onions	0.039	0.0094	0.067
Sugar beet	0.026	0.0062	0.084

Heat released in a bulk material increases its temperature. The heat-conduction coefficient of bulk materials is low, and so significant heat-conduction into the environment cannot be counted upon. No great error is committed if it is assumed that the total quantity of heat released is spent in increasing the temperature of the material. With this assumption, it is possible to determine the heating curve for a product pile.

The quantity of heat released during a time Δt is

$$dq = q_0 e^{b\vartheta} dt$$

which increases the temperature of the product by an amount $d\vartheta$ such that

$$dq = c d\vartheta,$$

where c is the product's specific heat. From the two above equations

$$dt = c d\vartheta / q_0 e^{b\vartheta}$$

integration of which yields the equation

$$t = (c/q_0 b)(1/e^{b\vartheta_1} - 1/e^{b\vartheta_2}) \tag{31}$$

The final temperature ϑ_2 may also be expressed using this equation as

$$\vartheta_2 = (1/b) \ln [c/(ce^{-b\vartheta_1} - q_0 bt)] \tag{31a}$$

As an example, suppose that potatoes are stored at 2 °C and the permissible maximum temperature is 5 °C. With $q_0 = 0.036$ kJ kg^{-1} h^{-1} and $b = 0.062$ °C^{-1}, $t = 239$ h $\simeq$ 10 days. This means that the potatoes must be aerated every 10 days in order to recool them.

In storing sugar beet, a close correlation exists between q_0 and the daily sugar loss: i.e.,

$$q_0 = 1.65\ m_c$$

where m_c is the daily sugar loss as a weight-percentage of the beet. Considering the value of q_0 in Table 4, the daily sugar loss is about 0.004%.

A certain quantity of heat is also released in piles during cooling, but this occurs at a decreasing temperature, and a correspondingly decreasing rate. The quantity of heat released during cooling may be calculated approximately by the following method.

The temperature at a given point in a body varies under the effect of cooling according to the equation

$$\Delta\vartheta/\Delta t = -(\alpha F'/c)(\vartheta-\vartheta_k)$$

where F' is the surface area of 1 kg of material, and ϑ_k the temperature of the cooling agent. The quantity of heat released in 1 kg of material during a time Δt is given by

$$\Delta q = q_0 e^{b\vartheta}\Delta t$$

By means of the above two equations, the following differential equation may be written:

$$dq = -(c/\alpha F')q_0 e^{b\vartheta}[(d\vartheta/(\vartheta-\vartheta_k)]$$

By integrating this equation, with the assumption of constant cooling temperature, the following equation can be derived:

$$q = (c/\alpha F')q_0[4\Delta\vartheta/(\Delta\vartheta+1)+(1/b)(2-1/\Delta\vartheta)](e^{b\vartheta_1}-e^{b\vartheta_2}) \qquad (32)$$

where ϑ_1 and ϑ_2 are respectively the initial and final temperatures of the material.

For example, on recooling a potato pile the following values may be counted upon: q_0=0.036 kJ kg^{-1} h^{-1}, b=0.062 °C^{-1}, c=3.56 kJ kg^{-1} °C^{-1}, α=20.93 kJ m^{-2} h^{-1} °C^{-1}, F=0.084 m^2 kg^{-1}, ϑ_1=8 °C, ϑ_2=2 °C and $\Delta\vartheta$=6 °C. Substitution of these values into eqn. (32) gives the quantity of heat released during recooling as q=1.226 kJ kg^{-1}. Recooling is realized in about 24 h. The calculated result may be corroborated by a simple method. Assume that a single cooling zone passes through the pile at a uniform velocity, and that the temperature ϑ_1 prevails before the zone and ϑ_2 beyond it. Then the quantity of heat released is

$$q = q_0[(e^{b\vartheta_1}+e^{b\vartheta_2})/2]\,t$$

where t is the cooling period.

Using the data of the above example, q=1.197 kJ kg^{-1}, which agrees well with the result obtained using eqn. (32).

7.14 Moisture exchange of fruits and vegetables with the air

For high-quality storage of fruits and vegetables two basic requirements must be met: uniform maintenance of the optimal temperature, and preservation of moisture content. The moisture content of fruits and vegetables is generally 80–85%. However, with the loss of only a few percent of the moisture content the turgor pressure decreases, the product begins to wither, and its resistance to damage and its shelf life decrease. Thus the preservation of moisture content is of paramount importance in maintaining quality. At the same time, condensation on the surface of the product must also be avoided, since it also decreases the preservability.

Observations have shown that withered products (e.g., potatoes, carrots, red beet, with 5–7% weight loss) do not regain their original turgor pressure even in an environment of 100% relative humidity. This may be explained by biological transformations (decomposition of organic materials, decreasing resistance to microorganisms) taking place in the material consequent upon the loss of moisture, whereby the material is no longer able to regenerate, i.e., to recover its original structural state.

The natural removal of water during the storage of agricultural products is a result of relatively slow diffusion processes, in which diffusion due to concentration differences plays the main role, while the thermal diffusion and other phenomena are generally negligible. The exchange of moisture between fruits or vegetables and the air may be described well by Dalton's law, which states that the moisture removed in unit time may be expressed as

$$G = \beta\gamma F(C_s' - C_s\varphi)$$

where C_s' is the concentration of saturated water vapor on the surface of the material, C_s the concentration of saturated water vapor in the air, φ the relative air humidity, γ the moisture exchange coefficient of the material. The evaporation coefficient β appearing in the equation is relative to a free-water surface; thus the coefficient γ expresses the proportion of the product's surface area which may be taken into account as a free water surface in terms of moisture exchange.

The coefficient γ is established experimentally. Table 5 lists γ values for the main fruit and vegetable products [24]. The data in the table show that differences as high as two orders of magnitude may exist between individual products. The high γ value for green peas indicates a very intense moisture exchange: green peas can dissipate water very rapidly. In contrast, onions have a low γ value and dissipate water very slowly, and may thus be kept for a long time, even under ambient conditions, without any moisture loss.

Table 5

Product	γ
Apples (summer)	0.025–0.030
Apples (winter)	0.011–0.014
Pears	0.015–0.020
Plums	0.020–0.025
Peaches	0.21 –0.27
Apricots	0.17 –0.24
Cherries, sour cherries	0.15 –0.18
Potatoes	0.009–0.012
Carrots	0.35 –0.40
Red beet	0.20 –0.30
Sugar beet	0.25 –0.30
Sugar beet (irrigated)	0.35 –0.45
Onions	0.002–0.003
Cabbages	0.37 –0.45
Grean peas in shell	0.7 –0.8

The moisture content of products grown on irrigated fields is generally higher, but they also discharge water more easily and so their γ values are higher. These products require more careful storage, and are less preservable.

It has been observed that for numerous products both heat production and moisture discharge are more intense in the initial phase of storage, decreasing once a certain time has passed. Figure 78 shows the moisture loss from potatoes

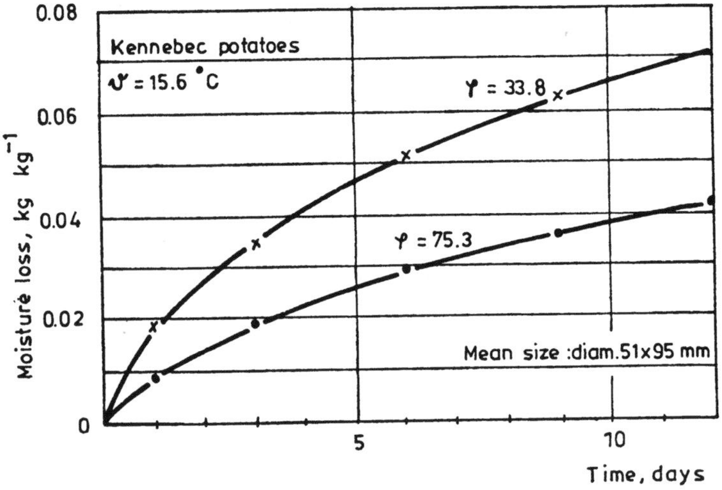

Fig. 78. Moisture loss from potatoes at beginning of storage.

during the first 10 days of storage at two relative humidities [27]. The equation of the curves may be expressed in terms of the partial-pressure difference and of time, in the form

$$G = 1.138\times10^{-4}[p_s(1-\varphi)]^{0.59}t^{0.65} \text{ (kg kg}^{-1}\text{)}$$

The moisture loss per hour is obtained by differentiating this equation:

$$dG/dt = 7.4\times10^{-5}[p_s(1-\varphi)]^{0.59}t^{-0.35} \text{ (kg kg}^{-1}\text{ h}^{-1}\text{)}$$

After 10 days the rate of moisture loss becomes stationary, under constant external conditions.

The heat introduced into a product during aeration is spent in changing the enthalpy of the material and in evaporation:

$$Q = \alpha F\Delta\vartheta+\beta\gamma F\Delta Cr$$

where r is the heat of evaporation. The ratio of the quantity of heat introduced and the amount of moisture removed is a quantity which can be used to characterize a conditioning process; its value is

$$Q/G = (\alpha/\beta)(1/\gamma)(\Delta\vartheta/\Delta C)+r = \Delta i/\Delta x \tag{33}$$

The ratio α/β appearing in the equation may be determined, by means of the analogy of heat and mass transfer, by the relationship $Nu/Nu'=1$, from which

$$\alpha/\beta = \lambda/D$$

and its value for the temperature range 0–20 °C is 1.21–1.25 kJ m^{-3} h^{-1} (0.29–0.30 kcal m^{-3} h^{-1}). The ratio $\Delta\vartheta/\Delta C$ affects the moisture exchange significantly. The concentration difference ΔC is a function of temperature and of relative humidity. The saturated concentration may be calculated in the interval -2 to $+4$ °C to good approximation using the equation

$$C_s = (4.9+0.35\vartheta)\times10^{-3} \text{ (kg m}^{-3}\text{)}$$

whereby the ratio $\Delta\vartheta/\Delta C$ can be determined (Fig. 79).

The amount of moisture removed during cooling may be calculated on the basis of eqn. (33) with appropriate substitution of the quantity of heat, Q:

$$G_0 = [c(\vartheta_1-\vartheta_2)+q]/[(\alpha\Delta\vartheta/\beta\gamma\Delta C)+r] \text{ (kg kg}^{-1}\text{)}$$

where ϑ_1 and ϑ_2 are respectively the initial and final temperatures of the material, and q is the quantity of heat released in the material, according to eqn. (32). By the end of the cooling process the equilibrium temperature ϑ_e of the material and the temperature ϑ_k of the medium may develop differently in relation to each other. For lower relative humidities, $\vartheta_k>\vartheta_e$, as the intense evaporation implies additional heat removal. At a certain relative humidity, $\vartheta_k=\vartheta_e$, while in the case of $\varphi=1.0$, generally $\vartheta_k<\vartheta_e$. The main task during the storage is the pres-

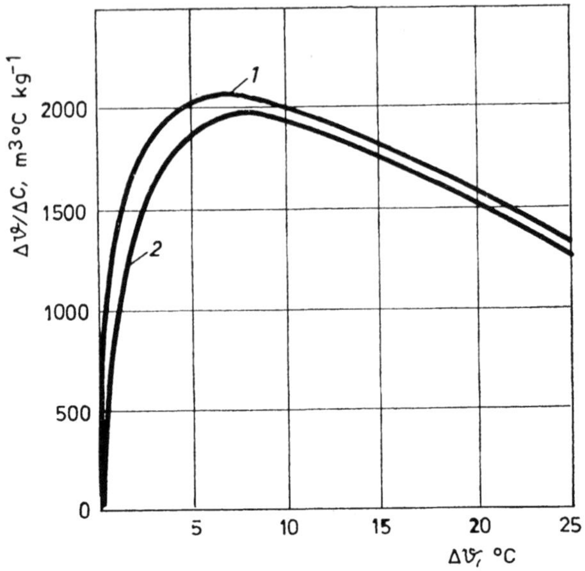

Fig. 79. Graphical representation of the ratio $\Delta\vartheta/\Delta C$. *(1)* $\varphi=0.94$; *(2)* $\varphi=0.85$

ervation of moisture in the material, so the first case is unfavorable and is to be avoided. The conditioning parameters are usually determined for the second case ($\vartheta_k=\vartheta_e$), since the moisture loss is then minimal. In the latter case, physiological heat is removed by cooling via evaporation, i.e.,

$$G = q_0 e^{b\vartheta}/r \text{ (kg kg}^{-1}\text{ h}^{-1})$$

The relative humidity φ corresponding to the condition $\vartheta_k=\vartheta_e$ may be determined on the basis of the heat balance

$$q_0 e^{b\vartheta} = \beta F'\gamma r(C_s - C_s\varphi)$$

from which

$$1-\varphi = q_0 e^{b\vartheta}/\beta F'\gamma r C_s \tag{34}$$

or

$$1-\varphi = q_0 e^{b\vartheta}/2069\alpha F'\gamma C_s \tag{34a}$$

If all the heat released is removed by evaporation cooling, the temperature of the air passing through the pile will not change. However, the air takes up moisture, and so it may occur that air of already high relative humidity will become saturated with moisture and condensation will appear in the upper layers. Therefore it is advisable to select parameters ensuring that the relative humidity of the cooling air remains nearly constant during its passage. This requires that the ratio of the sensible and latent heats has a preset value, such that the cooling air will be heated by a few degrees centigrade during its passage, and so its relative humidity will not be increased by any moisture taken up.

Related investigations have shown that the optimum value of the ratio Q/G determined by means of eqn. (33) is 6278 kJ kg^{-1} (1500 kcal kg^{-1}) at around 0 °C, from which the optimum ratio of the sensible and latent heats during the cooling process may be determined. The ratio of the sensible heat (Q') to the total quantity of heat may be written as

$$Q'/Q = (Q - Gr)/Q = 1 - (r/(Q/G)) = 1 - (600/1500) = 0.6$$

This means that 60% of the total quantity of heat is sensible heat and 40% latent heat in the optimal case. On the basis of this result eqn. (34) must be modified, since only 40% of the total physiological heat q is removed by evaporation. Now

$$1 - \varphi = 0.4 \cdot q_0 e^{b\vartheta}/2069\alpha F'\gamma C_s \tag{34b}$$

Table 6

Product	ϑ_k (°C)	d_c (m)	α (kJ m^{-2} h^{-1} °C^{-1})	γ	φ(%)	
					(34a)	(34b)
Carrots	2	0.035	19.88	0.35	99.5	99.8
Potatoes	3	0.045	17.58	0.012	86.4	94.5
Cabbages	2	0.18	9.63	0.37	96.9	98.7
Apples	2	0.06	15.36	0.014	79.2	91.6
Onions	2	0.05	16.74	0.003	38.0	75.0

Table 6 lists critical relative–humidity values for some characteristic products. The air velocity relative to the whole cross-sectional area is 0.05 m s^{-1} [24]:

It is seen from the table that the higher the value of γ for a product, the higher may be the relative humidity of the air used in ventilation. The φ values calculated from eqns (34a) and (34b) show the upper limits of the relative air humidity required to avoid condensation: the humidity chosen should be close to this limit in order to preserve moisture.

The equilibrium temperature of a material at the end of cooling may be calculated from the heat balance:

$$q_0 e^{b\vartheta} = \alpha F'(\vartheta_e - \vartheta_k) + \beta\gamma F'(C_s' - C_s\varphi)r$$

from which

$$\vartheta_e = [q_0 e^{b\vartheta}/\alpha F' + \vartheta_k - 9.97\gamma(1 - 204C_s\varphi)]/(1 + 0.7122\gamma) \tag{35}$$

The total loss of a stored product is the sum of the dry-substance loss due to oxidation, and the removed moisture.

8. THE BACKGROUND OF RHEOLOGY

The mechanical properties of materials govern their behavior under the effect of mechanical forces. Forces acting on a material cause deformation and flow (creep) in it, and the nature of the force is the primary factor deciding a given material's response. However, for many agricultural materials deformation and flow depends not only on the nature of the force (stress), but also on time. Such materials are termed rheological.

Rheology is thus the study of deformation and flow resulting from the application of forces, with time effects taken into account. Its main problems concern the relationships between time-dependent stresses and deformations, the phenomena of creep and stress-relaxation, and the study of viscosity.

In addition to rheological properties, there are other mechanical properties which concern the movement of a material under the effect of forces. Such properties are the drag coefficient, the terminal velocity, the friction coefficient, the flow characteristics of loose bulk materials, etc.

8.1 Characteristics of biological materials

Plants and agricultural products are living, biological materials, whose composition, moisture content and texture vary continuously during growth, ripening and even in the course of storage. The texture reacts sensitively during development to factors, such as the moisture content, temperature, the supply of oxygen and nutritive materials, etc. It suffices to refer to the fact that in dry, hot weather the stalk and crop of some plants (e.g., French beans) lose their elasticity: the plant becomes withered.

Consequently, the mechanical properties of biological materials depend on numerous factors. The majority of these relationships are still unknown today, especially with regard to their quantitative characteristics. The reason is that biological materials constitute biomechanical systems of very complex construction, whose behavior cannot be characterized by simple physical constants, as for example can that of steel.

Biological liquids may be regarded overwhelmingly as non-Newtonian media, whose rheology follows laws other than those applying to Newtonian media. This fact gives rise to considerable difficulties. One consequence of the complexity of biological materials is that relatively numerous assumptions must be applied in discussing of their mechanics, and the results obtained are valid only under the given conditions. Empirical methods must frequently be used to describe the observed phenomena.

Since purely theoretical considerations rarely lead to utilizable results, experimental investigations here have an especially important role. In evaluating test results it is very important to record accurately all the characteristics of the material which may influence the results (shape, size, moisture content, color, viscosity, etc.). In any rheological discussion, the characteristic structure of biological materials necessitates the introduction of certain concepts and definitions which

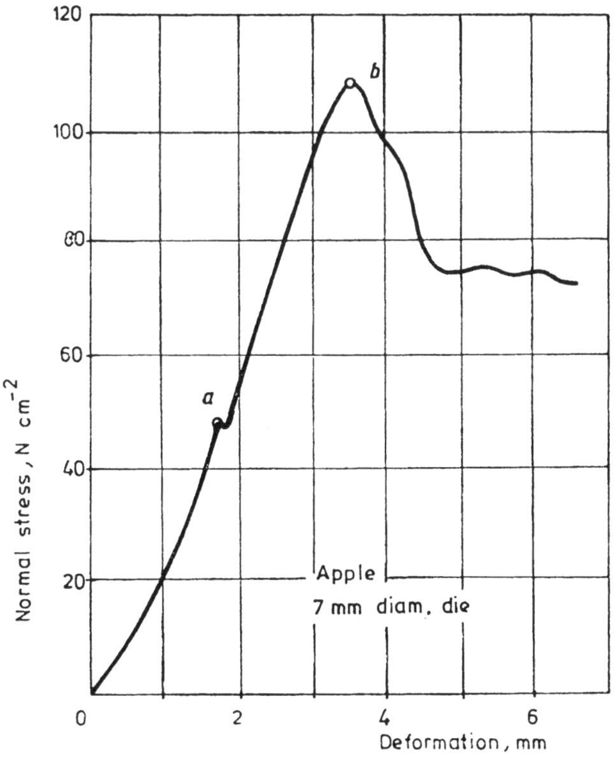

Fig. 80. Stress–deformation curve for biological materials.

are not usual in the mechanics of common elastic bodies. These are the following.

Biological yield point. (a) This is the point on the stress–deformation curve (Fig. 80) at which the stress decreases or remains constant with increasing deformation. This point indicates the appearance of initial cell rupture in a small volume of a cellular system. The yield point of biological materials plays an important part in determining their sensitivity to damage: if the load on a product does not reach the biological yield point, the cellular system will not be damaged, and spoiling of the product will not occur.

Rupture point. (b) This is the point on the stress-deformation curve (Fig. 80) beyond which the stress decreases rapidly and significantly with increasing deformation. This point indicates failure over a significant volume of material. In soft, tough materials rupture occurs only after considerable plastic deformation.

Rigidity. The rigidity of a material is characterized by the tangent to the initial, more or less linear section of the stress-deformation curve (this value is indeed nothing other than the modulus of elasticity). If the initial section of the curve is nonlinear, then either the initial tangent (shear) modulus, the secant modulus or the tangent modulus at a given point may be used (Fig. 81).

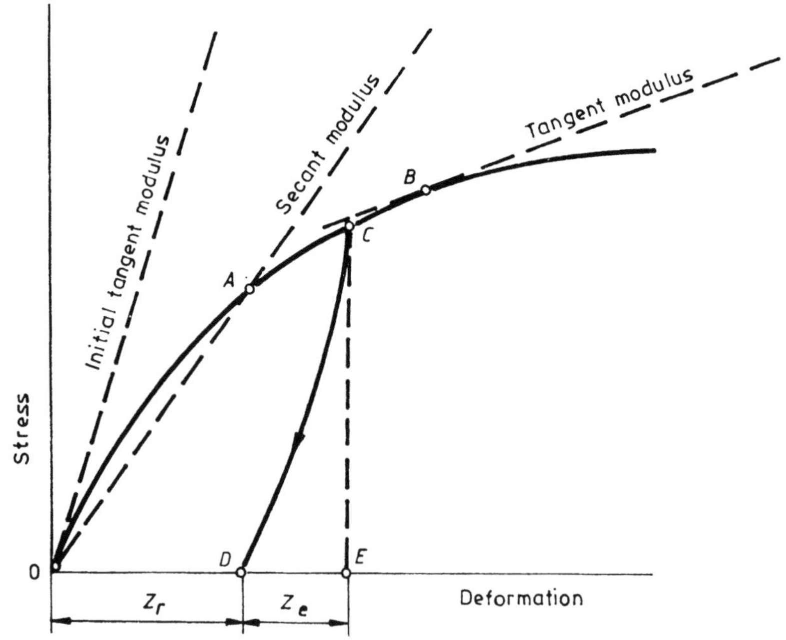

Fig. 81. Explanation of basic definitions

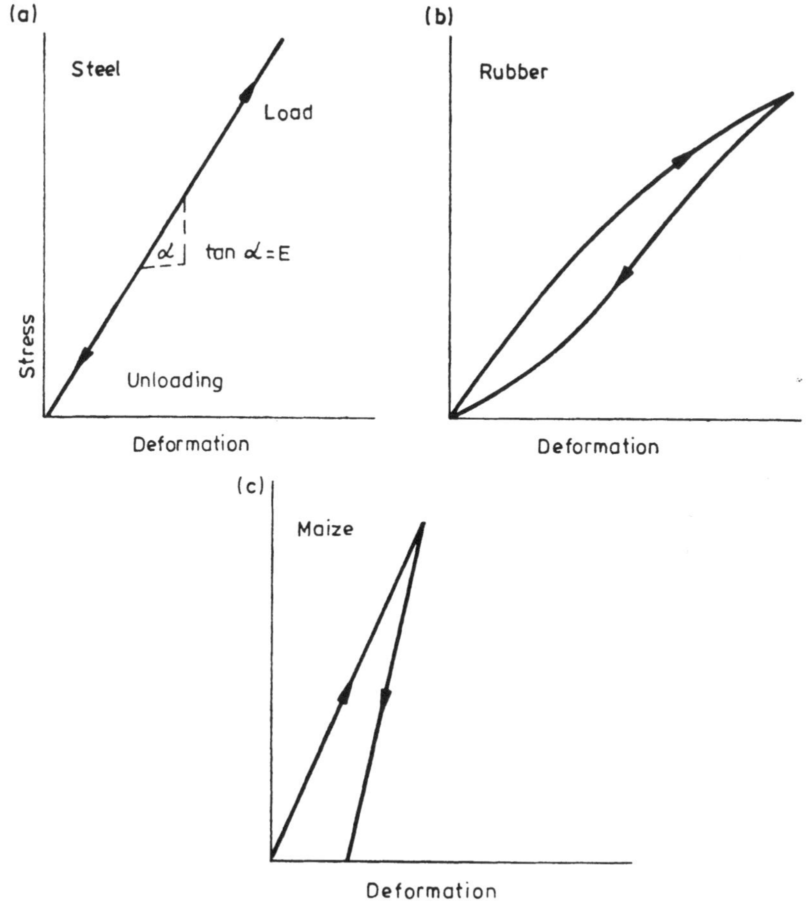

Fig. 82. Behavior of various bodies during the loading–unloading cycle

Degree of elasticity. This is the ratio of the elastic to the total deformation, when a material is loaded to a certain value and then unloaded (Fig. 81).

Toughness. The toughness is characterized by the work required to cause rupture in a material (mN m^{-3}), which is identical to the area under the stress–deformation curve.

Hardness. Hardness is characterized by the resistance of a material to penetration by an indenter.

Deformation work (resilience). The deformation work is a measure of the ability of a material to store energy (mN m^{-3}) in the range of elasticity. If the deformation is more or less elastic, the deformation work is given by the area under stress–deformation curve. If the deformation is inelastic, then the deformation work may be determined by plotting the loading–unloading cycle (Fig. 81): the work is given by area under the unloading curve *CD*.

Mechanical hysteresis. This is the energy absorbed by a material in the loading–unloading cycle: it is given by the area between the two curves. The mechanical hysteresis also characterizes the damping capacity of a material.

Energy recovery. This is the ratio of the energy recovered on unloading to the energy invested in loading.

8.2 Ideal materials and their properties

To facilitate the present discussion and survey it is convenient to divide the materials considered into groups according to certain basic properties. Extensive study has shown that the three fundamental properties by which the rheological behavior of a material can be characterized are its elasticity, plasticity and viscosity. The three ideal bodies showing these properties are termed the Hookean body, the St. Venant body and the Newtonian liquid. Real materials are never perfectly elastic or plastic, and so the three ideal bodies serve as standards or bases of comparison for evaluating real materials.

The behavior of an ideal elastic body is shown in Fig. 82(a). The stress is directly proportional to the elongation, as expressed by the well-known Hooke's law. On removing the stress, complete recovery of the strain takes place, and unloading occurs along the same line as loading. This behavior is termed linear elasticity.

Rubber recovers its original form on unloading, but the curve is not linear (Fig. 82(b)). In this case we speak of nonlinear elasticity. Compression testing of a wide variety of agricultural materials has shown that they do not follow Hooke properties even in the case of quite small deformations. At the end of the loading–unloading cycle a certain residual deformation always remains (Fig. 82(c)). For elastic bodies, the following basic relationships are valid. In the case of tension or compression, the modulus of elasticity is

$$E = \sigma/\varepsilon$$

where ε is the relative elongation or strain, given by $\varepsilon = \Delta l/l$. In the case of a torsional load, a given cross-section rotates by an angle Θ and the shear modulus of elasticity is

$$G = \tau/\gamma$$

where $\gamma = \tan\Theta$. When an elastic body is under hydrostatic pressure, the volumetric modulus of elasticity is given by

$$K = p/\varepsilon_v$$

where $\varepsilon_v = \Delta V/V$. The following relationships exist between the above modulus values with Poisson's ratio ν taken into account:

$$E = 3K(1-2\nu)$$

and

$$E = 2G(1+\nu)$$

or

$$\nu = (3K-E)/6K$$

and

$$\nu = (E-2G)/2G$$

Table 7

Poisson's ratio	E/G	K/E
0.0	2.0	0.333
0.1	2.2	0.417
0.2	2.4	0.556
0.25	2.5	0.667
0.30	2.6	0.833
0.35	2.7	1.111
0.40	2.8	1.667
0.45	2.9	3.333
0.5	3.0	—

Table 7 shows the ratios E/G and K/E as functions of Poisson's ratio.

Poisson's ratio for most materials varies between 0.2 and 0.5. The value $\nu=0.5$ is characteristic of liquids and rubber, and means that in closed spaces the horizontal pressure is identical to the vertical pressure. Another extreme case, for which $\nu=0$, is shown by cork. Poisson's ratios for a few materials are given in Table 8 [1].

Table 8

Material	Poisson's ratio
Cork	0.0
Sandstone	0.1
Concrete	0.19
Steel	0.3
Potatoes	
$X=250\%$	0.26–0.28
$X=300\%$	0.33–0.35
$X=350\%$	0.41–0.43
Apples	0.37–0.40

Figure 83 presents the force–deformation diagram for an ideal plastic body, showing that the material does not yield until the shear stress attains the yield value. However, the material can sustain no stress higher than the yield stress, and flows under the effect of this stress until some disturbing effect restricts it. The behavior of a plastic body may be compared to that of a body experiencing friction. A body placed on a surface fails to move until the force acting on it overcomes the static frictional force. As soon as the body starts to move, only dynamic friction of a constant value must be overcome. When the stress is removed, the flow stops immediately, and the material does not return to its original position but remains where it was at the instant when the stress was removed.

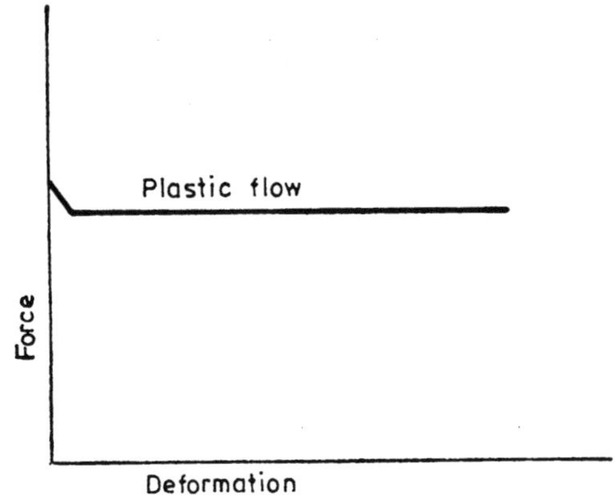

Fig. 83. Behavior of an ideal plastic body

An ideal viscous liquid starts to flow immediately when shear stress is applied. For liquids, deformation is a function not only of stress but also to a certain extent of time. For laminar flow, the velocity gradient may be expressed as

$$dv/dy = \tau/\eta$$

where η is the viscosity of the liquid. The velocity gradient here is actually the deformation rate. Use of the notation $dv/dy = \dot{\gamma}$ gives the well-known Newton relationship

$$\tau/\dot{\gamma} = \eta$$

illustrated graphically in Fig. 84. The mechanical model of a Newtonian liquid is the hydraulic dashpot element (Fig. 84(b)), for which the force P is proportional to the rate of displacement. The unit of viscosity is the poise, or its hundredth part, the centipoise. According to definition, 1 poise is the viscosity when 1 dyn

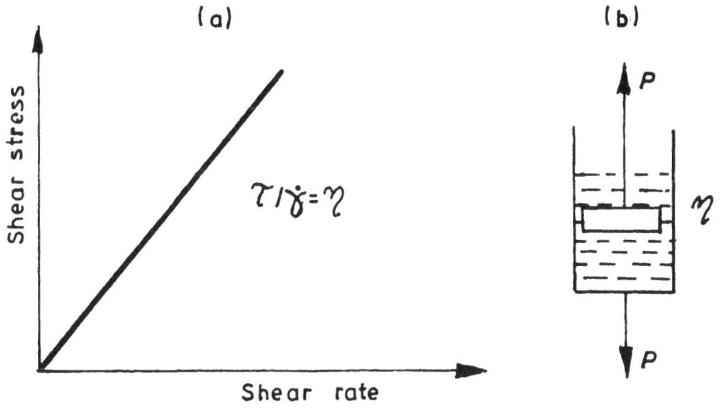

Fig. 84. Characteristic flow curve (a) and model (b) for an ideal liquid

force acting on 1 cm^2 surface area of plane plates at a distance of 1 cm results in a flow rate of 1 cm s^{-1}.

The viscosity of most materials depends greatly on temperature: with rising temperature it decreases. Table 9 lists viscosity values for some materials.

Table 9

Material	Temperature (°C)	Viscosity (centipoise)
Air	20	0.0186
Water	20	1.0
Water	0	1.79
Milk (skimmed)	25	1.37
Milk (whole)	20	2.12
Milk (whole)	0	4.28
Cream (20% fat content)	3	6.20
Cream (30% fat content)	3	13.78
Soybean oil	30	40.60
Olive oil	30	84.0

8.3 Time-dependent behavior of materials; viscoelasticity

The behavior of real materials always deviates to a greater or lesser extent from the behavior of ideal materials. This applies especially to agricultural materials: a striking property of biological materials is that the stress–deformation relationship also depends on the deformation rate. This means that a relationship must be found between not two factors (stress and deformation) but three (stress, deformation and time). Materials which show effects dependent on time are termed viscoelastic. These materials have partly the properties of solids and partly those of liquids.

For certain materials at relatively low loads the stress–deformation ratio is a function only of time, and does not depend on the magnitude of the stress. These are termed linearly viscoelastic materials. For many agricultural materials the stress–deformation ratio also depends on the magnitude of the stress in addition to time, as when a great part of the deformation caused by a load cannot be recovered on unloading. In this case we speak of nonlinear viscoelasticity. Unfortunately, a considerable proportion of agricultural materials must be assigned to this category. The general theory of nonlinear viscoelasticity has not yet been elaborated, and so it proves necessary in most cases to rely on assumptions and to apply the theory of linear viscoelasticity.

The time-dependent behavior of viscoelastic materials may be described by constitutive equations whose variables are stress, deformation and time. The constitutive equations for viscoelastic materials may be expressed by means of rheological models and with the aid of empirical relationships obtained by processing experimental data. The scope of the validity of such rheological models must also be established by experiment. The most frequently applied quasistatic experimental methods are creep and relaxation tests, as well as increasing the stress or deformation at a constant rate. Recently, certain dynamic methods such as cyclic loading at various frequencies have also been used, because their time requirements are significantly less and they permit simultaneous investigation of the changes (e.g., their softening) in a material appearing under repeated loading.

8.4 Creep

Creep is understood as the continuous deformation of a material under the effect of a constant stress. Generally three characteristic stages of the creep process are found (Fig. 85). In the first stage of creep the deformation rate is decreasing, and the process is termed primary creep; in the second stage the deformation rate is nearly constant, while in the third stage the deformation rate is increasing, and the process ends with rupture.

The total strain (relative elongation) at an arbitrary time t is composed of the instantaneous elastic and creep strain components:

$$\varepsilon = \varepsilon_e + \varepsilon_c$$

The strain rate is obtained by differentiation. Noting that ε_e=constant, the strain rate is

$$d\varepsilon/dt = d\varepsilon_c/dt = \dot{\varepsilon}$$

The time periods for the individual stages of creep depend decisively on the structure of the material and the stress. Therefore, in investigating a given problem the first step is to determine whether it concerns only the first stage or both the first and second stages of the creep curve.

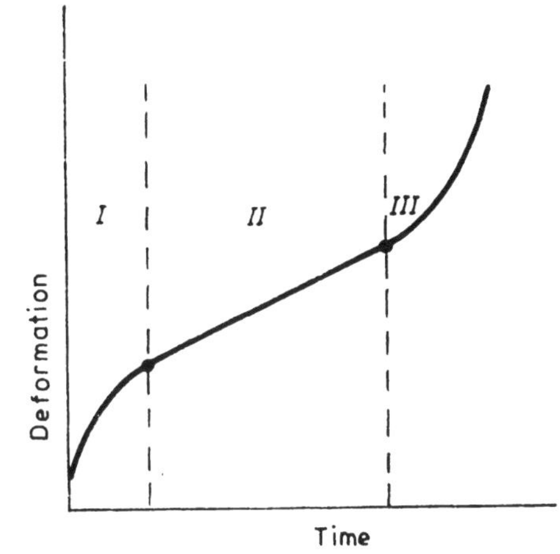

Fig. 85. Stages of creep

8.5 Recovery

At a given time in a creep test the load is removed and simultaneously the elastic deformation is recovered fully. The creep deformation decreases as a function of time, i.e., a time-dependent recovery process occurs (Fig. 86). The creep deformation does not vanish completely during recovery, even after a long time; the remaining value is the residual deformation.

The extent of recovery may differ for individual materials. Besides the structural properties of the material, the load (stress) also plays an important role: for agricultural materials the relative recovery decreases with increasing load. Recovery is also decreased by a temperature rise, and this property may be utilized with advantage, for example, in pelleting and wafering operations.

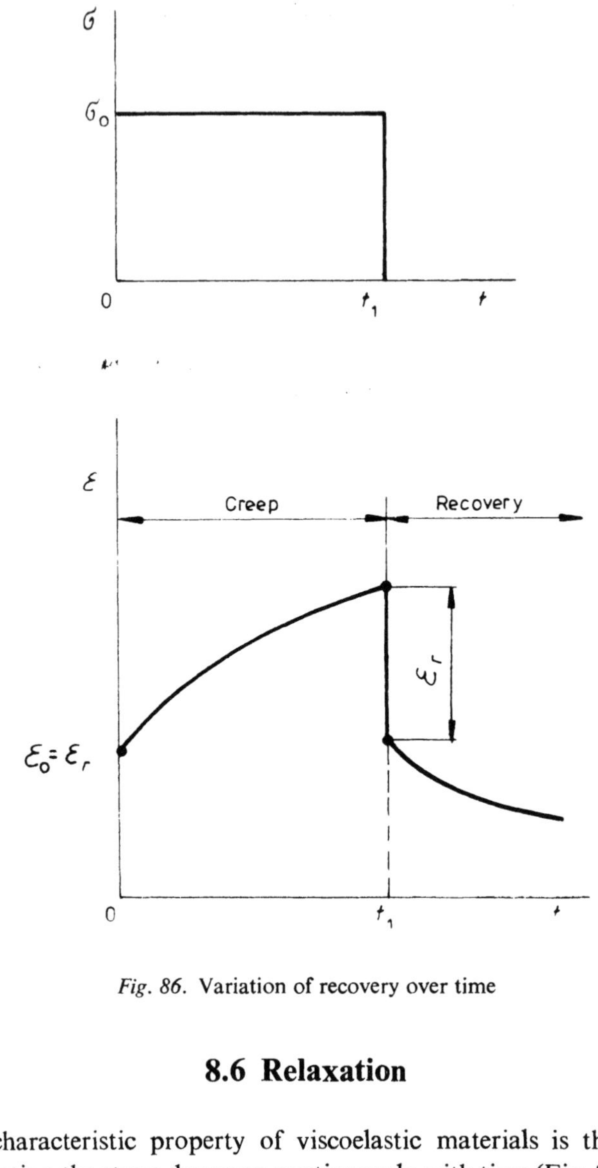

Fig. 86. Variation of recovery over time

8.6 Relaxation

Another characteristic property of viscoelastic materials is that under constant deformation the stress decreases continuously with time (Fig. 87). The extent and rate of stress reduction depend on the structure of the material and the amount of deformation. Generally, the decreasing stress tends asymptotically to a limiting value. The rate of stress decay is characterized by a relaxation time, i.e., the time period during which the stress decreases to $1/e$ (about 37%) of its original value.

8.7 Linearity

The behavior of viscoelastic materials is linear when the stress-deformation ratio is independent of stress and the principle of linear superposition then applies. These conditions may be expressed mathematically in the form [44]

$$\varepsilon[c\sigma(t)] = \varepsilon c[\sigma(t)]$$
$$\varepsilon[\sigma_0(t)+\sigma_1(t-t_1)] = \varepsilon\sigma_0(t)+\varepsilon\sigma_1(t-t_1)$$

Some agricultural materials may be treated as linearly viscoelastic, especially for smaller loads and for loads acting for a short time (e.g., during impact). In the case of greater or long-lasting loads, the majority of agricultural materials deviate from linear behavior, i.e., it is necessary to deal with non-linear materials. The description of nonlinear materials is more complex by far, since it requires the application of nonlinear constitutive equations.

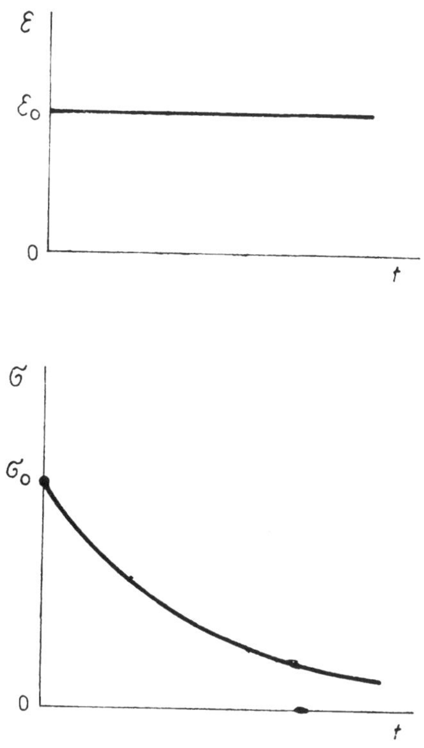

Fig. 87. The relaxation process

8.8 Rheological models

As was seen earlier, the behavior of an elastic material may be compared to that of a spring, while the behavior of a liquid is comparable to that of a dashpot element. Thus it is obvious to attempt to approximate the behavior of visco-elastic materials by some combination of a spring and a dashpot element. The mechanical models obtained in this way are termed rheological models.

The two simplest combinations of the spring and dashpot are the serial and parallel connections: i.e., the Maxwell and the Kelvin models (Fig. 88). These models each supply different stress–deformation relationships for different deformation rates.

In the Kelvin model the free ends of the spring and dashpot move together at a constant rate under the effect of an applied force. Therefore, the force absorbed by the dashpot has a constant value, independent of deformation, while the force absorbed by the spring increases from zero in a linear manner. As may be seen,

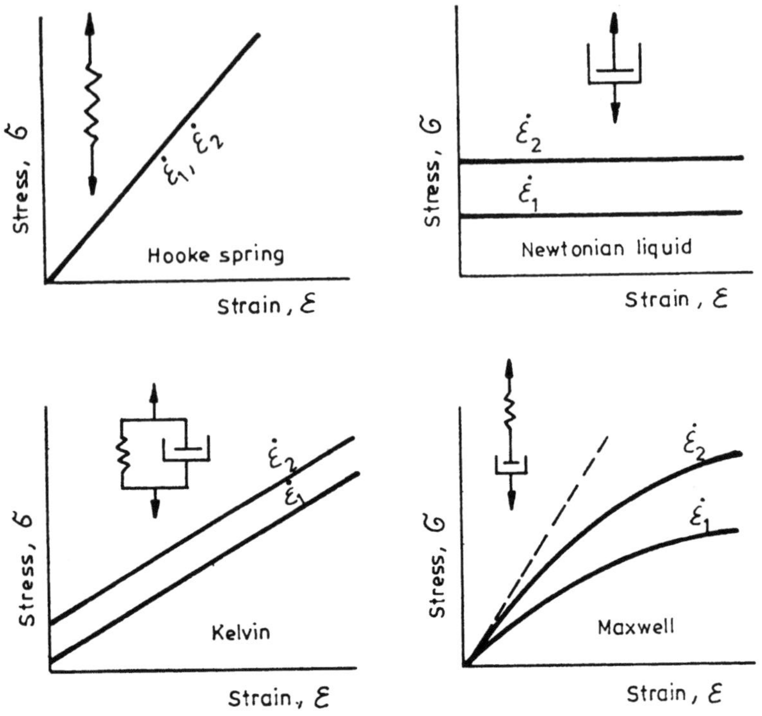

Fig. 88. The simplest rheological models and their characteristic curves

in this model the dashpot simply shifts the linear characteristics of the spring as a function of the deformation rate.

In the Maxwell model the whole stress is taken up initially by the spring, and this determines the initial tangent of the curve. On displacement of the spring the dashpot element starts to move as well with an increasing rate, and takes up a correspondingly increasing force. When the spring reaches its maximum compression, the whole force is taken up by the dashpot, moving at a constant rate: the deformation rate versus time curve becomes horizontal.

The above mechanical models may also be substituted by electrical models. In the electrical model the spring is replaced by a capacitor and the dashpot element by a resistor. Charging and discharging of the capacitor correspond respectively to tensioning and compression of the spring. The absorption of energy in the dashpot element is similar to the absorption of electrical power (its transformation into heat) by the resistor. The serially connected mechanical model is replaced by an electrical model connected in parallel. In the electrical model the voltage corresponds to the mechanical stress and the current to the deformation.

8.9 Rheological equations

In deriving the rheological equations for the above models, it is assumed that the spring obeys Hooke's law, and the dashpot element Newton's law, i.e.,

$$\sigma/\varepsilon = E$$

and

$$\sigma/\dot{\varepsilon} = \eta$$

Using the subscripts s and v for the values of σ and ε corresponding respectively to the spring and to the dashpot element, the spring in the Maxwell model may be represented by the equations

$$\sigma_s/\varepsilon_s = E$$

and

$$\dot{\varepsilon}_s = \dot{\sigma}_s/E$$

where a superscript point above any symbol indicates differentiation with respect to time. The equation for the dashpot element is

$$\eta = \sigma_v/\dot{\varepsilon}_v$$

In the Maxwell model, the strains of the two elements are added: i.e.,

$$\varepsilon = \varepsilon_s + \varepsilon_v$$

or, in differentiated form,

$$\dot{\varepsilon} = \dot{\varepsilon}_s + \dot{\varepsilon}_v$$

On substituting the values of $\dot{\varepsilon}_s$ and $\dot{\varepsilon}_v$ from the above equations and taking into account that the spring and dashpot element experience the same force, i.e., $\sigma_s = \sigma_v = \sigma$, the result is obtained that

$$\dot{\varepsilon} = \dot{\sigma}/E + \sigma/\eta$$

or, in another form,

$$d\varepsilon/dt = (1/E)d\sigma/dt + \sigma/\eta \tag{36}$$

If a given strain value is suddenly brought about in the model, and then remains constant over time, then $d\varepsilon/dt = 0$ in eqn. (36), i.e.,

$$d\sigma/dt + (E/\eta)\sigma = 0$$

from which, by integration,

$$\sigma = Ae^{-t/T} + C$$

where $\eta/E = T$ is the relaxation time, and A and C are constants. The integration constants must be determined from the boundary conditions: at $t=0, \sigma=\sigma_0=\varepsilon_0 E_0$, while at $t=\infty$, $\sigma=\sigma_e=\varepsilon_0 E_e$. With these values,

$$C = \varepsilon_0 E_e = \sigma_e$$

$$A = \varepsilon_0(E_0-E_e) = \sigma_d$$

Substituting the expressions for C and A, the variation of the stress σ over time is described by the equation

$$\sigma(t) = \sigma_d e^{-t/T}+\sigma_e \qquad (37)$$

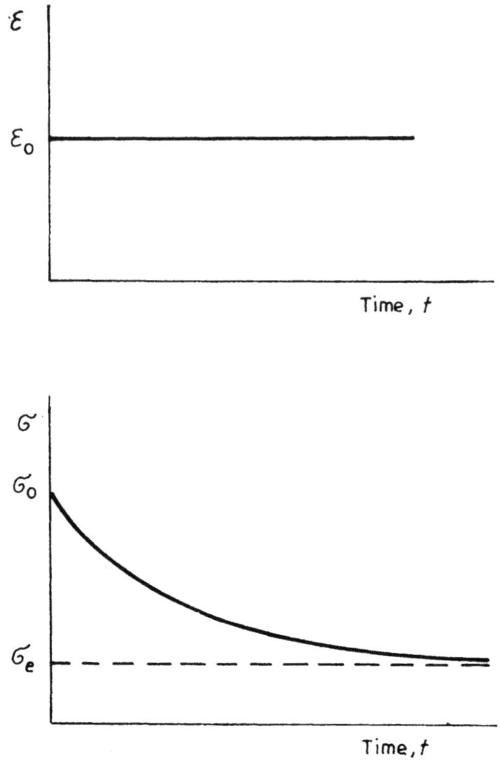

Fig. 89. Characteristic relaxation curve for the Maxwell model

In this equation σ_d is the stress drop during stress decay, and σ_e the residual stress when equilibrium has been attained (Fig. 89).

Observations of agricultural materials have shown that even after a long time a certain amount of deformation remains, which does not follow from the above Maxwell model. To eliminate this deficiency a spring with modulus E_e must be connected in parallel to the Maxwell model, as shown in the model illustrated schematically in Fig. 90. The Maxwell model is unsuited for a general description of linearly viscoelastic materials for other reasons also. For example, if a material is loaded suddenly by a given stress and this stress remains constant over time ($d\sigma/dt=0$), then the material is treated only as a Newtonian liquid by eqn. (36), while experiments show a continuously increasing deformation, similar to stress relaxation. To overcome this problem it is possible to use a large (infinite) number of Maxwell models, interconnected in parallel known as the generalized Maxwell model.

The generalized Maxwell model consists of n Maxwell elements and a spring, connected in parallel with the nth element (Fig. 90). The modulus of elasticity E_e of the spring corresponds to the equilibrium modulus of the material during stress relaxation. If $n=1$, i.e., the model consists of only a single Maxwell element and a spring connected in parallel with it, then the variation of the stress is described

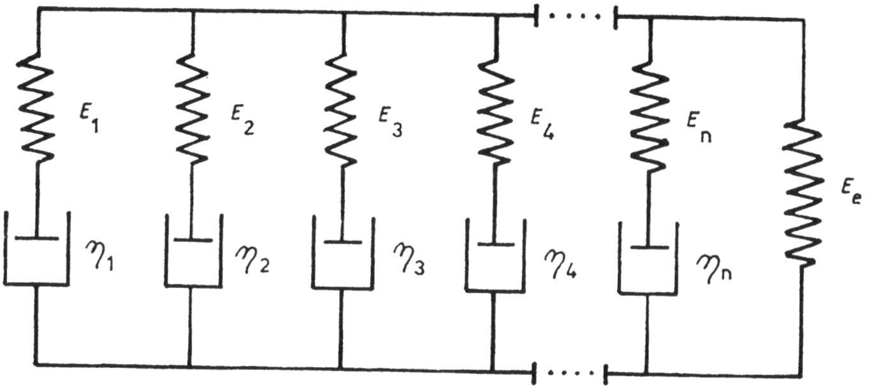

Fig. 90. The generalized Maxwell model

by the equation

$$\sigma = E_e\varepsilon + ET(d\varepsilon/dt) - T(d\sigma/dt) \tag{38}$$

where E is the instantaneous modulus of elasticity, and $T = \eta/(E-E)_e$ the relaxation time. The use of more Maxwell elements aggravates the calculations greatly, and this is done only for calculating the relaxation when $d\varepsilon/dt = 0$. In this case the total stress for a model consisting of n members, suffering deformation ε_0 at time moment $t=0$, is given by

$$\sigma = \sigma_1 + \sigma_2 + \sigma_3 + \ldots + \sigma_n + \sigma_e$$

and the reduction of the stress over time is

$$\sigma(t) = \varepsilon_0[(E_1 - E_0)e^{-t/T_1} + (E_2 - E_1)e^{-t/T_2} + \ldots + (E_n - E_{n-1})e^{-t/T_n} + E_e] \tag{39}$$

This equation indicates that the logarithmic stress-relaxation versus time curve is not linear, and so it cannot be given by eqn. (37). The curve must be replaced by a sufficient number of straight lines, each of which is described by an exponential function.

In the Kelvin model the total stress is distributed between the spring and the dashpot according to

$$\sigma = \sigma_s + \sigma_v$$

The deformations of the two elements are identical, i.e., $\varepsilon = \varepsilon_s = \varepsilon_v$. Substitution of the values of σ_s and σ_v gives

$$\sigma = E\varepsilon + \eta\dot{\varepsilon}$$

or, in another form,

$$\sigma/E = \varepsilon + T_r(d\varepsilon/dt) \tag{40}$$

where $T_r = \eta/E$ is the retardation time. Suppose that the material is loaded suddenly by a stress σ_0, which remains constant over time (dead load). Differentiation with eqn. (40) and $d\sigma/dt = 0$, gives:

$$T_r\ddot{\varepsilon} + \dot{\varepsilon} = 0$$

and, after integration,

$$\varepsilon(t) = \varepsilon_e - (\varepsilon_e - \varepsilon_0)e^{-t/T_r}$$

or

$$\varepsilon(t) = \varepsilon_0 + \varepsilon_d(1 - e^{-t/T_r}) \tag{41}$$

where $\varepsilon_d = \varepsilon_e - \varepsilon_0$. Equation (41) is shown graphically in Fig. 91. When $t = T_r$, the material has attained $1 - 1/e$, i.e., about 63%, of the retarded deformation, $\varepsilon_d = \varepsilon_e - \varepsilon_0$. Thus the retardation time T_r characterizes the rate of deformation during creep.

The Kelvin model, similarly to the Maxwell model, cannot be regarded as having general validity, since it fails to describe the behavior correctly for all

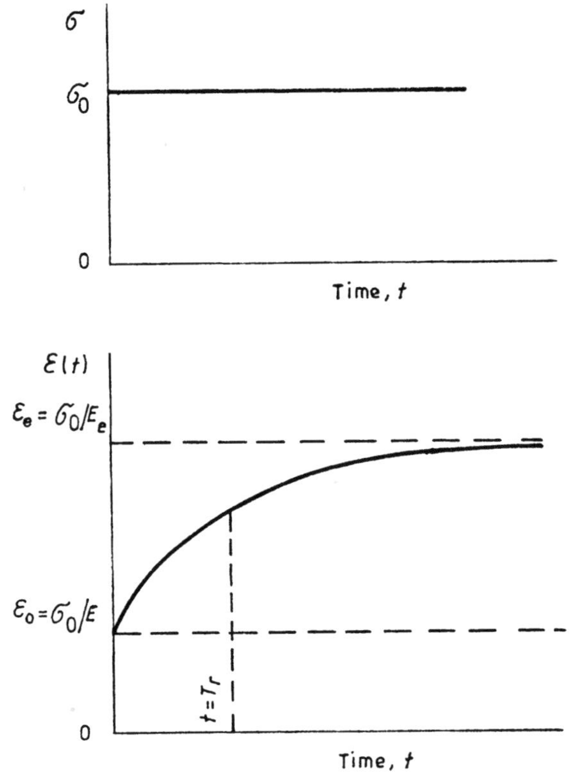

Fig. 91. Characteristic creep curve for the Kelvin model

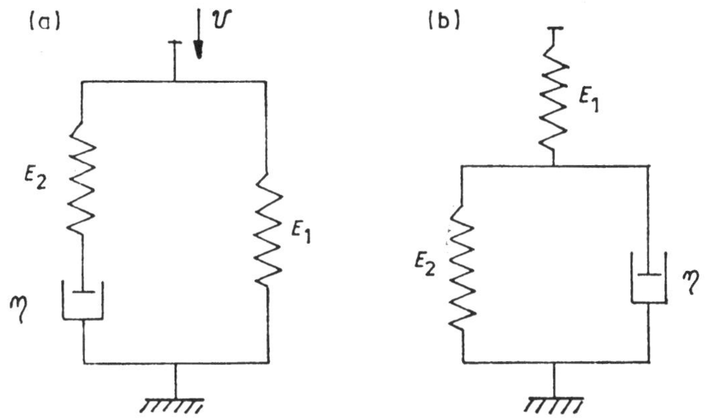

Fig. 92. Three-element models

loading modes; for example, it is not applicable to stress relaxation under constant stress. Therefore, the Kelvin model is also combined with other elements so as to obtain a model of more general validity.

The simplest model, which is applicable most frequently, is the three-element model, the two main types of which are shown in Fig. 92. The Maxwell and Kelvin models are respectively combined with a spring connected in parallel or serially. The model according to Fig. 92(a) may be described as

$$\sigma = E_1\varepsilon + T(E_1+E_2)\,d\varepsilon/dt - T(d\sigma/dt) \tag{42}$$

where $T=\eta/E_2$ is the relaxation time. In the case of a sudden load according to a step function, which is followed by constant deformation ($d\varepsilon/dt=0$), the relaxation of the initial stress σ_0 may be calculated from the equation

$$\sigma(t) = \sigma_0 e^{-t/T} + [E_1/(E_1+E_2)]\sigma_0(1-e^{-t/T}) \tag{43}$$

For a sudden load and a subsequently constant stress ($d\sigma/dt=0$), the variation of the deformation (i.e., the creep) is given by the equation

$$\varepsilon(t) = [\sigma_0/(E_1+E_0)]e^{-t/T} + (\sigma_0/E_1)(1-e^{-t/T}) \tag{44}$$

where $T=\eta/E_\infty$ and $E_\infty = E_1E_2/(E_1+E_2)$. A sudden load according to a step function cannot be realized in practice in many cases, and so the calculation may be performed instead with loading at a constant rate v_0, or with loading corresponding to the motion of a crankshaft. In the case of a constant deformation rate v_0, the instantaneous strain is

$$\varepsilon = (v_0/L)t = at$$

and

$$d\varepsilon/dt = a$$

where L is the length of material. With the latter relationship taken into account, the differential equation for the three-element model may be written as

$$d\sigma/dt+\sigma/T = (E_1/T)at+(E_1+E_2)a$$

of which the solution is,

$$\sigma(t) = E_1at+TaE_2(1-e^{-t/T}) \tag{45}$$

Equation (45) is valid in the time interval $0<t<t_1$, where t_1 represents the end of loading at a rate v_0. If the deformation remains constant after t_1, then the subsequent stress relaxation follows the equation

$$\sigma(t) = \sigma(t_1)e^{-t/T}+E_1\varepsilon(t_1)(1-e^{-t/T}) \tag{45a}$$

where $\sigma(t_1)$ and $\varepsilon(t_1)$ are the stress and strain at time t_1. After infinite time the stress tends to the value $\sigma_\infty=E_1\varepsilon(t_1)$.

The differential equation for the model shown in Fig. 92(b) is:

$$(d\sigma/dt)+\sigma/T = E_1(d\varepsilon/dt)+(1/T)[E_1E_2/(E_1+E_2)]\varepsilon \tag{46}$$

where $T=\eta/(E_1+E_2)$. In the case of a sudden load and subsequent constant deformation, the stress relaxation may be calculated from the equation

$$\sigma(t) = \sigma_0e^{-t/T}+[E_2/(E_1+E_2)]\sigma_0(1-e^{-t/T}) \tag{47}$$

For a sudden load and subsequent constant stress, the creep may be calculated from the equation

$$\varepsilon(t) = \sigma_0/E_1+(\sigma_0/E_2)(1-e^{-t/T}) \tag{48}$$

where $T=\eta/E_2$. In the case of a constant deformation rate v_0, the differential equation assumes the form

$$(d\sigma/dt)+\sigma/T = E_1a+(1/T)[E_1E_2/(E_1+E_2)]at$$

Introducing the asymptotic modulus $E_\infty=E_1E_2/(E_1+E_2)$, the stress-relaxation equation will be

$$\sigma(t) = E_\infty at+Ta(E_1-E_\infty)(1-e^{-t/T}) \tag{49}$$

Equation (49) is valid in the time interval $0<t<t_1$, where t_1 denotes the end of the loading period. The subsequent relaxation under constant deformation occurs according to the equation

$$\sigma(t) = \sigma(t_1)e^{-t/T}+E_\infty\varepsilon(t_1)(1-e^{-t/T}) \tag{49a}$$

The loading period t may be expressed in terms of the ratio of the deformation Δl and the loading rate v_0 and eqn. (49) may be brought to the form

$$\sigma(t) = \varepsilon(t)[E_\infty+(v_0T/\Delta l)(E_1-E_\infty)(1-e^{-\Delta l/v_0T})] \tag{49b}$$

In numerous cases the load is realized by a crankshaft, and the deformation then varies with time according to the expression

$$\Delta l = r(1 - \cos \omega t)$$

while the loading rate is

$$v = r\omega \sin \omega t$$

The differential equation for the model shown in Fig. 92(a) may now be written in the form

$$d\sigma/dt + \sigma/T = E_1 r/TL - (E_1 r/TL) \cos \omega t + (E_1 + E_2)(r\omega/L) \sin \omega t$$

where r is the radius and ω the angular velocity of the crank. The solution of this differential equation is [47]

$$\sigma(t) = (\varepsilon_0/2)\{E_1 + E_2(T\omega)^2 e^{-t/T}/((T\omega)^2+1) + E_2(T\omega)^2 \sin \omega t/((T\omega)^2+1) - \\ -(E_1 + E_2(T\omega)^2) \cos \omega t/((T\omega)^2+1\} \qquad (50)$$

At the upper (dead) point, $\cos \omega t = -1$ and $\sin \omega t = 0$; with these values, the stress σ is

$$\sigma(t_1) = (\varepsilon_0/2)\{2E_1 + [E_2(T\omega)^2/((T\omega)^2+1)](1 - e^{-t_1/T})\} \qquad (50a)$$

where t_1 is the half-rotation time of the crank, or $t_1 = 180/6n$ and ε_0 is the strain corresponding to the stroke of the crank. Equations (50) and (50a) are also valid for the model shown in Fig. 92(b), if E_∞ is substituted for E_1 and $(E_1 - E_\infty)$ for E_2.

A frequently used model is the so-called four-element Burgers model, shown in Fig. 93. The model consists of a Kelvin model connected in series to a spring

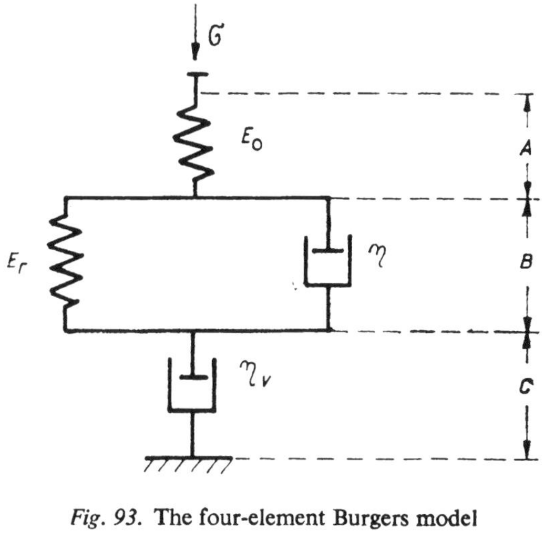

Fig. 93. The four-element Burgers model

and a dashpot element. The model may thus be divided into three parts. The total deformation is given by the sum of the deformations of the individual parts, i.e.,

$$\varepsilon = \varepsilon_A + \varepsilon_B + \varepsilon_C$$

The stress is identical in all the parts, i.e.,

$$\sigma = \sigma_A = \sigma_B = \sigma_C$$

The values of the individual stress components are:

$$\sigma_A = E_0 \varepsilon_A$$

$$\sigma_B = E_r \varepsilon_B + \eta \dot{\varepsilon}_B$$

and

$$\sigma_C = \eta_v \dot{\varepsilon}_C$$

The above three equations yield the differential equation [1, 44]

$$(d^2\varepsilon/dt^2) + (1/T_r)(d\varepsilon/dt) = (1/E_0)[d^2\sigma/dt^2 + (E_0/E_r T_r + E_0/\eta + 1/T_r)\, d\sigma/dt + (E_0/T_r \eta_v)\,\sigma] \tag{51}$$

Equation (51) is suitable for describing both creep phenomena under a dead load and stress relaxation in linearly viscoelastic materials. For a constant load, eqn. (51) becomes simpler, as then $d\sigma/dt = 0$. The differential equation may be written in the form

$$d^2\varepsilon/dt^2 + (1/T_r)(d\varepsilon/dt) = \sigma_0/T_r \eta_v$$

of which the solution is

$$\varepsilon(t) = \sigma_0/E_0 + (\sigma_0/E_r)(1 - e^{-t/T_r}) + \sigma_0 t/\eta_v \tag{52}$$

where $T_r = \eta/E_r$. The deformation rate is obtained by differentiating eqn. (52):

$$d\varepsilon/dt = (\sigma_0/\eta)e^{-t/T_r} + \sigma_0/\eta_v$$

Using this equation the deformation rate at $t=0$ is:

$$\dot{\varepsilon}(0) = (1/\eta + 1/\eta_v)\sigma_0 = \tan \alpha$$

while at $t = \infty$ the deformation rate tends asymptotically to the value:

$$\dot{\varepsilon}(\infty) = \sigma_0/\eta_v = \tan \beta$$

If the stress σ_0 is removed at time $t = t_1$, the elastic component of deformation ceases instantly, while the creep deformation decreases with time and tends asymp-

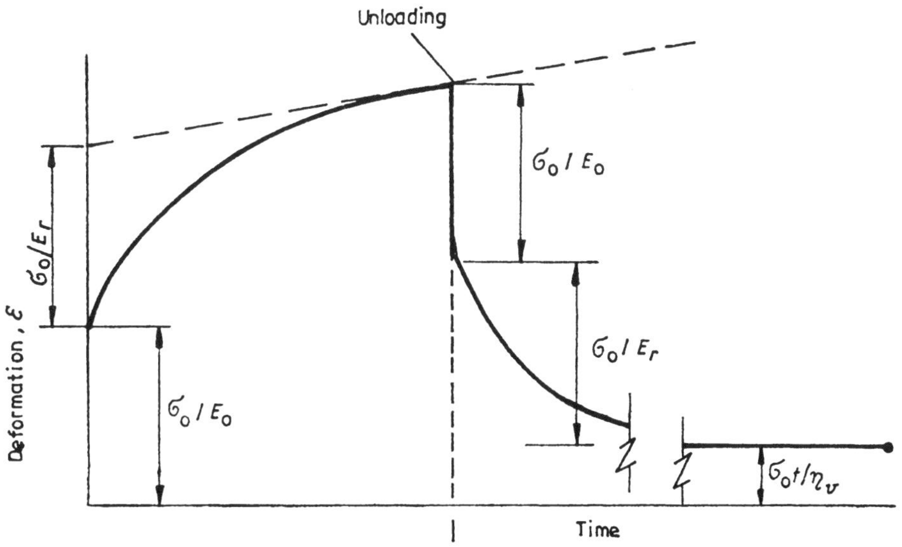

Fig. 94. Behavior of the four-element model

totically to the value $\sigma_0 t_1/\eta_v$. The decay of the deformation may be calculated by using the superposition principle such that a stress $\sigma = -\sigma_0$ is superimposed at $t = t_1$. Accordingly, the recovery of strain during time periods $t > t_1$ will be

$$\varepsilon(t) = (\sigma_0/\eta_v)t_1 + (\sigma_0/E_r)(e^{t_1/T_r} - 1)e^{-t/T_r}$$

Figure 94 shows the behavior of the four-element model. The elements of the model may be determined from the loading-unloading curve, using the values of $\tan\alpha$, $\tan\beta$ and the intercepts on the vertical axis.

The equations of the Kelvin and Burgers models discussed above contain an exponential function. If the strain ε is plotted using a logarithmic system of coordinates, then the exponential function will be represented by a straight line. However, the experimental data show that the relationship $\varepsilon = f(t)$ for a significant number of biological materials cannot be represented by a straight line even by logarithmic plotting. The problem may be solved in the same way as in the case of the Maxwell model discussed before (see Fig. 90). The curve is described by a finite number of equations and the equation will accordingly contain several exponential terms, with various retardation times $T_{r1}, T_{r2}, \ldots, T_{rn}$.

The mechanical model corresponding to the above requirement is obtained by connecting several Kelvin models in series. The model obtained in this way is termed the generalized Kelvin model (Fig. 95). The generalized Kelvin model consists of n Kelvin models and a spring, as well as a viscous element connected in series with them. The first spring takes into account the instantaneous elastic deformation of the material, the number n of Kelvin models characterizes the retained deformation, and finally the viscous element corresponds to the permanent flow. Equation (52) corresponding to the generalized Kelvin model may be written as

$$\varepsilon(t) = \sigma_0[1/E_0+(1/E_{r1})(1-e^{-t/T_1})+$$
$$+(1/E_{r2})(1-e^{t/T_2})+\ldots$$
$$\ldots+(1/E_{rn})(1-e^{-t/T_n})+t/\eta_v] \quad (53)$$

where $T_1, T_2, \ldots, T_n$ are the retardation times. Equation (53) may also be written in shorter form, as

$$\varepsilon(t) = \sigma_0[1/E_0+\sum_{i=1}^{n}\varphi_i(1-e^{-t/T_i})+t/\eta_v] \quad (53a)$$

where $\varphi_i=1/E_i$ is the reciprocal modulus. If the number of Kelvin elements is infinite, the symbol Σ may be written in integral form: i.e., excluding the E_0 and η_v terms,

$$\varepsilon = \sigma_0 \int_0^\infty \varphi(T)(1-e^{-t/T})\,dT$$

where $\varphi(T)$ is the distribution function of retardation time, or the retardation spectrum.

Investigation of the creep of various materials have shown that in many cases the experimental results may be approximated well by an equation of the form

$$\varepsilon(t) = \varepsilon_0+m't^n = \sigma_0(1/E_0+mt^n) \quad (54)$$

It can be demonstrated that this empirical function represents a defined

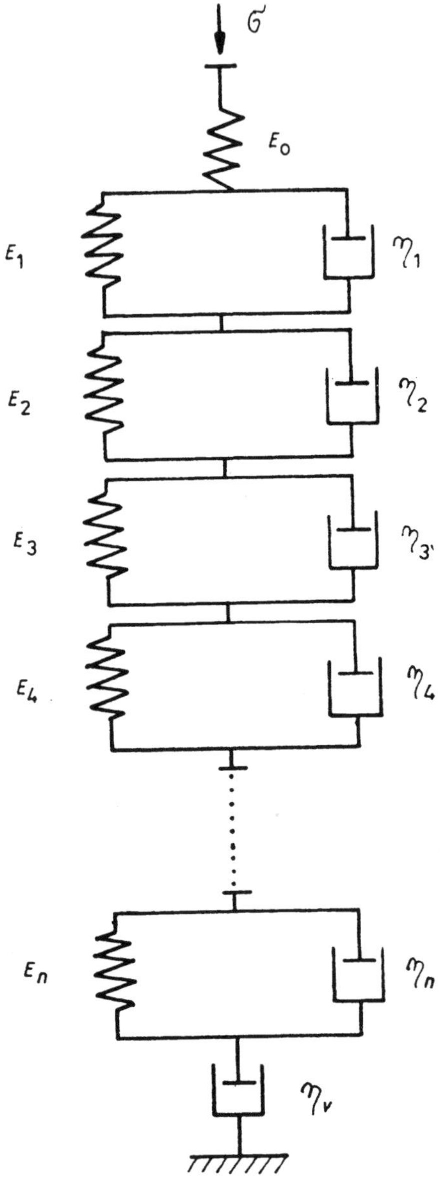

Fig. 95. The generalized Kelvin model

retardation spectrum. By comparing eqn. (54) and (53a), it may be seen that

$$mt^n = \int_0^\infty \varphi(T)(1-e^{-t/T})\,dT$$

from which

$$\varphi(T) = mnT^{n-1}/\Gamma(1-n) \tag{55}$$

The expression for the function Γ appearing in the denominator is

$$\Gamma(1-n) = \int_0^\infty (t/T)^{-n} e^{-t/T} d(t/T)$$

For most materials the exponent n is less than unity, and so in a logarithmic system of $\varphi(T)-T$ coordinates a straight line with a negative angle of inclination is obtained.

8.10 Integral representation of viscoelastic constitutive equations

Application of the models described in the preceding sections is not always permitted in cases of more complex constitutive equations or loading. The stress–deformation relationship may also be expressed in integral form, and this method is more universal with regard to its applicability. The ageing of biological materials, the effect of temperature, and nonlinearity can all be taken into account by integral representation.

We now introduce two time-dependent characteristics, the relaxation modulus and the creep modulus. For viscoelastic materials Hooke's law may be written formally as

$$\sigma(t) = E(t)\varepsilon_0$$

or

$$E(t) = \sigma(t)/\varepsilon_0$$

where the function $E(t)$ is termed the relaxation modulus. Similarly, the variation of deformation with time may be written as

$$\varepsilon(t) = J(t)\sigma_0$$

or

$$J(t) = \varepsilon(t)/\sigma_0$$

where $J(t)$ is the creep modulus or creep compliance. In the case of elastic materials, obviously $J=1/E$. The relaxation modulus may be written for the Maxwell and three-element models respectively in the forms:

$$E(t) = Ee^{-t/T}$$

and

$$E(t) = E_1 e^{-t/T} + E_\infty(1-e^{-t/T})$$

The values of the creep modulus for the Kelvin and three-element models are respectively

$$J(t) = 1/E + t/\eta$$

and

$$J(t) = 1/E_1 + (1/E_2)(1-e^{-t/T})$$

Now suppose that a viscoelastic material is loaded at $t=0$ by a stress σ_0, then at $t=\tau$ by a stress σ_1. Applying the superposition principle, the creep may be expressed as

$$\varepsilon(t) = J(t)\sigma_0 + J(t-\tau)\sigma_1$$

For any arbitrary course of loading, the above equation may be generalized in the form

$$\varepsilon(t) = \sum_{t=0}^{t} J(t-\tau)\sigma_i$$

or, by applying the symbol for integration for a continuous variation of the load,

$$\varepsilon(t) = \int_0^t J(t-\tau)(d\sigma/d\tau)\, d\tau \tag{56}$$

This integral equation describes with general validity the stress–strain relationship for any arbitrary course of loading. It is also usual to write eqn. (56) in other forms. If a sudden deformation appears, then

$$\varepsilon_0 = J_0\sigma$$

If the creep of a material increases continuously and does not tend to an asymptotic value, i.e.,

$$\varepsilon = \sigma t/\eta$$

then the integral equation may be written in the form

$$\varepsilon(t) = \int_0^t [J_0 + (t-\tau/\eta) + J(t-\tau)](d\sigma/d\tau)\, d\tau \tag{56a}$$

In this case the function $J(t)$ gives only the returning part of the deformation.

For a prescribed variation of deformation the stress-relaxation may be calculated in a similar way:

$$\sigma(t) = \int_0^t E(t-\tau)(d\varepsilon/d\tau)\, d\tau \tag{57}$$

If the relaxation modulus can be decomposed into a time-independent modulus E_0 and a time-dependent modulus $\psi(t)$, then

$$\sigma(t) = E_0\varepsilon(t) - \int_0^t \psi(t-\tau)(d\varepsilon/d\tau)\, d\tau \tag{57a}$$

On applying the Maxwell model

$$E(t) = Ee^{-t/T}$$

and, employing the integration rule by parts,

$$\sigma(t) = E\varepsilon(\tau) - (E/T) \int_0^t \varepsilon(\tau)e^{-(t-\tau)/T} d\tau \tag{58}$$

The relaxation modulus for the three-element model is

$$E(t) = E_1 e^{-t/T} + E_\infty(1-e^{-t/T})$$

and the stress relaxation for any arbitrary strain history becomes

$$\sigma(t) = E_1\varepsilon(\tau) - [(E_1 - E_\infty)/T] \int_0^t \varepsilon(\tau)e^{-(t-\tau)/T} d\tau \tag{59}$$

In the case of an arbitrary prescribed stress, the strain for the three-element model may be calculated from the equation

$$\varepsilon(t) = \sigma(\tau)/E_1 + (1/E_2 T) \int_0^t \sigma(\tau)e^{-(t-\tau)/T} d\tau \tag{60}$$

8.11 Behavior of viscoelastic materials under oscillating loads

A disadvantage of creep and stress-relaxation investigations is their relatively great time demand. During these investigations, sometimes lasting for several days, the mechanical properties of biological materials may vary considerably, and this circumstance may harmfully affect the evaluation of the results. Therefore, much less time-demanding dynamic tests have recently become more widespread.

An additional advantage of dynamic investigations is that by varying the frequency the effect of the deformation rate can also be studied in a simple way. Under the effect of a repeated load, biological materials often undergo softening, i.e., their modulus of elasticity E decreases. This means that the material "remembers" the course of loading which has preceded an instantaneous load. These phenomena may also be studied by the method of oscillatory loading. The essence of the dynamic testing method is that the material is loaded by periodically alternating stresses or deformations. The amplitude and frequency of the load may be varied arbitrarily. A periodic load of the frequency ν corresponds to a loading–unloading cycle of period $1/\nu$ s, from which the modulus of elasticity, the mechanical damping, and other material characteristics may be determined as functions of the frequency and number of cycles.

During the deformation of viscoelastic material, part of the energy applied is stored in the form of potential energy (i.e., energy stored in the spring of the model), while part is consumed (by the dashpot element of the model). Therefore, the deformation of the material alternates periodically, but with a certain phase retardation: the more energy consumed, the greater the phase retardation. A periodically alternating stress may be described by the equation

$$\sigma = \sigma_0 \cos \omega t$$

where σ_0 is the amplitude and ω the angular velocity of the stress. The deformation follows the variation of the stress with a phase retardation δ:

$$\varepsilon = \varepsilon_0 \cos (\omega t - \delta)$$

Utilizing the Euler relation, the deformation may be expressed as

$$\varepsilon = \varepsilon_0 e^{i(\omega t - \delta)} = (\varepsilon_0 e^{-i\delta}) e^{i\omega t} = e^* e^{i\omega t}$$

where ε^* is the complex strain amplitude given by

$$\varepsilon^* = \varepsilon_0 e^{-i\delta} = \varepsilon_0(\cos \delta - i \sin \delta)$$

Accordingly the complex creep modulus as introduced in Section 8.10 is

$$J^*(\omega) = \varepsilon^*/\sigma_0 = (\varepsilon_0/\sigma_0) e^{-i\delta} = (\varepsilon_0/\sigma_0)(\cos \delta - i \sin \delta)$$

i.e.

$$J^*(\omega) = J_1 - iJ_2$$

The absolute value of the complex creep modulus is

$$|J^*| = \sqrt{J_1^2 + J_2^2} = \varepsilon_0/\sigma_0$$

while the real and imaginary values are

$$J_1 = |J^*| \cos \delta$$

$$J_2 = |J^*| \sin \delta$$

The variation of the stress may be calculated similarly in the case of a periodically alternating strain. The input strain is expressed as

$$\varepsilon(t) = \varepsilon_0 e^{i\omega t}$$

in relation to which the stress shifts by an angle δ:

$$\sigma(t) = \sigma_0 e^{i(\omega t+\delta)}$$

or

$$\sigma(t) = \sigma^* e^{i\omega t}$$

where σ^* is the complex stress amplitude. The complex stress may also be divided into a real and an imaginary part:

$$\sigma^* = \sigma_0 e^{i\delta} = \sigma_0(\cos\delta + i\sin\delta)$$

Accordingly the complex relaxation modulus E^* is

$$E^* = \sigma^*/\varepsilon_0 = (\sigma_0/\varepsilon_0)e^{i\delta} = (\sigma_0/\varepsilon_0)(\cos\delta + i\sin\delta) = E_1 + iE_2 \qquad (61)$$

On the basis of the above, the frequency-dependent modulus E^* of a material may be divided into two parts: a real component, which is proportional to the stored energy, and an imaginary component, which is proportional to the energy loss (Fig. 96). The complex modulus E^* may be obtained from experimental results as the quotient of the stress and deformation peaks:

$$|E^*(\omega)| = \sigma_0/\varepsilon_0 = \sqrt{E_1^2+E_2^2}$$

The phase angle between the stress and strain is

$$\delta = \omega\Delta t$$

where Δt is the time shift between the curves for stress and strain. According to the vector diagram illustrated in Fig. 96,

$$\tan\delta = E_2/E_1$$

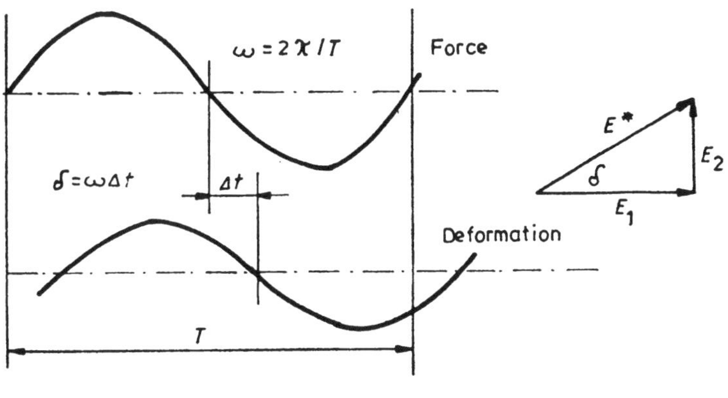

Fig. 96. Components of the elastic modulus E under oscillatory loading and deformation

The values of the components E_1 and E_2, based on eqn. (61), are

$$E_1 = (\sigma_0/\varepsilon_0)\cos\delta$$

and

$$E_2 = (\sigma_0/\varepsilon_0)\sin\delta$$

As mentioned before, the modulus E_1 is proportional to the stored and E_2 to the absorbed energy. Thus the ratio $E_2/E_1 = \tan\delta$ characterizes the damping capacity of the material; $\tan\delta$ is termed the loss modulus.

The damping energy is calculated as the ratio of the energy loss per cycle to the energy which may be stored maximally in a cycle ($\Delta W/W$). The individual quantities of energy are obtained by integration of the elemental energies $\sigma d\varepsilon$, and on this basis

$$\Delta W/W = 2\pi \sin\delta$$

It may be seen from this equation that the damping capacity of the linearly viscoelastic materials depends only on the phase angle δ which is a function of frequency, and is independent of load.

The complex relaxation- and creep modulus values may be determined easily for the various models. Since the expressions for σ and ε may be differentiated, the initial differential equations reduce to simple algebraic equations, permitting expressions for E^* and J^* to be obtained.

The differential equations for the Maxwell model, as seen before, may be written as

$$\sigma + T\dot{\sigma} = \eta\dot{\varepsilon}$$

Considering the complex relaxation modulus, the relationship $\varepsilon = \varepsilon_0 e^{i\omega t}$ yields $\dot{\varepsilon} = i\omega\varepsilon_0 e^{i\omega t}$ and $\sigma = \sigma^* e^{i\omega t}$ yields $\dot{\sigma} = i\omega\sigma^* e^{i\omega t}$. Substitution of these expressions into the differential equation yields

$$\sigma^*(1 + iT\omega) = i\eta\omega\varepsilon_0$$

from which

$$E^* = \sigma^*/\varepsilon_0 = i\eta\omega/(1 + iT\omega)$$

On separating the real and imaginary parts, the following expression is obtained:

$$E^*(\omega) = T\omega^2\eta/[1 + (T\omega)^2] + i\,[\omega\eta/[1 + (T\omega)^2]] \tag{62}$$

If the three-element model is considered (see Fig. 92(b)), a similar method yields the expression

$$E^*(\omega) = [E_1(T\omega)^2 + E_\infty]/[(T\omega)^2 + 1] + i[T\omega(E_1 - E_\infty)]/[(T\omega)^2 + 1] \tag{63}$$

The values of the loss modulus corresponding to $\tan\delta = E_2/E_1$ are for the Maxwell model

$$\tan\delta = E/\omega\eta$$

and for the three-element model,

$$\tan\delta = T\omega(E_1 - E_\infty)/[E_1(T\omega)^2 + E_\infty]$$

Expressions for the complex creep modulus are obtained by a similar calculation method; the relation for the three-element model may be written in the form

$$J^*(\omega) = [(T\omega)^2 E_1 + E_\infty]/[(T\omega)^2 E_1^2 + E_\infty^2] - i[T\omega(E_1 - E_\infty)]/[(T\omega)^2 E_1^2 + E_\infty^2] \quad (64)$$

8.12 Nonlinear constitutive equations

A considerable proportion of viscoelastic materials show nonlinear behavior under greater loads or strains; this must be taken into account for the purpose of more exact investigations. In describing nonlinear viscoelasticity, empirical and integral constitutive equations must be used. Some exponential function of stress generally appears in the empirical equations, for example, in the form

$$\varepsilon_c = k\sigma^m t^n$$

or

$$\varepsilon_c = k\sigma^m(1 - e^{-at}) + b\sigma^q t$$

where k, m, a n, q, and b are material constants at a given temperature. The integral form for nonlinear constitutive equations may be written by generalizing the linear form [44] as

$$\varepsilon(t) = J(0)\sigma(t) + \int_0^t F(t-\tau)(\partial/\partial\tau)f[\sigma(\tau)]\,d\tau \quad (65)$$

where F and f are time- and stress-dependent empirical functions.

As has been seen, the deformation of viscoelastic materials depends on the course of loading during the time preceding a given instant, i.e., the material "remembers" the history of loading over time. According to observations the time period immediately preceding a given instant is remembered by the material better than are earlier time periods. This also means that the material "forgets" the course of loading after the passage of a given time.

For a mathematical formulation of the above phenomenon, the stress or deformation function must be weighted for a given time interval by means of some weighting function. The weighted value of a given function may be expressed as

$$\overline{f(\tau)} = \int_{t_1}^{t_2} f(\tau)g(\tau)\,d\tau \Big/ \int_{t_1}^{t_2} g(\tau)\,d\tau$$

where $g(\tau)$ is the weighting function. One function which satisfies the requirements formulated above is the exponential function [46] i.e.,

$$g(\tau) = e^{a\tau}$$

where a is a material-dependent constant. The exponential function tends to zero as $\tau \to -\infty$. On substituting this weighting function and selecting $-\infty$ for the lower limit of integration, the weighted value of the function f may be written as

$$\overline{f(\tau)} = ae^{-at_2} \int_{-\infty}^{t_2} e^{a\tau} f(\tau)\, d\tau$$

or, in general, as

$$\overline{f(\tau)} = a \int_{-\infty}^{t} f(\tau) e^{-a(t-\tau)} d\tau \tag{66}$$

Equation (66) defines the exponentially mapped history of the function $f(\tau)$ for a given time interval. In dynamic investigations of cotton seed [46] it was assumed that the stress depends on the strain and the strain rate according to the equation

$$\sigma = C_1 \varepsilon + C_2 \dot{\varepsilon} + C_3 \varepsilon \dot{\varepsilon} + C_4 \int \varepsilon\, d\tau$$

On replacing the function $f(\tau)$ by this expression, the stress at a given time τ may be calculated, with the preceding course of deformation taken into account, from the equation

$$\sigma(\tau) = a \int_{-\infty}^{t} e^{-a(t-\tau)} (C_1 \varepsilon + C_2 \dot{\varepsilon} + C_3 \varepsilon \dot{\varepsilon} + C_4 \int \varepsilon d\tau)\, d\tau$$

This equation is the integral formulation of the nonlinear constitutive equation.

8.13 Temperature effects

The mechanical behavior of agricultural materials is almost always influenced by temperature. In principle, the temperature effect must always be taken into account as an additional variable besides time, i.e.,

$$J = J(t, \vartheta)$$

$$E = E(t, \vartheta)$$

etc. However, experiments have shown that for a great majority of materials the two variables may be reduced to a single variable by introducing the concept of "reduced time". This means that a material property determined at a given temperature may be applied at other temperatures, but the time must be reduced by a temperature-dependent constant, i.e.,

$$J(t, \vartheta) = J(t^*, \vartheta_0)$$

with

$$t^* = t/k(\vartheta)$$

where ϑ_0 is the temperature at which the material property was determined. Thus the effect of temperature is equivalent to extension or reduction of the effective time period, depending on whether the temperature is higher or lower than the reference temperature. Materials to which the above principle of time-temperature superposition is applicable are termed thermorheologically simple.

The factor $k(\vartheta)$ may vary between wide limits depending on temperature, and it is therefore advisable to plot $\log k(\vartheta)$ as a function of temperature. The curve obtained may be described well in numerous cases by an equation of the form [44]

$$\log k(\vartheta) = -k_1(\vartheta - \vartheta_0)/[k_2 + (\vartheta - \vartheta_0)] \tag{67}$$

where k_1 and k_2 are material constants. If the temperature varies with time, the reduced time may be used in integral form, as

$$t^*(t) = \int_0^t \{1/k[\vartheta(\tau)]\}\, d\tau$$

8.14 Non-Newtonian fluids; viscosimetry

In Section 8.2 we summarized briefly the basic properties of an ideal viscous liquid. A Newtonian liquid is characterized by the fact that the shear stress is proportional to the deformation rate, and the straight line describing the relationship passes through the origin. The behavior of non-Newtonian liquids shows essential deviations from this pattern. In the following the main characteristics of the non-Newtonian liquids are summarized.

8.14.1 Characteristic flow curves

We have characterized the behavior of Newtonian liquids by the relationship $\tau = f(\dot{\gamma})$, which also defines the viscosity. Since the viscosity is also the most important characteristic in the flow of non-Newtonian liquids, this mode of discussion is followed further.

Figure 97 shows the characteristic flow curve for ideal plastic flow. A minimum

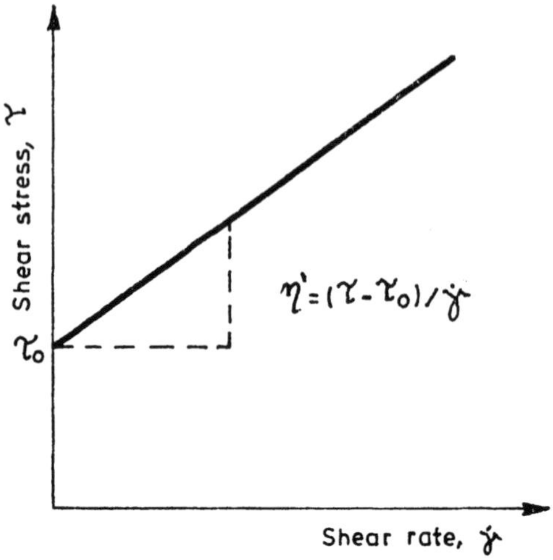

Fig. 97. Characteristic flow curve for an ideal plastic medium

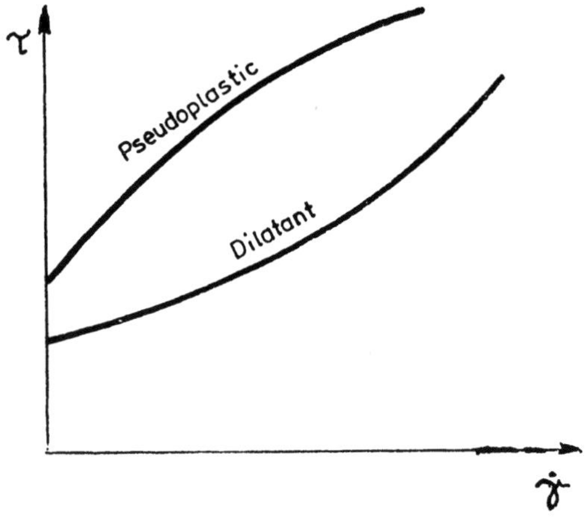

Fig. 98. Characteristic flow curve for a quasiplastic medium

stress τ_0 is required to initiate flow, and the behavior is similar to that of viscous liquids. The flow-rate gradient is given by

$$dv/dy = (\tau - \tau_0)/\eta$$

where η is the plastic viscosity coefficient. Liquids behaving as shown in Fig. 97 are also termed Bingham bodies. A typical material which behaves similarly is the oil paint. Figure 98 illustrates a plastic flow for which the relationship between shear stress and deformation rate is nonlinear. This type of flow is termed quasi-

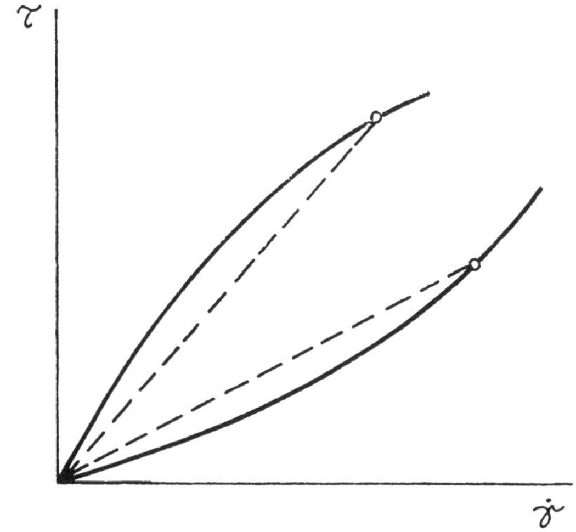

Fig. 99. Behavior of a quasiviscous liquid

plastic. The curve may be concave or convex, and accordingly we speak of pseudo-plastic and dilatant liquids. Figure 99 shows the behavior of a quasi-viscous liquid. The curves pass through the origin of the system of coordinates, but the

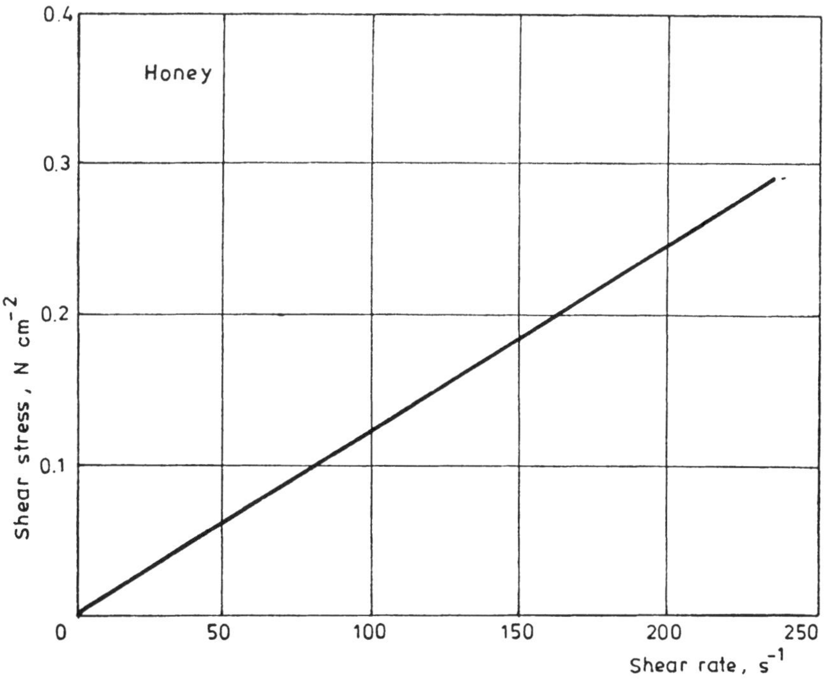

Fig. 100. Flow curve for honey

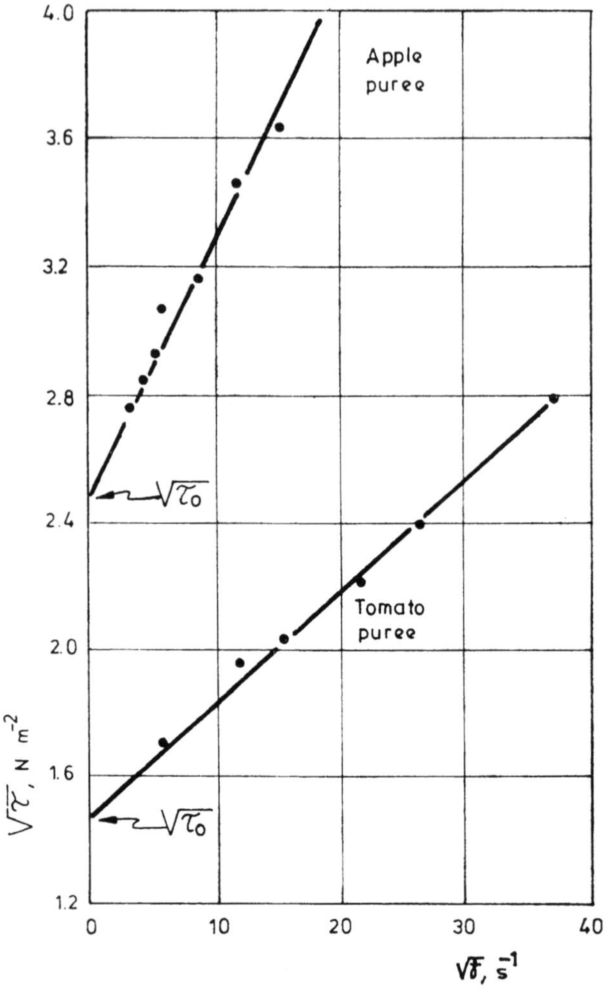

Fig. 101. Flow curves for apple and tomato purees

relationship is nonlinear. Liquids of the above types are termed non-Newtonian. Quasiplastic and quasiviscous liquids may be described respectively by the equations

$$dv/dy = (\tau - \tau_0)^n/\eta''$$

and

$$dv/dy = \tau^n/\eta''$$

where η'' is the apparent viscosity. The apparent viscosity is used in the case of all liquids for which the relationship $\tau = f(\dot{\gamma})$ is nonlinear. The apparent viscosity is identical to the viscosity of a Newtonian liquid which has the same resistance for the given shear stress and flow-rate gradient. The apparent viscosity is determined by the slope of the straight line connecting a given point of the curve with the origin.

The behavior of the liquids described above is time-independent. However, there exist liquids for which the shear stress for a given shearing rate decreases or increases with time. Increase of shear stress occurs rarely; it has been observed for certain material suspensions. Decrease of shear stress is observed frequently in mixing equipment, during the mixing of various industrial food materials (sauces, mashes). These are also termed thixotropic materials [126].

Figures 100 and 101 show characteristic curves for some agricultural materials. Honey behaves as a viscous material. Apple- and tomato purees show plastic behavior, requiring a defined stress τ_0 to initiate flow.

8.14.2 Viscosimetry

Knowledge of viscosity as a material property is necessary in all cases in order to calculate the flow of a liquid. The viscosity of agricultural materials is usually measured by means of a rotational viscosimeter, and less frequently by a capillary viscosimeter.

To calculate the viscosity it is necessary to establish regularity of flow in the viscosimeter. It is necessary to write the equation of motion for the flow purposely in a form permitting direct determination of viscosity. As has been seen before, the viscosity is the ratio of the shear stress to the shear rate, and so both these quantities must be measured. The shear stress is measured in the form of a torque (using a rotational viscosimeter) or a pressure drop (using a capillary viscosimeter), while the shear rate is a function of the flow rate.

On the basis of the above, the apparent viscosity in the case of the two types of the viscosimeter mentioned may be determined respectively by means of the equations

$$\eta'' = K_1(M/\omega)$$

$$\eta'' = K_2(\Delta p/Q)$$

where K_1 and K_2 are instrument constants, M is the torque, ω the angular velocity of flow, Δp the pressure drop, Q the volume flow rate. The constants K_1 and K_2 may be obtained by solving the equation of motion with the boundary conditions taken into account.

Figure 102 shows a schematic drawing of a rotational viscosimeter. A rotating cylindrical body is immersed in the liquid, held in a cylindrical container. The

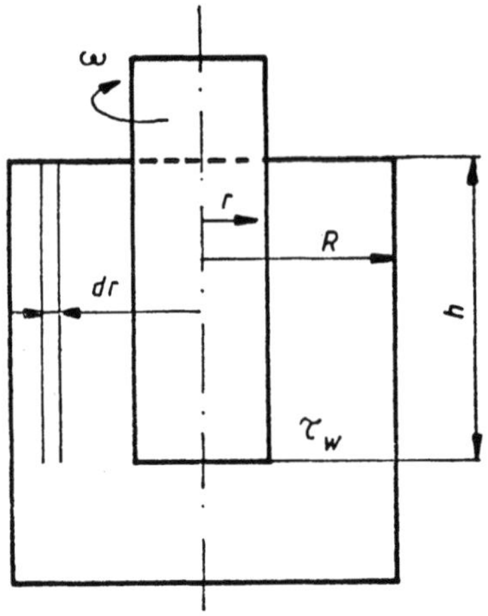

Fig. 102. Principal elements of a rotational viscosimeter

torque exerted on the rotating cylinder is balanced by the resistance due to the liquid: thus

$$M = 2\pi r h \tau r \tag{68}$$

The tangential velocity at a given radius r is $v = r\omega$, while at a radius $r+dr$ it is

$$v+dv = (r+dr)(\omega+d\omega)$$

from which, with small terms of second order neglected,

$$dv/dr = \omega + r(d\omega/dr)$$

The angular velocity of the rotating cylinder is ω, and so if there is no slippage between the cylinder and the liquid layer which is in contact with it, the shear rate is determined by the term $rd\omega/dr$ alone, i.e.,

$$dv/dr = r(d\omega/dr) \tag{69}$$

We now determine, using eqn. (69), the basic equations for the rotational viscosimeter for liquids showing various behaviors.

The initial equation for a Newtonian medium may be written as

$$\tau = \eta(-dv/dr) = \eta(-r(d\omega/dr))$$

Substitution of τ from eqn. (68) then gives

$$-d\omega = (M/2\pi h\eta)dr/r^3$$

The angular velocity along the external standing wall is zero; and along the inner wall it is ω. The radius of the liquid layer varies between r and R. Applying these integration limits,

$$\int_0^\omega -d\omega = \int_r^R (M/2\pi h\eta)dr/r^3$$

from which

$$\eta = (M/4\pi h\omega)(1/r^2 - 1/R^2) \tag{70}$$

Thus the angular velocity of the inner cylinder and the torque required for rotation must be known, and from these data the viscosity may be calculated. However, there also exist effects due to the construction of the viscosimeter (edge and surface effects) which cannot be allowed for in the calculations. Therefore viscosimeters are generally calibrated using a liquid of known viscosity and a correction to eqn. (70) is determined.

The basic equation for a plastic liquid may be written as

$$-dv/dr = -r(d\omega/dr) = (\tau - \tau_0)/\eta' \tag{71}$$

Substitution of τ from eqn. (68) then gives

$$-r(d\omega/dr) = (1/\eta')M/2\pi r^2 h - \tau_0/\eta'$$

or, on integration,

$$\int_0^{\omega} -d\omega = (M/2\pi h\eta') \int_r^R dr/r^3 - \tau_0/\eta' \int_r^R dr/r$$

which yields

$$\omega = (M/\eta' 4\pi h)(1/r^2 - 1/R^2) - (\tau_0/\eta') \ln (R/r) \tag{72}$$

The stress τ_w arising at the wall of the inner cylinder, according to the pattern of eqn. (68), is

$$\tau_w = M/2\pi r^2 h$$

which may be calculated knowing the measured torque M. The shear rate may be expressed on the basis of eqns (71) and (72) in the form

$$dv/dr = (\tau_w - \tau_0)\omega/[(M/4\pi h)(1/r^2 - 1/R^2) - \tau_0 \ln (R/r)] \tag{73}$$

where τ_w is identical to τ appearing in eqn. (71). The yield stress τ_0 must be determined experimentally. Knowing the values of τ_0 and dv/dr, the shearing diagram can be plotted; the plastic viscosity is given by the tangent to the curve.

The basic equation for non-Newtonian liquids, as has been seen, may be given by an exponential function of the general form

$$\tau = \eta''(-dv/dr)^n = \eta''(-r(d\omega/dr))^n$$

The equation for the torque, by means of eqn. (68) is

$$M = 2\pi r^2 h\tau = 2\pi r^2 h\eta''(-r(d\omega/dr))^n$$

Rearrangement and integration give

$$\int_0^{\omega} -d\omega = (M/2\pi h\eta'')^{1/n} \int_r^R dr/r^{1+2/n}$$

After integration, this yields

$$\omega = (n/2)(M/2\pi h\eta'')^{1/n}(1/r^{2/n} - 1/R^{2/n}) \tag{74}$$

In the case of Newtonian liquids, $n=1$, substitution of which gives eqn. (70). In eqn. (74) both η'' and n are unknown, and so a further transformation is necessary, to determine n. On expressing M in terms of the shear stress τ_w, eqn. (74) is modified to

$$\omega = (n/2)/(\tau_w/\eta'')^{1/n}[1-(r/R)^{2/n}] \tag{74a}$$

By taking logarithms of both sides of this equation, it may be seen that the slope of the straight line obtained on plotting $\log \tau_w$ versus $\log \omega$ supplies the required value of n. Knowing n, the apparent viscosity η'' can be calculated. In the case of pseudoplastic liquids the value of the exponent n is lower than unity, while in the case of dilatant liquids it is higher than unity.

9. CONTACT STRESSES

Contact between convex bodies, if they can be regarded as absolutely rigid, occurs at a single point. The contact surface area in this case is infinitely small, and thus the stress arising is infinitely great. In reality materials are not rigid, and so both bodies deform around the contact point and a contact surface is formed. The normal stress prevailing on the contact surface area is termed the contact stress.

Contact stress plays an important role in numerous technological processes. During handling, root bulb products, fruits and cereals will come into contact with their own and other surfaces. If the contact stress exceeds the biological yield limit of an agricultural product, the latter will be damaged and then spoilt.

Hertz has elaborated a theory of contact stress for elastic bodies. A general solution for viscoelastic materials is not yet available, but various approximate solutions may be applied usefully in numerous cases.

9.1 Contact stress in elastic bodies

In deriving relationships for contact stress it is necessary to rely on the following main assumptions: the material is homogeneous, the load is static, the surfaces are smooth, and the radii of curvature of the contacting bodies by far exceed the radius of the contact surface. This latter assumption means that only a normal (compressile) stress arises over the contact surface, and the effect of tangential stress is negligible.

Among the problems implied by contact are those of determining the contact surface area, the deformation of the contacting bodies, and the stress arising in the contact surface. In the general case the radii of curvature of the contacting bodies will differ in the various plane sections. In this case the shortest and longest radii of curvature found in mutually orthogonal planes are considered (Fig. 103). The angle Φ enclosed by the planes of the two shorter radii of curvature must also be given, as this may also affect the development of contact stress. In practice

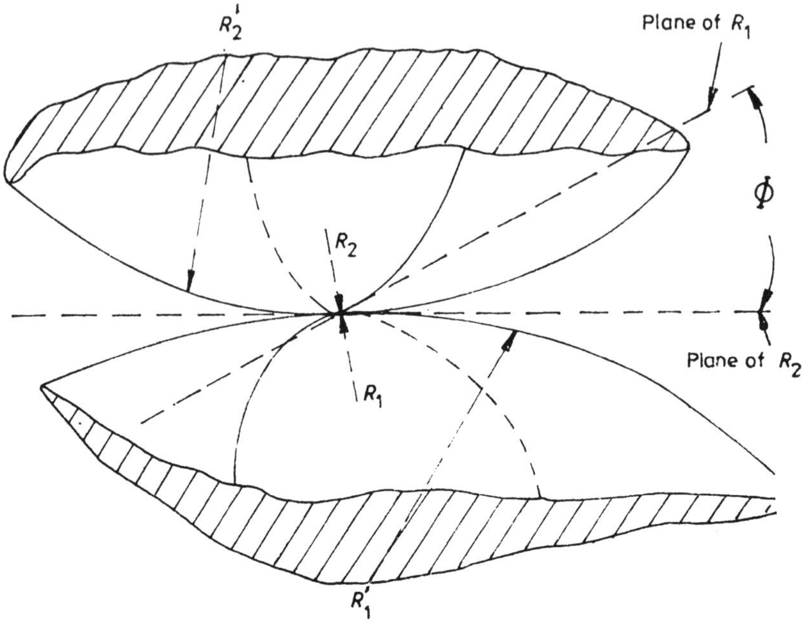

Fig. 103. Contact between convex bodies

two special cases occur most frequently: contact of two spherical bodies, and contact of a spherical body and a plane surface.

In the general case the contact surface is an ellipse, and the maximum stress arises in the center of the contact surface:

$$\sigma_{max} = 1.5(P/ab\pi) \tag{75}$$

where a and b are the semiaxes of the ellipse. As may be seen, the maximum stress is 1.5 times the mean value. The semiaxes of the ellipse may be calculated from the expressions [45]

$$a = m[3PA/2(1/R_1 + 1/R_1' + 1/R_2 + 1/R_2')]^{1/3} \tag{76}$$

$$b = n[3PA/2(1/R_1 + 1/R_1' + 1/R_2 + 1/R_2')]^{1/3} \tag{77}$$

and further

$$A = (1 - \nu_1^2)/E_1 + (1 - \nu_2^2)/E_2$$

where ν_1 and ν_2 are the Poisson ratios of the contacting bodies, E_1 and E_2 their moduli of elasticity, and m and n are constants, which are functions of the radii of curvature and of the angle Φ. The simultaneous deformation of two contacting bodies (the approach of the centers of the bodies towards each other) may be calculated from the equation [45]:

$$z_1+z_2 = (k/2)[(9P^2A^2/\pi^2)(1/R_1+1/R_1'+1/R_2+1/R_2']^{1/3} \tag{78}$$

where the constant k is a function of the radii of curvature and of the angle Φ.

The above relationships are simplified significantly in the two most frequent cases mentioned before. For the contact of a sphere with a plane surface, when $R_1=R_1'=\infty$ and $R_2=R_2'=d/2$, the following relationships are valid:

$$a = 0.721(PAd)^{1/3}$$

and

$$\sigma_{max} = 0.918(P/A^2d^2)^{1/3} \tag{79}$$

$$z_1+z_2 = 1.04(P^2A^2/d)^{1/3}$$

When two spherical bodies are in contact, $R_1=R_1'$ and $R_2=R_2'$; substitution then yields the equations

$$a = 0.721[PA/(1/d_1+1/d_2)]^{1/3}$$

$$\sigma_{max} = 0.918[P(1/d_1+1/d_2)^2/A^2]^{1/3} \tag{80}$$

and

$$z_1+z_2 = 1.04[P^2A^2(1/d_1+1/d_2)]^{1/3}$$

When agricultural materials come into contact with a steel body, the steel may be regarded as rigid; the expression for A then becomes much simpler, since the deformation concerns only the agricultural material. The radius of curvature of bodies deviating from sphericality may be determined by various methods. Appropriate measurement methods have been discussed by Mohsenin [1].

In testing, a material is generally loaded by a plane steel plate or a ball, and the loading force and deformation are measured. From these data the modulus of elasticity of the material studied may be calculated by means of the equations

$$E = [0.752P(1-\nu^2)/z^{3/2}](1/R)^{1/2} \tag{81}$$

for loading by a plate, and

$$E = [0.752P(1-\nu^2)/z^{3/2}](1/R+2/d)^{1/2} \tag{82}$$

for loading by a ball, where ν is the Poisson ratio of the material studied, and d the diameter of the ball.

Along the axis of the loading center the stress decreases rapidly beneath the contact surface, as may be seen from Fig. 104. The vertical axis represents the depth, relative to the radius of the contact surface. The maximum shear stress occurs at about $0.5z/a$, and its value is $\tau_{max} \simeq 0.27\sigma_{max}$.

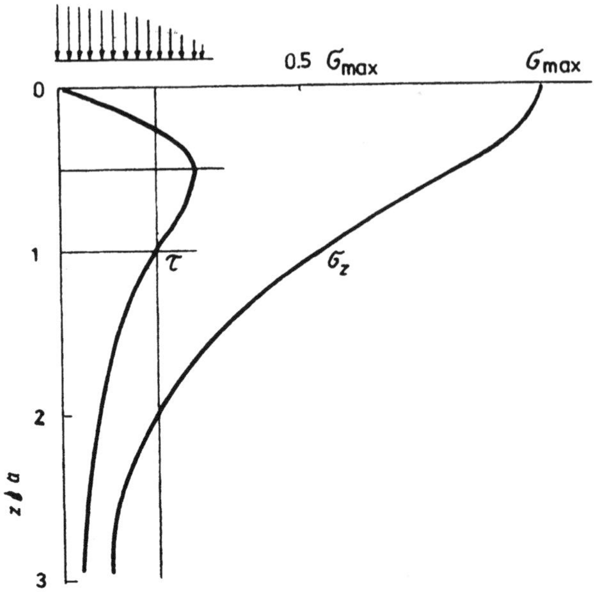

Fig. 104. Stress distribution beneath the contact surface

9.2 Contact stress in viscoelastic bodies

The contact-stress problems encountered in the case of viscoelastic bodies are generally solved by utilizing relationships which are valid for elastic bodies but with the time-dependent deformation properties of the former materials taken into account. Since the deformation varies as a function of time even under a constant load for viscoelastic materials, the distribution of pressure over the contact surface also varies.

Consider the case of loading on a plane surface of a viscoelastic material by a rigid steel ball. The pressure is given as a function of the radius of the contacting surface and time by the equation

$$p(r, t) = (4/\pi R)[G/(1-\nu)]\sqrt{a(t)^2 - r^2} \tag{83}$$

where G is the shear modulus of elasticity, $a(t)$ the radius of the contact surface at time t, and R the radius of the ball. The modulus of elasticity G in the case of viscoelastic materials is also time-dependent, and may be written for the Maxwell model as

$$G = G_0 e^{-t/T} \tag{84}$$

and for the three-element model as

$$G = G_0 e^{-t/T} + G_\infty (1 - e^{-t/T}) \tag{84a}$$

where G_0 is the instantaneous (initial) value of the shear modulus, G_∞ its asymptotic value, and T the relaxation time. Loading by the steel ball may be carried out in two ways: at a constant velocity v, or instantaneously for a given value of deformation. In the case of constant-rate loading, the approach of the ball center towards the compressed surface is given by

$$\alpha(t) = vt$$

while the instantaneous radius of the deformed surface is

$$[a(t)]^2 = Rvt$$

Figure 105 shows the distribution of contact pressure in dimensionless form at various times after contact [49]. If the time t is less than the relaxation time ($t/T<1$), the distribution curve is a semicircle, or of a form close to it, which means that the behavior of the material is elastic. If the time t is identical to the relaxation time, or exceeds it, the curve flattens, owing to creep of the material. The value a_0 appearing in Fig. 105 denotes the radius of the deformed surface at $t=T$. The term $G/(1-\nu)$ in eqn. (83) may be substituted by the modulus of elasticity E on the basis of the equation

$$G/(1-\nu) = E/2(1-\nu^2)$$

For instantaneous loading of the viscoelastic half-space by a steel ball (according to a step function), the radius of the contact surface varies with time (with the assumption that $E_0=2E_\infty$ [50]) according to the equation

$$a^2(t) = [(3RP/16E_0)(2-e^{-t/2T})]^{2/3} \tag{85}$$

where P is the load on the ball.

At $t=0$

$$a^2(0) = (3RP/16E_0)^{2/3} \tag{85a}$$

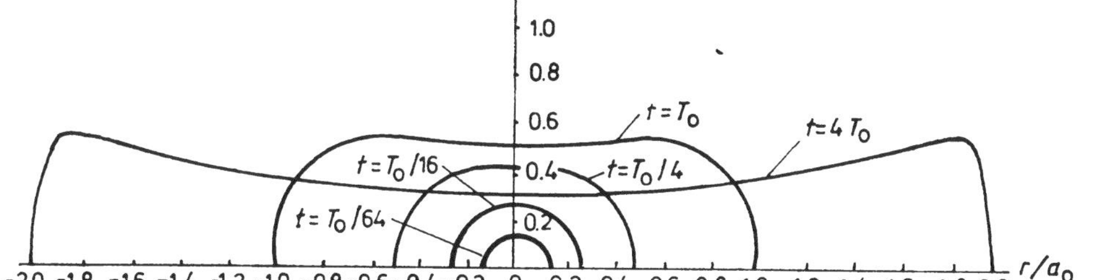

Fig. 105. Distribution of compressile stress in a viscoelastic body as a function of time (for loading at a costant rate)

The distribution and variation of pressure as a function of time may be calculated from eqn. (83), making use of eqns (85) and (84). Such a calculation is presented in Fig. 106, on relative coordinates [50]. The pressure distribution is initially of semicircular shape, meaning again that the material behaves elastically. As the time t approximates or surpasses the relaxation time, the plastic properties of the material become manifest: the deformation increases and the pressure distribution flattens. The pressure at the center of the deformed surface decreases considerably as a result of the stress relaxation.

One possible test method involves instantaneous deformation of a material by a rigid steel ball, maintaining the deformation at a constant value until a time t_1, and then lifting the ball instantaneously, i.e., removing the load (Fig. 107). On unloading (at time t_1), the viscoelastic material regains only a part

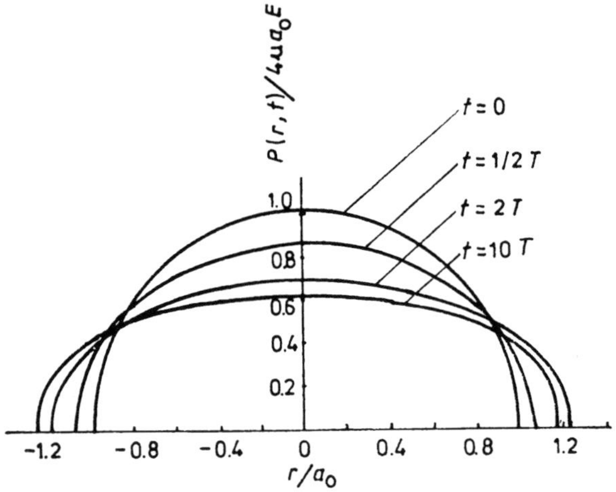

Fig. 106. Distribution of contact pressure during creep period, following instantaneous loading

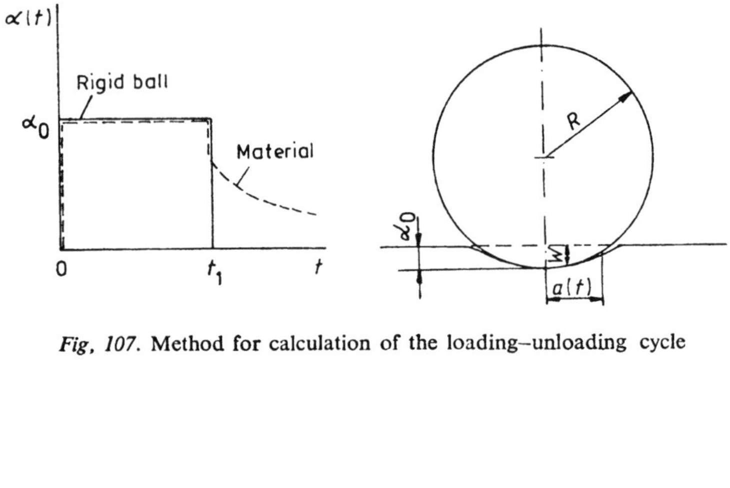

Fig. 107. Method for calculation of the loading–unloading cycle

of its deformation. It recovers a part of the residual deformation in the course of time, but a part of it remains even after the elapse of any arbitrarily long time. The pressure distribution under the ball varies during the loading period (with the viscoelastic half-space defined by the three-element model [50]) according to the equation

$$p(r, t) = (4a_0 E_0/\pi R)(1+e^{-t/T})(1-(r^2/a_0^2)) \tag{86}$$

where $0<t<t_1$. The change of deformation after unloading may be calculated from the equation

$$w(r, t) = (1/2)(e^{-(t-t_1)/2T} - e^{-(t+t_1)/2T})(\alpha_0 - (r^2/2R))$$

The deformation existing at the instant of unloading is obtained from the above equation by making the substitution $t=t_1$:

$$w(r, t_1) = (1/2)(1-e^{-t_1/T})(\alpha_0 - (r^2/2R))$$

The longer the loading period, i.e., the larger t_1, the higher the value of $w(r, t_1)$, i.e., the less the recovery of deformation at the instant of unloading.

9.3 The theory of the rigid die; the Boussinesq problem

The mechanical properties of agricultural materials are frequently investigated using various pressure dies. A pressure die is a cylindrical body of diameter d, which is pressed at constant velocity v into the material investigated while the force is measured as a function of indentation. The diameter of the die must be significantly smaller than the size of the body investigated (at most, one-tenth). Therefore, rigid dies may be applied only in testing products of relatively large size (apples, pears, peaches, potatoes, beet, etc.). The usual diameter of rigid dies is 5–10 mm. Smaller products (e.g., seeds) are investigated by loading with a plane plate.

Figure 108 illustrates the stress distributions under a plane plate and a rigid die. The respective stress distributions vary according to the relationships

$$\sigma = (3P/2\pi a^2)\sqrt{1-r^2/a^2}$$

and

$$\sigma = P/2\pi a\sqrt{a^2-r^2}$$

From the latter equation it follows that the stress under the outer edge of a rigid die is infinitely high. In practice the material continues to yield in this region,

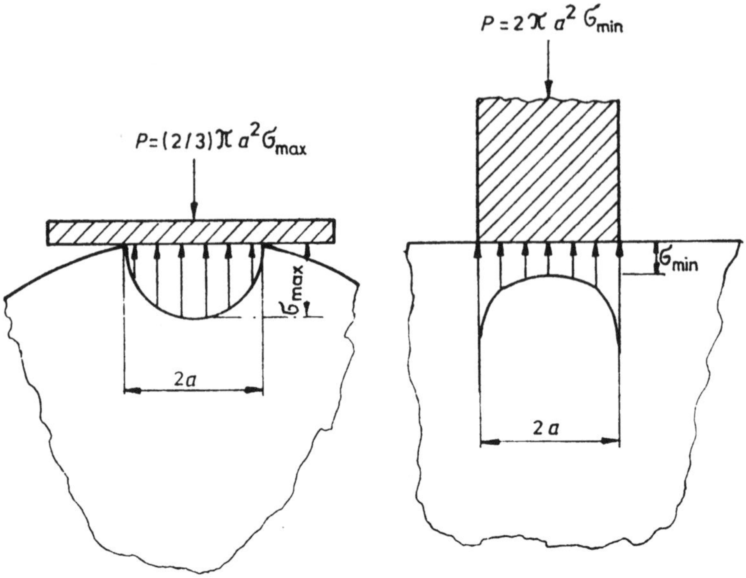

Fig. 108. Loading by a plane plate and a cylindrical die

and the stress has a finite value. The deformation under a plane plate may be calculated by means of the equation

$$z = [(9/16)(P^2/E^2R)(1-\nu^2)^2]^{1/3} \tag{87}$$

where R is the radius of the material investigated. The deformation under a rigid die may be calculated from the relationship [45]

$$z = P(1-\nu^2)/2aE \tag{88}$$

where $2a$ is the diameter of the die. In eqn. (88) the force P may be replaced by the mean pressure p; in this case,

$$p = [4E/\pi(1-\nu^2)]\,z/D \tag{88a}$$

where $D=2a$ is the diameter of the die. From this equation it follows that it is advisable to use the relative deformation z/D in graphical representations, as in this case the value of p is independent of the diameter of the die. Figure 109 shows loading curves for sugar beet and potatoes, plotted for dies of various diameters (between 8 and 25 mm). As may be seen, the points obtained for the various diameters fall practically on a single curve. The modulus of elasticity of the materials may be calculated, knowing the relevant experimental data, from eqn. (88a).

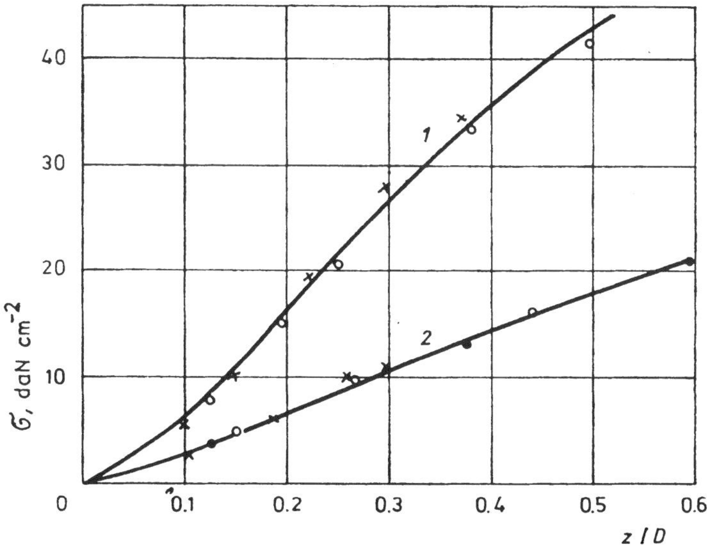

Fig. 109. Pressure–deformation curves for sugar beet and potatoes, measured using cylindrical dies of various diameters. *(1)* Sugar beet; *(2)* potato

In pressing convex (i.e., not plane) surfaces, the die behaves in the initial stage of deformation as a plane plate, until it reaches full contact over its whole surface area. The greater the diameter of the die in relation to the curvature of the body, the greater the deformation at which the whole surface area of the die comes into contact with the body. Figure 110 shows loading curves for pears, on a logarithmic system of coordinates, for two die diameters. The break points in the curves indicate full contact of the die surface. At larger loads the two curves practically coincide.

The relationships described above are valid for elastic bodies. The Boussinesq problem for viscoelastic bodies may also be solved, with certain approximations, [56]. This solution permits the mechanical properties to be determined also in the case of loads varying over time.

Assume that the viscoelastic half-space behaves according to the three-element model (Fig. 92(a)). If the load on the material,

$$P = \dot{P}t$$

or

$$\sigma = \dot{\sigma}t$$

is realized linearly, then the deformation under a cylindrical die during the period of loading may be described by the equation

$$z(t) = [(1-\nu^2)/2aE_1]P[1-(\dot{P}/P)(T_{\text{ret}}-T_{\text{rel}})(1-e^{-t/T_{\text{ret}}})] \qquad (89)$$

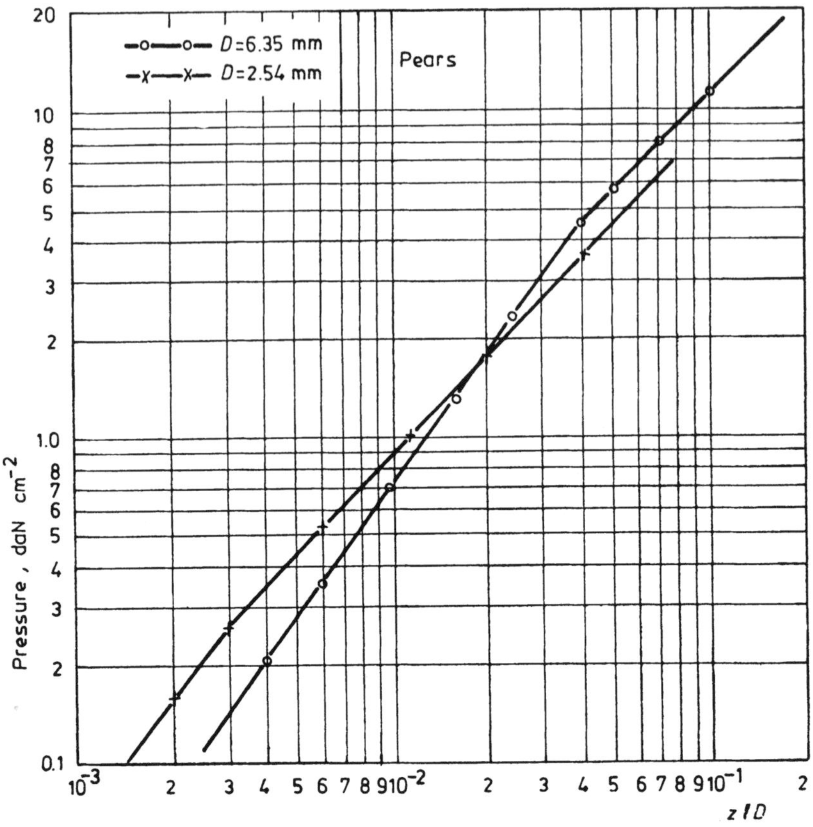

Fig. 110. Loading curves for pears, measured using dies of two diameters

where $T_{rel}=\eta/E_2$ and $R_{ret}=\eta/E_\infty$. In the case of an infinitely slowly varying load, obviously

$$z = [(1-\nu^2)/2aE_1]P \tag{89a}$$

For loading according to a step function, $\dot{P}\to\infty$ and $t\to 0$, and so in the above equation, using l'Hospital's rule,

$$\lim_{t\to 0} (\dot{P}/P)(T_{ret}-T_{rel})(1-e^{-t/T_{ret}}) = (T_{ret}-T_{rel})/T_{ret}$$

whence the indentation caused by the die is

$$z = [(1-\nu^2)/2a(E_1+E_2)]P \tag{89b}$$

or

$$z = [(1-\nu^2)P/2aE_1]T_{rel}/T_{ret}$$

In the initial phase of loading for convex bodies the die does not contact the material over its whole surface area, and so it is necessary again to assume contact between a plane face and a sphere, with eqn. (87) taken into account. As an approximation, the following simplified equations may be used:

$$z_{(\dot{P}\to 0)} = [3(1-\nu^2)/4aE_1]P$$

and

$$z_{(\dot{P}\to\infty)} = [3(1-\nu^2)/4a(E_1+E_2)]P$$

If the load remains constant after the loading period, the material creeps according to the equation

$$z(t) = (1-\nu^2)P_1/2aE_1 - [(1-\nu^2)P_1/2aE_1 - z_1]e^{-(t-t_1)/T_{ret}}$$

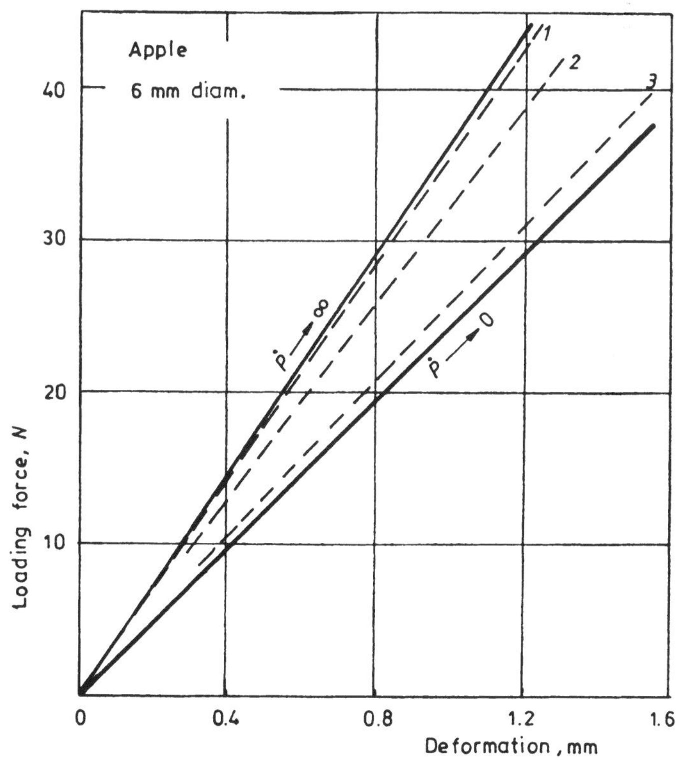

Fig. 111. Effect of loading velocity on deformation, of Jonathan apples. *(1)* $\dot{P}=1.0$ N s^{-1}; *(2)* $\dot{P}=0.1$ N s^{-1}; *(3)* $\dot{P}=0.01$ N s^{-1}

while under constant deformation the relaxation of the force is

$$P(t) = z_1 2aE_1/(1-\nu^2)+[P_1-z_1 2aE_1/(1-\nu^2)]e^{-(t-t_1)/T_{rel}}$$

where t_1, P_1 and z_1 are the time, loading force and deformation at the end of the loading period. Using eqns. (89a) and (89b), a zone may be defined in the $P-z$ system of coordinates within which curves plotted for an assumed finite loading rate must fall. Figure 111 shows test results for Jonathan apples. The mechanical properties appearing in the three-element model are as follows: $T_{rel}=133$ s, $T_{ret}=204$ s, $E_1=3.4$ N mm^{-2}, $E_2=1.66$ N mm^{-2} and $\eta=221$ N s^{-1} mm^{-2}. As may be seen, in the given case a loading rate of $\dot{P}=10$ N s^{-1} approximates the case $\dot{P}\rightarrow\infty$ sufficiently well.

In selecting the diameter of the compression die the size of the body investigated must be taken into account. On loading the infinite half space the displacements and stresses become zero theoretically only at infinity. In practical cases the extent of the bodies is finite, therefore the variation of the deformation and stress under the loading die must be examined as a function of depth. The dis-

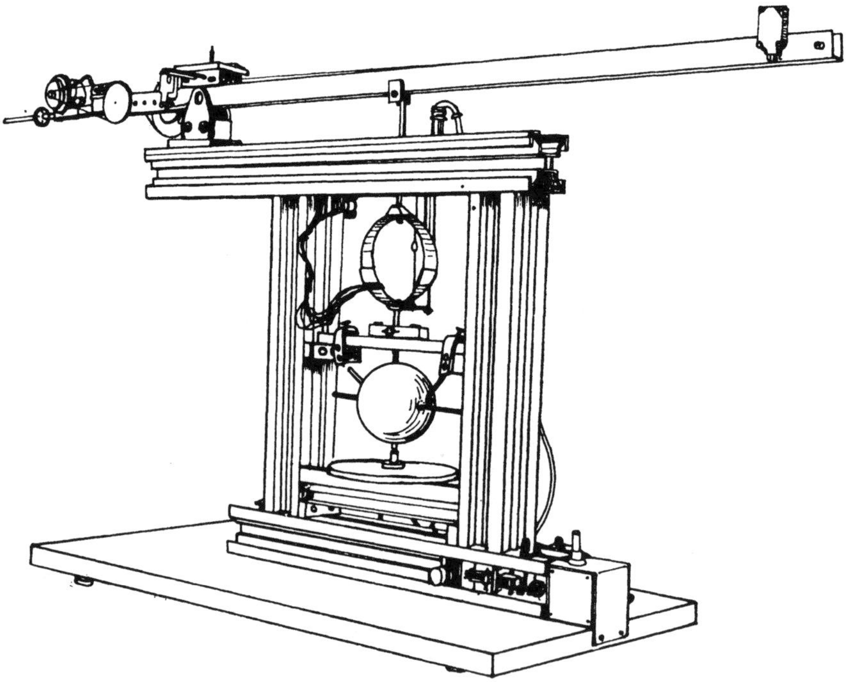

Fig. 112. Testing apparatus with moving weight (National Institute for Agricultural Engineering, Gödöllő, Hungary)

placement of internal points in the infinite half-space loaded by a die of diameter $D=2a$ may be calculated from the equation

$$w_{z,r} = (az_0/\pi)\arcsin\left[2a/\left(\sqrt{z^2+(a+r)^2}+\sqrt{z^2+(a-r)^2}\right)\right]$$

where z_0 is the displacement of the die, and z, r are the cylinder coordinates. At a depth of $z=6D$, $w_z/z_0=0.053$, and so if the dimensions of the body in the direction of the load exceed $(6\sim10)D$, then the distribution of internal stress is not influenced significantly by the size of the body.

Linear loading at various rates may easily be realized by applying a moving weight. Figure 112 shows a photograph of such a loading apparatus. The loading rate may be selected arbitrarily between 0 and 40 N s^{-1}. This measurement method is suited for determining the mechanical properties of a given agricultural product as function of various factors. As an example Fig. 113 shows for the three-element model the modulus of elasticity of Jonathan apples as a function of ripening time [56].

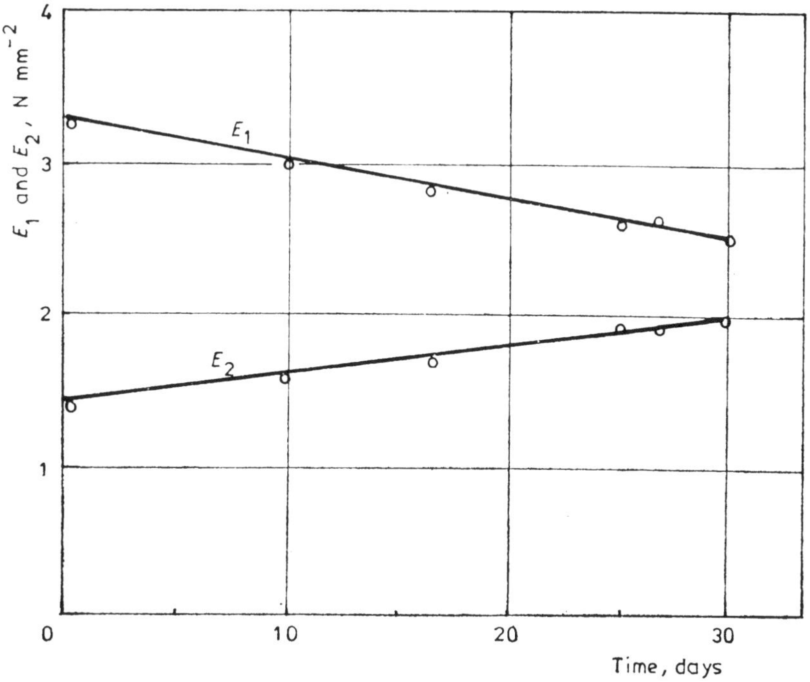

Fig. 113. Moduli of elasticity for the three–element model

10. IMPACT LOADING

The phenomenon of impact occurs frequently during the transport, handling and treatment of bulk agricultural materials. Products may impact either against each other or the surrounding walls during such operations. Fruit shaken off impacts against other fruit, against tree branches and finally against the catching surface. During the comminution of granular products and dried forage materials, the task of impact is to produce a force in the product which exceeds the rupture strength.

In the case of viscoelastic materials, the dynamic behavior under impact cannot be established safely on the basis of results obtained in static or low-velocity measurements, owing to the time effect. Therefore, special dynamic tests have been developed for applying an impact-type load to a material while the stress, deformation and duration of impact are measured.

Agricultural products are mostly convex bodies, and so impact appears together with the problem of contact stress.

10.1 Impact of elastic bodies

In agricultural practice the following types of impact occur, listed in order of importance:

(a) impact of viscoelastic materials against a stationary, rigid plane surface,

(b) impact of viscoelastic materials against stationary, rigid, oblique surfaces,

(c) impact of viscoelastic materials against each other, when one of the materials is stationary and its center of gravity cannot be displaced; and

(d) impact of viscoelastic materials against stationary surfaces coated by a cushioning material.

Naturally, other cases occur in addition to those listed above, for example, in hammer mills, where a rigid moving hammer impacts against practically stationary viscoelastic material.

For the sake of simplicity we first assume elastic bodies and examine the related general laws. The initial basic equation for impact problems is the expression for the variation of momentum:

$$mv_1 - mv_2 = \int P\,dt \tag{90}$$

where m is the mass of moving material, v_1 and v_2 are the velocities of the material at the start and end of impact, respectively, and P is the force arising during impact at an arbitrary time. In the majority of practical cases one of the bodies is stationary and is not displaced during impact. In such cases the velocity of the moving body is zero at the end of the impact process, i.e.,

$$v_2 = 0$$

Equation (90) may be solved in a relatively simple way by applying simplifying assumptions. Assume, according to Fig. 114, that during impact the force P increases linearly as a function of time. In a given case the value of the integral on the right-hand side is constant, so a larger P_{max} corresponds to a shorter impact period. With this assumption,

$$\int P\,dt = P_{max}\,\Delta t/2$$

and the maximum force appearing during impact is

$$P_{max} = 2mv/\Delta t$$

Consider the critical dropping height for agricultural products (e.g., sugar beet, potatoes) such that no damage results. The condition for avoiding damage is that the maximum stress arising during impact should not exceed the permissible value: i.e.,

$$\sigma_{perm} \leqq P_{max}/F$$

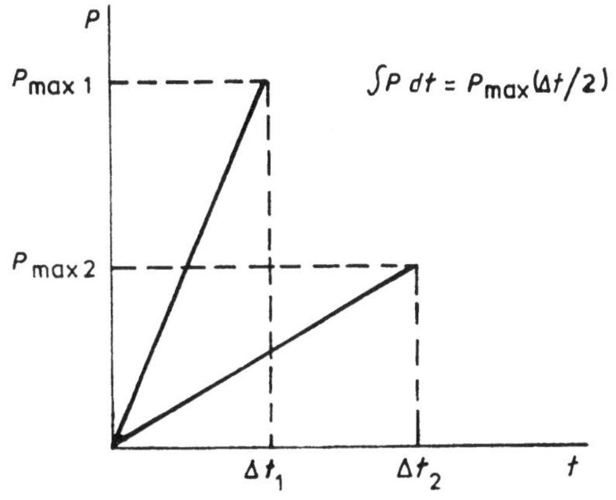

Fig. 114. Method for calculation of impact loading

where F is the area of the contact surface. The area of the contact surface according to eqn. (79) (with the expression squared and multiplied by π) is

$$F = 2.6(PRA)^{2/3}$$

Substitution of the values of P and F, for replacing the impact velocity by the expression for free fall, gives the permissible stress as

$$\sigma_{\text{perm}} \leqq 2m\sqrt{2gH}/2.6\,\Delta t\left[(2m\sqrt{2gH}/\Delta t)\,RA\right]^{2/3}$$

from which the value of H may be obtained as

$$H \leqq 0.04\sigma_{\text{perm}}^{6}(1-\nu^2)^4R^4\Delta t^2/m^2E^4 \tag{91}$$

As an example, let the weight of a sugar beet be 1.0 kg, its mean radius of curvature $R=5$ cm, $\Delta t=3\times10^{-3}$ s, $\sigma_{\text{perm}}=30$ daN cm^{-2}, $\nu=0.45$ and $E=180$ daN cm^{-2}. On substituting these values into eqn. (91), $H=61$ cm is obtained as the maximum dropping height.

The applicability of eqn. (91) is aggravated by the necessity of knowing the impact duration Δt. The impact duration may be determined experimentally by mounting an accelerometer in the moving material and determining the impact time from the graph of acceleration.

The maximum deformation occurring during impact may be determined by means of the energy equation

$$mv^2/2 = \int_0^{z_0} P\,dz \tag{92}$$

For impact by a cylindrical body, utilizing eqn. (88),

$$mv^2/2 = \int_0^{z_0} [2aE/(1-\nu^2)]\,z\,dz$$

from which

$$z_0 = \sqrt{(mv^2/2)[(1-\nu^2)/aE]} \tag{93}$$

For impact by spherical bodies the value of P must be expressed using eqn. (87) and substituted into eqn. (91). Then

$$mv^2/2 = \int_0^{z_0} [4E\sqrt{R}/3(1-\nu^2)]\,z^{3/2}\,dz$$

and, after integration,

$$z_0 = \left[(15/16)mv^2(1-\nu^2)/E\sqrt{R}\right]^{2/5} \tag{94}$$

Substitution of the data of the example attached to eqn. (91) into eqn. (94), yields the result that the maximum deformation of the beet during impact is $z_0=0.55$ cm.

Knowing the maximum deformation, the duration of impact may be determined from the equation [45]

$$t_0 = 2.94(z_0/v)$$

where v is the impact velocity. For the data of the above example (accepting $H=61$ cm as the height of free fall),

$$t_0 = 2.94(0.55/346) = 4.69\times10^{-3}\ \mathrm{s}$$

In the case of elastic bodies, the impact duration is symmetrical about the instant of reaching maximum deformation, and so the time Δt appearing in eqn. (91) is one-half the value of t_0. For viscoelastic materials the impact duration is asymmetrical in relation to the time of reaching z_0, and therefore a value slightly higher than $t_0/2$ must be substituted for Δt.

10.2 Impact of viscoelastic bodies

The solution of the impact problem for viscoelastic materials constitutes at present one of the most difficult tasks in this field. A solution may be obtained in closed form only for a few special cases, with the application of major simplifying conditions. The majority of cases require extensive mathematical apparatus and the application of a computer.

Experiments performed with viscoelastic materials have shown that in numerous cases these materials behave during impact similarly to the Maxwell model. For mechanical characterization of such materials, knowledge of the modulus of elasticity E, of the viscosity η and of the relaxation time T is necessary. These data may be assessed in the experiments described above.

Figure 115 shows the impact of a rigid body of mass m at velocity v_0 against

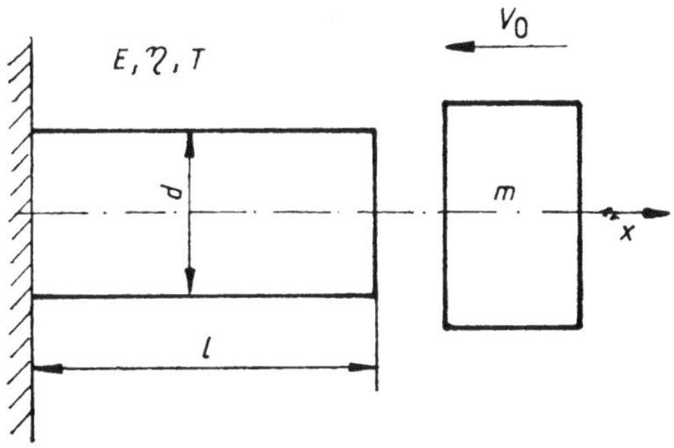

Fig. 115. Impact of a cylindrical viscoelastic body

a cylinder of viscoelastic material supported at one end. The mechanical characteristics E, η and T are known. Assume that the material behaves according to the Maxwell model, so that the rheological equation

$$\sigma = \eta(d\varepsilon/dt) - (\eta/E)(d\sigma/dt)$$

is valid. The dynamics of impact are described by Newton's equation of motion: i.e.,

$$m(d^2z/dt^2) = -P$$

where z is the deformation of the viscoelastic material (displacement of the free end). Simultaneous solution of the above equations is possible by applying the Laplace transformation [53], permitting the variations of force P and deformation z to be determined as functions of time:

$$z(t) = (mv_0 l/\eta F)\{1 - e^{-t/2T}[((1-B/2)/\sqrt{B-1})\sin(t/2T)\sqrt{B-1} + \\ +\cos(t/2T)\sqrt{B-1}]\} \qquad (95)$$

and

$$P(t) = (2v_0\eta F/l\sqrt{B-1})e^{-t/2T}\sin(t/2T)\sqrt{B-1} \qquad (96)$$

where F is the cross-sectional area of the viscoelastic material, and $B = 4\eta FT/ml$. By examining the extreme values of eqn. (95) it is possible to determine the values of the maximum and residual deformations:

$$z_{max} = (mv_0 l/\eta F)\{1 - [(1-B/2)/\sqrt{B-1}]\exp(-\pi/2\sqrt{B-1})\} \qquad (97)$$

$$z_r = (mv_0 l/\eta F)[1 + \exp(-\pi/\sqrt{B-1})] \qquad (98)$$

As a result of extreme value examination of eqn. (96) the maximum force and the duration of impact are obtained as

$$P_{max} = (2v_0\eta F/l\sqrt{B-1})\exp(-\pi/2\sqrt{B-1}) \qquad (99)$$

and

$$t_{max} = \pi T/\sqrt{B-1} \qquad (100)$$

Figures 116 and 117 show $P(t)$ and $z(t)$ curves for potato and sugar-beet test specimens. The points indicate the experimental results, while the continuous curve illustrates the results of the theoretical calculations. The deformation and impact duration for longer specimens are greater, while the resulting force effect is lower. It may also be observed that the maxima of force and deformation do not coincide. This is a consequence of stress relaxation [52, 53].

To describe the impact of viscoelastic materials against rigid bodies, approximate equations have been derived for the cases when all the materials are spherical or cylindrical, or when one of them is a plane plate [51]. The maximum force appearing during impact may be determined from the equation

$$P_{max} = 1.576K(mv_0^2/K)^{3/5}[1+0.09(1/T)(m/K)^{2/5}v_0^{-1/5}]^{3/5} \tag{99a}$$

The time required to attain maximum deformation is

$$t_{max} = 1.576(m/K)^{2/5}v_0^{-1/5}[1+0.09(1/T)(m/K)^{2/5}v_0^{-1/5}]^{2/5} \tag{100a}$$

where m is the mass of the moving body, v_0 the impact velocity, and T the relaxation time, while K may be determined from the equation

$$K = (4\pi/3)[E_1/(1-\nu_1^2)]\,q/\sqrt{(R_1+R_2)/R_1R_2}$$

The term q appearing in this equation is a constant whose value for spherical

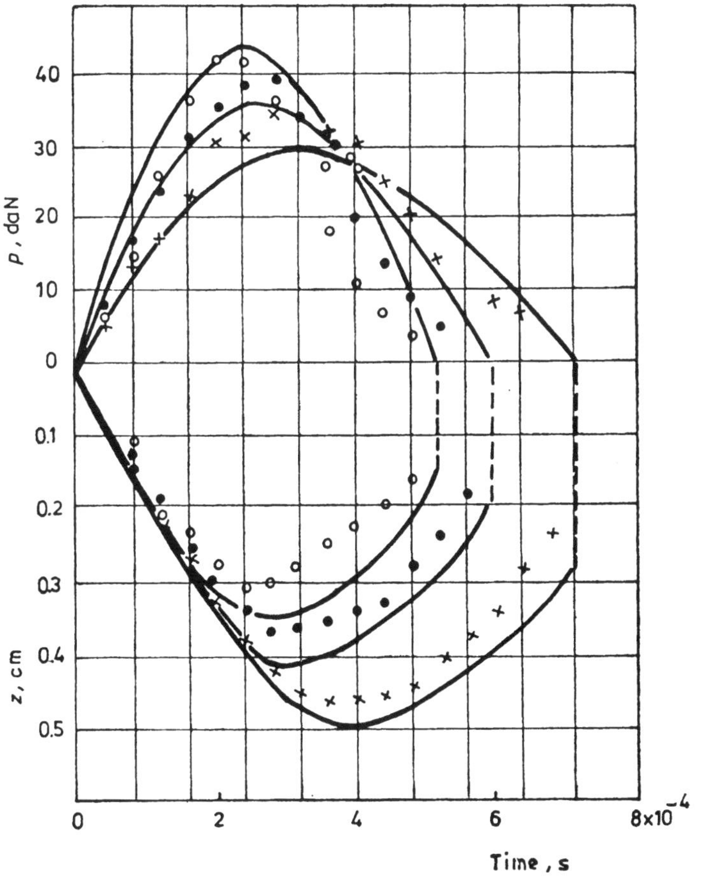

Fig. 116. Impact process for cylindrical specimens prepared from potatoes

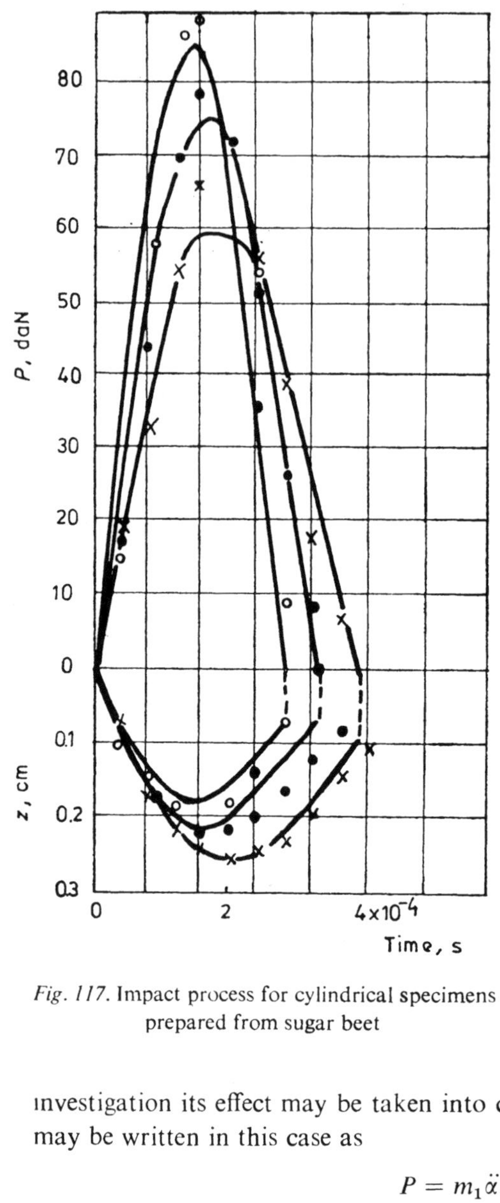

Fig. 117. Impact process for cylindrical specimens prepared from sugar beet

and plane bodies is 0.318. If two cylindrical bodies impact against each other in a mutually normal position, then the value of q may be taken from Table 10 as a function of the ratio R_2/R_1.

Intermediate values may be calculated by interpolation. In the case of a plane plate the radius is infinitely large, and then

$$K = 1.33([E_1/(1-\nu_1^2)]\sqrt{R}$$

where R is the radius of the spherical body.

Figure 118 shows the deformation occurring on impact for two spheres of viscoelastic material. In the general case the equation of motion for the impact process is

$$\ddot{\alpha} = \dot{v}_1 + \dot{v}_2 = -P(1/m_1 + 1/m_2)$$

where α is the approach of the centers of gravity of the two masses after contact.

If one of the masses remains stationary and is supported, its mass may be assumed to be infinite. The weight of the moving material itself generally affects the force arising in the contact surface only insignificantly but for more exact investigation its effect may be taken into consideration. The equation of motion may be written in this case as

$$P = m_1\ddot{\alpha} + G_1$$

For two spheres of identical radii, the instantaneous radius of the contact surface $a(t)$ is related to the instantaneous distance of approach $\alpha(t)$ by the relationship

$$[a(t)]^2 = (R/2)\alpha(t)$$

Table 10

R_2/R_1	q
1.00	0.318
0.70	0.321
0.49	0.332
0.33	0.350
0.22	0.382
0.13	0.430
0.031	0.666
0.0076	1.145

while the displacement of the surface over an arbitrary radius r (see Fig. 118) is

$$w_1 = w_2 = (1/2)[\alpha(t) - r^2/R] \qquad (101)$$

If the viscoelastic properties of the material are taken into account, the above equation of motion assumes the following form, under given boundary conditions [48]:

$$\ddot{\alpha}(t) + (1/T)\dot{\alpha}(t) + [\sqrt{2R}\, E_0/3m_1(1-\nu^2)][\alpha(t)]^{3/2} = v_1/T + g(1+1/T) \qquad (102)$$

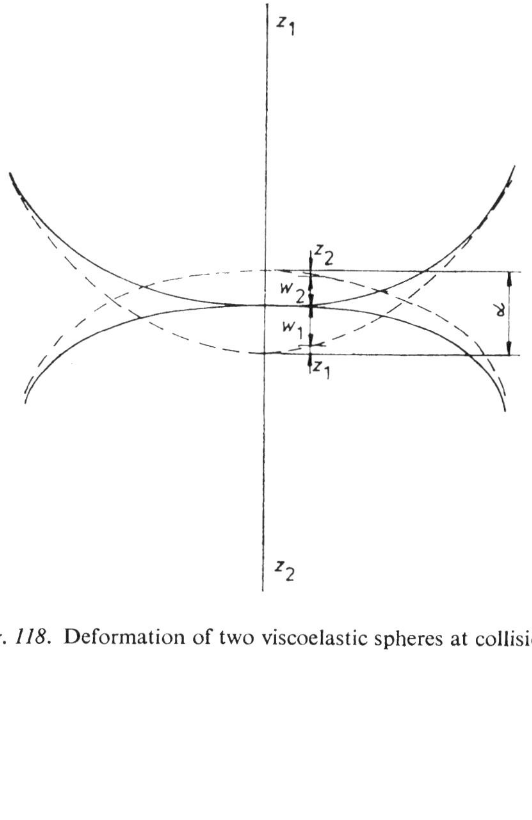

Fig. 118. Deformation of two viscoelastic spheres at collision

The compressile stress for any arbitrary radius of the contact surface at time t is

$$\sigma(r, t) = [0.637E_0/R(1-v^2)] \int_0^t e^{-(t-\tau)/T}\partial/\partial\tau[(R/2)\alpha(\tau)-r^2]^{1/2}\,d\tau \qquad (103)$$

where T is the retardation time, and τ the time coordinate of the process preceding the time t considered. The approach of the centers of the two spherical bodies is obtained as a function of time by solving eqn. (102). The displacement of the surfaces may be calculated subsequently from eqn. (101), and the pressure distribution at the contact surface from eqn. (103).

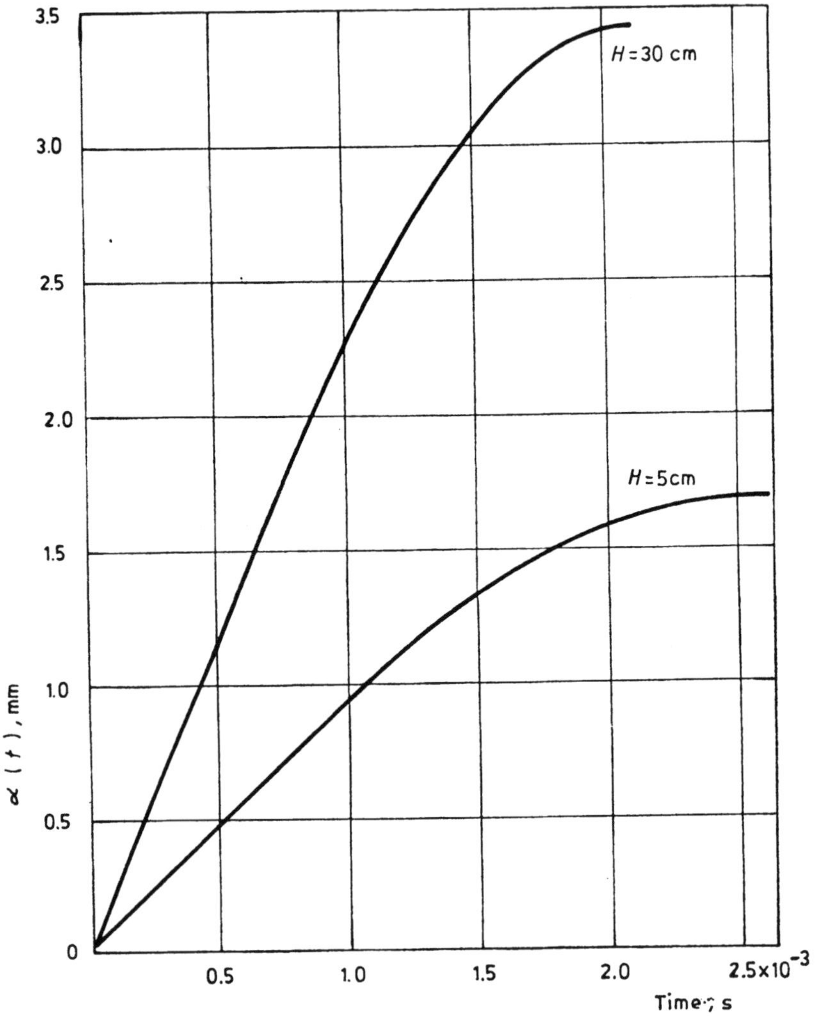

Fig. 119. Impact process for apples falling from various heights

Equations (102) and (103) cannot be solved in closed form, but they may be programmed on a computer. Thus concrete problems may be solved numerically. Figure 119 shows the impact process for two apples falling from heights of 5.0 and 30 cm, respectively, with the initial data [48] $G = 155$ g, $R = 3.66$ cm, $E_0 = 185$ daN cm^{-2}, $\eta = 1.26$ daNs cm^{-2}, $T = \eta/E_0 = 0.0068$ s and $\nu = 0.22$. The

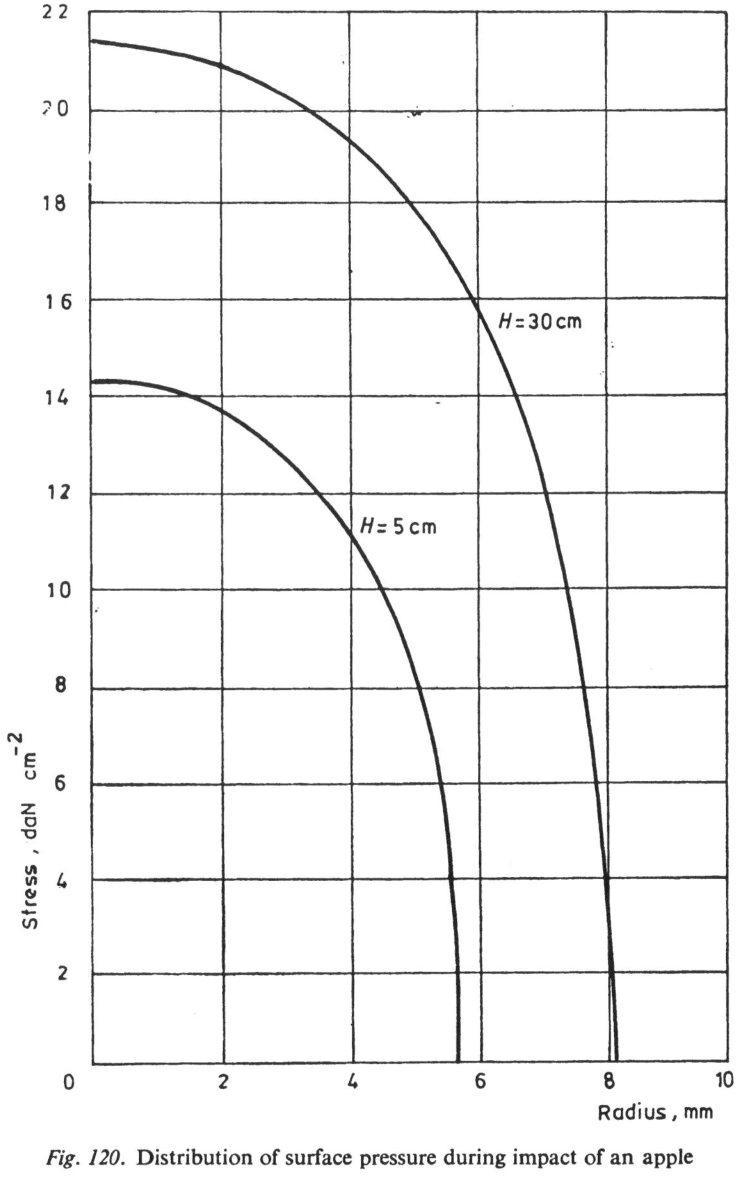

Fig. 120. Distribution of surface pressure during impact of an apple

time required to attain the maximum deformation is slightly less in the case of the greater falling height: the values of Δt are 2.13×10^{-3} and 2.62×10^{-3} s, respectively.

The distributions of surface pressure are shown in Fig. 120. The values of the maximum pressure are about 14 and 21.5 daN cm^{-2} in the two cases. If the biological yield limit of the apples is known, the critical falling height for which the load remains below the limit may be determined.

The distribution of stress over the space found under the contact surface (the compressile stress in the direction of the z-axis) is illustrated in Fig. 121 at the time of maximum deformation. Locations of constant stress are given by the contours.

The most advisable method of solution for viscoelastic impact problems is the method of finite elements. In more complex cases (e.g., with the peel of a fruit taken into account, or for shapes deviating from spherical, etc.), no other method is possible. The application of the finite-element method is discussed in the following Chapter.

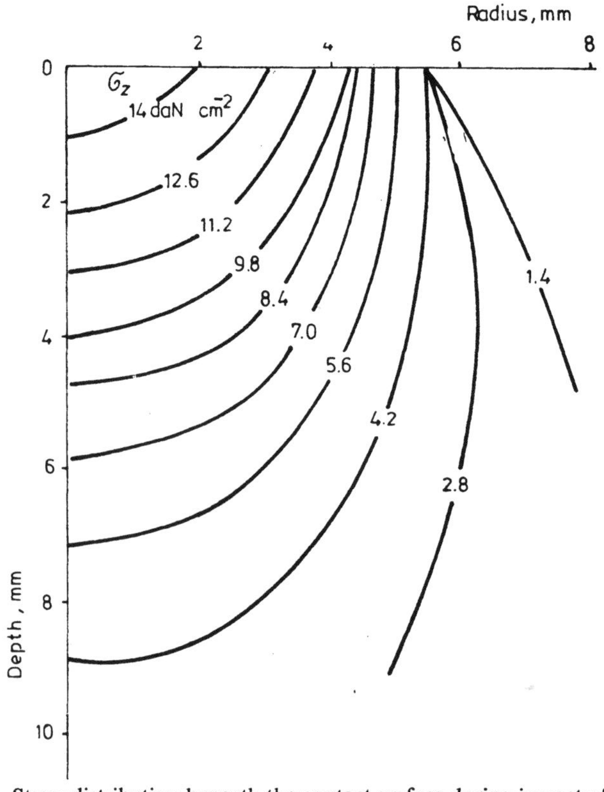

Fig. 121. Stress distribution beneath the contact surface during impact of an apple

10.3 Application of cushioning materials

During the treatment and manipulation of agricultural products, especially fruits, significant damage may occur. To diminish such damage, the relevant surfaces are often coated by soft cushioning materials. The task of the cushioning material is to absorb considerable energy at impact, to lengthen the duration of impact, and to increase the contact surface area considerably as a result of its own deformation. In this way the maximum force and stress arising during impact may be reduced considerably.

In designing a cushioning surface it is necessary to know the physicomechanical characteristics of the agricultural product considered, the permissible stress, the conditions of impact and the dynamic characteristics of the cushioning material. Investigations have shown that the duration of impact amounts to only some hundredths of a second even if cushioning materials are applied, and this is considerably shorter than the relaxation times obtained for most products. This means that to solve the impact problem the basic equations of the theory of elasticity may be utilized without committing any great error.

The maximum deformation occurring during impact may be determined by means of the energy equation, using the basic equations for contact stress due to Hertz. The general form of the energy equation is

$$mv^2/2 = \int_0^{z_0} P\,dz$$

where z is the deformation, P the impact force, and v the velocity of the impact. Utilizing eqns. (79) and (80) for the contact stress, and the relationship $v=\sqrt{2gH}$ for free fall, the impact characteristics (such as the deformations of the cushioning material and body, the maximum stress arising in the body, the duration of impact, etc.) for a spherical body falling on a cushioning material may be determined under various conditions.

With the assumption that the modulus of elasticity of the body is much higher than that of the cushioning material, the deformation of the cushioning material may be expressed [75] as

$$z_0/d = 1.1137[H\gamma(1-\nu^2)/E]^{0.4} \tag{104}$$

where ν and E refer to the cushioning material and γ is the specific weight of the falling body. Equation (104) expresses the deformation relative to the diameter of the body, which is illustrated in Fig. 122 for an apple with $\gamma = 0.8\ \mathrm{g\,cm^{-3}}$. The deformation assigned to a given falling height and modulus of elasticity may be determined for any arbitrary diameter from eqn. (104) and Fig. 122.

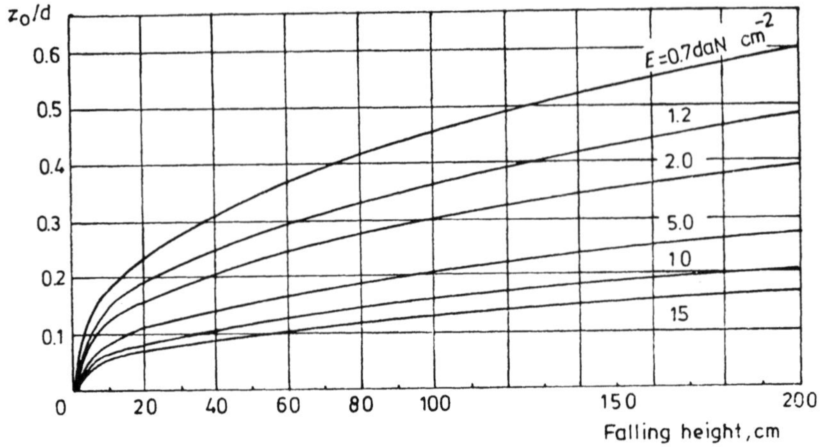

Fig. 122. Relative deformation of cushioning material

The damage to a material is determined under given circumstances by the maximum compressile stress arising in it, the value of which is given by

$$\sigma_{max} = 0.918(P/A^2d^2)^{1/3} = (0.9/A)\sqrt{z_0/d}$$

Using eqn. (104), the maximum stress is obtained as

$$\sigma_{max} = 1.094(GH)^{1/5}(1/A)^{4/5}(1/d)^{3/5} \tag{105}$$

where G is the weight of the product. The maximum shear stress occurs at a distance of $0.5R$ below the contact surface, where R is the radius of the contact surface. Its value is

$$\tau_{max} \simeq 0.27\sigma_{max}$$

Equation (105) shows clearly the effects of the individual factors on the maximum compressile stress. As may be seen, the effects of the weight and falling height are not significant, while the modulus of elasticity is an important factor; the diameter of the product also affects the stress significantly.

If the shape of a product is to a good approximation spherical, the weight G may be expressed in terms of its diameter, which equals the diameter of curvature of the contact surface. Equation (105) may now be written in the simpler form

$$\sigma_{max} = 0.96(H\gamma)^{1/5}(1/A)^{4/5} \tag{105a}$$

Figure 123 is a graphical representation of eqn. (105) for apples, with the assumption that the modulus of elasticity of apples ($E = 40$ daN cm^{-2}) is much

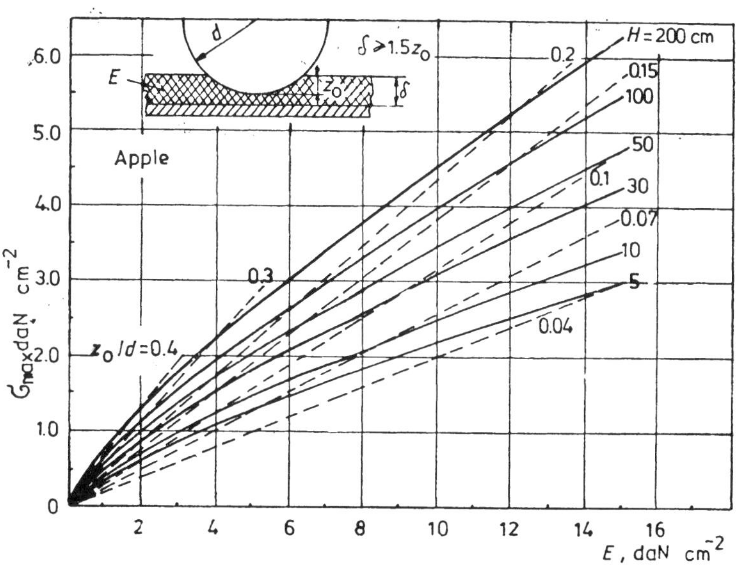

Fig. 123. Maximum compressile stress as a function of the modulus of elasticity of cushioning material

higher than that of the cushioning material. In the figure the falling height appears as a parameter, and constant z_0/d lines are also plotted.

If the permitted stress for a given crop is known, the required modulus of elasticity of the cushioning material for a given falling height (using the limiting condition that $\sigma_{max} \leqq \sigma_{perm}$) and also the ratio z_0/d may be read from a plot similar to Fig. 123. For example, for apples $\sigma_{perm} = 3.0$ daN cm^{-2}, and in the case of $H = 100$ cm; $E = 7$ daN cm^{-2} and $z_0/d = 0.19$. If the diameter of the largest apple is 70 mm, then the maximum deformation of the cushioning material is $z_0 = 0.19 \times 70 = 13.3$ mm. The thickness of the cushioning material must be selected to allow this value on the basis of the dynamic characteristic of the material. If the only cushioning material available has a modulus of elasticity $E = 2.0$ daN cm^{-2}, then $\sigma_{max} = 1.1$ daN cm^{-2}, $z_0/d = 0.3$, and so the maximum deformation is $z_0 = 0.3 \times 70 = 21$ mm, meaning that a much thicker layer of cushioning material will be required.

On shaking a crop (e.g., fruit), the individual products will fall against each other. This corresponds to the impact of two spherical bodies, and the maximum stress will be, similarly to eqn. (105),

$$\sigma_{max} = 0.72(GH)^{1/5}(1/A)^{4/5}[(R_1 + R_2)/R_1 R_2]^{3/5} \tag{105b}$$

Since in this case two bodies having identical moduli of elasticity are involved,

$A=2(1-\nu^2)/E$. With this in mind, the value of σ_{max} will be only 57% of the stress which would arise on impact against a rigid ball, and only 87% of that for impact against a rigid plane plate.

An important characteristic of impact processes is the duration of impact. The shorter the duration of impact, the faster the deceleration of the body, and the higher the load. According to the teachings of mechanics, the duration of impact for elastic bodies may be calculated from the relationship [45]

$$t_0 = 2.94(z_0/v)$$

Utilizing the relationships for deformation and falling velocity, the following expression is obtained:

$$t_0 = (0.09576/H^{0.1})[(1-\nu^2)G/E\sqrt{d}]^{0.4} \qquad (106)$$

which is illustrated in Fig. 124 for variously sized apples as a function of the modulus of elasticity of the cushioning material. It may be seen from the rela-

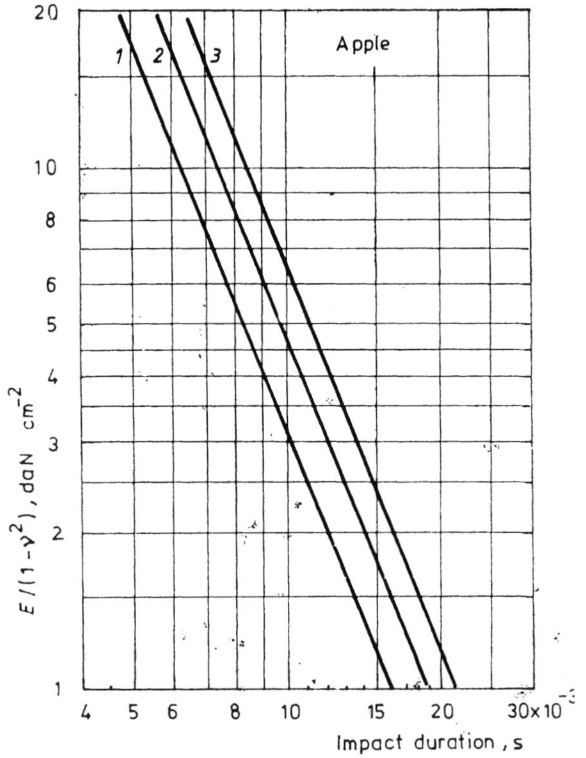

Fig. 124. Impact duration as a function of the modulus of elasticity of cushioning material *(1)* 5.5 cm diam.; *(2)* 6.5 cm diam.; *(3)* 7.5 cm diam.

tionship that the effect of the falling height is insignificant. The values calculated by means of eqn. (106) agree well with those obtained experimentally [95].

In selecting a suitable cushioning material the relevant characteristics must be known. The crucial characteristic of cushioning material of a given thickness is the variation of the quantity $E/(1-\nu^2)$ as a function of the relative deformation z/d, or of the dropping height. The relationship $E/(1-\nu^2)=f(z/d)$ may be determined under either static or dynamic loading conditions, of which only the dynamic characteristics can be used for dimensioning in the present case.

A few methodological remarks concerning the plotting of dynamic characteristics should be made. During impact investigations the force acting on a ball impacting against the surface of a cushioning material and the deformation are measured. The modulus of elasticity may be calculated from these data by means of eqn. (81). The theory of contact stress and also eqn. (81) assume that the modulus of elasticity is uniform everywhere in the material and is independent of deformation. However, this assumption is not always realized even approximately in the compressible cushioning materials. Laticel-type cushioning material can be compressed to one-tenth of its original thickness, but its modulus of elasticity will be higher by an order of magnitude as a result of the compaction. Since the radius of the ball (or product) and the thickness of the cushioning material are generally of the same order of magnitude, it may happen that E will be nearly constant over the greatest part of the contact surface, but will increase considerably at the location of maximum deformation. The calculations give only an average E value for the whole contact area, which may be much lower than that observed at the most compressed place.

The above points are illustrated clearly by the impact force versus time diagrams shown in Fig. 125. The impact–force curve over the whole contact surface area will be much steeper at point a, indicating hardening of the laticel. However, the peak force obtained in this way will not exceed the value which may be expected without hardening by a factor greater than two. However, when the force is measured over only a small surface showing the greatest deformation (i.e., is not an averaged value), the pattern of behavior is different. At the start of impact (after a running-up stage) the force remains practically constant. However, when the deformation exceeds 20 mm, the force increases to 10 times its initial value, owing to the compaction.

The modulus of elasticity of laticels is low; therefore in many cases a rubber sheet 1–2 mm thick is glued on their surface. Our experiments [75] showed that the averaged and local values of the modulus of elasticity may deviate from each other considerably in this case also. Finally, the modulus of elasticity is also influenced by whether the cushioning material lays freely or has its lower surface glued to a steel plate. In the latter case transverse elongations at great deforma-

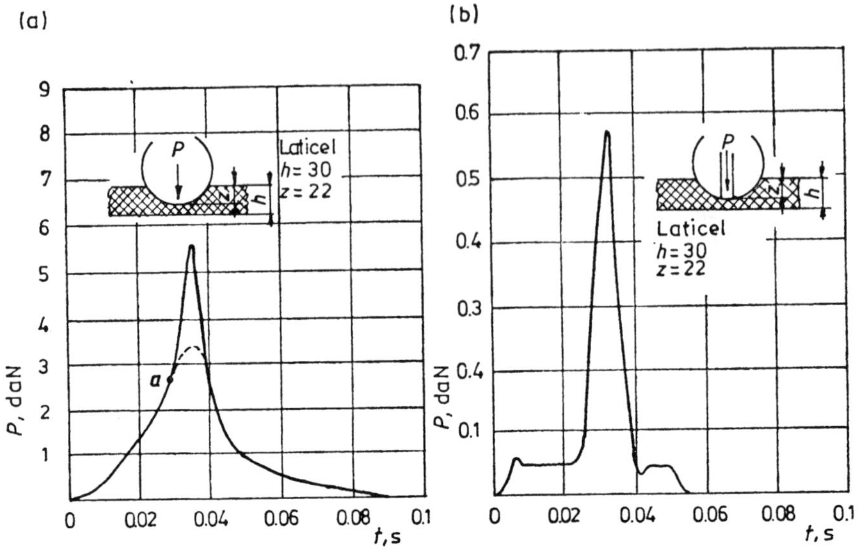

Fig. 125. Impact force–time diagrams relative to the total contact surface area (a), and the location of maximum deformation (b)

tions are restricted. Figure 126 shows the dynamic characteristics of laticels 20 mm and 40 mm thick for the following four cases: laticel lying freely; laticel glued on a steel plate; and laticels with rubber sheets 0.8 mm and 2 mm thick glued on their surfaces. The load-bearing capacity of laticels is higher in the case of slight deformations and remains nearly constant with increasing deformation, owing to their particular construction. When the material is compacted by great deformation, the modulus of elasticity increases rapidly. Figure 127 illustrates the modulus of elasticity measured at the location of greatest deformation as a function of relative deformation. On comparison with the preceding figure it may be seen that the values are much higher, since here the result is not distorted by averaging over surface regions each showing a different amount of deformation.

The modulus of elasticity may also be plotted as a function of deformation relative to the ball diameter. Lines of constant dropping height may also be indicated, as shown in Fig. 128. The thickness of the cushioning material appears in this case as a parameter.

In calculations for cushioning material of the above type the following method is recommended. First, the relative deformation should be determined using the modulus of elasticity relative to the complete ball, and then the maximum stress calculated using the modulus of elasticity relative to the location of greatest

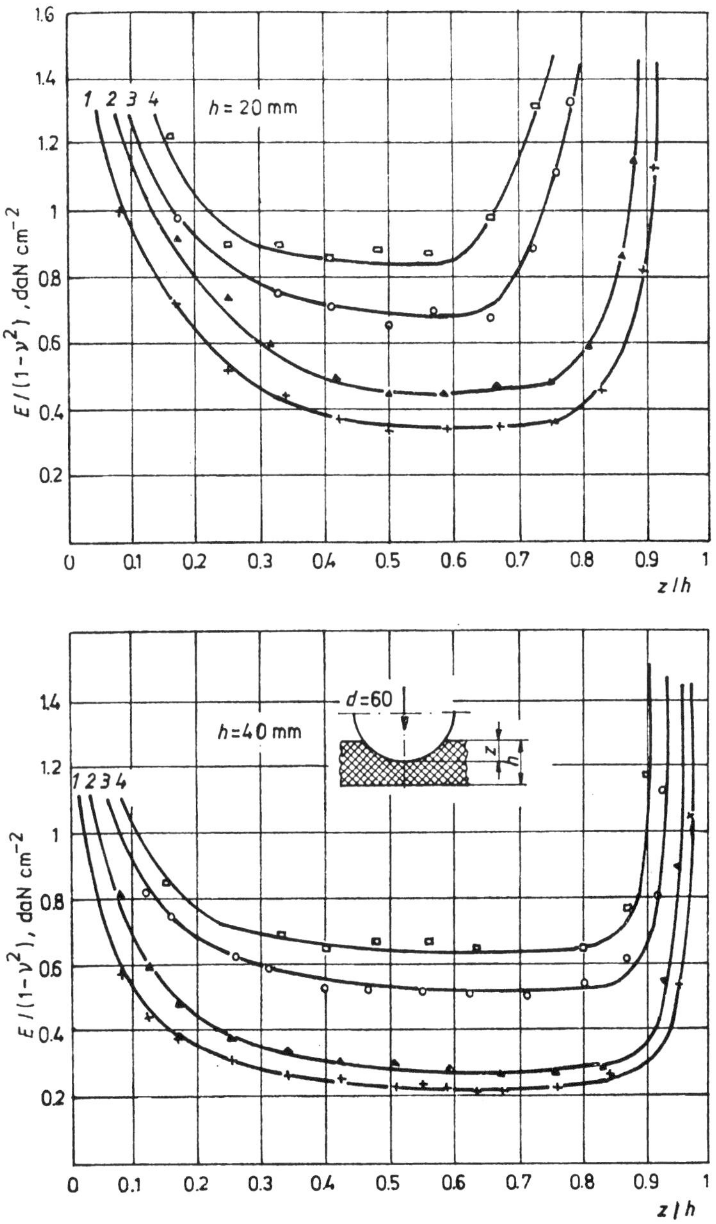

Fig. 126. Dynamic characteristic curves for laticels 20 and 40 mm thick. *(1)* Laticel; *(2)* laticel on steel plate; *(3)* laticel with rubber sheet of 0.8 mm thickness; *(4)* laticel with rubber sheet of 2 mm thickness

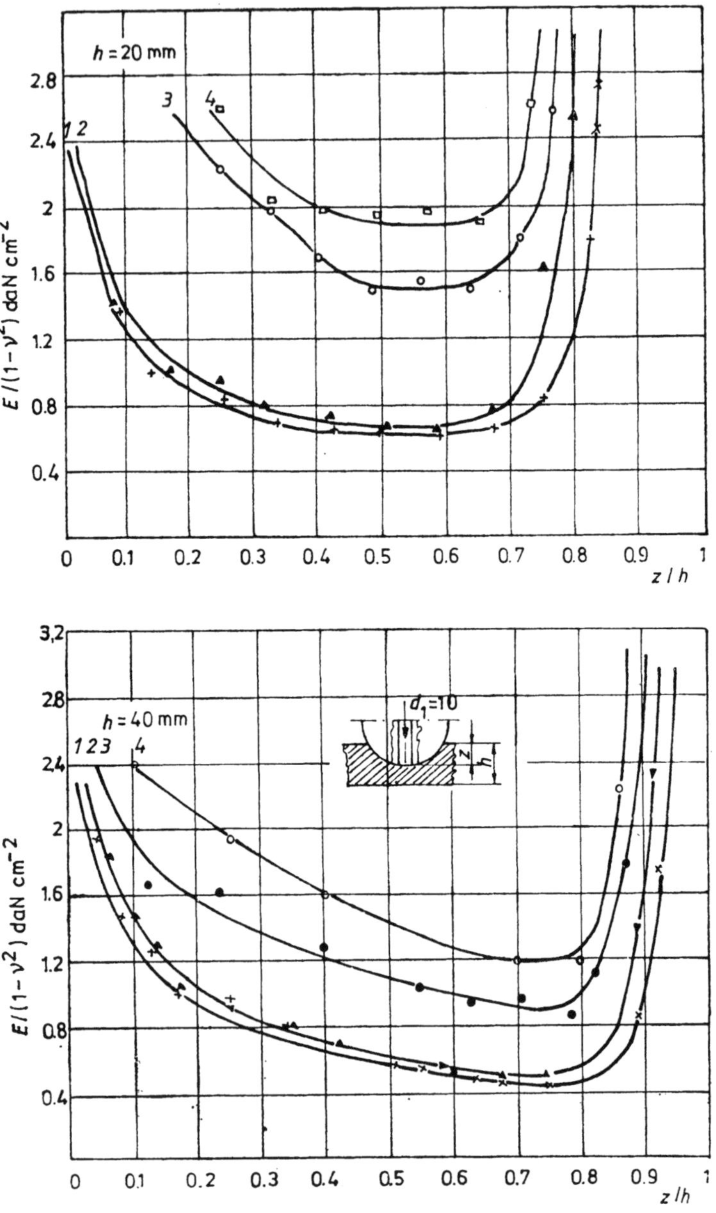

Fig. 127. Dynamic characteristic curves for laticels relative to the location of maximum deformation. *(1)* Laticel; *(2)* laticel on steel plate; *(3)* laticel with rubber sheet of 0.8 mm thickness; *(4)* laticel with rubber sheet of 2 mm thickness

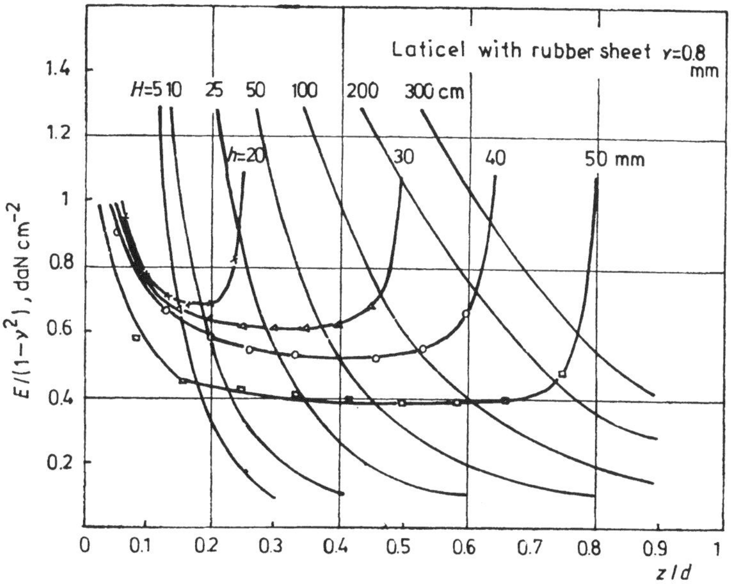

Fig. 128. Dynamic characteristic curves as functions of deformation relative to ball diameter

deformation. For example, if the falling height is $H=50$ cm and a laticel of thickness $h=40$ mm with a 2 mm rubber-sheet coating is employed, the mean modulus of elasticity is 0.65 daN cm^{-2}, the corresponding relative deformation is $z_0/d=0.365$ and the maximum stress is $\sigma_{max}=0.353$ daN cm^{-2}. However, if the local modulus of elasticity (1.25 daN cm^{-2}) is taken into account, then $\sigma_{max}=0.68$ daN cm^{-2}, i.e., nearly twice as high as the preceding value.

11. THE FINITE-ELEMENT METHOD

Analytical methods for viscoelastic stress analysis are available today only for a few simple cases, and so their application is limited. The practical cases occurring in agricultural engineering mostly represent complex problems which may be solved only by numerical methods. The most frequently applied numerical method is the finite-element method.

11.1 Concept of the method

The basic concept of the finite-element method is that a continuous function may be approximated in sections, by means of a system of piecewise continuous functions defined for a finite number of space elements. The piecewise continuous functions are defined by a finite number of points (nodal points) of the individual elements. The method is applied in the following main steps [80]:

(a) identify a finite number of points in space (these points are termed nodal points);

(b) prescribe the values of the continuous function at the nodal points;

(c) divide the space domain studied into a finite number of elements; the elements are connected to each other at the common nodal points, and in common characterize approximately the space studied; and finally

(d) approximate the continuous function describing the space studied by a system of functions describing the elements; each element function is continuous only within the corresponding element.

The main objective is then to determine the nodal-point values (e.g., the displacements) of the function $f(x)$ so as to obtain the best approximation to the required function $f(x)$. This task, as has been verified in several theoretical studies, may be reduced to the minimization of a given quantity. During the minimization process a system of equations is given, permitting the nodal-point values of the variable to be determined.

The finite-element method may be applied to one-, two- and three-dimensional spaces. A major proportion of three-dimensional problems are axisymmetrical, and this makes solution simpler. In the two-dimensional case the elements are triangular or quadrangular. The minimum number of nodal points is three for a triangle and four for a quadrangle. If the number of nodal points used is greater than the minimum, the side faces of the elements may be curved, and the element function will also describe a curved surface. Increasing the number of the nodal points makes the calculations more complicated, but the approximation may be greatly improved in this way. Therefore, by using a greater number of nodal points, satisfactory accuracy may be attained with a smaller number of elements.

An advantage of the finite-element method is that the material characteristics of the individual elements may differ, so that calculations may also be performed for anisotropic bodies (e.g., an apple with its peel). An irregularly shaped body may be approximated by a combination of elements having straight or curved sides. The sizes of individual elements may also differ, and thus smaller or larger elements may be used depending on the variation (gradient) of the required function. The greater the variation of the function, the smaller the elements which must be used to attain satisfactory accuracy.

11.2 Shapes of elements and the displacement function

The essence of the finite-element method, as has been seen, is the division of a given domain into a number of small elements and the approximation of the unknown parameter (e.g., displacement) within each element by a mathematical function. The approximate mathematical function used most frequently is the polynomial. The polynomial may be linear or parabolic, depending on the number of nodal points applied. Figure 129 shows the triangular and quadrangular elements used most frequently.

If the displacements of the nodal points are known, then the displacement of an arbitrary point within the element may be calculated according to the equation

$$\{f\} = [N]\{\delta\}^e = [N_i, N_j, N_m]\begin{Bmatrix}\delta_i \\ \delta_j \\ \delta_m\end{Bmatrix} \tag{107}$$

The displacement of the point involved is given by the components u and v, which are functions of the locus:

$$\{f\} = \begin{Bmatrix}u(x, y) \\ v(x, y)\end{Bmatrix}$$

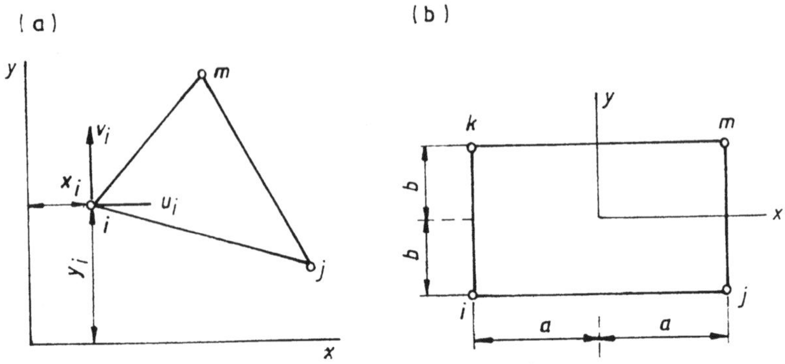

Fig. 129. Schematic drawing of triangular and quadrangular elements in finite-element method

The functions N_i, N_j and N_m must be selected so as to obtain the effective nodal-point displacement on substituting the coordinates of the nodal points. This condition is met if N_i, N_j and N_m are linear functions of the variables x and y. In this case the components of the displacement may be expressed by the linear polynomials

$$
\begin{aligned}
u &= \alpha_1+\alpha_2 x+\alpha_3 y \\
v &= \alpha_4+\alpha_5 x+\alpha_6 y
\end{aligned}
\tag{108}
$$

where the constants $\alpha_1, \ldots, \alpha_6$ are determined so as to satisfy the boundary conditions. For a triangular element with three nodal points, the following linear polynomials may be written:

$$
\begin{aligned}
u_i &= \alpha_1+\alpha_2 x_i+\alpha_3 y_i \\
u_j &= \alpha_1+\alpha_2 x_j+\alpha_3 y_j \\
u_m &= \alpha_1+\alpha_2 x_m+\alpha_3 y_m \\
v_i &= \alpha_4+\alpha_5 x_i+\alpha_6 y_i \\
v_j &= \alpha_4+\alpha_5 x_j+\alpha_6 y_j \\
v_m &= \alpha_4+\alpha_5 x_m+\alpha_6 y_m
\end{aligned}
$$

The values of $\alpha_1, \ldots, \alpha_6$ may be expressed from the above system of equations, and substitution into eqn. (108) then gives

$$u = (1/2\Delta)[(a_i+y_{jm}x+x_{mj}y)u_i+(a_j+y_{mi}x+x_{im}y)u_j+(a_m+y_{ij}x+x_{ji}y)u_m]$$

$$v = (1/2\Delta)[(a_i+y_{jm}x+x_{mj}y)v_i+(a_j+y_{mi}x+x_{im}y)v_j+(a_m+y_{ij}x+x_{ji}y)v_m]$$

where

$$\begin{aligned} a_i &= x_j y_m - x_m y_j \quad & y_{jm} &= y_j - y_m \\ a_j &= x_m y_i - x_i y_m \quad & x_{mj} &= x_m - x_j \\ a_m &= x_i y_j - x_j y_i \end{aligned}$$

and so on, and Δ is the area of the triangle. Applying matrix notation,

$$\begin{Bmatrix} u \\ v \end{Bmatrix} = \begin{bmatrix} N_i & 0 & N_j & 0 & N_m & 0 \\ 0 & N_i & 0 & N_j & 0 & N_m \end{bmatrix} \begin{Bmatrix} u_i \\ v_i \\ u_j \\ v_j \\ u_m \\ v_m \end{Bmatrix}$$

or, in shorter form,

$$\{f\} = [IN_i,\ IN_j,\ IN_m]\{\delta\}^e \tag{109}$$

where

$$N_i = (1/2\Delta)(a_i + y_{jm}x + x_{mj}y)$$

etc., and

$$I = \begin{bmatrix} 1 & 0 \\ 0 & 1 \end{bmatrix}$$

Equation (109) may also be applied in axisymmetrical three-dimensional cases, where the x coordinate is replaced by the radius r, and y by the coordinate z. Then in the above equation,

$$N_i = (1/2\Delta)(a_i + z_{jm}r + r_{mj}z)$$

and

$$a_i = r_j z_m + r_m z_j$$

etc. Figure 130 illustrates the subdivision of an axisymmetrical body into elements.

A quadrangular element has four nodal points in the simplest case, and the sides of the quadrangle are then straight. The displacement of an arbitrary point within the element is

$$u = \alpha_1 + \alpha_2 x + \alpha_3 y + \alpha_4 xy$$

$$v = \alpha_5 + \alpha_6 x + \alpha_7 y + \alpha_8 xy$$

The constants $\alpha_1, \ldots, \alpha_8$ are determined similarly as in the case of a triangular element. Introduction of the dimensionless coordinates $x' = x/a$ and $y' = y/b$

leads to

$$u = (1/4)\{(1-x')(1-y')u_i+(1+x')(1-y')u_j+(1+x')(1+y')u_m+ \\ +(1-x')(1+y')u_k\}$$

$$v = (1/4)\{(1-x')(1-y')v_i+(1+x')(1-y')v_j+(1+x')(1+y')v_m+ \\ +(1-x')(1+y')v_k\}$$

The following simpler equation may be written using matrix notation:

$$\{f\} = [IN_i, IN_j, IN_m, IN_k]\{\delta\}^e \tag{110}$$

where

$$N_i = (1/4)(1-x')(1-y'), \text{ etc.}$$

A linear interpolation polynomial gives an unsatisfactory approximation in numerous cases, unless very small elements are used. The accuracy may be improved greatly by using a second-order interpolation polynomial. Experiments have shown that for a given number of nodal points, a better approximation is supplied by a parabolic distribution function than by a simple triangular element, even with finer division [76]. Figure 131 shows the layout of a triangular element with six nodal points. The six nodal points permit the application of a second-order interpolation polynomial in the form

$$u = \alpha_1+\alpha_2 x+\alpha_3 y+\alpha_4 x^2+\alpha_5 xy+\alpha_6 y^2$$

$$v = \alpha_7+\alpha_8 x+\alpha_9 y+\alpha_{10} x^2+\alpha_{11} xy+\alpha_{12} y^2$$

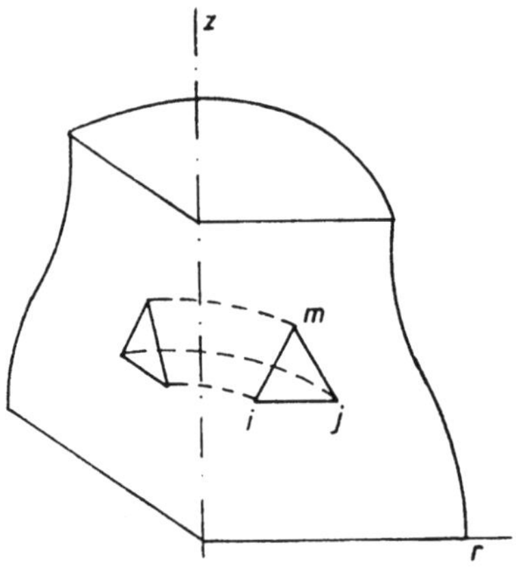

Fig. 130. Division of an axisymmetrial body into elements

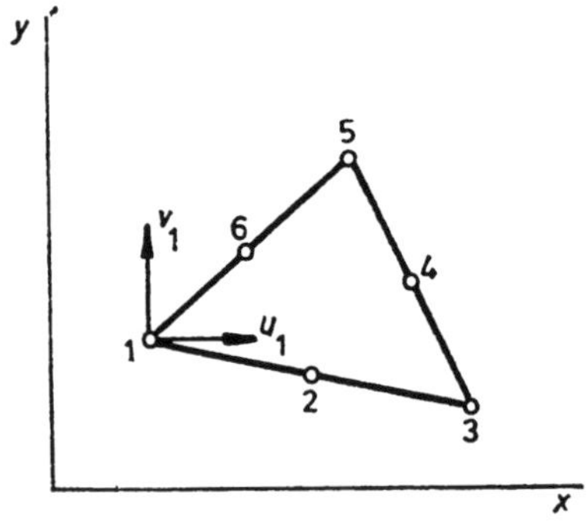

Fig. 131. Triangular element with six nodal points

The constants $\alpha_1, \ldots, \alpha_{12}$ are determined by a similar method as in the preceding cases. With regard to the large number of equations, it is advisable to determine the constants with the help of a computer.

The displacement of the nodal points may be written in matrix form as

$$\{\delta\}^e = \begin{Bmatrix} u_1 \\ v_1 \\ \vdots \\ u_6 \\ v_6 \end{Bmatrix} = \begin{bmatrix} 1, & x_1, & y_1, & x_1^2, & x_1y_1, & y_1^2, & 0, & 0, & 0, & 0, & 0, & 0 \\ 0, & 0, & 0, & 0, & 0, & 0, & 1, & x_1, & y_1, & x_1^2, & x_1y_1, & y_1^2 \\ 1, & x_2, & y_2, & x_2^2, & x_2y_2, & y_2^2, & 0, & 0, & 0, & 0, & 0, & 0 \\ 0, & 0, & 0, & 0, & 0, & 0, & 1, & x_2, & y_2, & x_2^2, & x_2y_2, & y_2^2 \\ \text{etc.} & & & & & & & & & & & \end{bmatrix} \begin{Bmatrix} \alpha_1 \\ \alpha_2 \\ \vdots \\ \vdots \\ \alpha_{12} \end{Bmatrix}$$

or, in shorter notation, as

$$\{\delta\}^e = [C]\{\alpha\}$$

The components of the displacements within the elements are

$$\{f\} = \begin{Bmatrix} u \\ v \end{Bmatrix} = \begin{bmatrix} 1, & x, & y, & x^2, & xy, & y^2, & 0, & 0, & 0, & 0, & 0, & 0 \\ 0, & 0, & 0, & 0, & 0, & 0, & 1, & x, & y, & x^2, & xy, & y^2 \end{bmatrix} \{\alpha\}$$

or

$$\{f\} = [P]\{\alpha\}$$

Using the preceding expression, the displacement components may be written as

$$\{f\} = [P][C]^{-1}\{\delta\}^e = [N]\{\delta\}^e \tag{111}$$

Frequently it is advisable to use area coordinates, as suggested by Zienkiewich [76]. The definition of area coordinates may be understood on the basis of Fig. 132.

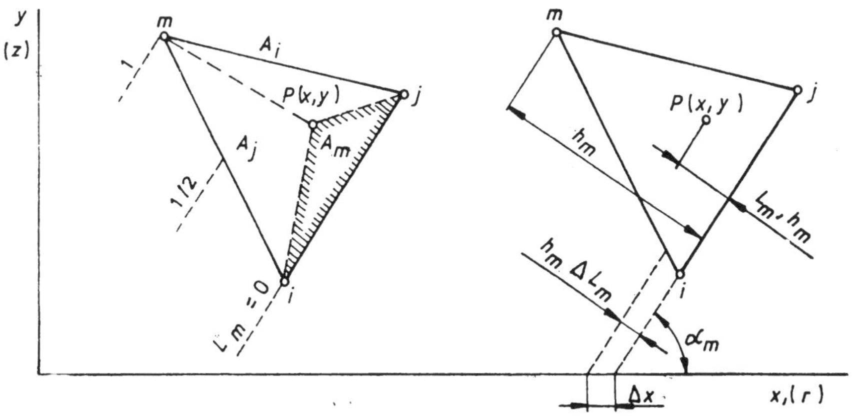

Fig. 132. Definition of area coordinates

A given point $P(x, y)$ divides the triangle into three component areas, and these areas, or their ratios relative to the total area, determine uniquely the location of the point P. On this basis, the area coordinates may be written as

$$L_i = A_i/\Delta, \quad L_j = A_j/\Delta, \quad L_m = A_m/\Delta \tag{112}$$

From the definition, it follows that

$$A_i+A_j+A_m = \Delta$$
$$L_i+L_j+L_m = 1$$

The constants L_i, L_j and L_m are straight lines which are parallel to the sides lying opposite the nodal points i, j and m. The line $L_i=0$ is the side of the triangle lying opposite the nodal point i; the line $L_i=1$ passes through the nodal point i.

Area coordinates may be expressed as a function of x and y coordinates as

$$\begin{aligned} L_i &= (1/2\Delta)(a_i+y_{jm}x+x_{mj}y) \\ L_j &= (1/2\Delta)(a_j+y_{mi}x+x_{im}y) \\ L_m &= (1/2\Delta)(a_m+y_{ij}x+x_{ji}y) \end{aligned} \tag{113}$$

where $a_i=x_j y_m-x_m y_j$ etc. The x and y coordinates may be expressed by means of the above equations, by utilizing the coordinates of the nodal points:

$$x = L_i x_i+L_j x_j+L_m x_m$$
$$y = L_i y_i+L_j y_j+L_m y_m$$

From comparison of eqns. (109) and (113), it follows that $N_i=L_i$, $N_j=L_j$ and $N_m=L_m$, permitting a much simpler treatment.

The displacement function for a triangular element with six nodal points may be written, applying area coordinates, as

$$\{f\} = \begin{Bmatrix} u \\ v \end{Bmatrix} = [IN_1, IN_2, IN_3, IN_4, IN_5, IN_6]\{\delta\}^e \tag{114}$$

where

$$\begin{aligned} N_1 &= L_1(2L_1-1) \\ N_2 &= 4L_1L_3 \\ N_3 &= L_3(2L_3-1) \\ N_4 &= 4L_3L_5 \\ N_5 &= L_5(2L_5-1) \\ N_6 &= 4L_1L_5 \end{aligned}$$

In certain cases knowledge of the partial derivatives of L_1, L_3 and L_5 as functions of the independent variables x and y is necessary. On the basis of Fig. 132,

$$\partial L_m/\partial x = (1/h_m)(h_m \Delta L_m/\Delta x) = (1/h_m)\sin\alpha_m = (1/h_m)(y_j - y_i)/L_m$$

Since $h_m L_m = 2\Delta$, the derivatives may be calculated as

$$\partial L_m/\partial x = (y_i - y_j)/2\Delta \quad \partial L_m/\partial y = (x_j - x_i)/2\Delta$$

$$\partial L_i/\partial x = (y_m - y_j)/2\Delta \quad \partial L_i/\partial y = (x_m - x_j)/2\Delta$$

$$\partial L_j/\partial x = (y_i - y_m)/2\Delta \quad \partial L_j/\partial y = (x_i - x_m)/2\Delta$$

In axisymmetrical cases the x and y coordinates must be replaced by r and z coordinates.

The sides of a quadrangular element with four nodal points may only be straight lines. If curved sides are to be applied in the interests of better approximation, then eight nodal points may be selected. Figure 133 shows the arrangement of a quadrangular element with eight nodal points, taking up the system of coordinates within the element. The latter is not orthogonal, and the axes of the system of coordinates pass through the nodal points assessed on the opposite sides. Within the element, the following relationship exists [81] between the common systems of x–y and η–ξ coordinates:

$$X = N_a X_a + N_b X_b + N_c X_c + N_d X_d + N_e X_e + N_f X_f + N_g X_g + N_h X_h,$$

$$Y = N_a Y_a + N_b Y_b + N_c Y_c + N_d Y_d + N_e Y_e + N_f Y_f + N_g Y_g + N_h Y_h,$$

where $X_a, \ldots, X_h$ and $Y_a, \ldots, Y_h$ are the coordinates of the nodal points, while the shape-functions $N_a \ldots N_h$ may be expressed in terms of the variables ξ and η

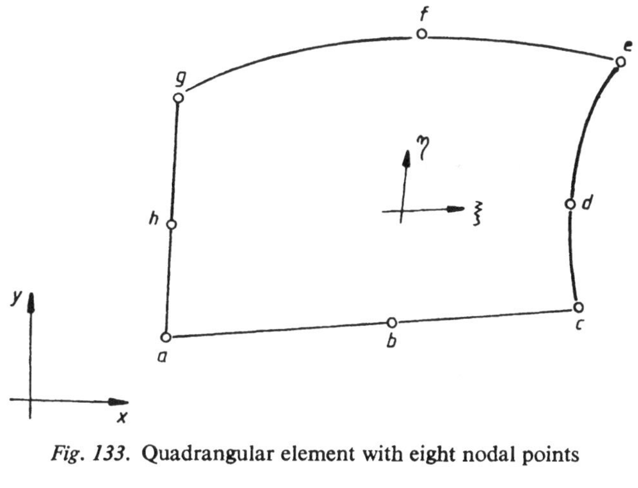

Fig. 133. Quadrangular element with eight nodal points

in the form:

$$N_a = -(1/4)(1-\xi)(1-\eta)(\xi+\eta+1) \quad N_b = (1/2)(1-\xi^2)(1-\eta)$$
$$N_c = (1/4)(1+\xi)(1-\eta)(\xi-\eta-1) \quad N_d = (1/2)(1-\eta^2)(1+\xi)$$
$$N_e = (1/4)(1+\xi)(1+\eta)(\xi+\eta-1) \quad N_f = (1/2)(1-\xi^2)(1+\eta)$$
$$N_g = (1/4)(1-\xi)(1+\eta)(-\xi+\eta-1) \quad N_h = (1/2)(1-\eta^2)(1-\xi)$$

The system of ξ–η coordinates is normalized, i.e., the absolute value of the coordinates cannot exceed unity. The coordinates of point a are $\xi=\eta=-1$, the coordinate of the side a–b–c is $\eta=-1$.

11.3 Embedding the elements into the continuum

In the preceding Section we have dealt with interpolation polynomials related to individual elements. The sizes and orientations of the individual elements may be arbitrary, and this is an important advantage in applying the finite-element method. Both regular and irregular configurations may be approximated well by an appropriate selection and arrangement of elements.

The task now is to embed the individual elements into the whole surface or body studied. This means that the interpolation functions must be expressed in terms of the coordinates and unknown variables (displacements) corresponding to the nodal points.

Consider a system which may be subdivided into five triangular elements (Fig. 134) [80]. The displacement components of the individual nodal points are

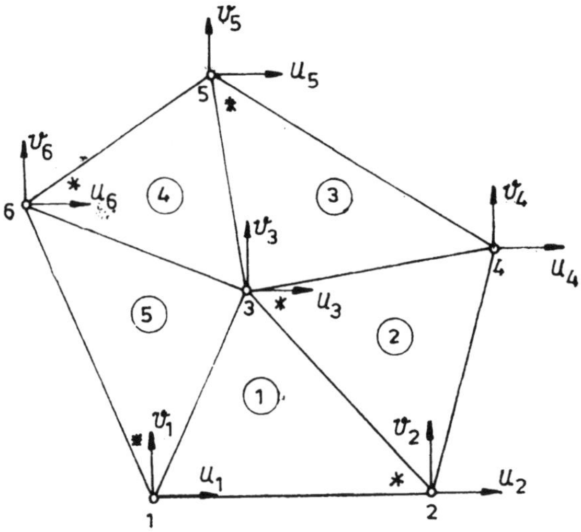

Fig. 134. Explanation of the assembly of elements

$u_1, v_1, \ldots, u_6, v_6$ and these are the unknowns. The coordinates $X_1, Y_1, \ldots, X_6, Y_6$ of the individual nodal points are known or can be determined by measurement. In the individual triangles the nodal points selected as i nodal points are marked by an asterisk. Accordingly, the suffixes i, j and m for the individual element are as follows.

first element: $i = 2, \quad j = 3, \quad m = 1$

second element: $i = 3, \quad j = 2, \quad m = 4$

and so on. According to eqn. (110), five interpolation functions may be written, with substitution of the corresponding suffixes i, j and m:

$$\{f\}^{(1)} = [IN_2^{(1)}, IN_3^{(1)}, IN_1^{(1)}]\{\delta\}^{(1)}$$

$$\{f\}^{(2)} = [IN_3^{(2)}, IN_2^{(2)}, IN_4^{(2)}]\{\delta\}^{(2)}$$

$$\{f\}^{(3)} = [IN_5^{(3)}, IN_3^{(3)}, IN_4^{(3)}]\{\delta\}^{(3)} \qquad 115)$$

$$\{f\}^{(4)} = [IN_6^{(4)}, IN_3^{(4)}, IN_5^{(4)}]\{\delta\}^{(4)}$$

$$\{f\}^{(5)} = [IN_1^{(5)}, IN_3^{(5)}, IN_6^{(5)}]\{\delta\}^{(5)}$$

where

$$\{\delta\}^{(1)} = \begin{Bmatrix} u_2 \\ v_2 \\ u_3 \\ v_3 \\ u_1 \\ v_1 \end{Bmatrix}$$

etc., and

$$N_2^{(1)} = (1/2\Delta^{(1)})(a_2 + Y_{31}x + X_{13}y)$$

$$a_2 = X_3 Y_1 - X_1 Y_3$$

$$Y_{31} = Y_3 - Y_1$$

$$X_{13} = X_1 - X_3$$

etc. The system of eqn. (115) may be summarized in a single matrix as a function of the total degree of freedom, in the form:

$$\begin{Bmatrix} f^{(1)} \\ f^{(2)} \\ f^{(3)} \\ f^{(4)} \\ f^{(5)} \end{Bmatrix} = \begin{bmatrix} IN_1^{(1)}, & IN_2^{(1)}, & IN_3^{(1)}, & 0, & 0, & 0 \\ 0, & IN_2^{(2)}, & IN_3^{(2)}, & IN_4^{(2)}, & 0, & 0 \\ 0, & 0, & IN_3^{(3)}, & IN_4^{(3)}, & IN_5^{(3)}, & 0 \\ 0, & 0, & IN_3^{(4)}, & 0, & IN_5^{(4)}, & IN_6^{(4)} \\ IN_1^{(5)}, & 0, & IN_3^{(5)}, & 0, & 0, & IN_6^{(5)} \end{bmatrix} \begin{Bmatrix} u_1 \\ v_1 \\ u_2 \\ v_2 \\ u_3 \\ v_3 \\ u_4 \\ v_4 \\ u_5 \\ v_5 \\ u_6 \\ v_6 \end{Bmatrix} \tag{115a}$$

11.4 Finite-element formulation

The main objective of the finite-element method is to calculate numerical values for the nodal quantities (i.e., displacement, temperature or concentration) that define the interpolation polynomials so that they provide the best possible approximation to the physical process under consideration. In most cases the nodal values will be determined by minimizing an integral quantity, termed as functional, related to the given physical process.

In solving field problems (e.g., heat transfer or diffusion), the problem is that of minimizing the integral of a functional having certain characteristic properties. The main property is that a given function, which provides the minimum value of the functional, must also satisfy the differential equation and boundary conditions for the process. The finite-element variational method can be applied to all scalar fields which satisfy either the Laplace or Poisson equations.

Consider a two-dimensional stationary temperature field, described in the axisymmetrical case, by the differential equation

$$\partial/\partial z(\lambda r(\partial\vartheta/\partial z)) + \partial/\partial r(\lambda r(\partial\vartheta/\partial r)) + Q = 0$$

Taking the boundary conditions into consideration,

$$\vartheta = \vartheta_s$$

(with known temperature values over the boundary surface);

$$-\lambda(\partial\vartheta/\partial n) = q$$

(with known heat flux over the boundary surface); and

$$-\lambda(\partial\vartheta/\partial n) = \alpha(\vartheta_m - \vartheta)$$

(with known heat transfer coefficient over the boundary surface). The variational formulation of the above differential equation involving the boundary conditions, may be expressed [77] as

$$\chi = \iint (1/2)\lambda r[(\partial\vartheta/\partial z)^2 + (\partial\vartheta/\partial r)^2]\,dz\,dr + \int (1/2)\alpha r(\vartheta_m - \vartheta)^2\,ds + \int qr\vartheta\,ds$$

In order to minimize this equation, it is necessary to make the derivative of the functional χ, with respect to the independent variable, equal to zero: i.e.,

$$\partial\chi/\partial\vartheta = 0.$$

Now consider a temperature field divided into a finite number of elements, in which the temperature is defined by the respective nodal-point values. The typical triangular element can be described, similarly to the displacement equations, by

$$\vartheta = [N_i, N_j, N_m]\{\vartheta\}^e = [N^e]\{\vartheta\}^e$$

where

$$N_i = (1/2\Delta)(a_i + z_{jm}r + r_{mj}z) = L_i$$

$$a_i = r_j z_m - r_m z_j$$

and further,

$$\{\vartheta\}^e = \begin{Bmatrix} \vartheta_i \\ \vartheta_j \\ \vartheta_m \end{Bmatrix}$$

$$\partial N_i/\partial r = (z_m - z_j)/2\Delta$$

$$\partial N_i/\partial z = (r_m - r_j)/2\Delta$$

etc. The functional formulation for an element can be written in matrix form as

$$\chi^e = \iint (\lambda r/2)\{\vartheta\}^T[B^e]^T[B^e]\{\vartheta\}\,dz\,dr + \int qr[N^e]\{\vartheta\}\,ds +$$

$$+ \int (\alpha r/2)\{\vartheta\}^T[N^e]^T[N^e]\{\vartheta\}\,ds - \int \alpha r\vartheta_m[N^e]\{\vartheta\}\,ds$$

Differentiation of the functional using the differentiation rules for matrix equations gives

$$\partial\chi^e/\partial\{\vartheta\} = \iint \lambda r[B^e]^T[B^e]\{\vartheta\}\,dz\,dr + \int qr[N^e]^T\,ds +$$

$$+ \int \alpha r[N^e]^T[N^e]\{\vartheta\}ds - \int \alpha r\vartheta_m[N^e]^T\,ds$$

where

$$[B^e] = \begin{bmatrix} \partial N_i/\partial r & \partial N_j/\partial r & \partial N_m/\partial r \\ \partial N_i/\partial z & \partial N_j/\partial z & \partial N_m/\partial z \end{bmatrix}$$

The minimization can be performed first for the individual elements, whence, after summing the results thus obtained and making them equal to zero,

$$\partial\chi/\partial\{\vartheta\} = \sum \partial\chi^e/\partial\{\vartheta\} = 0$$

The solution of problems in elasticity using the finite-element method is formulated in terms of minimizing an integral quantity related to the potential energy in the system. By assuming a displacement field and minimizing the potential energy of the system, the nodal values of the displacements are obtained. Knowing the displacement values, the strain and stress components can be obtained easily.

The components of deformation depend on the stress state. For example, in the case of a plane stress state the components of the strain may be written in the form

$$\{\varepsilon\} = \begin{Bmatrix} \varepsilon_x \\ \varepsilon_y \\ \gamma_{xy} \end{Bmatrix} = \begin{Bmatrix} \partial u/\partial x \\ \partial v/\partial y \\ \partial u/\partial y + \partial v/\partial x \end{Bmatrix}$$

while in the axisymmetrical case,

$$\{\varepsilon\} = \begin{Bmatrix} \varepsilon_z \\ \varepsilon_r \\ \varepsilon_\varphi \\ \gamma_{rz} \end{Bmatrix} = \begin{Bmatrix} \partial v/\partial z \\ \partial u/\partial r \\ u/r \\ \partial u/\partial z + \partial v/\partial r \end{Bmatrix}$$

The expression for strain as a function of displacement may be written in the general form

$$\{\varepsilon\} = [B]\{\delta\}^e = [B_i, B_j, B_m]\{\delta\}^e \tag{116}$$

where the construction of the matrix $[B]$ depends on the stress state. For a plane stress state,

$$[B] = (1/2\Delta)\begin{bmatrix} y_{jm}, & 0, & y_{mi}, & 0, & y_{ij}, & 0 \\ 0, & x_{mj}, & 0, & x_{im}, & 0, & x_{ji} \\ x_{mj}, & y_{jm}, & x_{im}, & y_{mi}, & x_{ji}, & y_{ij} \end{bmatrix}$$

and in the axisymmetrical case,

$$[B_i] = (1/2\Delta)\begin{bmatrix} 0 & r_{mj} \\ z_{jm} & 0 \\ a_i(r + z_{jm} + r_{mj}z)/r & 0 \\ r_{mj} & z_{jm} \end{bmatrix}$$

and so on. In the case of a plane stress state and a linear distribution function, the matrix $[B]$ is independent of the position of the point within the element, i.e., the deformation is constant within the element. This applies in the axisymmetrical case only to nodal point displacements, where the displacement u is proportional to the radius r.

Temperature or moisture variations will also cause deformation (thermal or moisture strain) in a material which will also depend on the stress conditions. Under plane stress condition the strain components for an isotropic material with expansion coefficient β will be

$$\{\varepsilon_0\} = \begin{Bmatrix} \beta \Delta U \\ \beta \Delta U \\ 0 \end{Bmatrix}$$

where ΔU is the moisture change. In the case of a plane deformation where stresses normal to the x–y plane are also produced, the Poisson's ratio ν will also influence the thermal or moisture strain. Therefore, in such instances,

$$\{\varepsilon_0\} = (1+\nu) \begin{Bmatrix} \beta \Delta U \\ \beta \Delta U \\ 0 \end{Bmatrix}$$

In axisymmetrical cases the strain components may then be expressed as

$$\{\varepsilon_0\} = \begin{Bmatrix} \beta \Delta U \\ \beta \Delta U \\ \beta \Delta U \\ 0 \end{Bmatrix}$$

Thus thermal or moisture expansion does not seem to cause any shear strain at all.

Assuming elastic behavior of the material, a linear correlation will be observed between stress and strain, in the general matrix form

$$\{\sigma\} = [D]\{\varepsilon\} - [D]\{\varepsilon_0\}$$

where the elasticity matrix $[D]$ may be constructed on the basis of the usual stress–strain relationship. For example, the above relationships may be written in the case of a plane stress state as

$$\sigma_x = [E/(1-\nu^2)](\varepsilon_x + \nu\varepsilon_y)$$

$$\sigma_y = [E/(1-\nu^2)](\nu\varepsilon_x + \varepsilon_y)$$

$$\tau_{xy} = [E/(1-\nu)]\gamma_{xy}$$

and the elasticity matrix assumes the form:

$$[D] = E/(1-\nu^2) \begin{bmatrix} 1 & \nu & 0 \\ \nu & 1 & 0 \\ 0 & 0 & (1-\nu)/2 \end{bmatrix}$$

In the axisymmetrical case, the stress components are σ_z, σ_r, σ_φ and τ_{rz} and the elasticity matrix may be written in the form (for clarity, only the half-matrix is shown; the matrix is symmetrical about the diagonal)

$$[D] = E(1-\nu)/(1+\nu)(1-2\nu)\begin{bmatrix} 1 & \nu/(1-\nu) & \nu/(1-\nu) & 0 \\ & 1 & \nu/(1-\nu) & 0 \\ & & 1 & 0 \\ \text{symmetric} & & & (1-2\nu)/2(1-\nu) \end{bmatrix}$$

The general expression for the stress may be written, using eqn. (116), in the form

$$\{\sigma\} = [D][B]\{\delta\}^e \tag{117}$$

If an external distributed load is also acting on the element, this must be added to the stress caused by the displacement:

$$\{\sigma\} = [D][B]\{\delta\}^e + \{\sigma\}_p^e. \tag{117a}$$

In determining the nodal values of displacement, as explained above, the total potential energy of the elastic system should be minimized. The total potential energy can be separated into two components: the first obtained from the strain energy in the body, and the second related to the potential energy of the loads applied.

The strain energy is given by integrating over the volume, as

$$\pi_i = (1/2)\int (\{\sigma\}\{\varepsilon\} - \{\sigma\}\{\varepsilon_0\})\,dV$$

Taking eqns. (116) and (117) into consideration, then

$$\pi_i = (1/2)\int (\{\delta\}^T[B^e]^T[D^e][B^e]\{\delta\} - \{\delta\}^T[B^e]^T[D^e]\{\varepsilon_0\})\,dV$$

The work done by a concentrated load is given by the product of the concentrated load and a displacement, or

$$\pi_{a1} = \{P\}^T\{\delta\}$$

The work due to a distributed load acting on the surface is

$$\pi_{a2} = \int [N^e]\{\delta\}\{p\}\,ds$$

The total potential energy can now be written as

$$\pi = \pi_i - (\pi_{a1} + \pi_{a2})$$

To minimize the total potential energy of the system, π should be differentiated and made equal to zero. For an element,

$$\partial\pi^e/\partial\{\delta\} = \int [B^e]^T[D^e][B^e]\,dV\{\delta\} - \int [B^e][D^e]\{\varepsilon_0\}\,dV - \int [N^e]^T\{p\}\,ds - \{P\} \tag{118}$$

or, in shorter form,

$$\partial\pi^e/\partial\{\delta\} = [k^e]\{\delta\} + \{f^e\} \tag{118a}$$

where the term

$$[k^e] = \int [B^e]^T [D^e][B^e]\, dV$$

is termed the element stiffness matrix. For the total system the minimization procedure is given by

$$\partial\pi/\partial\{\delta\} = \sum \partial\pi^e/\partial\{\delta\} = 0$$

In the case of plane problems, when the thickness t of the element is constant, integration must follow the area of the triangle:

$$[k^e] = \int [B^e]^T [D^e][B^e] t\, dx\, dy$$

When a linear distribution function is used, the matrix $[B^e]$ is independent of the position within the element, and so the stiffness matrix will read simply

$$[k^e] = [B^e]^T [D^e][B^e] t\Delta$$

The global stiffness matrix and the global force vector are given by summing the element matrices, as

$$[K] = \sum [k^e]$$

and

$$\{F\} = \sum \{f^e\}$$

11.5 Viscoelastic stress analysis; numerical method

The finite-element method is applicable to stress analysis for various viscoelastic materials, but an appropriate numerical method must be used. Herrmann and Petersen [79] have described such a numerical method, which was subsequently applied successfully [78].

In elaborating the method it is assumed that the material is linearly viscoelastic, thermorheologically simple, and shows elastic behavior under hydrostatic pressure. The components of the stress and strain tensor are calculated, with application of the superposition principle, in the usual form as

$$\sigma_{ij} = \delta_{ij}\sigma_v + S_{ij} \tag{119}$$

where δ_{ij} is the Kronecker delta, defined as

$$\delta_{ij} = \begin{cases} 1 & \text{if} \quad i = j \\ 0 & \text{if} \quad i \neq j \end{cases}$$

and where σ_v is the average normal stress, whose value is

$$\sigma_v = (\sigma_{11}+\sigma_{22}+\sigma_{33})/3$$

The S_{ij} (deviatoric stress) components are obtained by subtracting the value σ_v from the components arranged in the diagonal σ_{ij}. The tensor for deformation may be used in a similar form:

$$\varepsilon_{ij} = \delta_{ij}\varepsilon_v + e_{ij}, \tag{120}$$

where

$$\varepsilon_v = (\varepsilon_{11}+\varepsilon_{22}+\varepsilon_{33})/3 = \varepsilon_{kk}/3$$

In the case of hydrostatic compression, the relationship for elastic stress-deformation is

$$\sigma = K3\varepsilon_v$$

where K is the volumetric modulus of elasticity. The viscoelastic stress–strain relationship may be written in integral form similarly to eqn (57), as

$$\sigma_{ij} = \delta_{ij} \int_0^t \varphi_1(t-\tau)(\partial\varepsilon_{kk}/\partial\tau)\, d\tau + 2 \int_0^t \varphi_2(t-\tau)(\partial e_{ij}/\partial\tau)\, d\tau \tag{121}$$

where φ_1 and φ_2 are the shear-relaxation functions, which must be given in the form

$$\varphi(t) = a_0 + \sum_{i=1}^{m} a_i e^{-\beta_i t} = a_0 + \varphi^*(t) \tag{122}$$

In eqn. (121), stresses caused by temperature variations may also be allowed for through reduction of the time coordinate. The reduced time is obtained, as described in Section 8.13, on the basis of the integral

$$t^*(x, t) = \int_0^t \{1/k[\vartheta(x, \tau)]\}\, d\tau$$

To determine the deviatoric stress, the time interval $0-t$ is divided into N parts. The individual intervals are designated as $t_0=0$, $t_1=t_0+\Delta t_1$, ..., $t_N=t_{N-1}+$ $+\Delta t_N=t$. At the end of the time interval, $S_{ij}(x, t_N)=S_{ijN}$ and $e_{ij}(x, t_N)=e_{ijN}$. Assuming a deformation- and stress-free state at the beginning of the time interval, the deviatoric stress may be written on the basis of eqn. (121) as

$$S_{ijN} = 2a_0 e_{ijN} + 2\int_0^{t_{N-1}} \varphi^*(t-\tau)(\partial e_{ij}/\partial\tau)\, d\tau + \int_{t_{n-1}}^{t_N} \varphi^*(t-\tau)(\partial e_{ij}/\partial\tau)\, d\tau \quad (123)$$

The variations of the individual stress components within each time interval are approximated as

$$\partial e_{ij}(x, t)/\partial t = (e_{ijN} - e_{ijN-1})/\Delta t_N$$

etc. With the above approximation, eqn. (123) assumes the form

$$S_{ijN} = 2a_0 e_{ijN} + 2\int_0^{t_{N-1}} \varphi^*(t-\tau)(\partial e_{ij}/\partial\tau) + 2[(e_{ijN} - e_{ijN-1})/\Delta t_N] \int_{t_{N-1}}^{t_N} \varphi^*(t-\tau)\, d\tau$$

or, in shorter notation,

$$S_{ijN} = 2\mu_N e_{ijN} + L_{ijN} \quad (124)$$

where

$$\mu_N = (1/\Delta t_N) \int_{t_{N-1}}^{t_N} \varphi^*(t-\tau)\, d\tau + a_0$$

and

$$L_{ijN} = 2\left[\int_0^{t_{N-1}} \varphi^*(t-\tau)(\partial e_{ij}/\partial\tau)\, d\tau - (\mu_N - a_0)e_{ijN-1}\right]$$

In the above equation μ_N is the shear modulus at the given point x, at time t_N; L_{ijN} is the effect of the preceding deformation (in a given case also of the temperature history) at time t_N. The values of μ_N and L_{ijN} must now be determined. The equation for μ_N may be written, with the expression for φ^* taken into account, as

$$\mu_N = \sum_{i=1}^{m} a_i \left[(1/\Delta t_N) \int_{t_{N-1}}^{t_N} e^{-\beta_i(t_N-\tau)}\, d\tau\right] + a_0$$

or, in shorter form, as

$$\mu_N = \sum_{i=1}^{m} a_i J_{Ni} + a_0 \quad (125)$$

where

$$J_{Ni} = (1/\Delta t_N) \int_{t_{N-1}}^{t_N} e^{-\beta_i(t_N-\tau)}\, d\tau \quad (126)$$

The expression for L_{ijN} in the shorter notation is

$$L_{ijN} = 2[A_{ijN} - (\mu_N - a_0)e_{ijN-1}] \quad (127)$$

where

$$A_{ijN} = \int_0^{t_{N-1}} \varphi^*(t_N - \tau)(\partial e_{ij}/\partial \tau)\, d\tau$$

or

$$A_{ijN} = \sum_{i=1}^{m} a_i C_{ijNi} \tag{128}$$

with

$$C_{ijNi} = \int_0^{t_{N-1}} e^{-\beta_i(t_N - \tau)}(\partial e_{ij}/\partial \tau)\, d\tau$$

On again approximating the differential quotient in the above expression by a finite difference, the following equation is obtained:

$$C_{ijNi} = e^{-\beta_i \Delta t_N}[C_{ijN-1,i} + (e_{ijN-1} - e_{ijN-2}) J_{N-1,i}] \tag{129}$$

It may be seen from the form of eqn. (129) that the effect of the earlier deformation history may be calculated by a recursive method, based on the individual parameters for the preceding intervals.

The relaxation function $\varphi(t)$ is determined most frequently by uniaxial compression tests of cylindrical specimens. In this case the stress σ_{yy} is zero, and the stress σ_{xx} may be expressed, using eqn. (121), as

$$\sigma_{xx}(t) = 2 \int_0^t \varphi_2(t-\tau)(\partial \varepsilon_{xx}/\partial \tau - \partial \varepsilon_{yy}/\partial \tau)\, d\tau + 2[\varepsilon_{xx}(0) - \varepsilon_{yy}(0)]\, \varphi_2(t) \tag{121a}$$

allowing for a finite jump in strain at time zero. Assuming the bulk modulus to be elastic, it is possible to write

$$\sigma(t) = K\varepsilon_{kk}(t)$$

and in the case of uniaxial compression, this equation becomes

$$\sigma_{xx}(t) = 3K[2\varepsilon_{yy}(t) + \varepsilon_{xx}(t)]$$

Solving this equation for ε_{yy} and substituting into eqn. (121a) gives

$$\sigma_{xx}(t) = 3 \int_0^t \varphi_2(t-\tau)(\partial \varepsilon_{xx}/\partial \tau)\, d\tau - (1/3K) \int_0^t \varphi_2(t-\tau)(\partial \sigma_{xx}/\partial \tau)\, d\tau +$$

$$+ 3\varepsilon_{xx}(0)\, \varphi_2(t) - (1/3K)\, \sigma_{xx}(0)\, \varphi_2(t)$$

which can be solved for φ_2 at discrete time increments. Equation (121a) can also be solved by means of the Laplace transformation, which yields [78].

$$\bar{\varphi}_2 = (2K\bar{\sigma}_{xx}/S)/[6K\bar{\varepsilon}_{xx} - (2/3)\bar{\sigma}_{xx}]$$

During the investigations ε_{xx} and $\sigma_{xx}(t)$ are known as the input and output values. The volumetric modulus of elasticity K may be determined by special tests. The deformation ε_{xx} is generally selected as a step function, i.e.,

$$\varepsilon_{xx} = \varepsilon_0 H(t),$$

where $H(t)$ is the Heaviside step function. The measured function $\sigma_{xx}(t)$ is processed in the form of eqn. (122); three exponential terms result in satisfactory accuracy in most cases. The required relaxation function is obtained by substituting the transformed values of σ_{xx} and ε_{xx} into the equation for $\bar{\varphi}_2$ and retransforming the function obtained.

For example, in testing the relaxation of apples, a cylindrical test specimen of diameter 12.7 mm and 19 mm high was used and the value of the deformation selected according to the step function $\varepsilon_{xx}(t)=0.05H(t)$ [78]. The test was continued for 1.5 min and the following stress and relaxation functions were obtained:

$$\sigma_{xx}(t) = 0.0877e^{-56.5t}+0.0885e^{-3.39t}+0.675e^{-0.0768t}$$

for the stress, and

$$\varphi_2(t) = 0.627e^{-56.9t}+0.628e^{-3.41t}+4.56e^{-0.0799t}$$

for the relaxation. The volumetric modulus of elasticity was K=36.8 daN cm^{-2}.

11.6 Application of the finite-element method to flow fields

The finite-element method is also suited to solving complex, multidimensional flow problems. Such problems are encountered in the aeration of large bales, stacks built of bales, or loose forage materials, granular piles of complex form, etc. Most of these are two- or three-dimensional and axisymmetrical problems. Some three-dimensional tasks may be reduced to simpler two-dimensional tasks by the application of correction factors.

Agricultural granular and forage piles are characterized by a nonlinear relationship between pressure gradient and flow velocity; generally, the pressure gradient increases in proportion to velocity to the power of 1.4–1.6. In addition, the porosity or volumetric weight plays an important role. Accordingly, the expression for the pressure gradient may be written as

$$\operatorname{grad} p = -C\gamma^m \vec{v}^n = -K'\vec{v}^n$$

where $\vec{v}$ is the flow velocity vector, and γ the volumetric weight of the material. The partial differential equation for two-dimensional flow must be written with the function grad $p=f(v)$ taken into account. It is advisable to use the above exponential relationship in the form

$$\text{grad}\, p = -K\vec{v}$$

where K is now generally a function of velocity, and may be written [106] as

$$K = K'^{1/n}[(\partial p/\partial x)^2 - (\partial p/\partial y)^2]^{(n-1)/2n}$$

With the above taken into consideration, two-dimensional flow may be described by the differential equation

$$\partial/\partial x[(1/K)(\partial p/\partial x)] + \partial/\partial y[(1/K)(\partial p/\partial y)] = 0 \qquad (130)$$

or, in the axisymmetrical case, using polar coordinates r and z as

$$\partial/\partial r[(r/K)(\partial p/\partial r)] + \partial/\partial z[(r/K)(\partial p/\partial z)] = 0 \qquad (130a)$$

In applying finite elements to solve the differential equation, the variational principle is generally used. Accordingly, the p function minimizing the integral

$$\chi = \iint g(x, y, p, \partial p/\partial x, \partial p/\partial y)\,dx\,dy$$

must satisfy the following equation

$$\partial/\partial x[\partial g/\partial(\partial p/\partial x)] + \partial/\partial y[\partial g/\partial(\partial p/\partial y)] - \partial g/\partial p = 0 \qquad (131)$$

Subdividing the space domain into a number of smaller elements, the functional χ must be determined for each element and then summed for all the elements:

$$\chi = \sum \chi^e$$

The minimization condition is satisfied if the derivative with respect to p of the functional χ equals zero:

$$\partial\chi/\partial p = \sum \partial\chi^e/\partial p = 0$$

Equations (130) and (131) may be used to determine the function g. From comparison of the two equations,

$$\partial g/\partial(\partial p/\partial x) = (1/K)(\partial p/\partial x)$$

$$\partial g/\partial(\partial p/\partial y) = (1/K)(\partial p/\partial y)$$

and

$$\partial g/\partial p = 0$$

If K is taken as constant within the individual elements, then the first two equations give

$$g^e = (1/2K^e)[(\partial p/\partial x)^2 + (\partial p/\partial y)^2]$$

and the functional χ^e to be minimized is

$$\chi^e = \iint g^e dx\, dy \tag{132}$$

To determine the interpolation polynomials for the finite elements, it is again assumed that the pressure values prevailing within each of the elements may be expressed by the pressure values at the respective nodal points:

$$p = \{N\}\{p\}^e \tag{133}$$

The objective is to find nodal-point pressures giving the minimum of the integral χ. This may be achieved by differentiation with respect to each nodal-point pressure and making the resulting function equal to zero. For a single element, the derivative of the functional χ^e is

$$\partial\chi^e/\partial\{p\}^e = [h]^e\{p\}^e$$

where the elements of the rigidity matrix $[h]^e$ are obtained on the basis of the following considerations. Differentiation of eqn. (132) with respect to the pressure at one nodal point gives

$$\partial\chi^e/\partial p_i = \iint (1/K)[(\partial p/\partial x)(\partial/\partial p_i)(\partial p/\partial x)+(\partial p/\partial y)(\partial/\partial p_i)(\partial p/\partial y)]dx\, dy$$

The partial derivative of the pressure according to eqn. (133) is given by

$$\partial p/\partial x = \{\partial N_i/\partial x,\ \partial N_j/\partial x,\ \partial N_m/\partial x,\ \ldots\}\{p\}^e$$

Comparison of the above equations for the components of the matrix $[h]^e$ gives the integral

$$h_{ij}^e = \iint (1/K)\ [(\partial N_i/\partial x)(\partial N_j/\partial x)+(\partial N_i/\partial y)(\partial N_j/\partial y)]dx\, dy$$

The derivative of the functional χ for the whole system after summation of the element functional is

$$\partial\chi/\partial\{p\} = [H]\{p\} = 0$$

where the elements of the matrix $[H]$ are obtained by summing the corresponding components for the individual elements:

$$H_{ij} = \sum h_{ij}^e$$

The pressure gradient is calculated from the pressure drops in the x and y directions

$$\text{grad}\ p = \sqrt{(\partial p/\partial x)^2+(\partial p/\partial y)^2}$$

from which the flow velocity may be calculated simply.

12. APPLICATION OF RHEOLOGY

Relatively numerous experiments have recently been performed investigating the physicomechanical properties of agricultural materials. These have included plotting load–deformation (stress–strain) relationships for compression, tension, shear, bending and hydrostatic compression, and investigations with the aim of characterizing viscoelastic properties. During these investigations all of the state characteristics influencing the mechanical properties appear as variables. The most important of these state characteristics are the moisture content, temperature, the type of product, stage of ripening, etc.

Investigations may be performed with materials in their natural state, or with specimens prepared from them. Since most agricultural materials are processed in their natural state (e.g., whole fruits with peel), measurements should generally be carried out under circumstances corresponding as closely as possible to this state. In certain cases the preparation of test specimens, with shapes facilitating evaluation and reducing disturbing effects, may be of advantage.

12.1 Force–deformation relationships

During investigations of agricultural materials the force–deformation relationship is that is tested most frequently. The load may be applied by cylindrical dies, by spherical indenters or by plane plates. Steel loading dies may be regarded as rigid in relation to the material tested, and so they may be considered not to undergo deformation.

From the force–deformation relationship it is possible to determine the modulus of elasticity E, the biological yield point, the rupture point, and even, by complementary measurements, the Poisson's ratio. During measurements the state characteristics of the material (moisture content, temperature, ripeness, period of storage after harvesting or threshing, etc.) must be recorded exactly, and must not change during the measurements.

Figure 135 shows the force–deformation relationships for pears and peaches

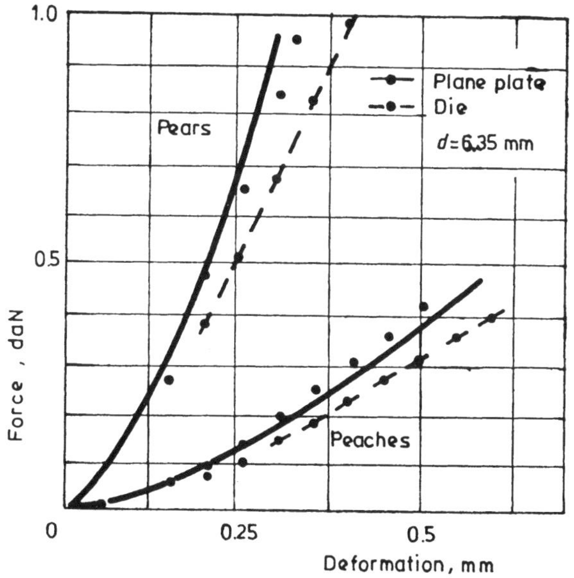

Fig. 135. Force-deformation relationships for pears and peaches

[57], for loads applied by a plane plate and by a die of diameter 6.35 mm, at a load velocity of 3.3 cm min^{-1}. As may be seen, there is no essential difference between the two modes of loading. The modulus of elasticity E is determined on the basis of the initial section of the curve, at relatively low deformations. For this purpose, only the section of the curve lying under the biological yield limit may be used.

Figure 136 shows the force-deformation curve for raspberries, loaded between plane plates [69]. The first section of the curve is nearly straight and the linear

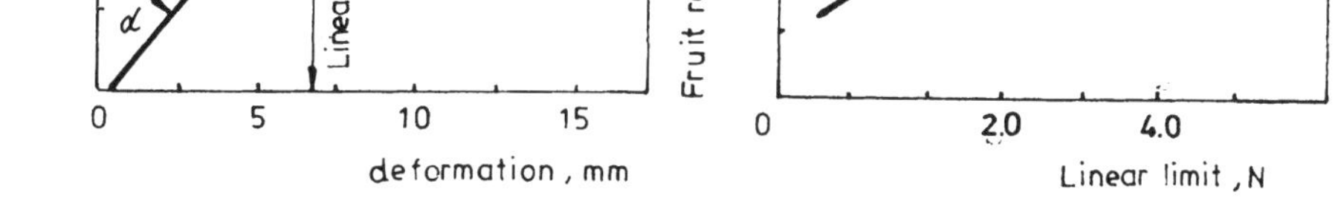

Fig. 136. Force-deformation curve for raspberries, loaded between parallel plates

limit may be established visually. On reaching the linear limit (the biological yield limit), a small volume of the material yields, and under continued load the whole volume collapses. During ripening the linear limit and the slope of the curve vary considerably. With variation of the linear limit the force required for plucking the fruit also varies; these two variables are in fact closely correlated. In the interests of subsequent transport and handling, fruit should be plucked when the linear limit is still sufficiently great. However, it cannot then be shaken off so easily, because the force is still relatively high. The task is to find the optimal compromise, possibly considering also the external or internal color.

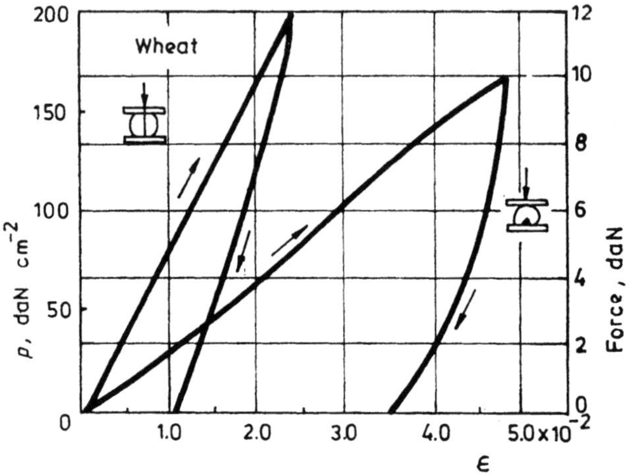

Fig. 137. Deformation of wheat grains with cut-off ends and lying flat

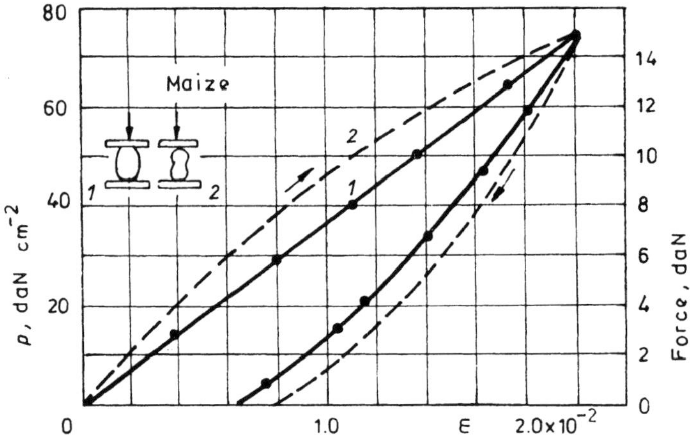

Fig. 138. Deformation of maize kernels with cut-off sides *(1)* and in the natural state *(2)*

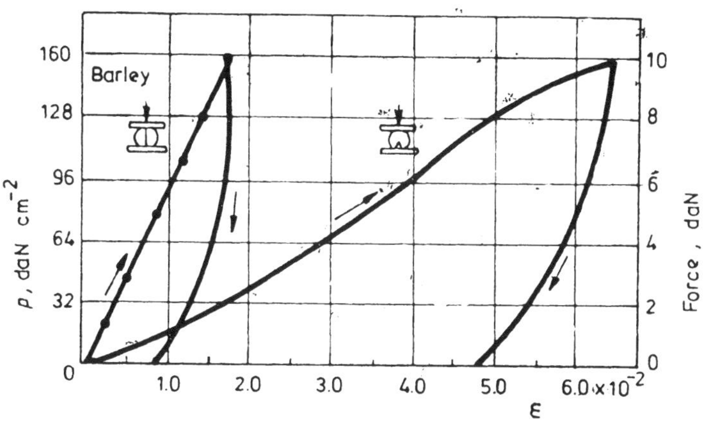

Fig. 139. Deformation of barley grains with cut-off ends and lying flat

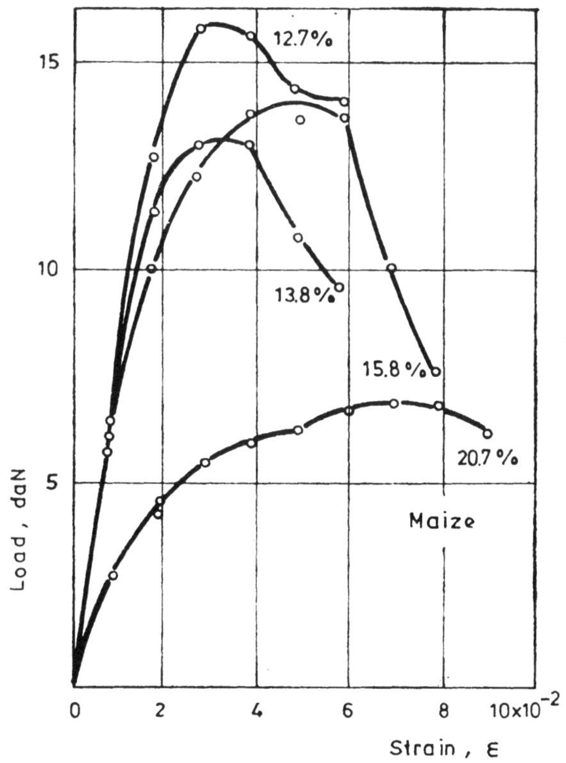

Fig. 140. Loading curves for maize kernels with various moisture contents

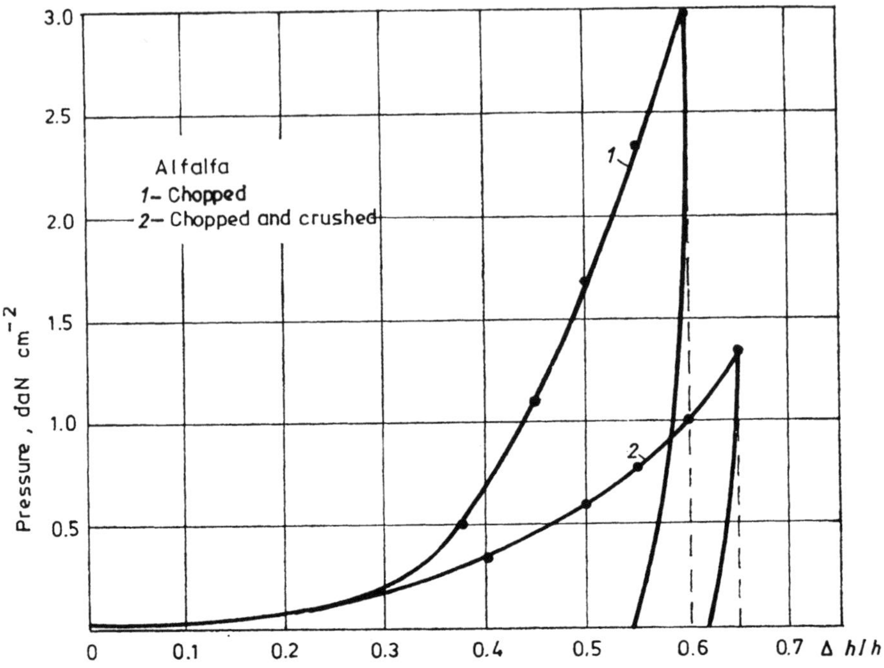

Fig. 141. Compressibility of chopped and crushed alfalfa. *(1)* Chopped; *(2)* chopped and crushed

The force–deformation relationship in the case of cereals depends greatly on the position in which the grains are loaded. The grains may be loaded either in a lying position or set on their edge, or with their ends cut off. Figure 137 presents loading curves for wheat grains loaded in two different positions [67]. The modulus of elasticity E is 8300 daN cm^{-2} for wheat grains loaded with cut-off ends, while that for flat-lying grains is 3300 daN cm^{-2}. Figure 138 shows loading curves for cut-off and whole maize kernels in the edge position. The high strength properties of the hard crust may be utilized better in the case of whole grains, whose initial modulus of elasticity is therefore somewhat higher: the value of E varies between 3750 and 5000 daN cm^{-2}. Figure 139 shows loading curves for barley grains. The difference between the two loading positions is more striking in this case. The modulus of elasticity of grains with cut-off ends is around 9000 daN cm^{-2}, while that of flat-lying grains is 2200 daN cm^{-2}. The three preceding figures also show the unloading curves. The moisture content of the cereal grains varies between 12 and 13% in the three cases.

The load-bearing capacity of cereal grains is greatly influenced by their moisture content. Figure 140 illustrates loading curves for maize kernels set on their

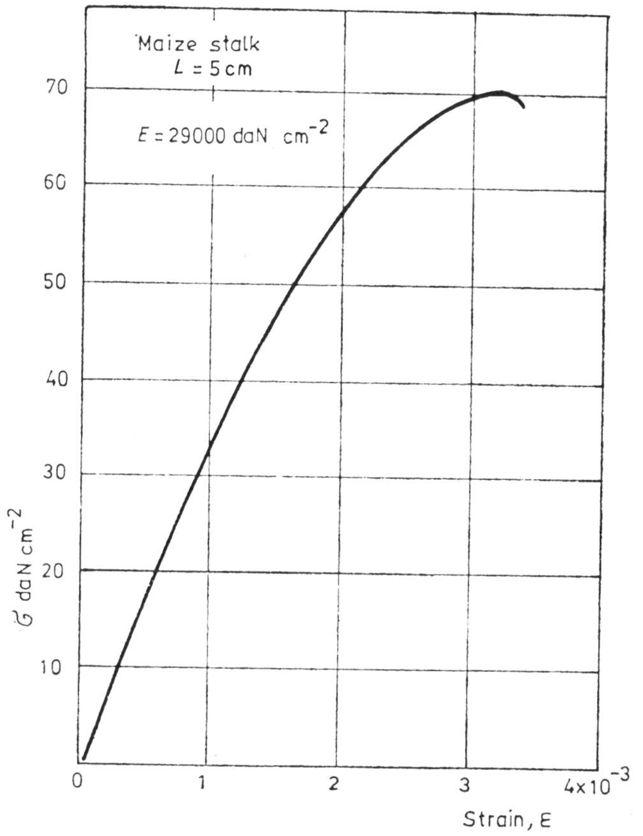

Fig. 142. Stress–strain curve for dry maize stalks

edges, for various moisture contents [58]. The initial modulus of elasticity does not decrease visibly for moisture contents up to 18%, but for higher moisture contents it decreases significantly.

In agricultural technology, compaction of the chaff components of various plants (maize, alfalfa) occurs frequently. The quality of chopping is influenced by the mechanical behavior of chaff, in particular by its compressibility. The quality of chopping may be characterized by the dimensional distribution of the chaff and the degree of crushing. The compressibility is greatly increased by crushing, as proved by the measurement results given in Fig. 141. Alfalfa chaff obtained by exact forage harvesting requires for the same compaction twice the force as does chopped and shredded chaff [181].

The compressile strength of plant stalks (alfalfa, maize, etc.) is tested by special measuring devices. For thicker stalks strain gages may also be employed to measure longitudinal and transverse deformations, from which the Poisson's ratio may be calculated. Figure 142 shows the stress–strain curve for dry maize stalks. The modulus of elasticity is relatively high, 29 000 daN cm^{-2}. The Poisson's ratio obtained during these experiments was 0.37 [59]. The Poisson's ratio is determined using a cylindrical test specimen cut from the material investigated, by measuring its axial and transverse deformations. The Poisson's ratio generally depends on the moisture content, on the magnitude of the stress and even on time. Figure 143 shows the variation of Poisson's ratio for apples over time [70]. During the axial creep following loading the transverse deformation also increases within about 0.5 s, and then remains practically constant. At the same time the axial deformation increases, which causes a decrease of the Poisson's ratio. Generally, with increasing load the value of ν increases.

For several agricultural materials the rupture stress on loading depends also on the temperature of the material. As an Example, Fig. 144 shows the rupture force as a function of temperature during the compression of potato tubers [68]. From the figure it follows that on harvesting in cold weather the tubers will be cracked open by a lower force, i.e., they are more sensitive to damage.

It has been seen while examining the rheological models (Fig. 88, eqns. (45) and (49)) that the deformation rate affects the force–deformation and stress–strain relationships. Figure 145 shows the force–deformation relationship of the

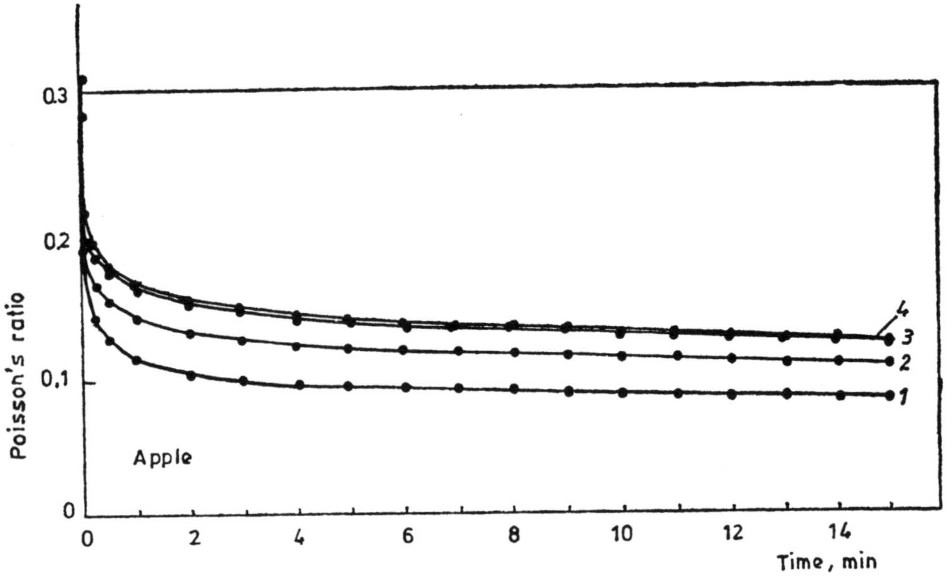

Fig. 143. Poisson's ratio of apples as a function of time for various loading levels. *(1)* 0.61 daN cm^{-2}; *(2)* 0.92 daN cm^{-2}; *(3)* 1.27 daN cm^{-2}; *(4)* 1.62 daN cm^{-2}

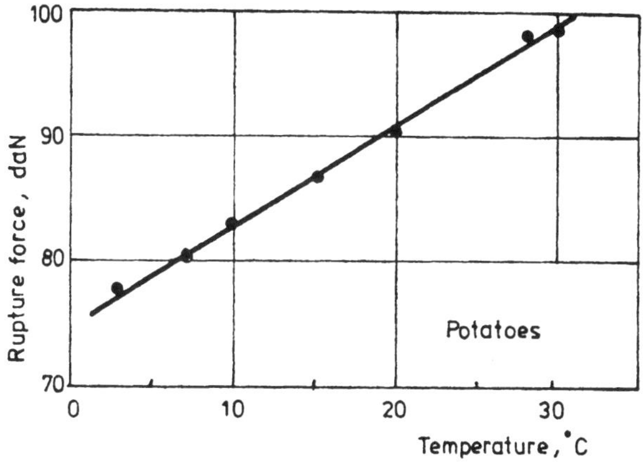

Fig. 144. Rupture force for potatoes as a function of temperature

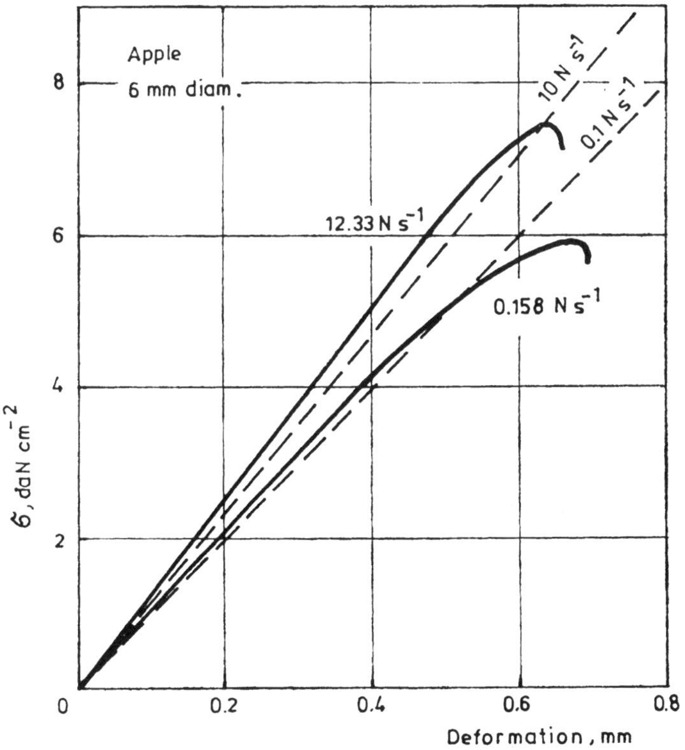

Fig. 145. Effect of loading rate on the force–deformation relationship for a fresh apple

fresh apple at various loading rates [56]. For small deformations the curves run close to each other, indicating that a great part of the deformation is elastic. For greater deformations the curves deviate from each other greatly. This is an indication of the effects of loading rate and magnitude of deformation on the mechanical behavior of the material, which must always be kept in mind, since greatly differing forces may otherwise be obtained for the same deformation. The effect of deformation rate may be calculated by applying the three-element model, using eqns (45), (49) or (89). When applying other models a similar method is used.

12.2 Stress-strain relationships

Knowing the loading relationship, stress–strain relationships may be constructed simply if the cross-sectional area and original length of the material loaded are known. However, this will be the case only for axially loaded cylindrical materials. Plant stalks may be regarded as more or less cylindrical, and so compression or tension tests give rise to no problems in this respect.

The shapes of many agricultural products resemble spheres or ellipsoids. If the internal texture of a product can be regarded as more or less homogeneous, then cylindrical test specimens may be cut from it for the purposes of various tests. The results obtained using cylindrical test specimens may be useful in studying general regularities, but they are not always suitable for evaluating the behavior of a whole product. In numerous cases the peel plays an important role in the strength-related behavior of a whole product, and the effect of the peel cannot be investigated by considering a test specimen. In addition, test specimens cannot be prepared at all from several ripe products (tomatoes, raspberries, etc.).

12.2.1 Uniaxial compression

One of the modes of loading applied most frequently to test specimens is uniaxial compression. Several requirements must be met in carrying out these experiments to ensure the validity of the results:

(a) the load must be accurately axial, to avoid the appearance of bending loads;

(b) friction between the end face of the test specimen and the compression plate must be kept as low as possible, so as not to limit transverse elongation of the specimen; and

(c) a length/diameter ratio must be selected at which no risk of deflection yet exists.

In uniaxial compression tests it is primarily the modulus of elasticity of a material which is determined; in certain cases the Poisson's ratio may also be investigated. Test specimens for fruits and root bulb products are typically prepared with a diameter of 1–2 cm, in lengths of 2–3 cm. For cereal grains both ends are cut off to obtain test specimens having a more or less constant cross-sectional area. The test specimen is deformed at a constant rate, generally of a few cm min^{-1}. In testing cereal grains, lower deformation rates are preferred, owing to the small deformation.

The modulus of elasticity is calculated from the theory of elasticity:

$$E = (P/F)/(\Delta L/L) = \sigma/\varepsilon$$

In the calculations the nearly linear initial section of the curve is utilized. Since the materials involved are usually viscoelastic, E may be termed the apparent modulus of elasticity corresponding to a given loading rate, or the modulus of deformation.

A problem in determining the Poisson's ratio is that the test specimen is deformed to a barrel shape, owing to the presence of friction between the pressure plate and the end face of the specimen (Fig. 146) [61]. In this case the equiv-

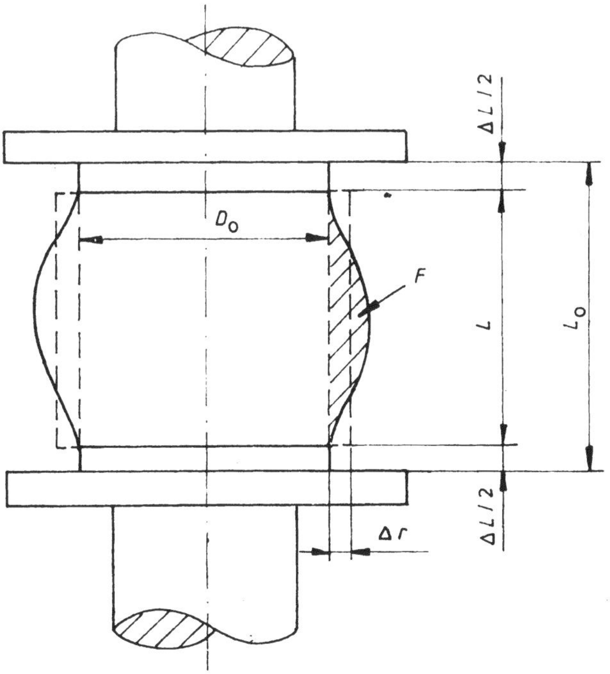

Fig. 146. Determination of Poisson's ratio

alent radial deformation must be determined, whose value is

$$\Delta D = 2\Delta r = 2F/L$$

and with it

$$\nu = (\Delta D/D_0)/(\Delta L/L_0) = \\ = 2L_0 F/D_0 L/(L_0 - L)$$

12.2.2 Uniaxial tension

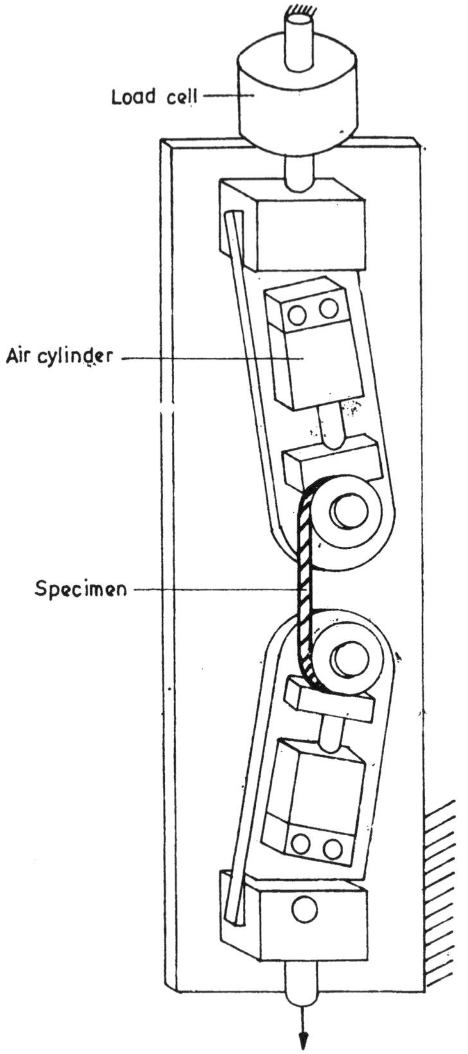

Fig. 147. Device for measuring the tensile strength of fibrous products

The tensile strength of agricultural materials is relatively less important during the various technological processes, and so is measured less frequently. The measurements are also aggravated by several problems, the most important of which is the gripping and holding of a biological specimen so as to avoid pressing the material to a degree which may cause essential changes in its texture. The development of stress collecting regions and bending moments must also be avoided.

Measuring the tensile strength of forage materials is a relatively simple task. Exploiting frictional forces, the long stalks of such materials can be clamped over an appropriate length. Figure 147 shows the principles of construction of a device designed for testing forage materials [59]. Each end of the test specimen is folded over a roller, and clamping jaws are pressed against the test specimen with constant force by a pneumatic cylinder. Figure 148 shows the breaking tensile strength of alfalfa as a function of moisture content. The tests were performed during flowering. The lower moisture content was attained by artificial drying.

The sensitivity to damage of numerous agricultural products is influenced by the tensile strength of their peel, so testing the latter may be of importance in many cases. Figure 149 shows the tensile strength of apple peel as a function of

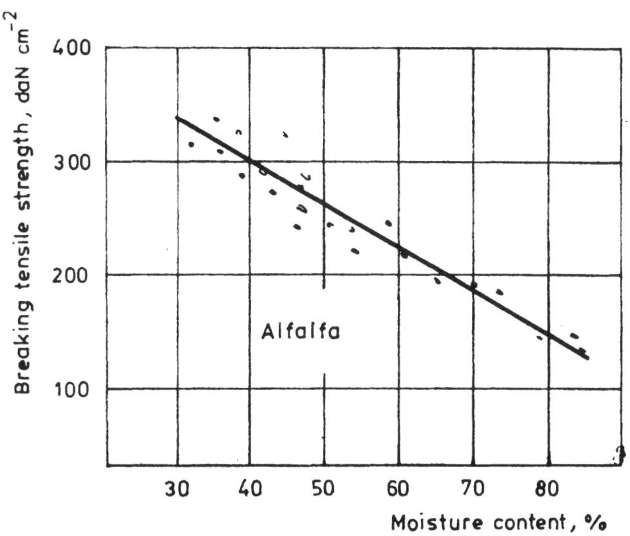

Fig. 148. Tensile strength of alfalfa as a function of moisture content

strain [63]. The relationship is not linear, the apparent modulus of elasticity calculated on the basis of the initial section of the curve is about 140 daN cm^{-2}. The Poisson's ratio of the peel is obtained as 0.32.

Tests of specimens cut from various parts of potato tubers have shown that

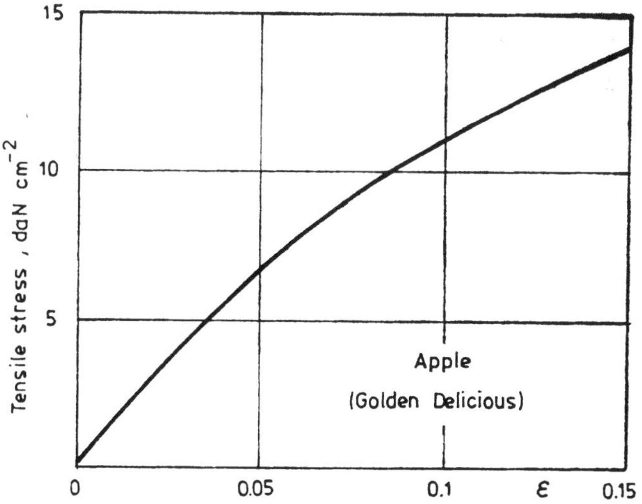

Fig. 149. Stress–strain relationship for apple peels

the modulus of elasticity relative to a given tension is not everywhere identical. The value of E is higher in the central part of a potato tuber than close to the peel. This is probably explained by differences in moisture content between the center and near the peel. This difference in moisture content and the resulting difference in tensile strength may increase during storage.

In testing the tensile strength of maize kernels the apparent modulus of elasticity appears identical to the value obtained under compression ($E = 4100$ daN cm^{-2}), but the breaking tensile strength ($\sigma_{tensile} = 90$ daN cm^{-2}) is far lower than the breaking strength under compression ($\sigma_{break} = 340$ daN cm^{-2}).

12.2.3 Shear

Shear of agricultural materials occurs frequently during various technological processes. In the harvesting of cereals the stalk of the plant is separated by shearing from the root, and shear plays an important role also during comminution processes. When a material is sheared, the direction of shear stress is parallel to the plane of shearing. Shear stress arises as a result of opposing forces acting in the plane of shearing. The condition of pure shear is rather difficult to realize in practice, because of the appearance of bending moments. Shearing may be studied using various devices, depending on the size and shape of the test material. Schematic drawings of the equipment most frequently used are given in Fig. 150.

The shear strength of fruit peels is measured using dies 5–10 mm in diameter. The shorn cross-section is a cylindrical mantle, and the shear stress is given by

$$\tau = P/d\pi h$$

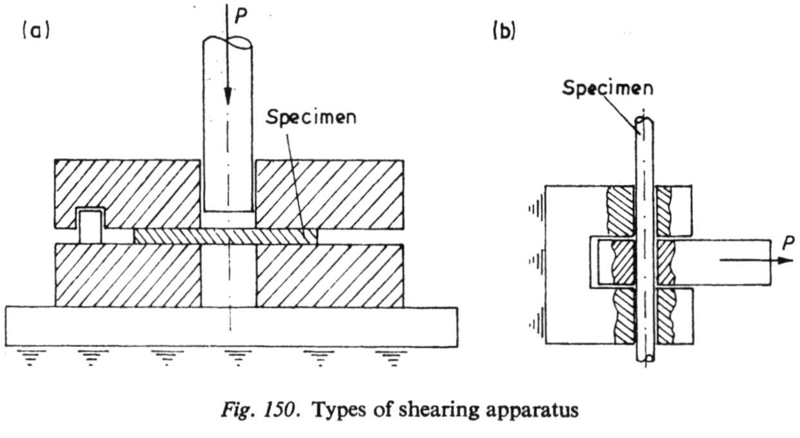

Fig. 150. Types of shearing apparatus

where d is the diameter of the die, and h the thickness of the test specimen. The shear strength of thin slices cut from the flesh of fruits and other materials may be measured in the same way. The shear strength of fruits varies considerably during ripening, and so the shear strength may also be used to indicate the stage of ripening.

The shear strength of cylindrical materials (e.g., plant stalks) is tested by means of a double shearing apparatus. The material to be tested is tucked into variously sized boreholes, and a moving central part shears the material over two cross-sections.

Grains are shorn by flat tools with appropriately formed edges. Brittle fracture may also occur on shearing dry grains, and in this case calculations of shear strength are conditional. The shear strength of grains varies greatly with moisture content. Figure 151 presents the shear strengths of maize and wheat grains and of peas as functions of moisture content. It should be noted that the data for wheat relate to a soft variety.

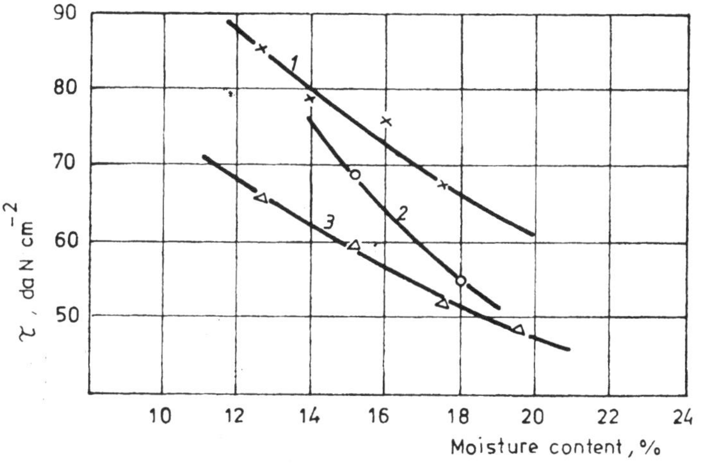

Fig. 151. Shear strengths of wheat, maize and peas as functions of moisture content. *(1)* Maize; *(2)* pea; *(3)* wheat

On shearing forage materials in a greater mass the shear strength depends on the volumetric weight of the material. The greater the volumetric weight, the more dry material must be shorn over a given cross-section, and the higher the shear strength. Figure 152 illustrates the mass shear strength of alfalfa silage as a function of dry content per unit volume [64].

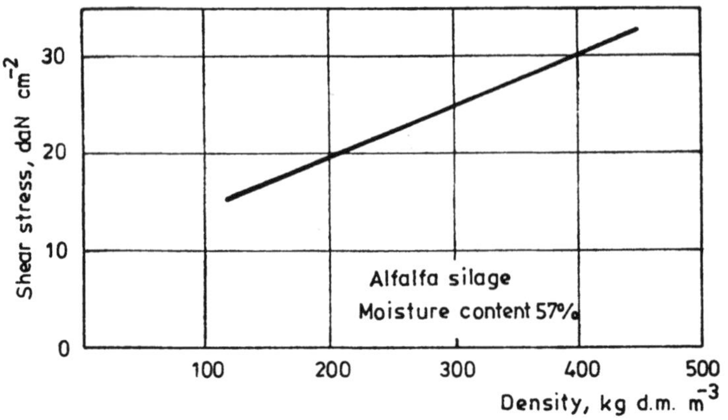

Fig. 152. Shear strength of alfalfa as a function of dry-material content per unit volume

12.2.4 Bending

During the mechanical harvesting of plants with thick stalks the mechanical load on the cutting device is determined decisively by the bending strength of the stalk. The stability of the stalk (its resistance to bending or breaking), which affects harvesting losses, is also related to the bending strength. Plant stalks may be treated, to good approximation, as solid cylindrical bodies or as tubular. Figure 153 illustrates the most important loading cases. The bending strength of stalks is usually determined by the deflection of a two-support beam, by means of the relationship

$$z = PL^3/48IE$$

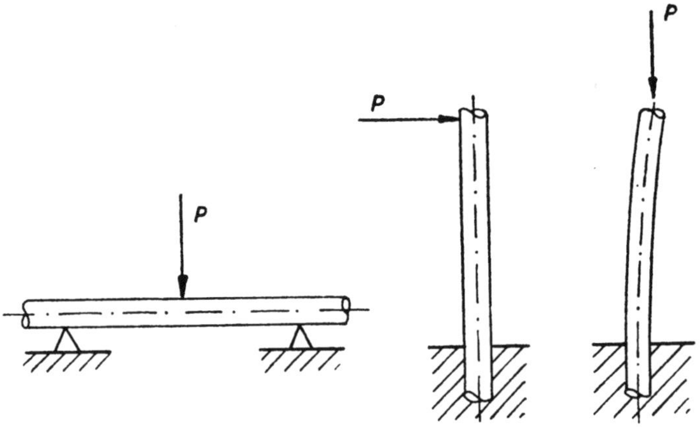

Fig. 153. Various cases of bending

where z is the deflection of the beam, L the length of the beam between the supports, I the moment of inertia of the cross-sectional area relative to the neutral axis, and E the modulus of elasticity of the material.

The bending strength of maize stalks may be tested using the simple device shown in Fig. 154 [65]. The stalk is supported by rollers, the load is applied by a screw spindle, and the displacement may be read directly from the scale found on the spindle shaft. The force is measured by strain gages attached to the beam. The modulus of elasticity E may be obtained from the above equation: in the case of a test specimen of a given length and cross-sectional area, it is

$$E = (P/z)L^3/48I = \text{const } P/z$$

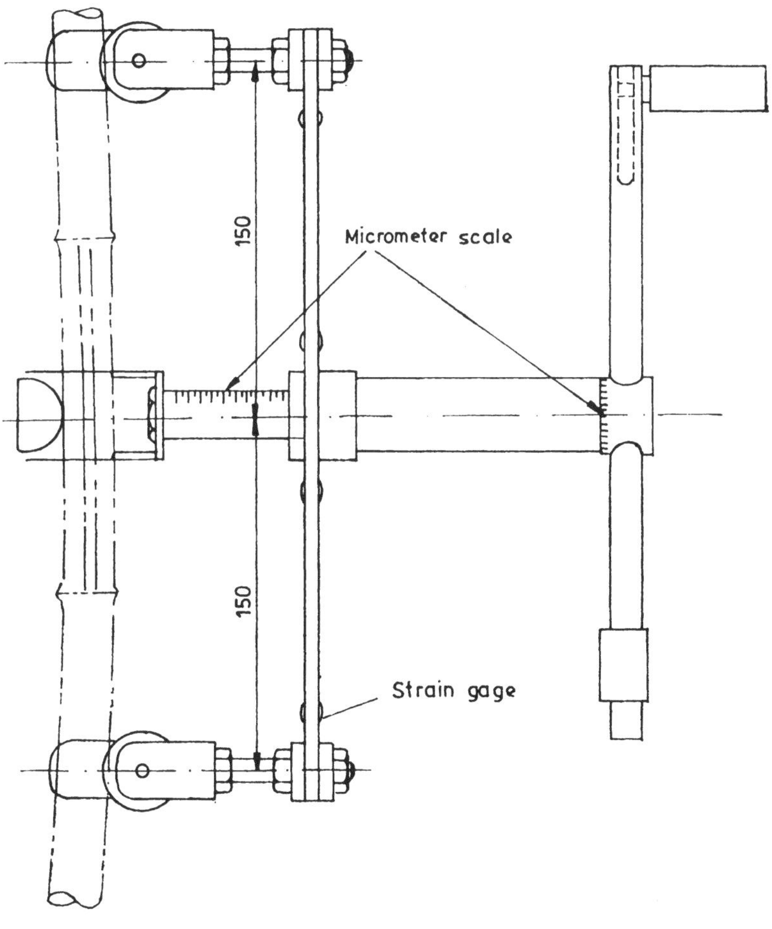

Fig. 154. Apparatus for measuring bending strength

i.e., the modulus of elasticity is proportional to the quotient of the directly measurable force P and the deflection z. Figure 155 shows the quotient P/z for a maize-stalk test specimen of length $L=30$ cm, as a function of the stage of growth. The diameter of the maize stalk varied between 22.5 and 28.6 mm; the measurement results have been reduced to the smaller diameter [66].

In cutting thick plant stalks the part of the stalk remaining in the soil is bent by the moving cutting mechanism. The force required for this bending loads the cutting mechanism, and must be taken into account during estimation. Figure 156 shows the horizontal force required to bend maize stalks cut at a height

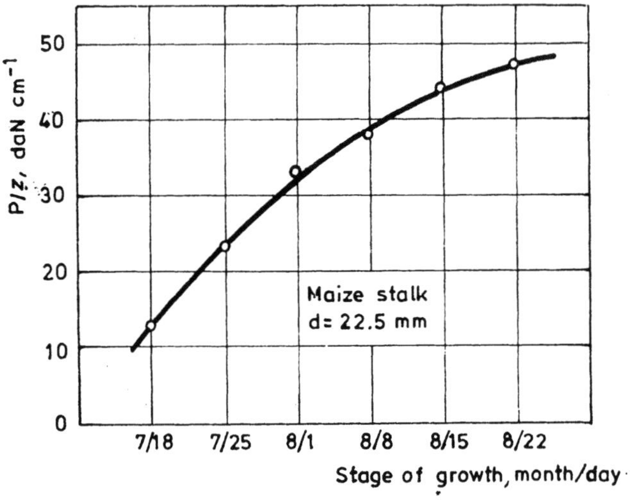

Fig. 155. P/z as a function of growth period for maize stalks

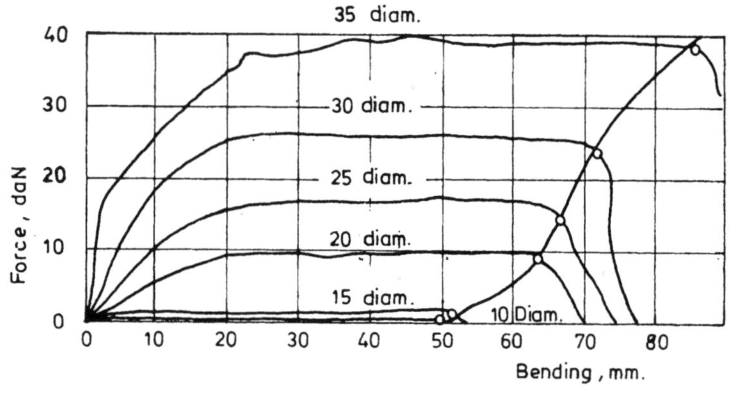

Fig. 156. Bending force as a function of deflection for maize stalks of various thicknesses

of 10 cm above the soil, as a function of the degree of bending. For stalk thicknesses up to 15 mm the force is insignificant, but in the case of stalks 30–35 mm thick a force of 30–35 daN is encountered. In bending sunflower stalks of a similar thickness, the force is greater by 20–30%.

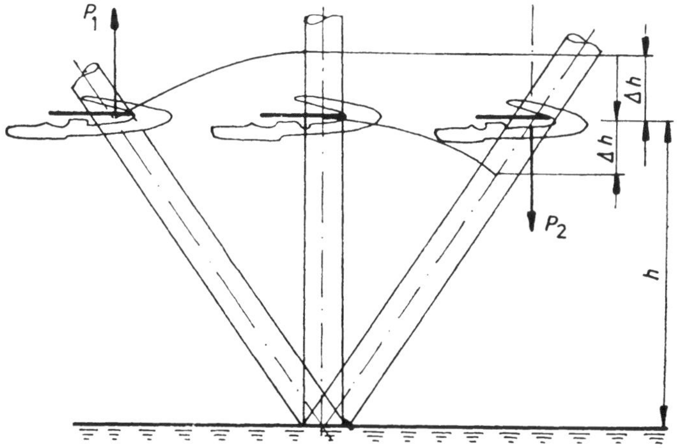

Fig. 157. Displacement of a stalk during cutting

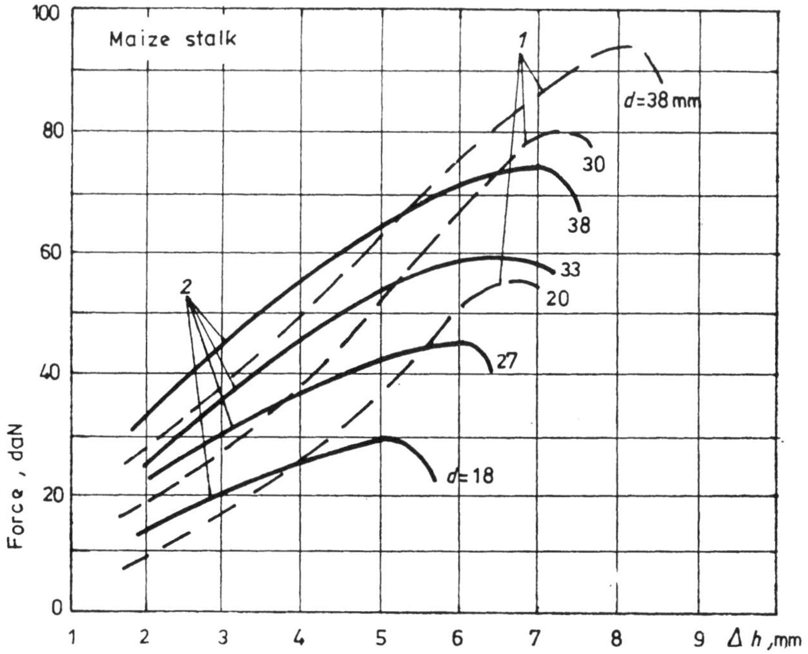

Fig. 158. Vertical force during cutting of maize stalks as a function of deformation. *(1)* Tension; *(2)* compression

The upper end of a stalk during bending will move along a circular arc (Fig. 157), and so the stalk is exposed to the action of a vertical force in addition to the horizontal one. The vertical force may be directed downwards or upwards, depending on the position of the stalk when engaged by the cutting mechanism. Figure 158 shows the vertical force as a function of deformation for various stalk thicknesses [66]. It may be seen that forces of 60–100 daN may be encountered if the stalk thickness is 30 mm or more.

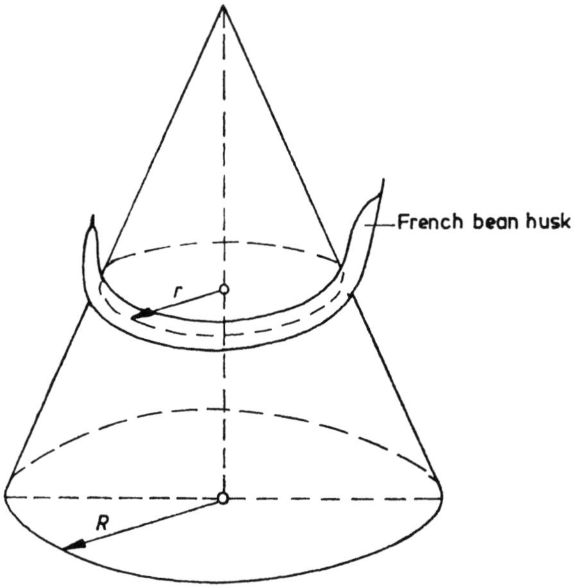

Fig. 159. A simple apparatus for determining the critical bending radius

The physical state of certain vegetables may be assessed by their behavior under a given bending load. It is known that fresh bean pods burst easily on bending (i.e., the load easily exceeds the tensile strength), while bursting is more difficult if they are withered. If a string-bean pod is bent over a conical surface and pushed over the mantle in the direction of smaller radius of curvature until it breaks (Fig. 159), than the radius of curvature corresponding to breaking may be used to characterize its physical state. Knowing this radius of curvature the critical tensile stress arising in the outside fiber may be determined if the modulus of elasticity E is known.

12.2.5 Elastic–plastic behavior of agricultural materials

The loading–unloading curves for agricultural materials show that their deformation is partly elastic and partly plastic. As a result of the plastic deformation the unloading curve does not return to the starting point before loading, i.e., permanent deformation remains. The degree of elasticity of a material is characterized by the ratio of the elastic and total deformations (sum of the elastic and plastic deformations).

The residual plastic deformation is due primarily to the presence of pores, microscopic cracks and discontinuities in the texture. Consequently, the ratio of the elastic and plastic deformations will not remain constant during

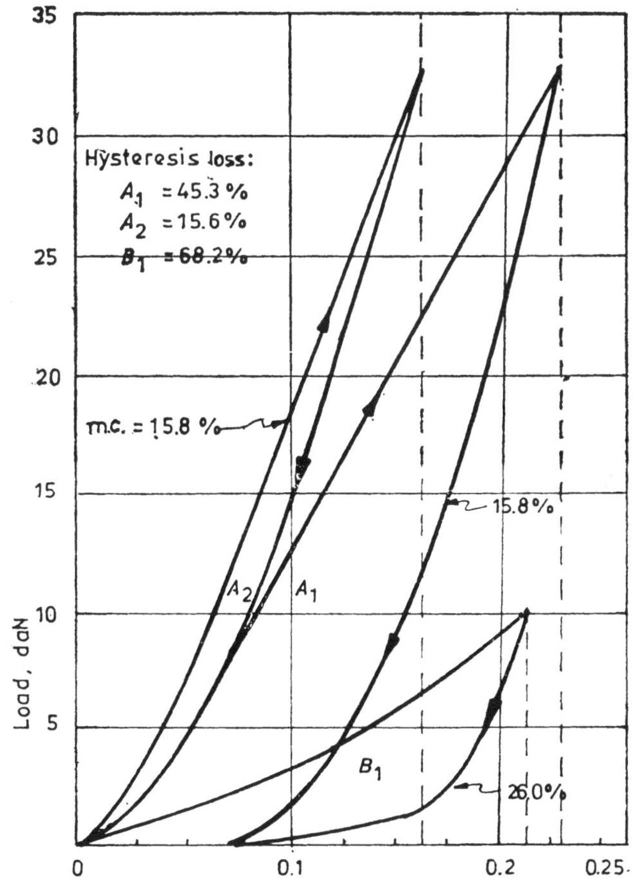

Fig. 160. Hysteresis losses for maize kernels in the first and second cycles of loading

subsequent loading–unloading cycles. Plastic deformation is greatest during the first cycle and decreases in subsequent cycles, as may be seen clearly from Fig. 160, which gives the first and second loading–unloading curves for maize kernels. An influence of moisture content may also be observed here: the plasticity increases with an increase in the moisture content [58].

A loaded material absorbs energy during the loading–unloading cycle as the result of mechanical hysteresis. The energy absorbed is given by the area between the two curves. The damping capacity may also be characterized by the hysteresis loss. The more plastic a material, i.e., the greater the hysteresis loss, the greater its capacity to damp mechanical oscillations. In many cases the plastic properties of agricultural materials are related to their stage of growth or ripening, etc.

The degree of elasticity of agricultural materials also depends greatly on the magnitude of the load or deformation. Figure 161 shows the degree of elasticity of potatoes as a function of relative strain [60]. With increasing deformation, ever more microscopic cracks are formed in the material, whereby its plasticity is increased.

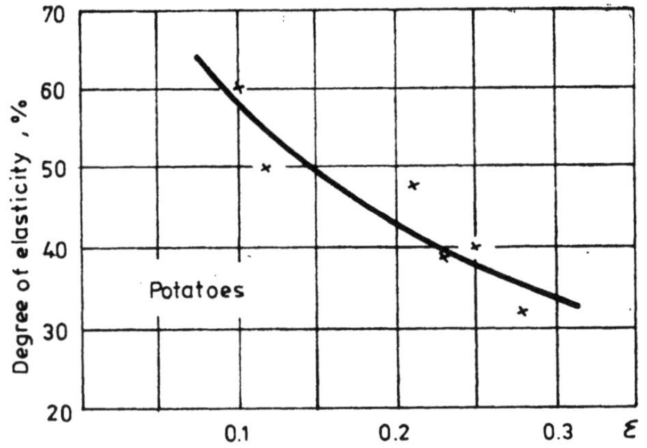

Fig. 161. Degree of elasticity as a function of strain for potatoes

12.2.6 Hydrostatic compression

We speak of hydrostatic compression when a material is exposed to identical pressure on all sides. Under the effect of the pressure the volume of the material decreases by ΔV, and the pressure and change in volume are related according to

$$p = K(\Delta V/V_0) = K\varepsilon_v \tag{134}$$

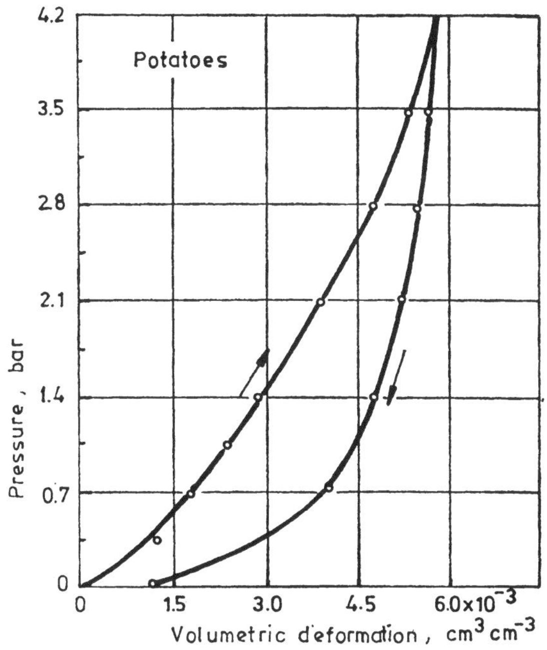

Fig. 162. Hydrostatic compression curve for potatoes

where K is the volumetric modulus of elasticity of the material, and V_0 the original volume. Equation (134) is identical in form to Hooke's law, but here a three-dimensional stress state is involved, where the values of the three main stresses are identical.

Figure 162 shows the loading curve for hydrostatic compression of a potato. The curve becomes increasingly steep with increasing deformation, and so the value of K is not constant but increases with the load. In the central section of the curve $K=700$–800 daN cm^{-2} [60]. For ripe apples a value of $K=35$ daN cm^{-2} has been found: the considerably lower value of K here may be explained by the fact that apples contain a relatively large amount of air [54].

Significant mechanical hysteresis during unloading is also found under hydrostatic compression, as in the case of uniaxial pressure. If the volumetric and linear moduli of elasticity are known from experimental results, the Poisson's ratio ν may be determined by means of the relationship

$$K = E/3(1-2\nu)$$

from which

$$\nu = (3K-E)/6K = 0.5-(E/6K)$$

In certain cases hydrostatic pressure may also be produced within a material and utilized for certain investigations. To test the strength of eggshells water may be injected through a thin needle into an egg, and the pressure required to crack the shell open measured. This critical internal pressure shows close correlation with the critical dropping height at which the egg breaks. Similar tests have also been carried out for tomatoes.

12.2.7 Viscoelastic properties

The behavior of agricultural materials is time-dependent, and they must therefore be studied using rheological methods to determine their elastic and viscous properties. These investigations have been introduced on a wide scale only recently, and so for a relatively great number of materials adequate data are not yet available. The most frequent test methods are stress-relaxation, creep and dynamic tests.

Stress-relaxation tests are performed under constant deformation, during which the reduction of stress over time is measured. The deformation must be effected suddenly (according to a step function) in the material; otherwise, a certain amount of relaxation will occur (before the maximum deformation is attained) which cannot be measured. In practice, sufficiently instantaneous loading is hard to realize if the relaxation time is short.

The test results are plotted on a system of logarithmic σ–t coordinates and the curve obtained is generally not straight, which means that the stress-relaxation cannot be expressed by a single exponential term. The necessary multiterm expression is obtained by the successive residual method.

The essence of the method is illustrated in Fig. 163. The first exponential term is given by the tangent drawn to the original curve for larger value of time. The first residual is not yet straight, and so again a tangent is drawn to it, which gives the second term. The second residual is already sufficiently straight, and this gives the third exponential term. The curve may now be expressed by the equation

$$\sigma = 3.72e^{-t/76}+2.3e^{-t/3.2}+1.0e^{-t/0.5}$$

The relaxation time T appearing in the denominator of the exponent may be calculated from the slopes of the straight lines:

$$T = (t_2-t_1)/(\ln \sigma_1-\ln \sigma_2) = (t_2-t_1)/2.3(\log \sigma_1-\log \sigma_2) \qquad (135)$$

where t_1 and t_2 are two arbitrarily selected times, and σ_1 and σ_2 the stress values corresponding to them. The multiplication factors appearing in the individual terms are the intercepts of the straight lines at $t=0$.

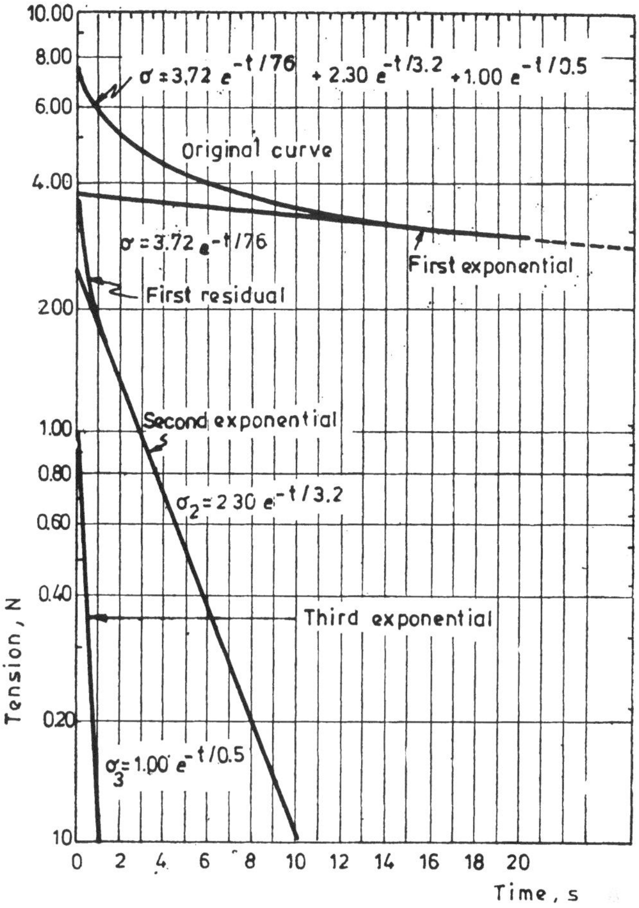

Fig. 163. Explanation of successive-residual method

In practice, in many cases it is not necessary to know the exact course of relaxation over longer time intervals. When pressing hays and straws, the duration of compaction is less than 1 s, and so it suffices to examine the relaxation for a time interval extending to 1 s. In other cases the required interval may be considerably longer, but then an accurate description of the relaxation occurring during the first few seconds is of no interest. In such cases the best-fitting straight-line method may be utilized well.

The essence of the method is now presented in the framework of an example. We have examined the rheological properties of apples when compressed by a die, and measured, after realization of a given deformation, the decrease of stress as a function of time. After a certain time the stress does not decrease any further, and the stress relaxation, with the residual stress σ_e taken into account, is described by eqn. (37):

$$\sigma(t) = \sigma_d e^{-t/T} + \sigma_e$$

Instead of the stress the modulus of elasticity E may be used, which is proportional to the former. Then

$$E(t) = E_d e^{-t/T} + E_e$$

or

$$E(t) - E_e = E_d e^{-t/T}$$

On the basis of measured values the logarithm of $E(t) - E_e$ may be plotted as a function of time (Fig. 164). The curve is not exactly straight, and so the straight line best approximating the curve must be found by the method of least squares. The relaxation time T may be determined from the slope of the straight line, and the ordinate corresponding to time $t = 0$ gives the modulus of decay, E_d. The equation of the straight line is

$$E(t) = 4.95 e^{-t/13.8} + 2.94$$

where the time t must be substituted in minutes (i.e., the value of T is 13.8 min). It may be seen from Fig. 164 that for longer time periods the difference between the slopes of the actual and calculated curves is not great, and so there will not be any significant deviation between the effective and calculated relaxation times. However, in the case of shorter periods (less than 1 min), the deviation is considerable. If for a given technological process only this part of the plotted curve is to be approximated, then it is necessary to adopt the values $T = 3.5$ min and $E_d = 5.6$ daN cm^{-2}. Conversely, incorrect results are obtained for long periods using these data (see the dashed line in Fig. 164).

To describe stress relaxation it is also possible to utilize the equations of the four-element Burgers model or of the three-element model. The four-element model, obtained by connecting the Maxwell and Kelvin models in series, cannot generally be applied for longer relaxation times, since the dashpot element η_1 permits total relaxation of the system. This is not attained in practice, and so the three-element model, which does not contain the dashpot element η_1, is used more frequently; the corresponding equations are also simpler. In practice, the loading preceding relaxation occurs at a finite rate in many cases. This results in a certain amount of relaxation also taking place during loading. The longer the loading time, the more relaxation occurs during loading. Consequently the relaxation after loading depends on the loading rate. The effect of a finite loading rate on relaxation may be examined by means of eqns. (45), (45a), or (49), (49a),

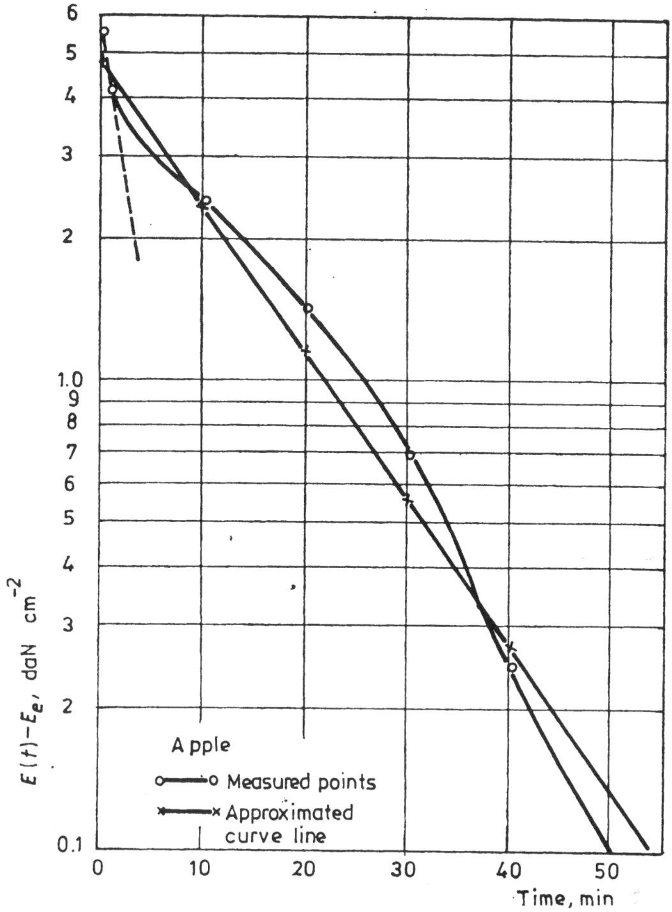

Fig. 164. Determination of best-fitting straight line, in calculation of the relaxation time

and (89). The deviations calculated on the basis of these equations are supported by experimental data. Figure 165 shows the relaxation curve for a potato subsequent to loading at various rates. It may be seen that the loading rate greatly affects the measured relaxation curve. Disregarding the effect of loading rate may lead to considerable error. The figure also shows that the effect of loading rate asserts itself mainly during the first few seconds, after which the curves run nearly in parallel [60].

The phenomenon of creep occurs during numerous technological processes (the settling of silage and granular piles, the deformation of fruits by their dead weight, etc.); therefore viscoelastic characteristics are required for dimensioning in these cases. During creep tests the material is loaded suddenly (according to a step

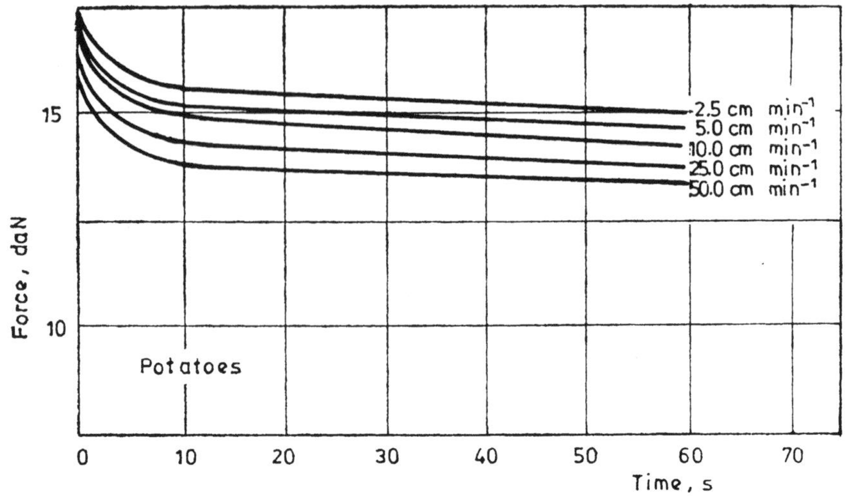

Fig. 165. Relaxation of potatoes for various loading rates

function) and the load kept constant for the duration of the test. The increase of the deformation is measured as a function of time, and the results are processed in the form of the time-dependent parameters $E(t)$, $G(t)$ and $K(t)$, depending on whether tension or compression, torsion, or volumetric compression is involved.

The rheological models of creep are the Kelvin model, the three-element and the four-element (Burgers) model. The behavior of these has been described in detail in Section 8.9.

Besides the behavior of the material, the time interval for which the model is valid (i.e., for which it describes the creep of the material with satisfactory accuracy) also greatly affects the applicability of the individual models. This time interval is less than 1 min in the case of many technological processes, but settling periods during storage may be as long as 6–12 months.

In the case of shorter time intervals the four-element model, whose elements may be determined on the basis of Fig. 94, may often be applied conveniently. To determine the retardation time T_r, eqn. (52) may be written in the simpler form

$$A = B(1-e^{-t/T_r})$$

where

$$A = \varepsilon(t)-\sigma_0/E_0-\sigma_0 t/\eta_v$$

and

$$B = \sigma_0/E_r$$

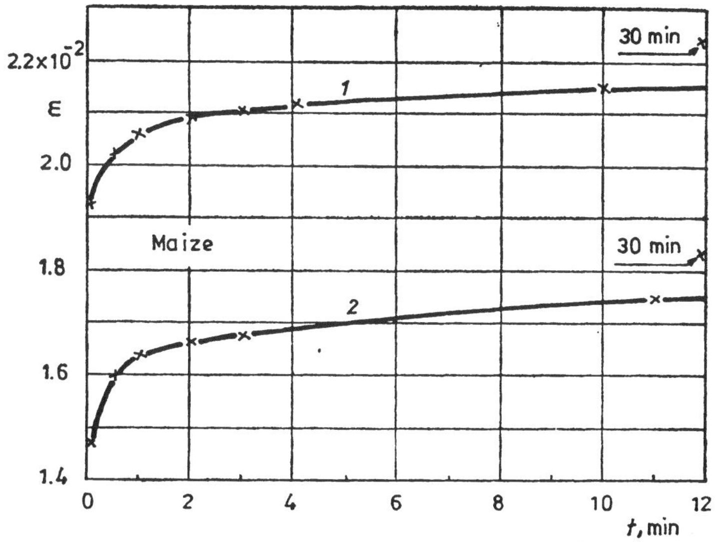

Fig. 166. Creep of maize kernels after multiple loading–unloading *(1)* (14 daN) and during the first loading cycle *(2)* (11 daN)

Taking logarithms of this equation gives

$$\log(1 - A/B) = t/2.3T_r$$

Values of A are plotted, utilizing the data of the curve, in a semilogarithmic system of coordinates as a function of time. The slope of the straight line obtained is $1/2.3T_r$, from which T_r may be calculated. Since $T_r = \eta/E_r$, η can also be obtained by calculation.

Figure 166 shows creep curves for maize kernels in the edge position under constant loads of $P = 11$ and 14 daN, as functions of time [67]. The deformation increases within the first 30 min to 1.25 times its initial value, further increase is slower. The curve may be described approximately by eqn. (41):

$$\varepsilon(t) = \varepsilon_0 + \varepsilon_d(1 - e^{-t/T_r})$$

where T_r is the retardation time. $T_r = 4$ min for the curves given in Fig. 166, and as an example, the equation for curve *(1)* may be written as

$$\varepsilon(t) = 1.92 \times 10^{-2} + 0.32 \times 10^{-2}(1 - e^{-t/4})$$

Cereals are generally stored and treated in the loose bulk state; the phenomenon of creep must therefore be studied for this case also. Figure 167 shows creep curves for loose bulk wheat and maize under a constant load of $p = 1.65$ daN cm^{-2}, as functions of time [67]. The retardation time T_r is 2.25 h

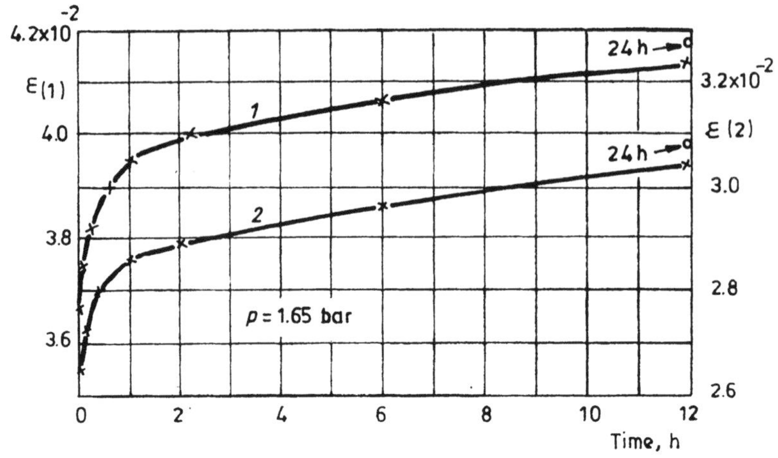

Fig. 167. Creep curves for *(1)* loose bulk wheat and *(2)* maize

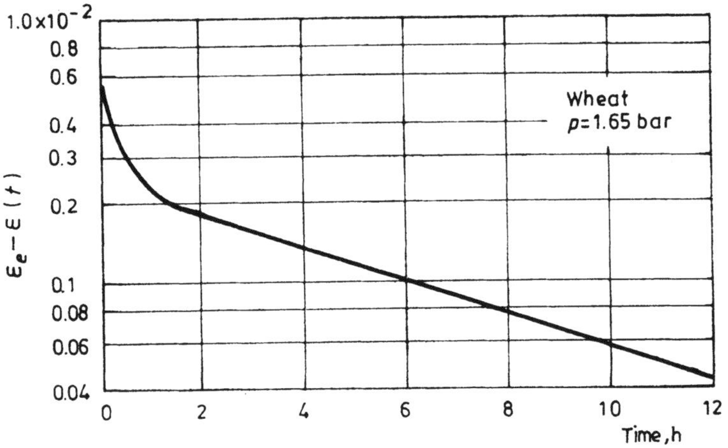

Fig. 168. Plot of $\varepsilon_e - \varepsilon(t)$ as a function of time

for wheat, and 3 h for maize. As may be seen, the retardation time for individual grains is much shorter than that for a grain pile. One of the reasons for this is most certainly the friction between the grains during creep (settling) in a pile.

Equation (41) describes the creep curve only approximately. Significant deviations are found, primarily at the beginning of the process. Figure 168 shows $\varepsilon_e - \varepsilon(t)$ plotted as a function of time on a system of semilogarithmic coordinates. The fact that the relationship is not linear means that it may be described only by a multiterm exponential function, similar to eqn. (53) [67].

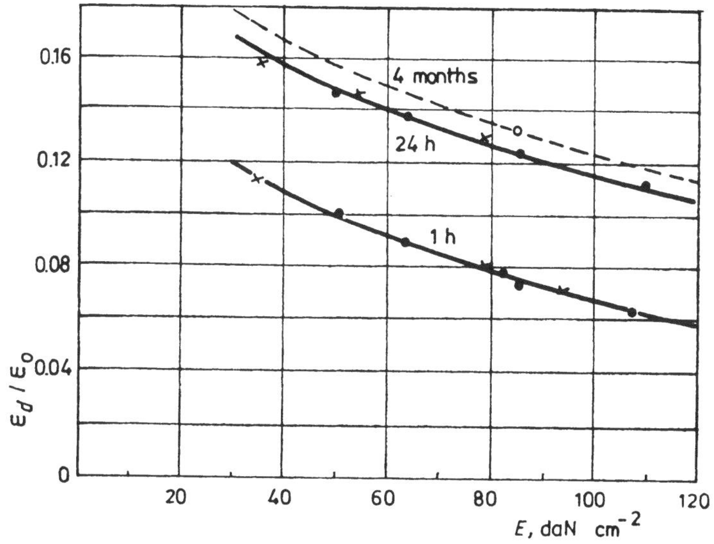

Fig. 169. General relationship between ε_d and ε_0 for creep of cereals in bulk

Tests performed for various bulk cereals (wheat, maize and barley) have shown that ε_d has a defined relationship to the initial deformation ε_0 and the modulus of elasticity E of the material. This general relationship is shown in Fig. 169 for creep processes of various time periods [67]. The figure permits important conclusions to be drawn. The cereals can be characterized as concerns creep by their modulus of elasticity E relative to a given pressure. The curves are thus valid for wheat, maize and barley alike, under any arbitrary load. A determined relationship also exists between the amounts of creep occurring in 1 h and during an infinitely long time period: the 1 h creep amounts to about 60% of the maximum value. This offers the possibility of reasonably accurate conclusions on the basis of the 1 h value as to the expected maximum value.

The modulus of elasticity of bulk cereals increases under the effect of a load, and this fact must be taken into account in dimensioning storage operations. Figure 170 shows the modulus of elasticity of bulk piles of maize, and wheat as a function of pressure after a sufficiently long consolidation period.

The creep properties of maize silage with 68% moisture content have been examined using a triaxial apparatus. On processing the experimental results the volumetric and shear creep compliances were determined, and it was found that they could be approximated by the equations [82]

$$J_v(t) = p^{-0.4527}(0.3282+0.112\times10^{-5}t-0.0714e^{-6.526t})$$

and

$$J(t) = 1.1846+0.2255\times10^{-2}t-0.2e^{-1.989t})$$

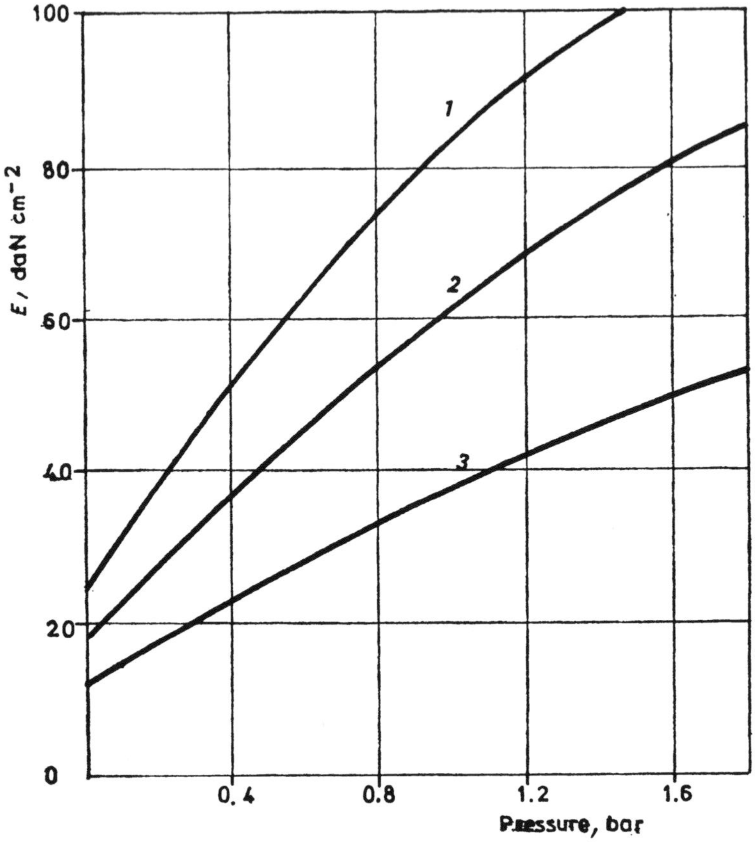

Fig. 170. Young's modulus of various crop products in bulk as a function of pressure. *(1)* Maize; *(2)* wheat; *(3)* barley

where p is the hydrostatic pressure (daN cm^{-2}), and t is time (days). The expressions for the corresponding relaxation moduli are

$$G_v(t) = p^{0.4527}(3.046e^{-0.0034t} + 0.848e^{-8.343t})$$

and

$$G(t) = 0.8439e^{-0.0019t} + 0.1725e^{-2.39t}$$

The above functions apply to specimens of 770–800 daN m^{-3} initial volumetric weight. In practice, values of 250–300 daN m^{-3} occur in a silo at the start of filling, and so the relaxation modulus values must be reduced in the corresponding calculations. The reduction may be performed by introducing a density coefficient of the form

$$k = (\gamma/\gamma_0)^n$$

where γ and γ_0 are the volumetric weights respectively in a given case and in the tests. The value of the exponent n may be accepted, on the basis of a limited number of data, as 2.75. However, additional research is still required. The above creep and relaxation moduli have been used to calculate the pressure distribution in storage silos by the finite-element method.

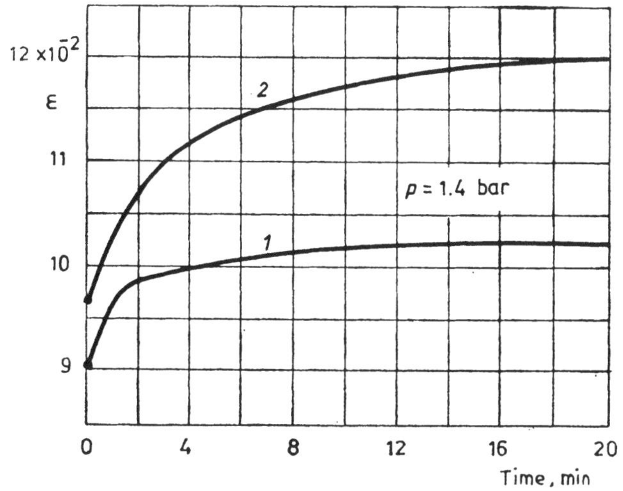

Fig. 171. Creep curves for granular superphospate *(1)* and potassium fertilizer *(2)*

The creep properties of chaff are greatly influenced by its quality. According to investigations, the initial volumetric weight of shredded chaff is greater than that of cut chaff and subsequent settling (creep) occurs more rapidly in the former case, which is favorable from the point of view of storage. Therefore, attention has recently been focussed increasingly on the shredding of maize stalks.

Various loose bulk fertilizers also settle considerably, i.e., creep appears in them also. Figure 171 shows creep curves for granular superphosphate and potassium fertilizers under a constant pressure of $\sigma_0 = 1.4$ daN cm^{-2}. The granular superphosphate has coarser grains which are nearly spherical. The initial modulus of elasticity is $E_0 = 15.7$ daN cm^{-2}, the retardation time is $T_r = 1.3$ min. The potassium fertilizer is fine-grained, and the grains are not spherical. The initial modulus of elasticity is $E_0 = 14.5$ daN cm^{-2}, the retardation time is $T_r = 3.4$ min. From the above it may be established that a pile consisting of relatively large grains of spherical shape, consolidates more rapidly than a pile made up of tiny, nonspherical grains.

12.2.8 Dynamic testing

Recently, dynamic testing of agricultural materials has propagated rapidly, since a considerable proportion of the loads encountered in processing have a dynamic or repeated dynamic form. Viscoelastic properties may also be determined from dynamic tests (see Section 8.11).

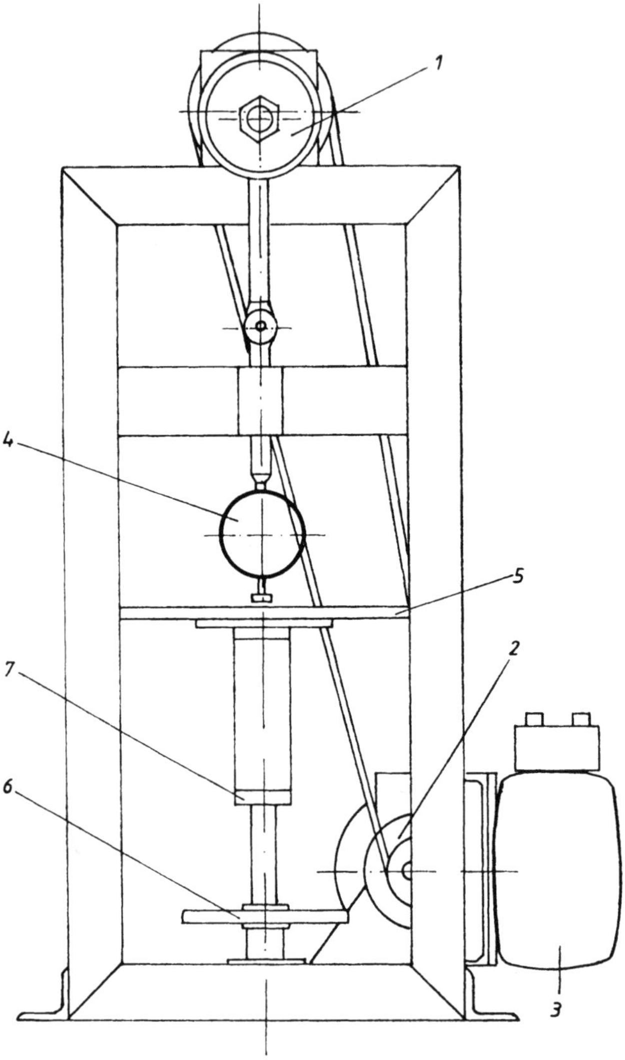

Fig. 172. Dynamic test apparatus for oscillatory loading. *(1)* Excentric drive; *(2)* countershaft; *(3)* motor; *(4)* measuring ring; *(5)* table; *(6)* adjusting wheel; *(7)* fixing nut

A highly important function of dynamic tests is the determination of deformation and damage for fruits under the effect of dynamic oscillating forces. At the Agricultural College of Körmend (Hungary) a special dynamic test apparatus has been developed for these tests [71]: its construction is illustrated schematically in Fig. 172. The stroke of the loading device may be adjusted continuously between 0 and 12 mm, its maximum speed is 900 rpm. Spherical indenters of various diameters, or plane plates, are applied as the loading elements. The apparatus is equipped with the necessary electrical measuring instruments.

The measurements thus far have primarily concerned dynamic oscillation tests of various apple species, using a rigid ball 60 mm in diameter or a plane plate, and the rigid ball could be replaced by an apple (i.e., tests of impact between two apples). During the tests the deformation or force of impact was kept constant; the number of cycles varied between 1 and 10^4.

For constant deformation, the theoretical value of the deformed volume during contact of a plane plate with a sphere is

$$\Delta V_{th} = (\pi/3)z^2(3R-z) \tag{136}$$

and for contact of two spheres (one of which is rigid) it is

$$\Delta V_{th} = (2\pi/3)(z/2)^2(3R-z/2) \tag{136a}$$

where z is the deformation, and R the radius of the sphere.

The volume deformed in reality, i.e., in which biological yield occurs will be larger than the theoretical value and also depends on the number of loading cycles. Figure 173 shows the theoretical value of ΔV_{th} and the volume actually damaged for Jonathan and Golden apples as functions of deformation for $i=1000$ loading cycles. The loading was performed using a sphere 60 mm in diameter [72, 73].

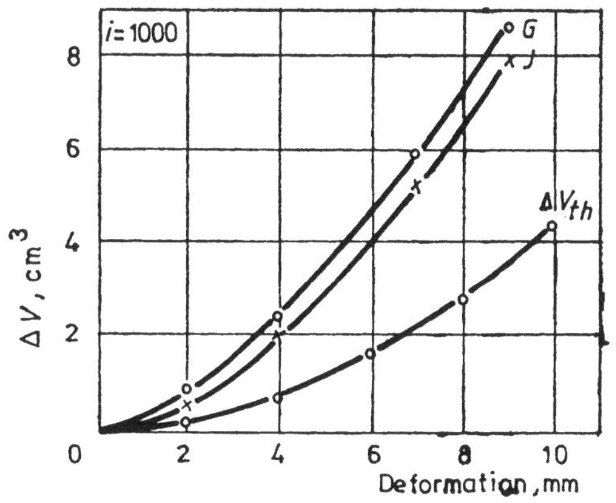

Fig. 173. Theoretical deformed volume and volume actually damaged during 1000 loading cycles, for Jonathan and Golden apples as functions of deformation

The magnitude of the force appearing during the first loading cycle may be determined from the equation

$$P = [E/(1-\nu^2)]\, z^{1.5}/0.752(1/R+2/d)^{1/2}$$

where E is the modulus of elasticity, ν is Poisson's ratio, d the diameter of the loading ball, and R the radius of curvature of the apple. It is convenient to determine the modulus of elasticity together with Poisson's ratio. For Jonathan and Golden apples, respectively, we have measured at the beginning of the harvesting season values of $E/(1-\nu^2)=32$ and 28 daN cm^{-2}. By the end of the harvesting season, reductions in these values by 8–10% and 15–20%, respectively, were observed.

In cycles subsequent to the first, the stress and force decrease continuously for viscoelastic materials owing to relaxation. Figure 174 shows the variation of the

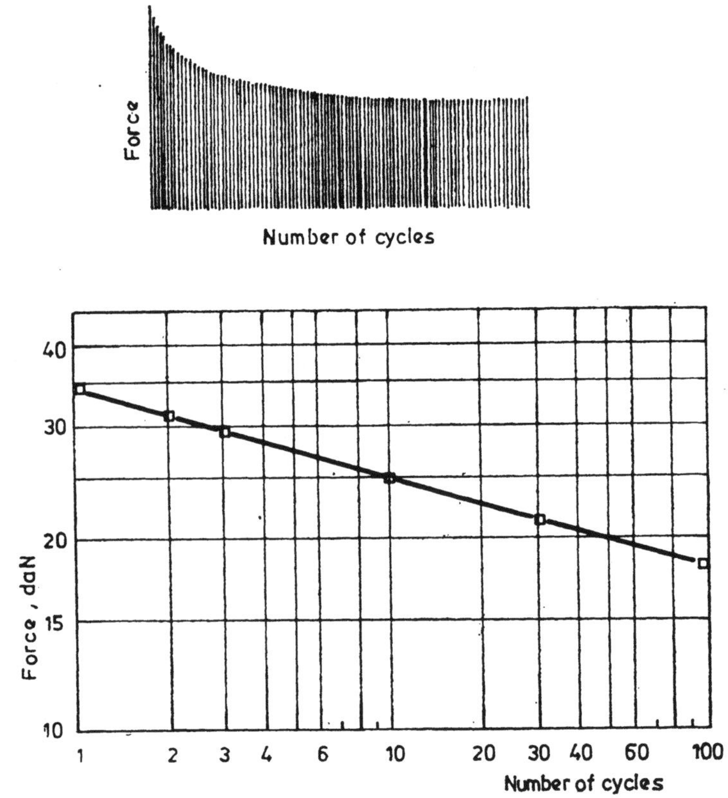

Fig. 174. Peak value of force during first 100 loading cycles at constant deformation for Jonathan apples

peak force value for Jonathan apples during the first 100 cycles ($P_{max}=34$ daN, $P_{min}=18$ daN, $z=9$ mm). The relaxation of the peak force value may be obtained from the equation

$$P(i) = \sum_{k=1}^{n} a_k e^{-(i-1)/I_k} \tag{137}$$

where i is the number of cycles, and I_k the cycle number of relaxation. For the curve in Fig. 174, three terms supply sufficient accuracy for 100 cycles:

$$P(i) = 22e^{-(i-1)/443} + 8e^{-(i-1)/10.5} + 4e^{-(i-1)/1.92}$$

The relative reduction of the peak force value as a function of the number of cycles depends greatly on the magnitude of the deformation. Figure 175 shows the relative decrease of the peak value for Jonathan apples as a function of deformation, for various numbers of cycles. The curves may be approximated by the empirical formula

$$P(i)/P_1 = 1-k(i)\,z^n$$

where the value of the exponent n is obtained as 0.6–0.65, while the constant k is a function of the number of cycles.

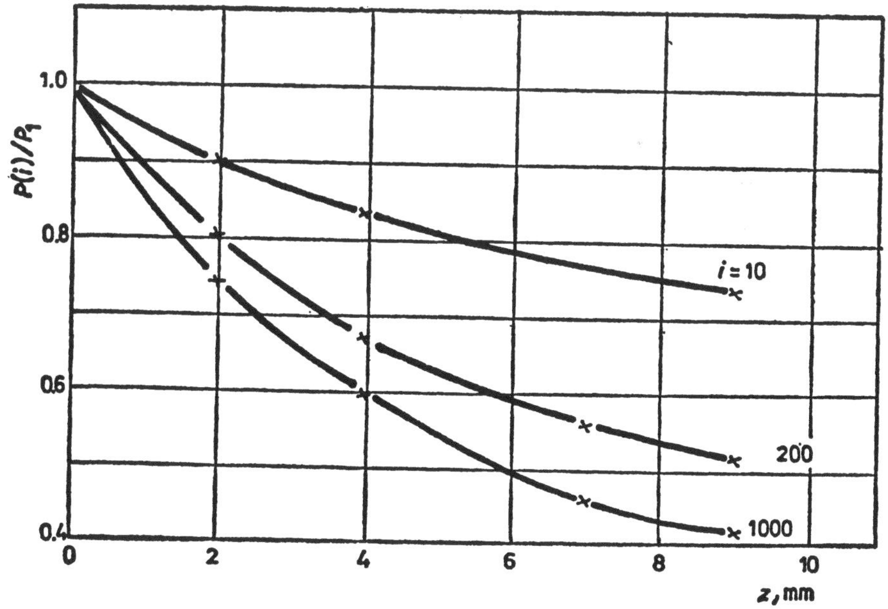

Fig. 175. Relative reduction of peak value of force as a function of deformation, for Jonathan apples

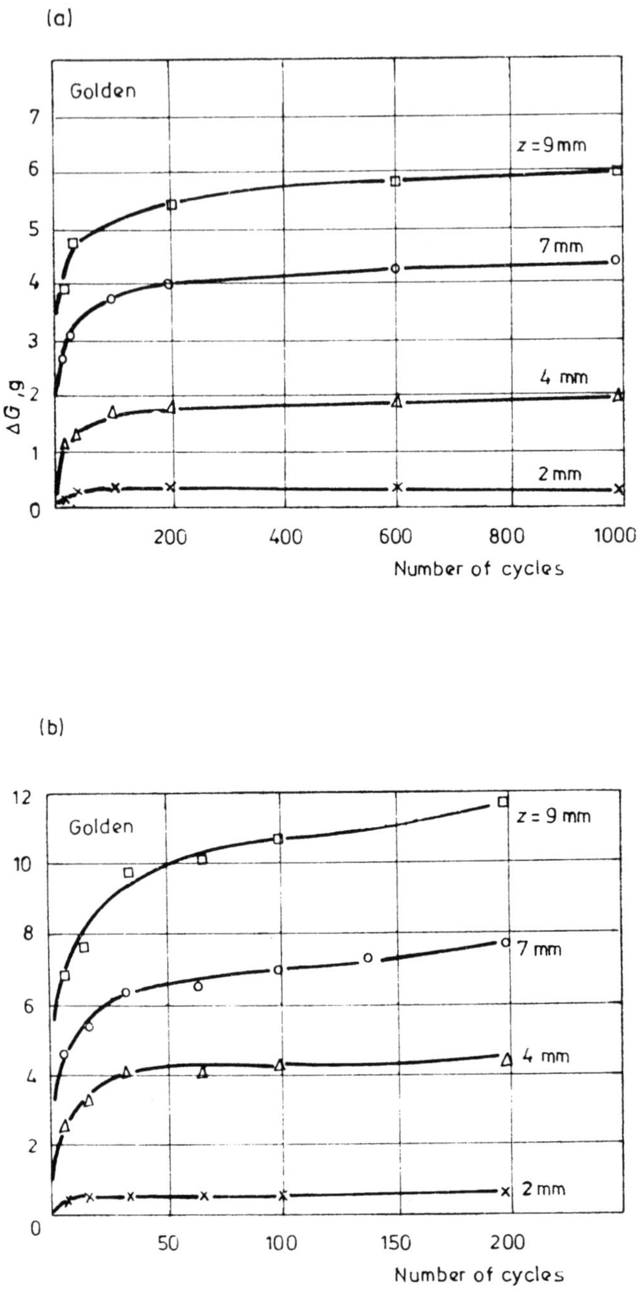

Fig. 176. Weight of damaged volume as a function of number of loading cycles for Jonathan apples at the beginning *(a)* and end *(b)* of ripening

Application of eqn. (137) is followed by complications in many cases, and so it is advisable to use an equation of the form

$$P(i) = P_1 i^{-n}$$

where the exponent n depends on the deformation; its value in the majority of cases varies between 0.06 and 0.1.

For constant deformation, the damaged volume attains nearly its maximum value during the first 100 cycles, and increases only slightly during subsequent cycles. Figure 176 shows the weight of the damaged region as a function of the number of cycles, for various deformations [72].

In practical applications, the case where the load remains constant during subsequent cycles is of the most interest. In this case the deformation occurring in the first cycle increases during subsequent cycles, owing to the creep of the material. Figure 177 presents deformation curves for Jonathan apples as functions of the number of cycles, for various loads. In the first cycle the deformations are 1.6, 2.8 and 6 mm. In the following cycles the deformations increase rapidly; the curves then become flat. The deformation may be divided into two parts: that occurring in the first cycle, and that occurring due to creep in the following cycles:

$$z = z_1 + z_c$$

The deformation in the first cycle for contact between two spheres is

$$z_1 = \{[0.752(1-\nu^2)P/E](1/R+2/d)^{1/2}\}^{2/3}$$

while the deformation due to creep, since the material involved is not linearly

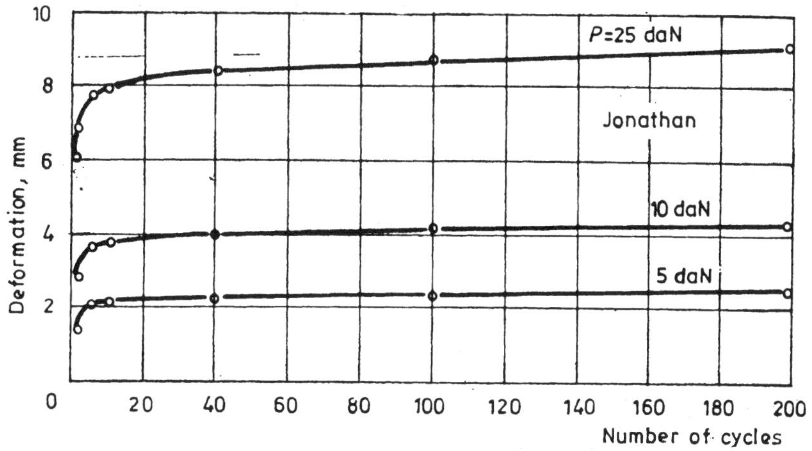

Fig. 177. Creep curves for Jonathan apples as functions of the number of cycles for various loads

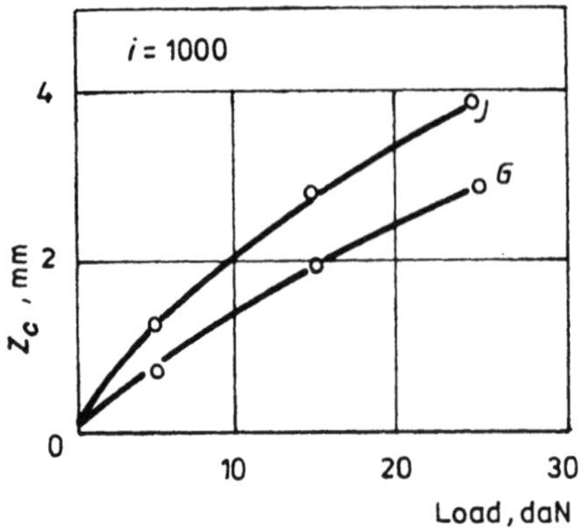

Fig. 178. Creep deformation as a function of load for Jonathan and Golden apples

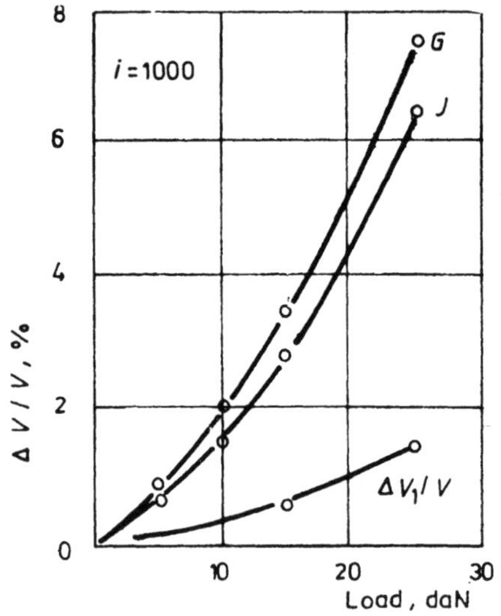

Fig. 179. Relative damaged volume as a function of load for Jonathan and Golden apples

viscoelastic, may be best approximated by the empirical relationship

$$z_c = A(i-1)^n P^m$$

By appropriate processing of the experimental data, we obtained for Jonathan and Golden apples at the start of harvesting the respective equations

$$z_c = 0.0127(i-1)^{0.15}P^{0.75}$$

and

$$z_c = 0.0095(i-1)^{0.15}P^{0.75}$$

where i is the number of cycles, and P the loading force. Figure 178 shows the variations of the creep deformation with load. An interesting result is that the creep deformation of Golden apples is slightly lower than that of Jonathan apples, although the moduli of elasticity of the two species are similar, (in fact somewhat lower in the case of Golden apples).

The damaged volume is almost proportional to the square of the deformation, and so the relative value of the damaged volume is composed, similarly to the deformation, of two terms:

$$\Delta V/V = \Delta V_1/V + B(i-1)^{2n}P^{2m} \quad (138)$$

where $\Delta V_1/V$ is the theoretical value of the relative volume damaged (deformed) during the first cycle. On processing the experimental data the following equations were obtained for Jonathan apples:

$$\Delta V/V = \Delta V_1/V + 7.18\times10^{-3}(i-1)^{0.25}P^{1.5}$$

and for Golden apples,

$$\Delta V/V = \Delta V_1/V + 8.75\times10^{-3}(i-1)^{0.25}P^{1.5}$$

Figure 179 illustrates the above equations graphically. As may be seen, the damage to Golden apples is somewhat more extensive than that to Jonathan apples, despite the lower creep deformation of the former. This phenomenon may be explained by other differing properties of the texture, namely, the compressibility and energy-absorption properties.

The form of the deformation zone is more favorable on loading by a plane plate than by a sphere, and so the volume damaged in the case of identical loads will be less extensive in the former case. Figure 180 shows the damaged volume as a function of number of cycles, for identical loadings by a rigid sphere and a plane plate. The experimental results support the preceding assumption.

During the progress of ripening, the mechanical properties of apples vary. As has been seen earlier, the modulus of elasticity for Golden apples decreases by 15–20% during the harvesting season of about 5 weeks. A proportional increase in sensitivity to damage might be expected. However, the investigations show that the damaged volume increases to a much greater extent. Figure 175 presents measurement results obtained at the end of the season for Golden apples, as a function of the number of cycles, for various deformations. On comparing these with similar data obtained at the beginning of the season, it may be seen that the volume (or weight) damaged approximately doubles in value by the end of the season. Such a great increase in damage is not observed in the case of Jonathan apples.

The investigations also supply another result of interest. After a great number of cycles ($i > 100$), the damaged volume under the spherical indenter becomes completely soft, and so the stress is greatly equalized over the deformed surface area. The mean stress developing after a great number of cycles is the characteristic of the load-bearing capacity of the sound tissue found under the deformed

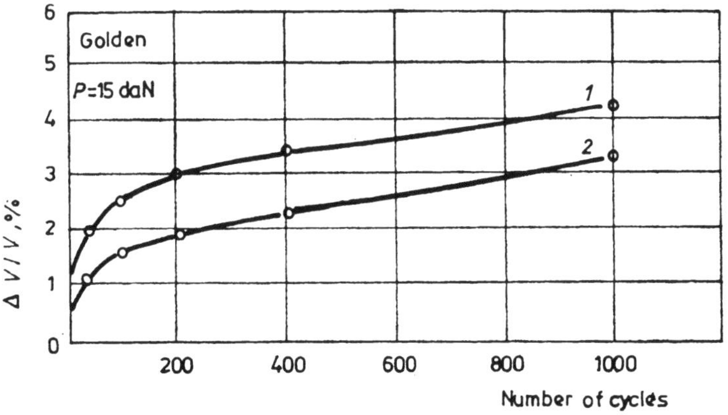

Fig. 180. Relative damaged volume as a function of number of cycles for loading by a ball *(1)* and a flat plate *(2)*

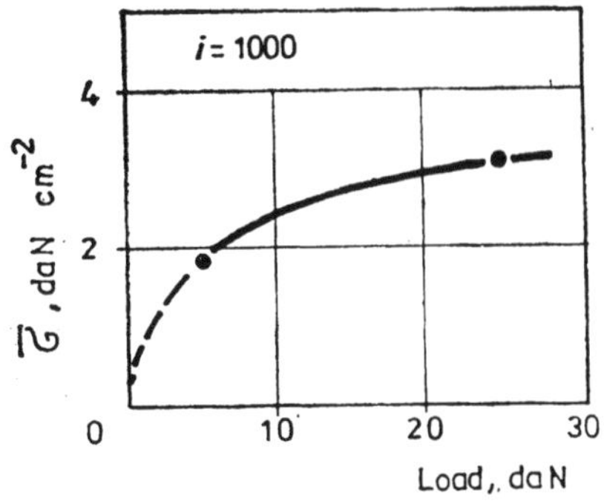

Fig. 181. Mean compressile stress as a function of load after 1000 cycles, for apples

part. We determined the mean stress as a function of loading force for $i = 1000$ cycles on the basis of the measured results (Fig. 181). As may be seen, the mean stress varies between 2 and 3 daN cm^{-2} for a loading interval of 5–25 daN.

From earlier static tests it is known that the biological yield limit of apples is found at stresses of 4–5 daN cm^{-2}. The values obtained above permit the conclusion to be drawn that in the case of dynamic oscillating loads, the permissible stress in apples is lower than the static biological yield limit.

During oscillating impact between two apples, the total damaged volume and the ratio of the two damaged volumes depend on the moduli of elasticity of the individual fruits. If the moduli of elasticity are identical and the stages of ripening are also similar, then the two damaged volumes are practically of equal magnitude. If the moduli of elasticity differ, then the damaged volumes may differ significantly. If the modulus of elasticity of one apple is 1.7–2.0 times higher than that of the other, then the apple with the lower modulus of elasticity will only be damaged even after a great number of cycles. A difference in the radii of curvature around the contact surfaces will also contribute to inequality of the damaged volumes: the fruit with the smaller radius of curvature will be in a more favorable stress state (considering the tangential stresses on the surface), and so will be damaged less or remain undamaged; furthermore, the loading conditions for the less damaged apple improve in subsequent cycles. Our experiments showed that if an apple is not damaged during the first few cycles, it will remain sound even after 10^4 cycles, and only the damage to the other apple increases with the number of cycles.

Consequently, the total deformation due to creep during oscillating impact depends on the ratio of the deformations of the two bodies involved. If the two individual fruits are deformed equally, the total deformation will be maximum; if to all intents only one of them is deformed, the total deformation will be minimum. Figure 182 shows the creep deformation of Golden apples as a function of load for two numbers of cycles. The band indicates the possible values of the total creep deformation depending on the ratio of the partial deformations. The creep deformation increases in proportion to load to the power of 0.7–0.75. If two spherical bodies are in contact, then

$$z/R = 1.04(PA/R^2)^{2/3}$$

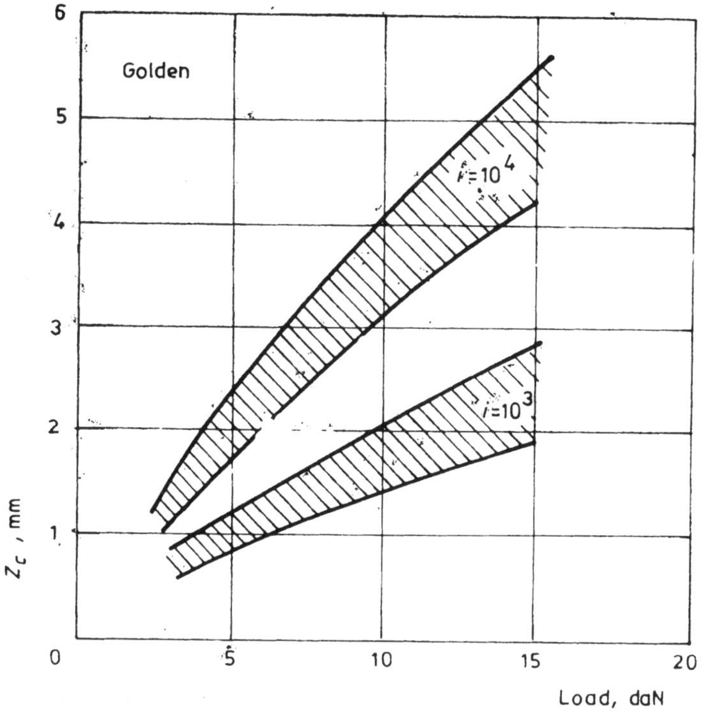

Fig. 182. Creep deformation during impact of two apples as a function of load

on the basis of Hertz's theory, while in the case of i cycles an equation of the following form may be used:

$$z/R = \text{const}\,(PA/R^2)^m i^n \tag{139}$$

On processing our experimental results the following equation was obtained for Golden apples:

$$z/R = 1.06(PA/R^2)^{0.7} i^{0.08}$$

Figure 183 shows the damaged volume as a function of number of cycles during repeated impact of two apples against each other. Curve *(1)* refers to the case when both are damaged to the same extent, while in the case of curve *(2)* only one of the apples is damaged. Since impact between two apples is involved, the total volume damaged is in both cases given relative to the total volume of the two impacting fruits.

As may be seen, when the two apples are damaged identically, the relative damaged volume is nearly twice as high as the value obtained when only one apple is damaged. Cases when both fruits are damaged but not to the same extent fall between the two curves.

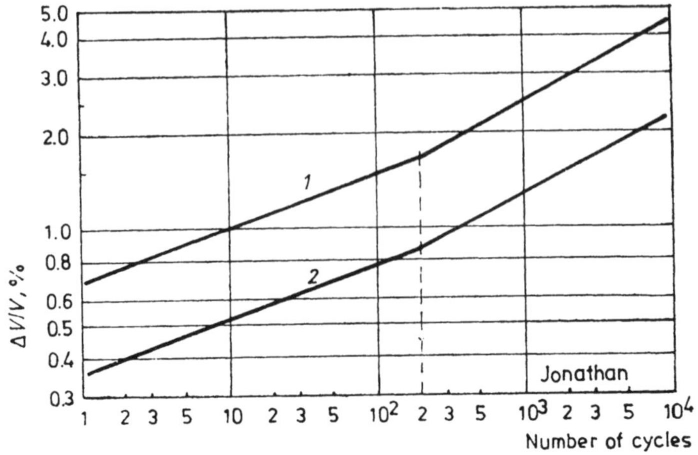

Fig. 183. Increase of damaged volume during impact of two apples as a function of number of cycles. *(1)* $E_1 = E_2$; *(2)* $E_1 = 2E_2$

In the above curves describing damage a break may be observed in the range of 200–1000 cycles. The greater the load and the riper the fruit, the lower the number of cycles before the break occurs. An observation of interest is that the relaxation of the loading force shows a similar break point in the same range of number of cycles (Fig. 174).

The value $\Delta V_1/V$ appearing in eqn. (138) describing the damaged volume may be obtained as follows. According to eqn. (136), the volume damaged during the first cycle is proportional to z^2R, while the volume V is proportional to R^3. Accordingly,

$$\Delta V_1/V \simeq z^2R/R^3 = z^2/R^2$$

If the expression for z is substituted from eqn. (80), the damaged volume may be written in the form

$$\Delta V_1/V = \text{const} \cdot (AP/R^2)^{4/3}$$

On the basis of this equation, it may be recognized that the second term of eqn. (138) may also be made more general by adding the missing term A/R^2 to the force P. The exponent of 4/3 found theoretically does not differ significantly from that obtained experimentally (1.5), and so eqn. (138) may be used in the form

$$\Delta V/V = k_1(AP/R^2)^{4/3} + k_2(i-1)^{2n}(AP/R^2)^{2m}$$

or, more simply,

$$\Delta V/V = (AP/R^2)^{2m}[k_1 + k_2(i-1)^{2n}] \tag{140}$$

On the basis of the data of Fig. 183 the values of the exponents are $2m=1.5$ and $2n=0.3$, while those of the constants are $k_1=0.34$ and $k_2=0.075$. The expression AP/R^2 appearing in the above equations is dimensionless and may be treated as the similarity criterion for relative damaged volume.

A significant proportion of cereal crops are dried after harvesting, and the mechanical strength of the grains is reduced by drying. If the reduction of strength is significant, the quality suffers owing to the presence of damaged and cracked grains, and storage losses are increased. Cereal grains are generally loaded dynamically (e.g., by impact), and so it is advisable to select some dynamic test method to characterize their mechanical strength. The dynamic test apparatus shown in Fig. 172 is suitable for testing cereal grains according to the following method. A cereal grain is loaded cyclically at a given deformation until it ruptures. Rupture is indicated by a sudden decrease of the loading force. If the number of cycles before rupture is plotted versus deformation, a uniquely defined relationship is found between the two variables. The relationship is hyperbolic, and in a system of log–log coordinates straight lines are obtained. Figure 184 shows the number of cycles before rupture for maize kernels, as a function of strain, for various drying temperatures [71, 74]. As may be seen, the straight lines corresponding to different temperatures are shifted: the higher the temperature, the lower the number of cycles before rupture for a given deformation. The curve corresponding to 20 °C refers to hand-collected and naturally dried maize; the greatest mechanical strength is obtained in this case.

It is of interest to plot the deformation resulting in rupture after 100 cycles (i.e., the related deformation or strain) as a function of temperature or other relevant variables. Such a plot is presented in Fig. 185. As may be seen, a unique relationship exists between the related strain and drying temperature. Rapid cooling after drying makes the grains more brittle; i.e., the material can support 100 loading cycles only at a lower deformation. On the basis of the above, the related strain may be regarded as a measure of dynamic mechanical strength. This form of representation is useful for expressing numerically the effect of various factors (species, harvesting period, drying, etc.) on the mechanical strength.

The experimental results presented in Fig. 184 may also be explained theoretically. Applying the relationship observed during the comminution of cereals, i.e., the proportionality of rupture energy to surface area newly created [188], the energy equation may be written for cyclical loading in the form:

$$\sum_{i=1}^{i_r} \int P(i)\,dz = vd^2$$

where d is the equivalent grain diameter, v the specific energy of rupture

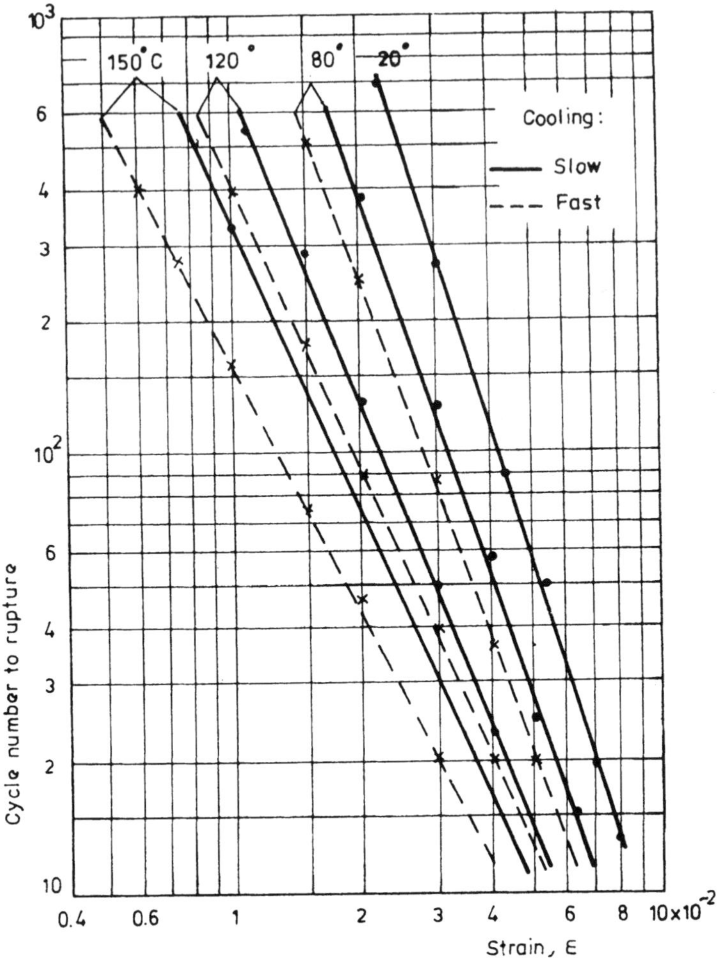

Fig. 184. Number of loading cycles required to rupture maize kernels as a function of deformation

(cm N cm^{-2}), and i_r the number of cycles before rupture. In the case of constant deformation, the relaxation of the force may be taken into account using an equation of the form

$$P(i) = P_1 i^{-n}$$

The force P may be expressed using the basic equations for contact stress due to Hertz, i.e., by means of eqn. (79), as

$$P \simeq [\sqrt{d}E/(1-\nu_1^2)]\, z^{3/2}$$

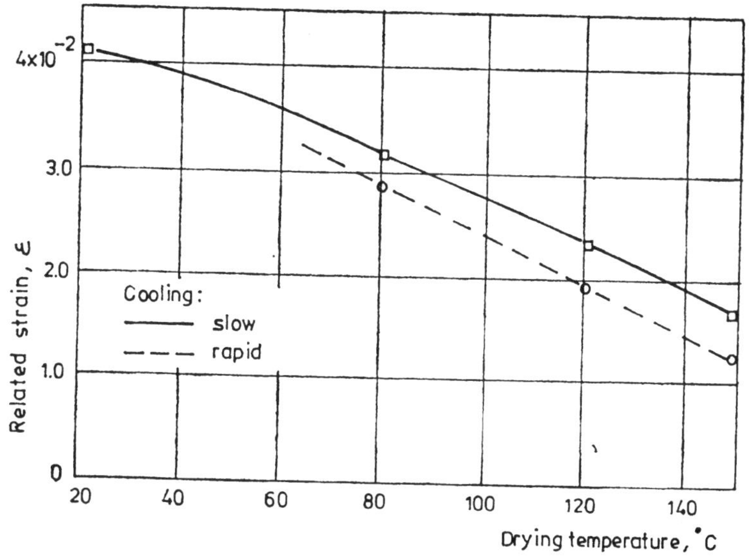

Fig. 185. Relationship between related strain and drying temperature for maize kernels

Considering this expression, the energy equation may be written as

$$\sum_{i=1}^{i_r} i^{-n}[\sqrt{d}E/(1-v_1^2)] \int z^{3/2}dz = vd^2$$

Integration yields the relationship [189]

$$i_r = [(1-n)vd^{1.5}(5/2)(1-v_1^2)/Ez^{5/2}]^{1/(1-n)}$$

If a constant force is applied, the increase of the deformation may be described by an equation of the form

$$z(i) = z_1 i^n$$

and, by a similar method as in the preceding case, the following expression is obtained:

$$i_r = [(1+5n/2)(5/2)vd^{1.5}(1-v_1^2)/Ez_1^{5/2}]^{2/(2+5n)}$$

In the case of elastic bodies, $n=0$, and so the above equations may be written in the general form

$$i_r = K(1/z^{5/2})$$

According to theory, the number of cycles before rupture is inversely proportional to deformation to the power of 5/2. In the case of relaxation, the exponent increases slightly, while in the case of creep it decreases in relation to the value which is valid for elastic bodies. Thus, for example, if the calculations are performed with the value $n=0.1$ and constant deformation is assumed, then the exponent of the deformation is 2.77, which agrees well with the value 2.6–2.8 found experimentally for temperatures below 100 °C. For drying temperatures of 120–150 °C the exponent values found experimentally (2.2–2.4) are lower, and so the agreement with theory is unsatisfactory. The explanation is that cereal grains become more brittle at high temperatures and their mechanical properties vary during the loading period. The decrease of the exponent is caused by the continuous deterioration of the mechanical properties.

Another important application of dynamic tests is determination of the vibration properties of fruit species, in order to assess their sensitivity to damage during transport. Fruit is generally transported in containers by motor vehicles. If during transport the resonance frequency of a fruit column packed into a container coincides with the excitation frequency of the road or vehicle, then the acceleration of the fruit will increase considerably owing to resonance, and it will be damaged by impact [89, 90]. The resonance frequency of fruit in a container may be calculated approximately from the equation

$$f_r = (1/4L)\sqrt{Eg/\gamma}$$

where L is the height and E the modulus of elasticity of the fruit column in the container, and γ the volumetric weight of the fruit.

The resonance frequency of fruit packed into a container and the dynamic states (displacement, acceleration) of the individual layers may be examined using special test equipment (Fig. 186). A container full of fruit is fixed on an oscillating table and excitation is effected by means of eccentric weights rotating in opposite

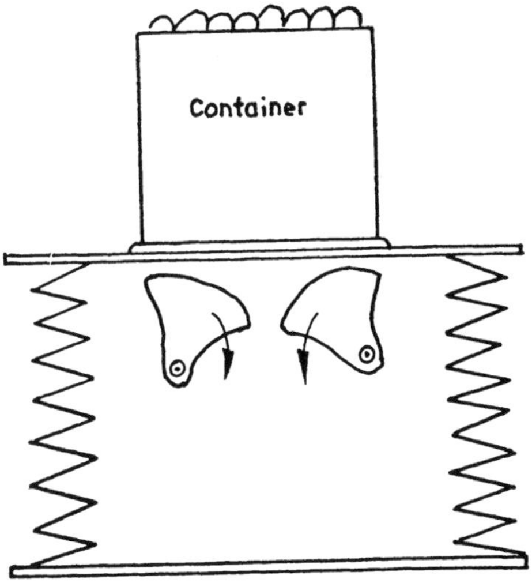

Fig. 186. Apparatus for investigating the state of motion of fruit packed in a container

directions. The magnitude and angular velocity of the rotating masses may be varied.

Observations show that the upper 2–3 layers of fruit become unstable if the acceleration exceeds 0.7*g*. The individual fruits move freely, both vertically and around their axes (rotational motion). Figure 187 illustrates the acceleration in various (upper, central and lower) layers of oranges in a 60×60×60 cm container, as a function of excitation frequency [90.] Up to a frequency of 10 Hz the curves nearly coincide, and then the acceleration increases rapidly, especially in the upper layer, and attains its maximum value at the resonance point *A*. The resonance frequencies are practically identical for the individual layers. At point *B* the acceleration decreases suddenly, and then increases again. This phenomenon is related to resonance of the oscillating table used in the experiments. The acceleration of the central layer is greatest at point *C*, while that of the lower

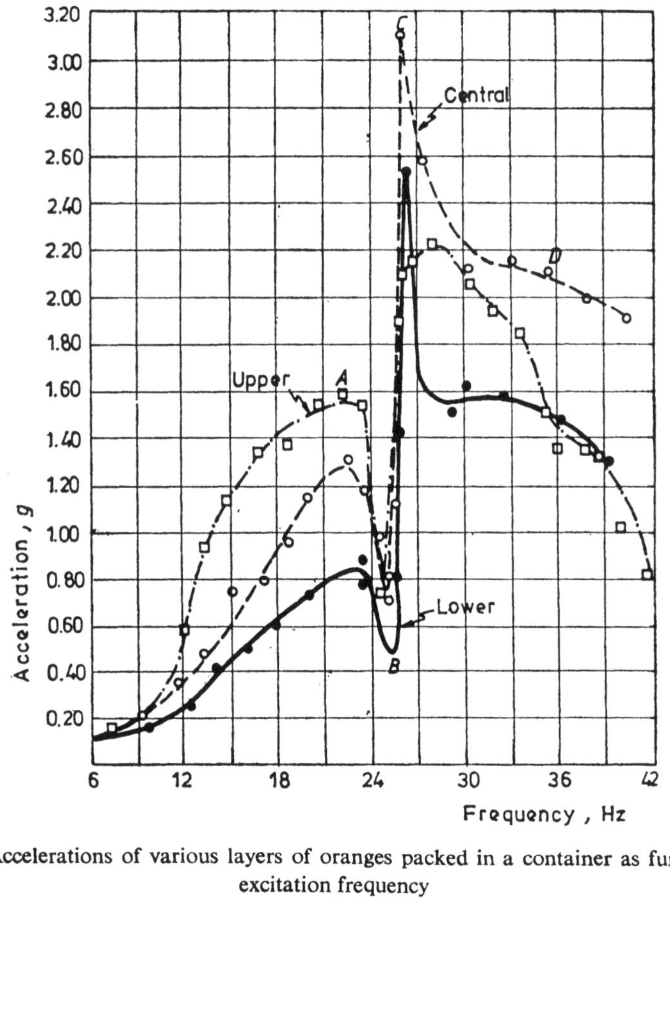

Fig. 187. Accelerations of various layers of oranges packed in a container as functions of excitation frequency

and upper layers is less. This phenomenon is related to the form of the oscillation. At point *A*, the nodal point of oscillation occurs at the bottom, permitting the greatest acceleration in the upper layer. At point *C*, nodal points occur in the upper and lower layers, and maximum acceleration becomes possible in the central layer. The acceleration in lower layers is naturally always less than in upper layers, because the bottom layer also supports the weight of the column found above it.

Another application of dynamic tests is the determination of intrinsic and extrinsic (e.g., as a result of air resistance) damping coefficients. For example, both coefficients play a role in the shaking of fruit trees and through knowledge of their values the design of tree shakers may be improved significantly.

The intrinsic damping of a material is characterized by its logarithmic decrement, or damping coefficient, which may be determined in the following way. One end of a prismatic bar cut from the material to be investigated is clamped, and the other end undergoes free oscillation after deflection. The deflection of the free end decreases continuously owing to damping, and this decreasing deflection is plotted as a function of the number of cycles (Fig. 188) [91]. In a semi-

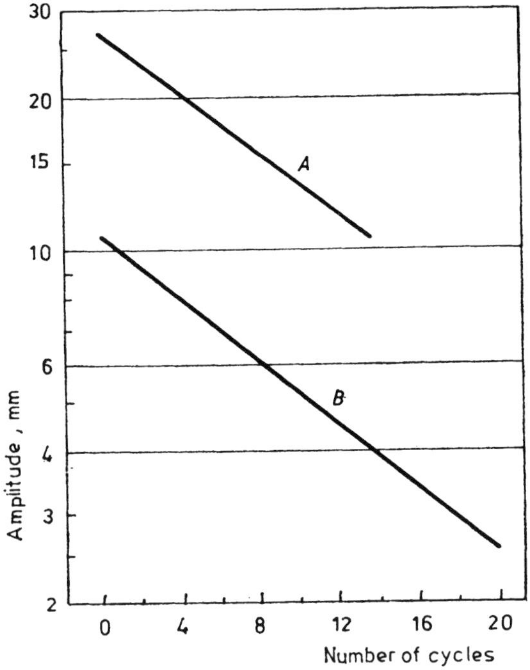

Fig. 188. Reduction in amplitude of free end as a function of number of cycles during oscillation of a bar. *A*, *B* two different specimen

logarithmic system of coordinates straight lines are generally obtained, and the logarithmic decrement may be calculated on the basis of measurement results from the expression

$$\delta' = (1/n) \ln (y_0/y_n) \tag{141}$$

where y_0 is the initial maximum deflection, and y_n the deflection in the nth cycle. The damping coefficient may be calculated from the logarithmic decrement using relation

$$\xi = \delta'/2\pi \tag{142}$$

The damping coefficient for trees is practically independent of moisture content for the range encountered in living specimens. However, if the free water (corresponding to about 25–30% moisture content) is completely removed from the cellular cavities, the damping decreases. Figure 189 shows the logarithmic decrement for almond-tree test specimens as a function of moisture content [91].

The loss tangent ($\tan \delta = E_2/E_1$), characterizing also the energy-absorption properties of a material, shows a defined correlation with the above factors δ' and ξ: i.e.,

$$\tan \delta = 4\pi\delta'/(4\pi^2 + \delta'^2) = 2\xi/(1+\xi^2) \tag{143}$$

The loss tangent for tree bark is about four times that for living wood, and so tree bark is able to absorb a considerable amount of energy during dynamic loading, thus protecting the inner parts of the tree. However, the modulus of

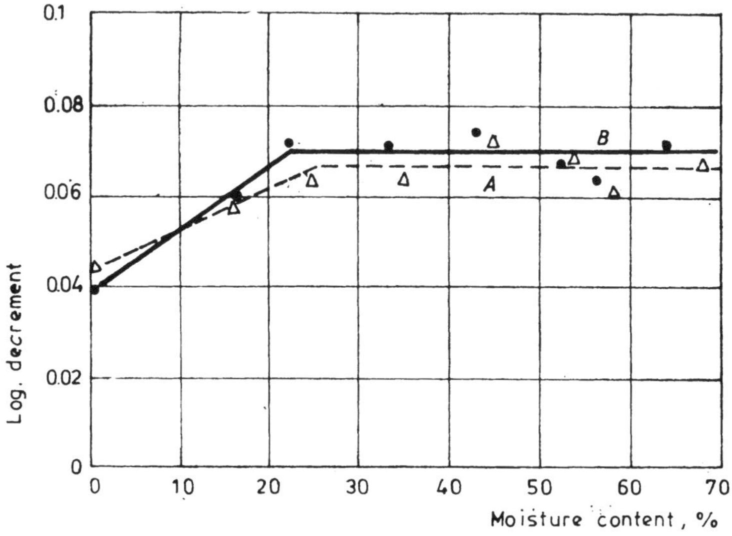

Fig. 189. Logarithmic decrement as a function of moisture content for almond-tree test specimens

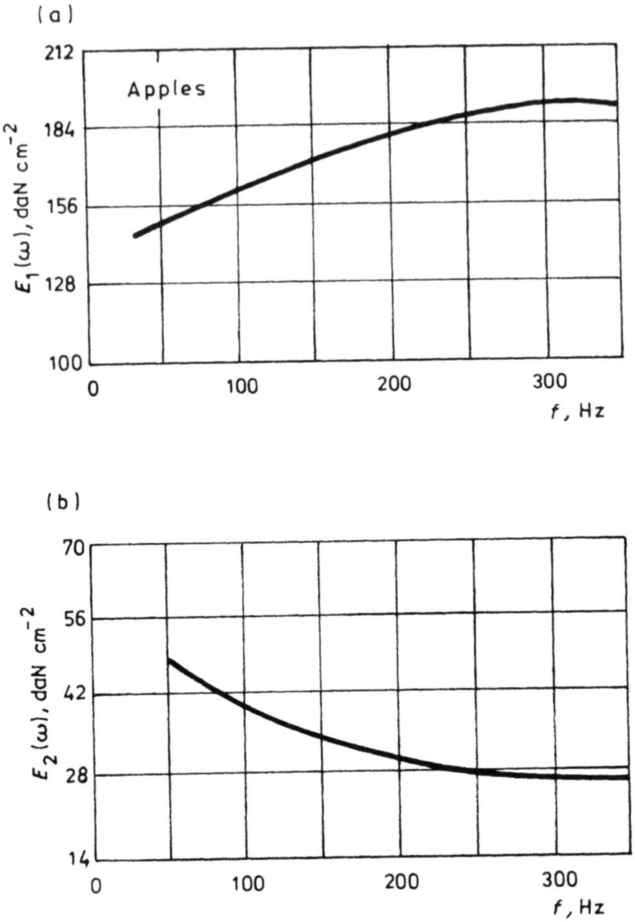

Fig. 190. Storage (a) and loss moduli (b) of an apple specimen as functions of loading frequency

elasticity of bark increases with frequency, and so it behaves with increasing hardness and its sensitivity to damage increases at higher frequencies.

Extrinsic damping originates mostly from air resistance. Air resistance may be expressed generally by the equation

$$P = c_w(\varrho/2)Fv^2$$

where c_w is the air-resistance (drag) coefficient, ϱ the density of the air, F the projected area of the body normal to its direction of movement, and v the velocity difference between the body and the air. Experiments performed with leafy tree branches have shown that the value of the drag coefficient may be accepted approximately as 1.5. However, the drag coefficient also depends on the orientation of the leaves during movement. This latter factor cannot as yet be supported by appropriate experimental data.

Dynamic tests are also suitable for determining viscoelastic characteristics as a function of frequency (see Section 8.11). The test specimen is loaded by a sinusoidally varying force, and the relationship between force and deformation is measured. For materials inclined to softening, the deformation must be slight, in order to prevent additional variation of the properties during the measurements as a result of variations in the material. Figure 190 presents storage (a) and loss moduli (b) for apple test specimens 25 mm in diameter [55]. It may be seen that the material apparently becomes more elastic on increasing the frequency.

The elements of the Maxwell model may be calculated on the basis of experimental results using the relationships

$$E_0 = (E_1^2 + E_2^2)/E_1$$

and

$$\eta = (E_1^2 + E_2^2)/\omega E_2$$

13. MECHANICAL DAMAGE

The mechanization of various harvesting and subsequent manipulation operations has an unfavorable consequence in that it leads to an increase in damage to the material processed. In every case the quality of the product is directly lowered as a result, and in numerous cases mechanical damage is followed by rapid spoiling, whereby the material deteriorates completely. In the course of longer storage, spoiled material also endangers sound material which is in contact with it. Thus it is understandable that the reduction of mechanical damage is of high economic importance.

The possibilities for reduction of mechanical damage may be divided into three main groups:

(a) designing the elements of the machine performing an operation in a way such that the forces acting on the material are the lowest possible;

(b) breeding product species of improved quality which are able to support relatively great loads without suffering mechanical damage; and

(c) carrying out harvesting or manipulation operations when the state of the product is such that its mechanical strength is sufficient to render it less sensitive to damage.

13.1 Causes and forms of appearance of damage

Damage to agricultural products may appear in greatly differing forms. The form of damage depends on the physical and biological construction of the product and on the type of load. The various grains are damaged primarily during threshing and mechanical transport (e.g., by transport screws). Damage occurs here in the form of rupture or cracking under the effect of excessive deformations or forces appearing at impact. Excessive deformations may appear if a product is forced to pass through a given gap, or during the impact of grains on treshing or filling into storage towers, etc. The damage may appear in very diverse forms, from hair cracks to total rupture. Damage decreases the germination capability of cereal grains and increases the rate of oxidation during storage, whereby losses increase considerably and the quality of the flour made from such grains is lowered.

Mechanical picking of fruit implies significant mechanical damage. On shaking a tree the fruit impacts against the tree branches, against other fruits and finally against the catching surface. The tissue beneath the skin is deformed by impact. If the deformation surpasses the biological yield limit, the tissue will brown within a short time and be spoilt. In certain cases browning under the skin is not visible from the outside (e.g., in pears), the affected fruit cannot be sorted out and processing is thereby aggravated.

Certain fruits (e.g., cherries and sour cherries) fall without their stems on shaking the tree. Where the stem is torn out of the fruit, juice appears, representing a loss on the one hand and promoting deterioration on the other. Fruit collected in this way is suitable mainly for fast processing. However, by selecting the correct shaking frequency it may be possible to ensure that most of the fruits fall together with their stems.

In harvesting root bulb products (beet, potatoes), significant damage must be counted upon during loading into and unloading from the means of transport. The main forms of damage result from impact, bruising and cuts. The extent of damage also depends on the species, on the stage of ripening and on temperature. It is observed that later harvesting of potatoes, in cold weather, implies greater damage than harvesting in warm weather.

The examples listed above show that agricultural products are generally damaged by static or dynamic external forces, and more rarely by internal forces. Mechanical damage resulting from internal forces is caused by physical variations taking place inside a product, for example, variations of the temperature and moisture content, or chemical and biological variations. In the case of cherries and tomatoes, cracking of the skin due to an increase in internal pressure may be observed frequently.

The mechanisms of damage in the case of agricultural products are not fully known at present, but the occurrence of rupture or tears in the outer or inner cellular system during damage is certainly implicated. For biological materials the beginning of rupture in the cellular system is indicated by the biological yield limit. Thus to a first approximation it may be stated that damage occurs when the load exceeds the biological yield limit. However, under loads repeated many times, biological yield may occur even if the individual loads are smaller than that corresponding to the biological yield limit. The reason is that many materials soften under repeated loads and their strength decreases.

To determine the permissible load for an individual material it is necessary to know the mechanical properties of the material in simple stress states (under tensile, compressile and shearing loads, both static and dynamic). Knowing these mechanical properties, the complex cases of loading occurring in practice may be evaluated and approximate prediction made of whether or not damage is liable to occur.

13.2 Biological and chemical reactions after damage

Damage to agricultural materials is generally followed by infections caused by various fungi. As a consequence of damage, the skin of a product and the layers under the skin are torn, whereby the possibility arises of penetration by bacteria into the product. The damaged parts are in direct contact with the air, and so the rate of oxidation is increased. This latter phenomenon occurs mainly in the case of cereals.

The oxidation of cereals increases in direct proportion to the extent of damage; if 30% of the grains become damaged, the rate of oxidation approximately doubles. During the storage of large volumes of damaged cereals, the storage pile will heat up more rapidly and must be aerated or turned over more frequently. Hair cracks, invisible to the naked eye, frequently form on the surface of grains permitting the penetration of bacteria and oxygen and increasing the rate of oxidation. Wet grain is apparently able to accept greater deformation without external rupture, but the protective tissues under the skin may be damaged and bacteria may then penetrate through the hair cracks unopposed into the internal tissue.

Damaged fruit tissue browns within a short time. The chemistry of browning has not yet been solved completely, but enzymes are sure to play an important role in certain cases, while in other cases they may not be involved. Enzymatic browning appears when damaged tissue comes into contact with air. As examples, the browning of apples, pears, peaches, sour cherries and strawberries may be mentioned. The oxygen required for browning is provided not only by the ambient air, but also by intracellular air. For example, apples contain a relatively large volume of air in the microvolumes found between the cells, and so internal browning may occur in apples in the absence of apparent external damage. Nonenzymatic browning may occur in certain processed and preserved food products, for example, in fruit juices, purees and dried fruits.

A black core or spot may form inside or on the surface of potatoes as a result of mechanical damage. Biological yield may occur in a potato tuber compressed between plane plates, without any visible traces of damage on the surface. The deformed inner part blackens in a few days. This internal blackening is explained by deficiency of oxygen. The respiration of the cells becomes more intense in the deformed tissues, involving greater oxygen consumption. These assumptions are supported by the observation that no blackening occurs when a potato is cut directly after deformation. The surface of a potato blackens within 24 h after impact. This type of blackening is caused by fermentation-type oxidation of the tyrosine found in the cell sap. The product of oxidation is a black pigment which causes the discoloration.

13.3 Establishing and measuring damage

A generally accepted method for establishing the extent of damage to agricultural products has not yet been developed. The reason is that mechanical damage takes a great many forms. The same damage may also be judged differently, depending on whether a product is to be processed directly or is destined for storage. The methods of assessment established in practice are described in the following, from which that most appropriate to given circumstances must be selected.

The main forms of appearance of mechanical damage are as follows [1].

Abrasion. Here the skin of a product is damaged or partly separated from the tissues beneath it. Abrasion is sometimes hardly visible directly after harvesting, but will become apparent after storage for 1–2 weeks.

Bruising. In this case, damage to plant tissue occurs as a result of external forces causing physical changes, in certain cases discoloration and change of taste. Bruising need not imply rupture of the skin.

Cracking. This category is limited to cracking of the skin or tissue due to impact or pressure, without causing the product to fall apart into several pieces.

Cutting. Cutting may be defined as the penetration of a sharp tool into the product, without any significant crushing effect.

Puncture. This type of wound is caused by pointed needle-type tools, plant stems, or thorns penetrating the surface of a product and the tissue beneath.

Shatter cracking. This takes the form of multiple cracks starting radially from the point of impact.

Skin cracking. This category concerns cracks restricted to the outer skin alone.

Splitting. Splitting occurs when a product divides into several parts.

Tearing. Tearing is usually caused by stem ends, i.e., when the skin of a fruit is torn on removing the stem.

Swell cracking. This cracking is caused by an increase in the internal osmotic pressure.

Distortion. Distortion concerns changes of form caused by loads acting on a product.

The classification of mechanical damage according to the above list is useful not only as concerns the further treatment of a product, but also from the point of view of detecting the causes of damage. Within the individual categories of damage some index must be found to characterize the extent of damage. For example, in the case of abrasion the percentage of products damaged in relation to the total quantity and the mean value of the abraded surface area as a percentage of the total must be given. In the case of bruising the damaged volume must be determined, which may again be expressed as a percentage of the total volume. Some research workers have also used the mean diameter of the damaged surface, but this may lead to erroneous results, since the depth of damage may differ. In the case of cuts in a product, the length and depth are the critical parameters. If a cut product falls apart into two pieces, i.e., a free cut surface is formed, then the cut surface area may be related to the total surface area.

The time delay between damage and processing is also of importance. A crushed volume develops browning and then spoils, and therefore damaged products must be processed as soon as possible. In numerous cases the damage is at first invisible to the naked eye, becoming visible only later. Such damage may lead to internal browning, blackening and deterioration, or in the case of seeds to a reduction in germination capacity.

Internal damage and changes may be demonstrated by nondestructive test methods. The measurement methods applied most frequently are based on the fact that the optical properties (e.g., absorption, reflectivity) of a damaged product change. Sound maize kernels show high reflectivity. The quantity of light reflected is, however, diminished by the presence of hair cracks, bruises or ruptures, and the absorption of light is correspondingly increased.

Knowing the appropriate experimental data, optical methods are suitable for rapid determination of the extent of damage for numerous products [1]. Internal changes (e.g., browning, the presence of a water core or black core) in fruits and potatoes may also be detected by optical methods. A universal instrument (the IQ Analyzer) suitable for detecting the above changes with the use of monochromatic light at two purposely selected wavelengths has recently become available commercially. The extent of invisible damage in sowing seeds may also be determined by germination trials. The germination capacity of damaged seeds is considerably reduced, and so conclusions may be drawn from germination capacity as to the extent of damage. The respiration process in many products is enhanced by damage, and so a measure of damage may also be derived from the rate of evolution of carbon dioxide. The respiration rate for damaged potato tubers is about double the value found for sound tubers [1].

Classes of damage are often established according to the extent of damage. Damaged products are assigned to various classes and their quantities expressed as percentages of the total crop. The definitions of the individual classes are determined primarily by the intended subsequent utilization of a product. As an example, French beans picked mechanically can be classified in the following way: the first class consists of perfectly sound legumes, the second class contains those, where a small amount of tissue has broken off with the stem; and the third contains broken legumes.

13.4 Effect of moisture changes on damage

The physicomechanical properties of agricultural materials are determined decisively by their moisture content, especially when it is high (as for fruits and

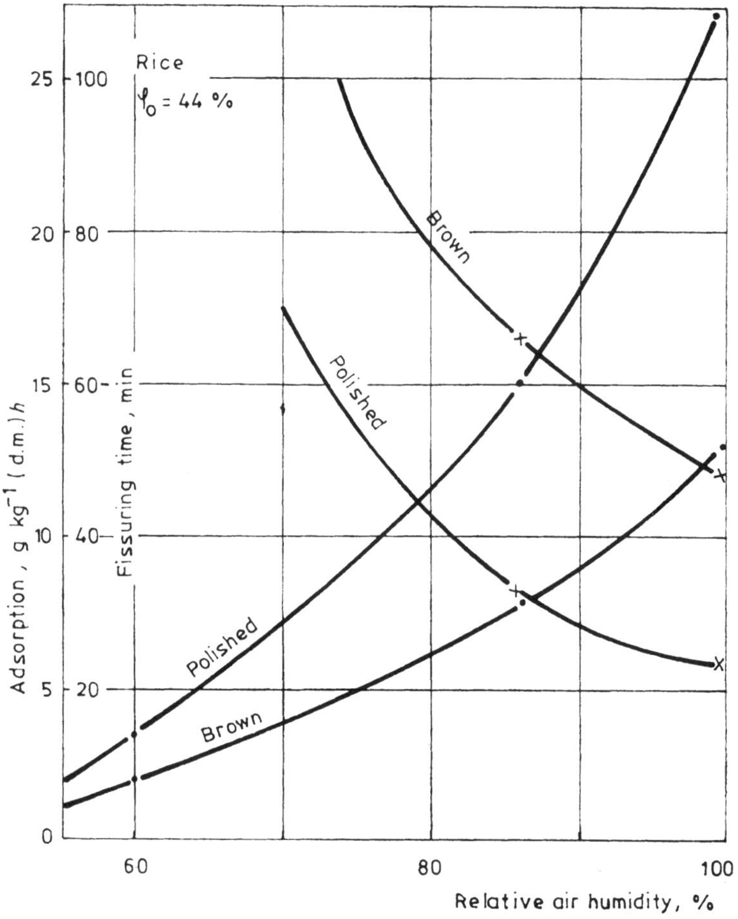

Fig. 191. Time required to obtain initial fissuring of polished and brown rice as a function of air humidity

vegetables). Materials with a high moisture content lose water easily, whereupon the turgor pressure in the cells decreases. The turgor pressure greatly influences the elastic properties and together with them the stress–strain relationship for plant tissues. As an example, it may be mentioned that the modulus of elasticity of potatoes varies linearly with water potential according to the equation [20]

$$E = 3.3\psi + 7.1 \quad \text{(MPa)}$$

The effect of moisture content on sensitivity to damage varies. In materials with a high moisture content (tomatoes, French beans, etc.), loss of moisture decreases the turgor pressure and the modulus of elasticity. With reduction of the modulus of elasticity the sensitivity to damage generally decreases. The internal flow of moisture is rapid in these materials, and so no significant moisture gradients occur in them.

For other agricultural materials, such as cereals, moisture adsorption or desorption implies a significant moisture gradient. The moisture gradient results in the

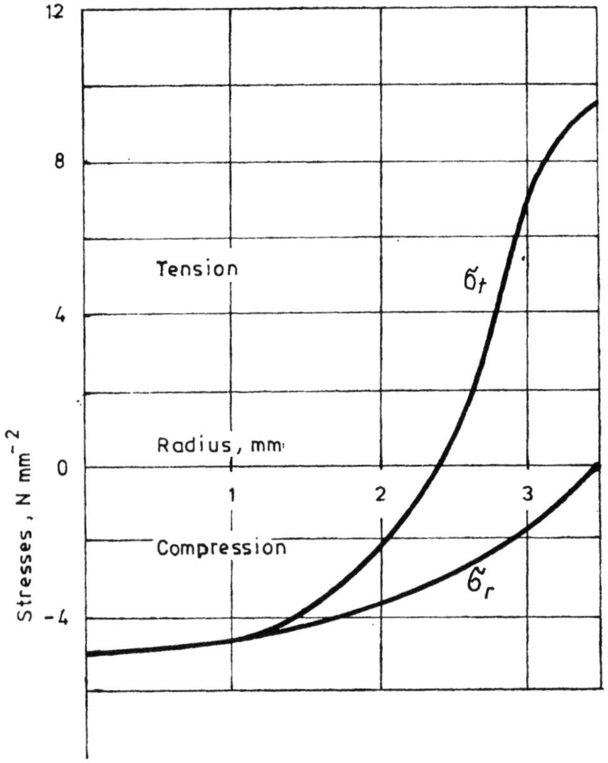

Fig. 192. Stress distribution in a soybean kernel during drying

appearance of stresses in the material, causing surface fissures. Figure 191 shows the time required for the development of initial fissuring in shelled and brown rice as a function of relative air humidity [87]. During the experiments rice grains in equilibrium at a relative air humidity of $\varphi_0 = 44\%$ were placed into a medium of higher humidity, where they adsorbed moisture. The greater the relative humidity difference $\Delta\varphi$, the more rapid the adsorption and the greater the moisture gradient under the surface.

In drying granular products considerable tangential tensile stresses appear close to the surface. Figure 192 shows the distribution of the radial and tangential stresses caused in soybeans by drying, as a function of the distance from the center [84, 85, 86]. In the central region, only compressile stresses appear, but close to the surface the tangential stress is tensile and may be responsible for fissuring.

In drying forage materials in the field, the main source of loss is the tearing of

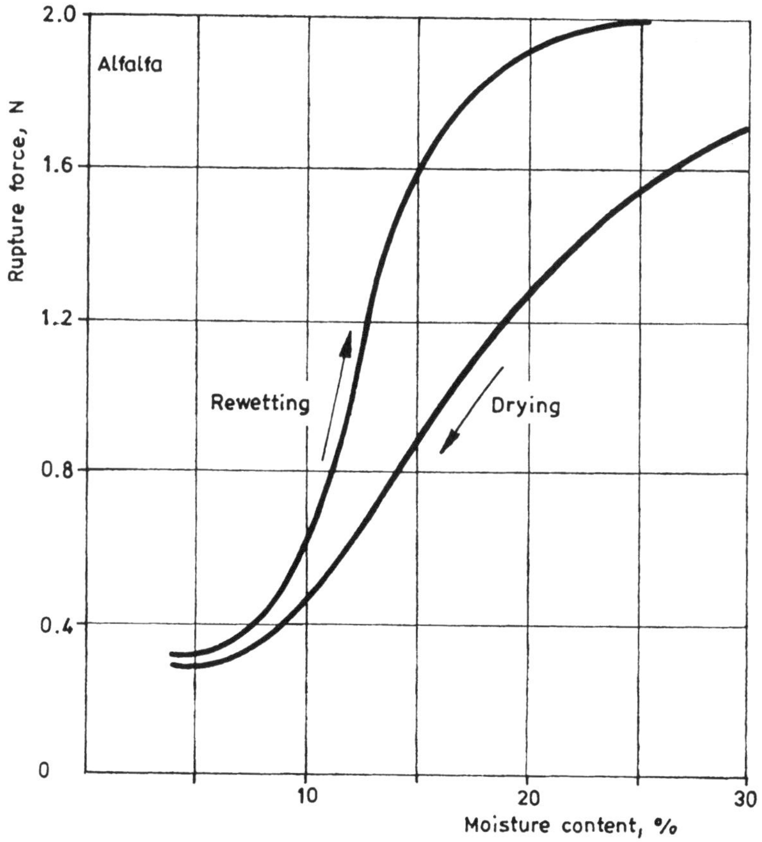

Fig. 193. Leaf retention force as a function of moisture content during drying and rewetting

leaves from the stalk. The force required to tear the leaves off depends greatly on the moisture content, as may be seen from Fig. 193 [62]. The experimental results also show that the tearing force depends on whether drying or rewetting occurs.

13.5 Impact damage

Agricultural materials are damaged most frequently by impact. During harvesting, transport and various handling and treatment processes, moving products impact against stationary or moving mechanical parts. The surface of the product is convex in most cases, and so during impact a Hertz contact stress is produced.

Calculation methods for Hertz stresses have been discussed in Chapter 9, and impact problems in Chapter 10. The solutions are relatively simple for elastic bodies, whereas in the case of viscoelastic materials solution is frequently possible only by means of computers.

Four phases may be distinguished during impact:

(a) at the start of impact, elastic deformation arises which ceases completely on unloading; this stage of impact may be calculated on the basis of contact stresses (considering the deformation, mean pressure and duration of impact);

(b) plastic deformation then begins in the course of which the mean pressure exceeds the dynamic yield stress of the material; thus a part of the deformation remains after removal of the load;

(c) fully plastic deformation sets in, during which deformation continues and the mean pressure drops to below the dynamic yield stress; and

(d) finally there is an unloading phase, during which the elastic stresses and deformations stored in the material cease to exist; the plastic deformations remain permanently.

It follows from the above that the maximum pressure and maximum deformation fail to coincide in time (see Figs. 116 and 117) and the course of the impact force is asymmetrical as a function of time. The volume damaged as a result of impact may be studied theoretically using the methods of stress analysis, assuming damage to occur in the volume where the equivalent stress exceeds the biological yield limit. Determinations of the equivalent stress with respect to damage have not yet been elaborated for biological materials, and so an approximate method must be applied. Examination of the individual stress components appearing at impact shows that the compressile stress σ_{zz} is considerably higher than the radial (σ_{rr}), tangential ($\sigma_{\varphi\varphi}$) and τ_{max} components, and so it may be accepted that the damage at the surface is caused mainly by the compressile stress σ_{zz}, according to the condition $\sigma_{zz} \geqq \sigma_{perm}$. The calculation methods for stress distributions assume a small deformation in relation to the dimensions of the body.

This condition is not always fulfilled during the loading of agricultural materials. In the case of greater deformations the material is compacted and yields plastically, the texture of the material is destroyed, and its mechanical properties are changed. In such cases damage may be expected over volume differing from that calculated.

Figure 194 illustrates typical cases of impact. In the first case [(a) and (b)] one body may be regarded as rigid relative to the second impacting body. The rigid body may be a sphere or a plane plate. In the second case [(c) and (d)] the moduli of elasticity of the two bodies are identical, or may deviate to a slight extent. The shape and volume of the part deformed immediately, and the stress distribution under the contact surface, vary depending on the form of loading.

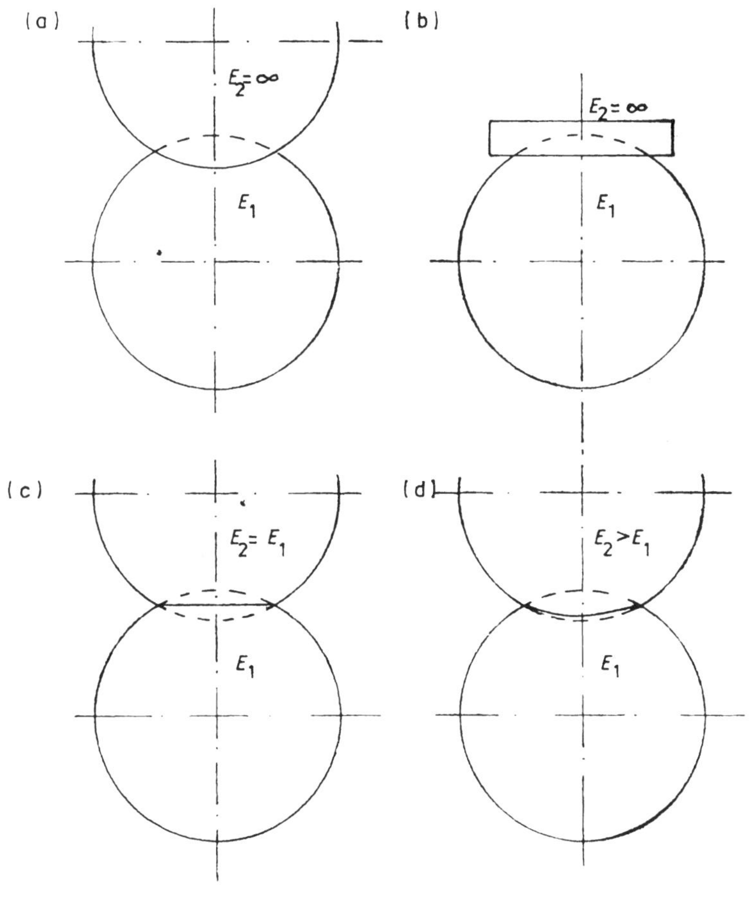

Fig. 194. Typical cases of impact

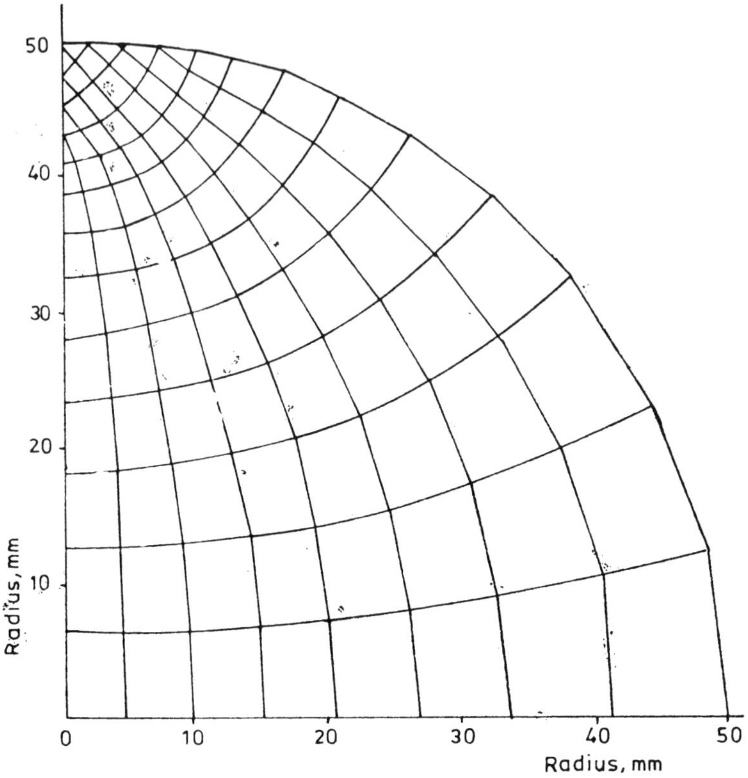

Fig. 195. The 92-element model for calculating stress distributions

Rumsey [78] has investigated the load caused by impact and the stress distribution in spherical viscoelastic bodies (apples) using the finite-element method. The numbers of finite elements chosen were 92, 116 and 140. According to the investigations, application of 140 elements supplied adequate results; the model with 92 elements showed at certain places (where the variation of stress was greater) errors extending to 5–10%. Figure 195 shows the discretization of the body in the case of 92 elements. The nodal points of the finite grid of elements were calculated by applying bipolar coordinates, with the exception of the upper 20 elements. The latter would have been too small, and were therefore developed by hand. The shear-relaxation function was applied in the form of the single term expression

$$\varphi_2(t) = 75.7e^{-179.8t}$$

The volumetric modulus of elasticity for apples was taken as 109.4 daN cm^{-2}, the radius of the apple was $R=3.657$ cm, the Poisson ratio was $\nu=0.219$, the modulus of elasticity of the peel was $E=133.4$ daN cm^{-2}, its Poisson ratio was $\nu=0.32$ and its thickness was 0.5 mm (these values of the modulus of elasticity are valid for apples having a high flesh strength such as found at the beginning of ripening; generally, much lower values are found).

During the investigations four calculation methods were compared: for a homogeneous elastic body, for a homogeneous viscoelastic body, for a viscoelastic body with an elastic peel and finally for a viscoelastic body with a viscoelastic peel.

Figures 196 and 197 illustrate the solutions obtained for a viscoelastic body

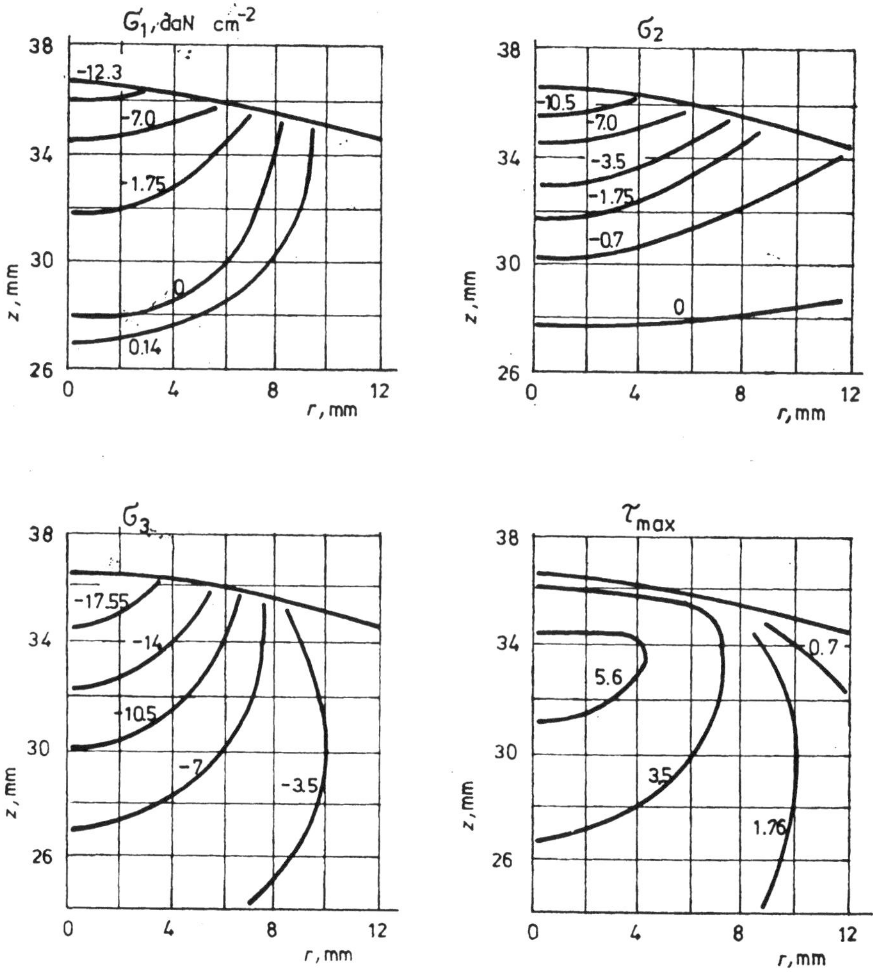

Fig. 196. Stress distribution in an apple after free fall from a height of 5 cm (homogeneous viscoelastic body)

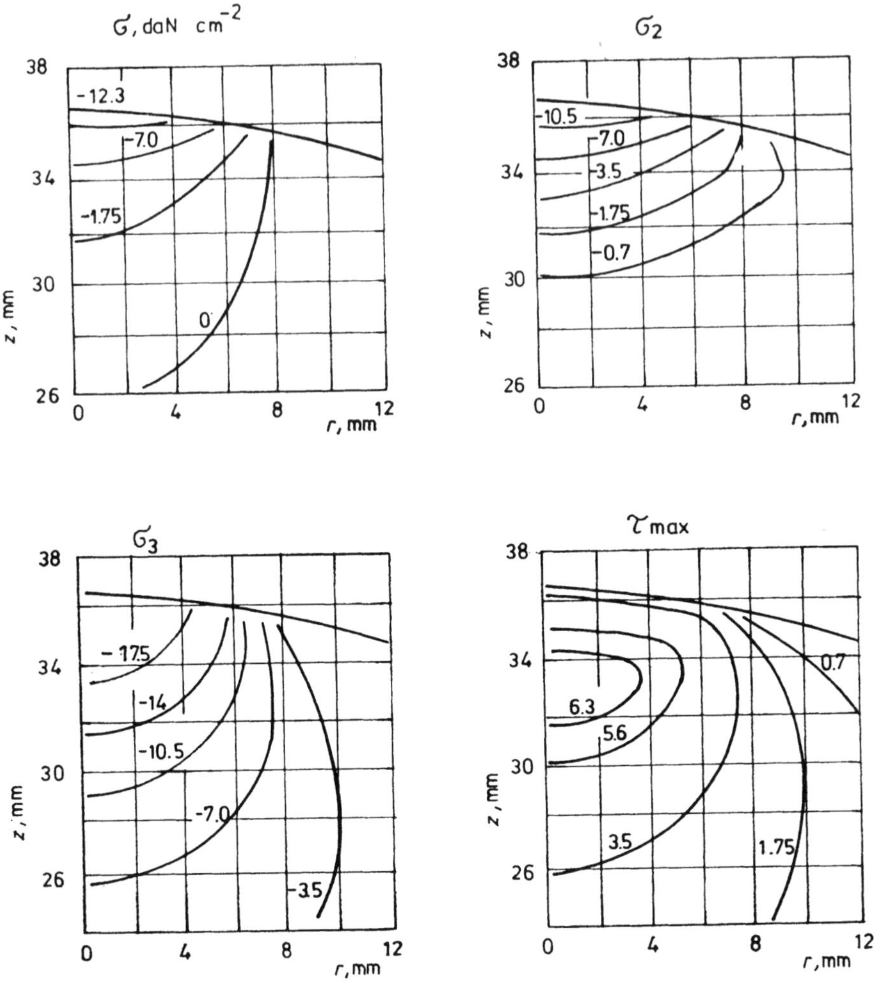

Fig. 197. Stress distribution in an apple after free fall from a height of 5 cm (elastic peel and viscoelastic inner part)

and for a viscoelastic body with an elastic peel in the case of impact after free fall from a height $H=5$ cm. The maximum deformation is 1.14 mm, which is attained 16×10^{-4} s after contact. The maximum compressile stress is obtained after $\Delta t=14\times10^{-4}$ s, i.e., the maximum deformation is delayed in relation to the maximum stress. The main stress σ_3 is the compressile stress in the direction of the z axis; σ_1 is the tangential and σ_2 the radial stress. Comparing the two figures, it may be seen that the main stresses σ_3 do not differ significantly, whereas the distribution of the stresses σ_1 and σ_2 deviate. The deviation of the radial stress σ_2 is particularly striking in the vicinity of the surface at a radius $r=9$ mm, at the boundary of the deformed surface.

The following facts are established through the above comparison of the individual calculation models. The total stress distribution is obtained with the required accuracy only if the peel is also taken into account. Application of a viscoelastic peel does not alter the distribution of stresses as compared to the case of an elastic peel, except that the tangential and radial stress components are greater by 5–10% in the case of elastic peels. The distribution of the compressile stress σ_3 is similar for all four models, and the absolute values do not differ significantly. Since it is primarily the compressile stress σ_3 which determines the damaged volume, a simpler model may be applied. If the impact duration is not longer than one-quarter of the relaxation period of the material, then the elastic model gives results almost identical to those obtained using the viscoelastic model. In the opposite case, higher stresses are supplied by the elastic model.

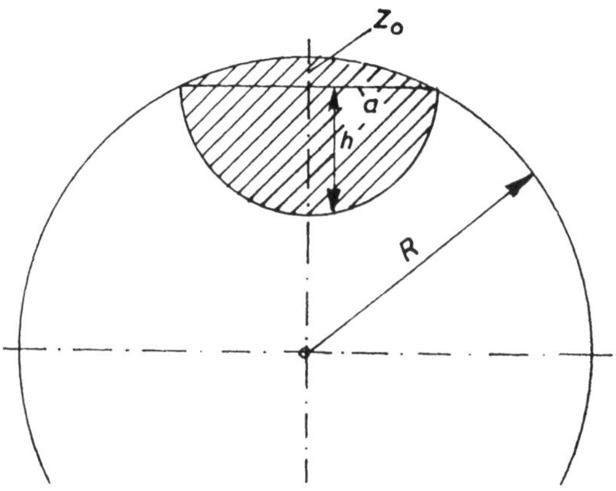

Fig. 198. Calculation of damaged volume

Knowing the stress distribution, the damaged volume may be calculated as follows. The damaged volume is composed of the volume deformed immediately and the volume in which the stress exceeds the biological yield limit (Fig. 198). The latter may be approximated by a spherical section of height h, and then

$$\Delta V = \Delta V_1 + \Delta V_2 = (\pi/6)\, z_0(3a^2 + z_0^2) + (\pi/6)\, h(3a^2 + h^2)$$

where the radius of the deformed surface (for impact with a plane plate) is

$$a = \sqrt{2Rz_0 - z_0^2}$$

On the basis of the available calculation data, the relative damaged volume may be approximated as a function of falling height by the empirical equation

$$\Delta V/V = kH^n$$

where the value of the exponent n for apples has been obtained as 0.7–0.8. However, the calculated damage greatly exceeds the experimental value, and therefore the calculated results for the stress distribution must be treated with corresponding caution. Significant error may result from the fact that at higher loads the modulus of elasticity decreases, as a consequence of destruction of the texture. Thus at the point of maximum load (on the axis of contact) the stress will decrease, while in regions experiencing lower loads it increases. Further significant error may originate from the fact that the stress theory of Hertz is valid only for small deformations, and this condition is generally not fulfilled.

By applying the previous assumption that the relative damaged volume is proportional to z^2/R^2, dimensionless similarity numbers may be derived which facilitate correct processing and generalization of experimental results. The relative damaged volume for one impact cycle, depending on whether the falling height H, or the impact rate v, or impact force P is known, may be expressed in the respective forms

$$\Delta V/V = K_1(H\gamma A)^{n_1} \tag{144}$$

$$\Delta V/V = K_2(v^2A\gamma/g)^{n_2} \tag{144a}$$

and

$$\Delta V/V = K_3(AP/R^2)^{n_3} \tag{144b}$$

where the theoretical values of the exponents are $n_1=4/5$, $n_2=4/5$ and $n_3=4/3$, while K_1, K_2 and K_3 are constants. In the case of repeated loads the number i of cycles must also be allowed for, and the similarity equations may then be used in the forms

$$\Delta V/V = K_1'(v^2A\gamma/g)^n i^m \tag{145}$$

or

$$\Delta V/V = K_2'(AP/R^2)^n i^m \tag{145a}$$

The above similarity numbers and equations are valid only for the impact of two spherical bodies having identical radii of curvature. If the difference is not great, the mean radius of the curvature may be substituted. In the case of large differences, more complex equations must be used.

On the basis of Hertz's theory, the value of K_1 appearing in eqn. (144) should be 3.74. Measurement results for apples give $K_1=3.8$–4.2, depending on the species and the ripeness. The theoretical value of the constant K_3 appearing in eqn. (144b) is 0.2 for two deformable spherical bodies in mutual impact (e.g., two apples,

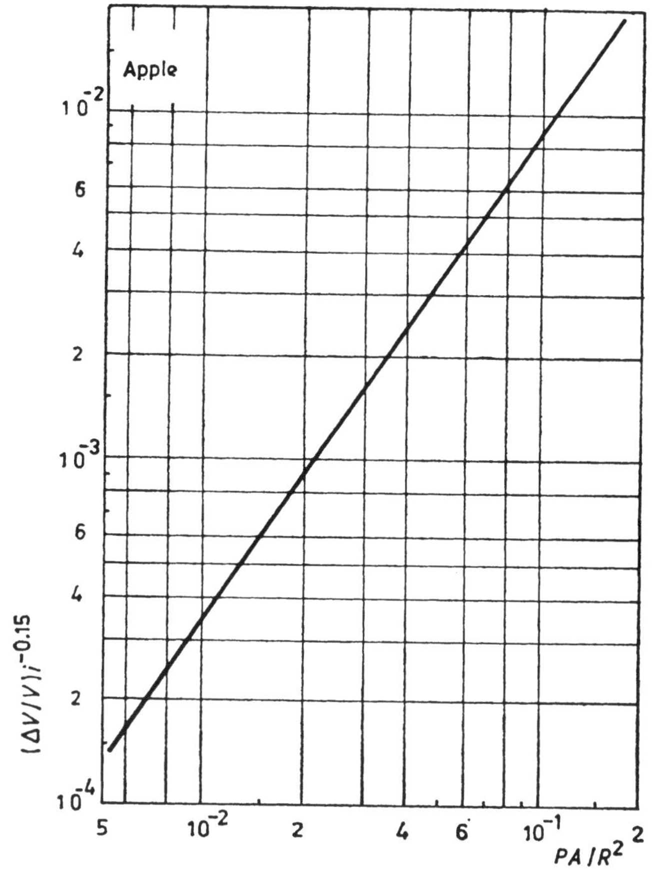

Fig. 199. Plot of relative damaged volume for apples as a function of dimensionless quantities

peaches, etc.). In practice, the values of K_3 are slightly higher. For the oscillating impact of two apples the following similarity equation is obtained according to the model of eqn. (145a) [190]:

$$\Delta V/V = 0.22(AP/R^2)^{1.4}i^{0.15}$$

as illustrated graphically in Fig. 199. According to this equation the volume damaged during the first cycle is only a little larger than the theoretical value obtained by taking into account only the effectively deformed volume as the damaged volume.

The experimental results quoted demonstrate that in the case of materials whose texture is destroyed completely by deformation such that water is segregated, calculations of the stress in regions beneath the deformed zone using conventional methods may lead to completely erroneous results. The destroyed tissue embedded into water transfers a nearly equalized compressile stress to sound regions beneath it, and at the same time it has significant energy-absorbing capacity. The latter statement is also supported by the following theoretical considerations. Assume that a specific energy v_v (cm N cm^{-3}) is required for the destruction of tissue; the energy equation for repeated loading in the case of constant deformation may then be written as

$$\sum_{i=1}^{i} \int P(i)dz = v_v \Delta V$$

The relaxation of the force is again described by an equation of the form $P(i) = P_1 i^{-n}$, and integration of the above equation then yields

$$\Delta V/V = 0.764[i^{1-n}E/(1-n)v_v d^{2.5}(1-v_1^2)]\, z^{5/2} \tag{146}$$

For constant force, application of the equation $z(i) = z_1 i^n$ for creep yields

$$\Delta V/V = 1.53[i^{(5n+2)/2}E/(5n+2)v_v d^{2.5}(1-v_1^2)]\, z_1^{5/2} \tag{146a}$$

or

$$\Delta V/V = 1.68[i^{(5n+2)/2}A^{2/3}/(5n+2)v_v d^{20/6}]P^{10/6}$$

The above equations are based on the theory of elasticity, although relaxation and creep are also taken into account. This means that the energy applied is spent entirely in destroying the sound tissue.

The preceding experimental considerations indicated that the volume destroyed increases as the number of cycles to the power of 0.15 during repeated loading of apples. However, in eqn. (146a) the exponent of the number of cycles calculated with $n = 0.1$ will be 1.25, which far exceeds the value found experimentally. This difference may be explained by the significant energy-absorbing capacity of the destroyed volume after a few cycles, whereby the extent of destruction of additional volume is decreased. Nonetheless, the exponent of the force P obtained theoretically (1.66) does not deviate basically from the experimental values of 1.4–1.5.

Appropriate selection of a cushioning material serving to reduce damage caused by impact may be achieved as described in Section 10.3.

13.6 Effects of various parameters on sensitivity to damage

Sensitivity to damage is influenced by numerous factors. One group of parameters concern the physical and biological state of the material (e.g., temperature, moisture content, stage of growth, ripeness), while others are related to the load characteristics (static, dynamic, oscillating, loading rate, etc.). In most cases, temperature greatly affects the mechanical properties of agricultural products and thereby their sensitivity to damage. With variation of temperature the turgor pressure of cellular material and together with it the elasticity both vary.

Elasticity is often measured by means of a pendulum-type impact device; the ratio of the impact and rejection velocities of a weight is used as an index, or

$$e = v_2/v_1 = \sin(\beta/2)/\sin(\alpha/2)$$

where β is the rebound angle and α the starting angle of the weight.

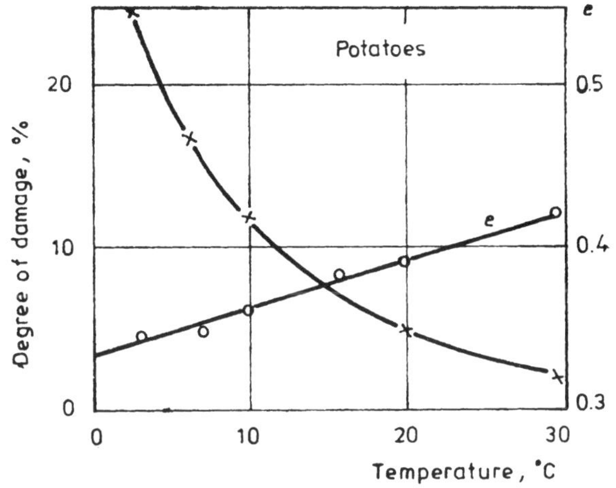

Fig. 200. Elasticity and extent of damage as functions of temperature for potatoes

Figure 200 shows the variations of *e* (characterizing the elasticity) and of the extent of damage as functions of temperature for potato tubers [68]. With decreasing temperature the elasticity decreases and the damage increases. The overall damage was assessed on the basis of the proportions of tubers damaged to various extents as follows:

$$\text{degree of damage} = 0.1m_1 + 0.3m_2 + 1.0m_3$$

where m_1 is the percentage of tubers damaged not deeper than 1.7 mm, m_2 the percentage damaged to between 1.7–5.0 mm, and m_3 the percentage showing damage to a depth greater than 5.0 mm.

The sensitivity to damage of other products with a high moisture content (e.g., fruits) is influenced similarly by temperature; therefore the effect of temperature must be taken into account when harvesting in late autumn.

During high-temperature treatment (e.g., drying) the texture of many agricultural materials changes; the elasticity decreases and the material becomes

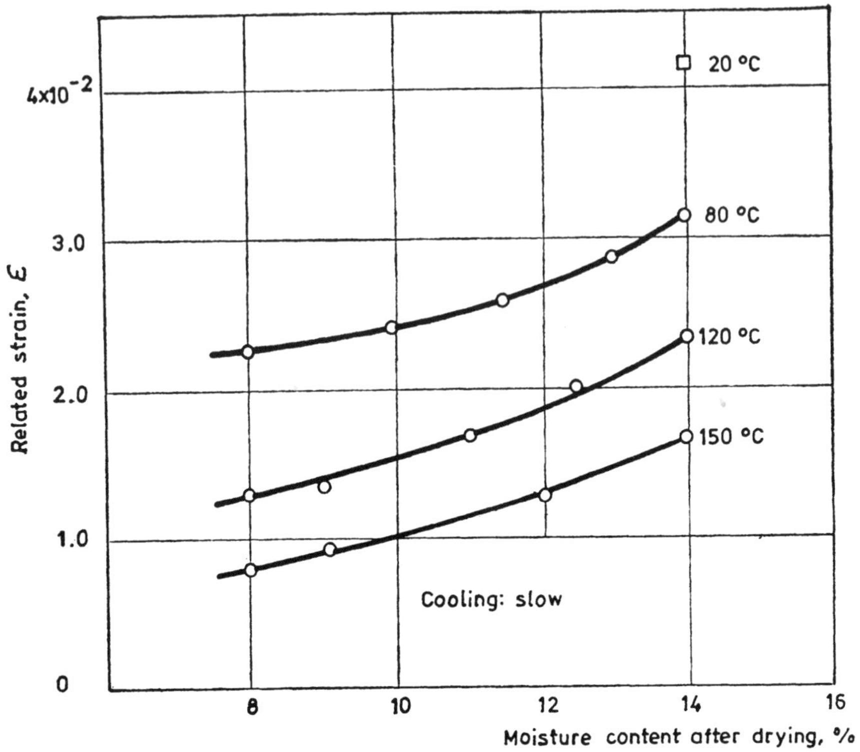

Fig. 201. Rupture strain for overdrying of maize as a function of drying temperature

brittle, whereby its susceptibility to damage increases and the permissible load decreases. For example, the force required to break maize kernels after drying at 140–160 °C is less than one-half the value required to break grains dried in ambient air.

For many products the mechanical strength is greatly lowered and the sensitivity to damage increased by overdrying and by rapid cooling after drying (see Figs 184 and 185). Figure 201 shows the variation of the rupture strain for 100 loading cycles as a function of overdrying for various drying temperatures [74]. The sensitivity to damage increases in direct proportion to the reduction of the related strain.

Moisture content and mechanical properties are in general closely correlated, and so sensitivity to damage is also moisture-dependent. In numerous cases the moisture content is also related to the stage of ripening. In this case, variations in the texture and moisture content exert effects in common on the mechanical properties and the sensitivity to damage.

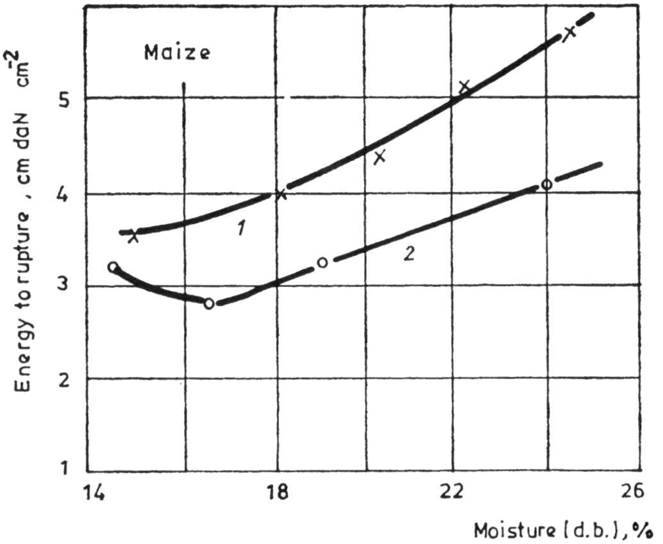

Fig. 202. Static and dynamic energy required to rupture maize as a function of moisture content *(1)* Dynamic; *(2)* static loading

Figure 202 illustrates the energy required to rupture maize kernels as a function of their moisture content, during static and dynamic loading [58]. The rupture energy varies for the given moisture range in a nearly linear fashion. With increasing moisture content the elasticity decreases, and so the energy required for rupture increases. It may also be observed that the dynamic rupture energy is always higher than the static value.

The cause of breaking during the threshing of cereals is high-velocity impact. Considerable grain breakage must generally be taken into consideration for circumferential velocities above 30 m s^{-1}. The germination capacity of apparently sound grains may be reduced by impact, the more so, the higher their moisture content (Fig. 203).

The stage of ripeness influences the sensitivity to damage decisively for certain fruits and vegetables (peaches, apricots, raspberries, pears, tomatoes). The mechanical characteristics of these fruits vary by orders of magnitude over the few days preceding full ripeness, and their susceptibility to damage varies accordingly. For other fruits and products the sensitivity to damage does not vary significantly or may even decrease during the ripening period. For example, in harvesting maize mechanically, unripe grains which have a higher moisture content, are damaged to a greater extent than are ripe kernels.

In avoiding damage to fruit species the permissible falling height and permissible static pressure are of great importance. The former is important in plan-

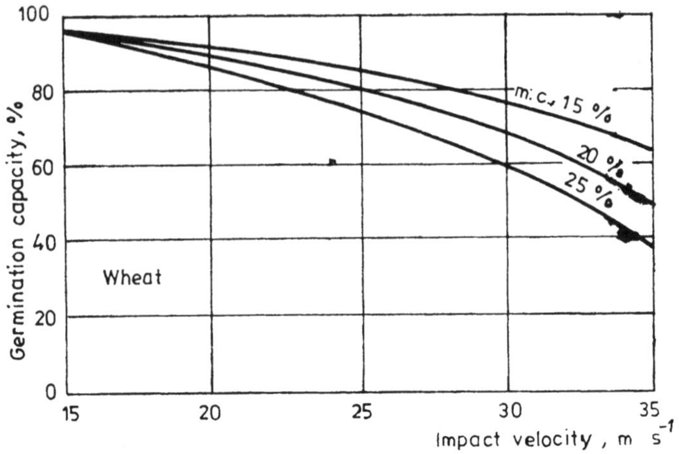

Fig. 203. Relationship between germination capacity and impact velocity during threshing of wheat

ning harvesting and handling operations, the latter in selecting the height of transport containers. Table 11 offers orientating values of the permissible falling height for various fruits. These values refer to the conditions prevailing during the optimum harvesting period. The strength of some overripe fruits (e.g., peaches, apricots, pears) decreases to such an extent that treatment without deformation is impossible. The permissible falling height for a fruit species depends also on the weight. A smaller apple can sustain a greater falling height. It is advisable to

Table 11

Fruit	Surface	Permissible falling height (cm)
Apricots	Tree	20
	Fruit	40
Peaches	Tree	10–20
	Fruit	20–40
Pears	Tree	5–10
	Fruit	10–20
Plums	Tree	60
	Fruit	120
Tomatoes	Stalk	25
	Fruit	45
Apples	Tree	3–6
	Fruit	5–10

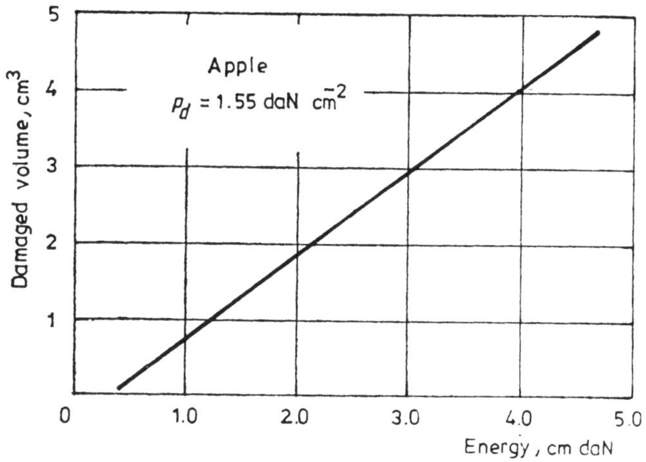

Fig. 204. Effect of impact energy on damaged volume for apples

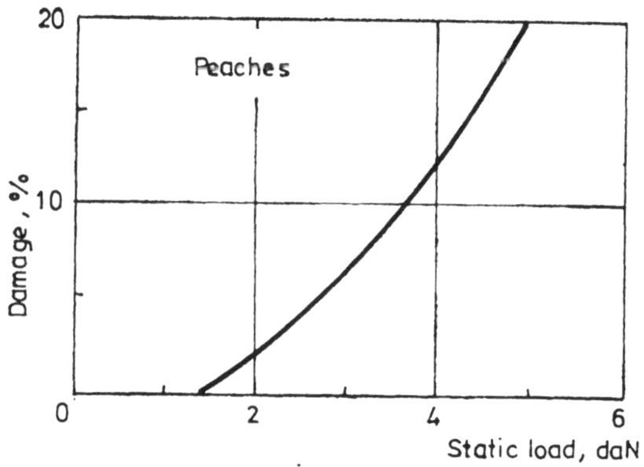

Fig. 205. Relationship between static load and damage for peaches

determine the maximum energy absorption capacity of a given fruit and to calculate the permissible falling height from it, knowing the weight. Figure 204 shows the relationship between impact energy and bruised volume for apples [94]. The damaged volume increases linearly with the energy absorbed.

Fruits are generally transported in containers, and the static and dynamic forces which then act on the fruit will cause damage if they exceed a given value. The static force may be calculated from the weight of the fruit column being transported while the dynamic load is a consequence of vibration caused by transport. The permitted static load for a given fruit may be determined experimentally. Figure 205 presents experimental static-load results for peaches

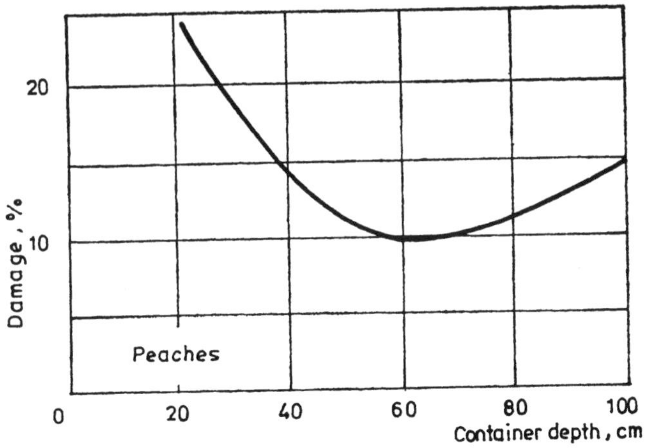

Fig. 206. Relationship between depth of container and extent of damage to peaches during transport

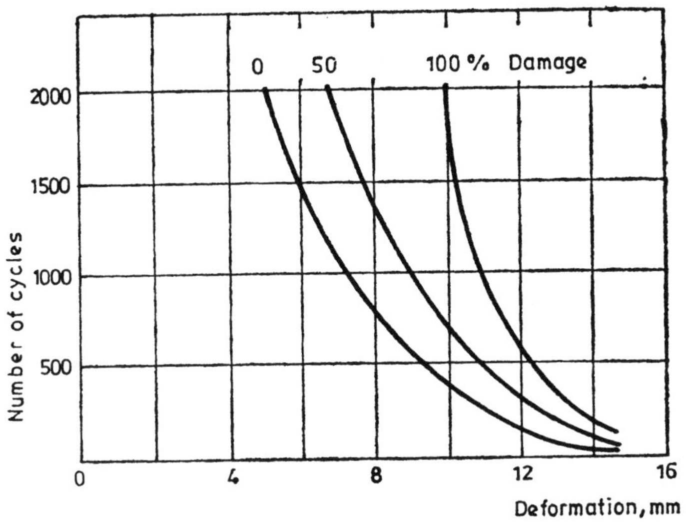

Fig. 207. Relationship between the number of loading cycles, damage and deformation for potatoes

showing that peaches can support about 1.5 daN static load without damage. This corresponds to the weight of a column of fruit approximately 70 cm high [92]. During transport the upper layer of fruit is exposed to the maximum acceleration. The deeper the container, the lower the volume ratio represented by the upper layer. Thus the proportion of fruit damaged may be reduced significantly by increasing the depth of the container up to a certain point. Figure 206 shows the proportion of fruit damaged as a function of container depth during transport over a distance of 160 km [92]. According to the investigations the optimal container depth is 60 cm. For greater depths, damage increases in the lower layers, owing to the static pressure. The damage caused by transport may be reduced by appropriate design of the suspension (generally air springs) of the transport vehicle. In addition, it is advisable to cover fruit packed into a container by a spongy layer 2–3 cm thick, in order to reduce the acceleration of the upper layer. In this way the damage may be reduced considerably [93].

The damage caused by oscillating dynamic loads may be calculated from eqns. (145) and (145a), or from other empirical equations. In certain cases relationships may exist among the number of cycles, the damage and the deformation caused by the load. Figure 207 shows such a relationship for potatoes, for a load applied by a rigid ball of diameter 60 mm. The force corresponding to a given deformation may be calculated knowing the modulus of elasticity ($E/(1-\nu^2) \simeq 65$ daN cm^{-2} in the case being considered), and on this basis the damage caused by various loads during transport or other operations may be judged.

14. AERODYNAMIC AND HYDRODYNAMIC PROPERTIES AND PHENOMENA

During the treatment of agricultural materials air or water is often used as the transport medium. Pneumatic transport and cleaning of various agricultural products have been known for a long time, and recently water has also been utilized as a means of transport, as it may involve lower energy consumption and less damage. During these proceses aerodynamic and hydrodynamic properties play an important role and must be known for optimum design and operation of the equipment.

The two most important aerodynamic characteristics of a body are its aerodynamic drag coefficient and terminal velocity.

14.1 Aerodynamic drag coefficient

A body placed into a flowing medium is subject to the action of frictional forces and a force due to the asymmetrical pressure distribution, termed the aerodynamic or drag force. The drag force on a given body may be calculated from the equation

$$W = c_w F(\gamma/2g)v^2 \tag{147}$$

where c_w is the drag coefficient, F the cross-sectional area of the body normal to the flow, and v the flow rate. Values of the drag coefficient have been determined for bodies of regular shape (spheres, cylinders, flat discs), and are generally plotted on a system of logarithmic coordinates as a function of the Reynolds number Re (Fig. 208). At low flow rates ($Re<1$), Stoke's law is valid and the drag coefficient is given by

$$c_w = 24/Re$$

The drag coefficient decreases with increasing Reynolds number. In turbulent flow the drag coefficient is nearly constant up to $Re=2.5\times10^5$. In the transition zone, an equation of the form

$$c_w = \text{const}/Re^n$$

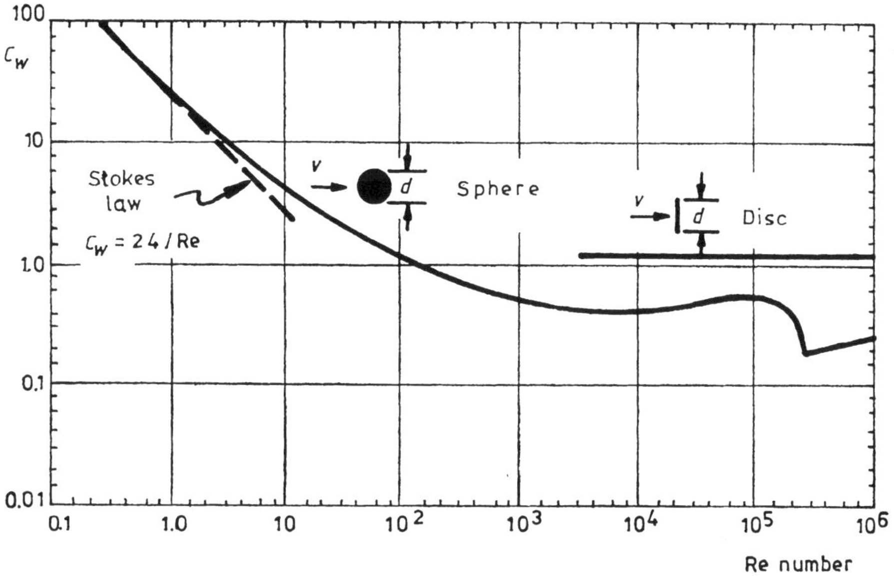

Fig. 208. Relationship between drag coefficient and Reynolds number for various bodies of regular shape

is used for the calculation, where the constant and the exponent n are valid only for a narrow interval of Reynolds numbers.

A great proportion of agricultural products are irregular bodies; in this case the drag coefficient depends also on the shape and instantaneous position (orientation), and so standard processing of data obtained experimentally is a difficult task. Generally, an equivalent diameter is calculated on the basis of the volume of the product, and this equivalent diameter is used to calculate the Reynolds number. The projected cross-sectional area is sometimes also calculated by means of the equivalent diameter, but this generally leads to considerable error, for example in the case of flat maize kernels.

Available drag-coefficient values for various crops are summarized in Table 12 [1, 96].

14.2 Terminal velocity

The acceleration of a freely falling body ceases after a time owing to air resistance, and it then falls further with a constant velocity. In this case the resisting force is identical to the weight of the body, i.e.,

$$W = c_w F(\gamma/2g) v_{cr}^2 = G$$

from which, for a spherical body,

$$v_{cr} = \sqrt{2Gg/c_w F\gamma} = \sqrt{4d\gamma_m g/3c_w\gamma} \tag{148}$$

where d is the diameter of the body, and γ_m its specific weight. For turbulent flow a constant value $c_w=0.44$ may be substituted, and then

$$v_{cr} = 1.74\sqrt{d\gamma_m g/\gamma} \tag{148a}$$

In the interval $2<Re<1000$, the equation

$$c_w = 18.5/Re^{0.6}$$

may be used to give approximate values of the drag coefficient, and the terminal velocity for spherical or nearly spherical bodies may be calculated using this value in eqn. (148).

The terminal velocity of irregularly shaped crops cannot be calculated theoretically with sufficient accuracy, and so it is better determined experimentally. The test equipment comprises essentially a dropping tube with a dropping head at the upper end and a receiver cup at the bottom (Fig. 209) [96]. The dropping head holds the product to be tested by vacuum, and drops it, when the vacuum is released. The product passes in front of a photocell, which starts an electronic timer. The product falls onto the membrane of a microphone built into the

Table 12

Crop	Drag coefficient	Terminal velocity (m s^{-1})	*Re*
Wheat	0.50	9.6	2700
Barley	0.50	7.6	2300
Maize	0.56–0.7	11.4	5700
Soybean	0.45	14.5	6300
Oats	0.47–0.51	6.6	2000
Wheat ears		2.4	
Wheat straw			
0.6 cm	0.84	5.15	
2.5 cm	0.80	4.25	
7.5 cm	0.90	3.0	
25 cm	0.91	2.7	
Potatoes	0.64	32.0	
Round stones		40.0	
Apples		42.0	1.9×10^5
Apricots		34.0	
Cherries		24.0	
Peaches		42–44	
Plums		32–34	8×10^4

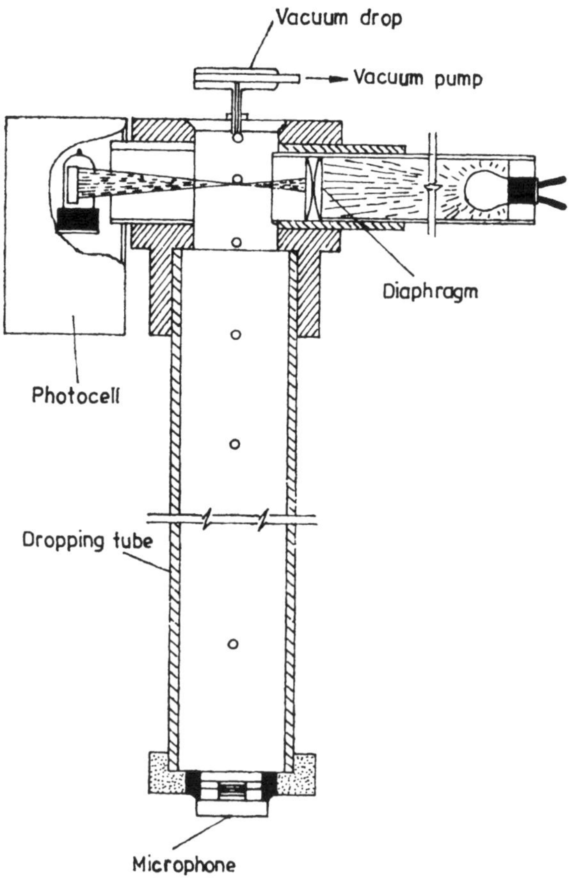

Fig. 209. Equipment for measuring terminal velocity

receiver cup, and the resulting signal stops the timer. The dropping head and receiver cup may be mounted easily on dropping tubes of various lengths (generally from 1.0 to 10 m). The measurements are performed with dropping tubes of various lengths, and the duration of fall is plotted as a function of the distance fallen. The curves constructed on the basis of the measured points may be expressed as

$$S = (v_{cr}^2/g) \ln \cosh (g/v_{cr})t$$

from which the value of v_{cr} may be determined. If the dropping tube is sufficiently long, the body will attain its terminal velocity. This is indicated by the straight section of the distance–time curve. The terminal velocity may be calculated simply in this case from the slope of the straight section. Table 12 lists terminal velocities for various crops. Given an experimental value of the terminal velocity, the drag coefficient c_w may be calculated from eqn. (148). The projected cross-sectional area F appearing in the equation may be calculated from the two longest orthogonal axes of the product:

$$F = L_1 L_2 \pi/4$$

In addition to the value of c_w determined as above, the Reynolds number characterizing the nature of the flow (laminar, turbulent, transitional) must be known.

14.3 Aerodynamic resistance of granular bulk materials

A considerable number of agricultural materials are stored or treated for various periods in the loose bulk state. During treatment or storage, air is often blown through such materials to dry, aerate or cool them. For correct selection of a fan having the required capacity, the aerodynamic resistance of the product must be known. The shapes of individual grains are irregular, and so the grains take up random positions relative to each other in the loose bulk state. The aerodynamic resistance of such piles cannot be calculated purely theoretically, but the task may be solved by determining experimentally certain constants characterizing individual products.

The aerodynamic resistance of layers of bulk granular and root bulb crops may be calculated using the equation [97]

$$\Delta p = k\xi(h\gamma/\varepsilon^4 d_k 2g)v^2 \tag{149}$$

where k is a dimensionless material characteristic, ξ the resistance coefficient of a loose bulk pile of spheres, which is a function of the Reynolds number, ε the void ratio and h the height of the bulk material, d_k the reduced diameter, γ the specific weight of air, and v the air velocity relative to the whole cross-sectional area.

The reduced or equivalent diameter of the product is calculated on the basis of the effective volume of a large number (100~1000) of individual grains or bulks, from the equation

$$d_k = \sqrt[3]{6V/\pi}$$

The true volume of a test pile may be determined by direct measurement or by calculation using the weight and volumetric weight:

$$V = G/z\gamma_{vw}$$

Table 13

Crop	Moisture content (%)	γ_{vw} (kg m^{-3})
Wheat	12	1260
	18	1220
Rye	12	1230
	18	1230
Barley	12	1180
	18	1170
Oats	12	1060
	18	1040
Maize	13	1240
Peas	16	1360
Potatoes	85	1100
Fodder beet	88	1020
Sugar beet	73	1030

where G is the weight of z grains, and γ_{vw} the volumetric (specific) weight of the individual grains.

Values of γ_{vw} for various crops are contained in Table 13.

The porosity ε of a bulk material may be calculated from the expression

$$\varepsilon = V_t - V/V_t = 1 - \gamma_b/\gamma_{vw}$$

where V_t is the total volume of the bulk material, V the volume filled by the individual grains, and γ_b the volumetric weight of the bulk material. Equation (149) shows that the aerodynamic resistance is greatly affected by the porosity. Different methods of filling a storage structure (pneumatic, gravitational, etc.) will result in differing porosities, and so the aerodynamic resistance will vary correspondingly. The aerodynamic resistance of a bulk product is greatly increased by contamination, which reduces the porosity and clogs the continuity of the pores in many places. It is therefore not advisable to store contaminated cereals.

The resistance coefficient ξ appearing in eqn. (149) is valid for a bulk pile consisting of regular spheres of identical size, and varies as a function of the Reynolds number (Fig. 210) [97]. The Reynolds number may be calculated on the basis of the equation

$$Re = d_k v/\nu$$

where ν is the kinematic viscosity of air. Individual sections of the ξ curve plotted in Fig. 210 may be replaced by straight lines and corresponding equations written.

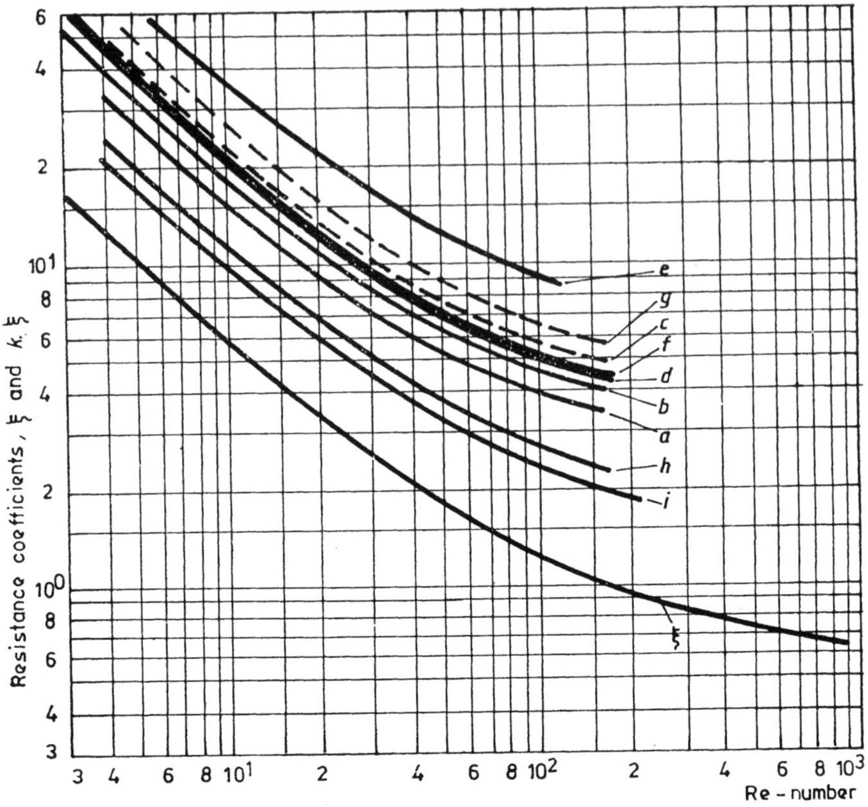

Fig. 210. Resistance coefficients ξ and $k\,\xi$ as functions of Reynolds number for bulk materials: *(a)* Spring barley; *(b)* autumn barley (cleaned); *(c)* autumn barley (uncleaned); *(d)* oats; *(e)* oats (low quality); *(f)* barley–oat mixture (cleaned); *(g)* barley–oat mixture (uncleaned); *(h)* autumn rye; *(i)* wheat

For example, the following equations may be used in the given intervals:

$$\xi = 32\ Re^{-0.74} \qquad 5 < Re < 30$$

$$\xi = 19\ Re^{-0.57} \qquad 15 < Re < 80 \quad (d_k = 3\text{–}5\ \text{mm})$$

$$\xi = 11.5\ Re^{-0.44} \qquad 30 < Re < 150 \quad (d_k = 6\text{–}10\ \text{mm})$$

The material characteristic k appearing in eqn. (149) is essentially a multiplication factor for the aerodynamic resistance of a pile consisting of spheres of identical size, and is used to characterize the resistance of individual crops. Table 14 summarizes values of k for various crops, showing that the aerodynamic resistance of cereals harvested by combines and uncleaned exceeds considerably that of cleaned cereals. The variation of $k\xi$ for various crops is also given in Fig. 210.

To render the calculations more simple, experimental results are frequently obtained in the form of the pressure drop through a layer 1 m thick as a function of the air velocity v. Figure 211 shows the aerodynamic resistance of 1 m thick layers of various granular and root bulb products as a function of air velocity [97].

Forage materials (e.g., hay or straw) are dried by blowing cold or hot air through them. To select an appropriate fan it is necessary to know the aerodynamic resistance of the material considered, possibly as a function of all the factors which may affect it.

Physical characterization of forage materials (i.e., concerning their porosity and characteristic dimensions) has not yet been achieved, and thus only empirical equations based on measured results can be established. In processing the measured results, the specific resistance $\Delta p/h$ is generally plotted as a function of air velocity (Fig. 212) [99], but the curve obtained is valid only for conditions similar to those of the measurements.

The aerodynamic resistance of forage materials is influenced by numerous factors. The most important of these are the structure of the fibers and foliage, the arrangement of the stalks, the moisture content and the volumetric weight. The effect of the latter is especially important, therefore it is also usual to present the specific resistance as a function of the volumetric weight (Fig. 213) [98]. The thinner the stalk and the more leaves on it, the steeper the curve.

The specific aerodynamic resistance may be treated on the basis of the equation

$$\Delta p/h = C\gamma_{vw}^{m} v^{n}$$

Table 14

Crop	k
Peas	1.05
Rape seed	1.2
Beans	1.4
Potatoes	1.4
Wheat (clean)	1.8
Wheat (combine-harvested)	3.0
Rye (clean)	2.7
Rye (combine-harvested)	4.4
Barley (clean)	2.7
Barley (combine-harvested)	4.4
Maize (with damaged kernels)	3.2
Sugar beet	3.3
Oats (clean)	3.8

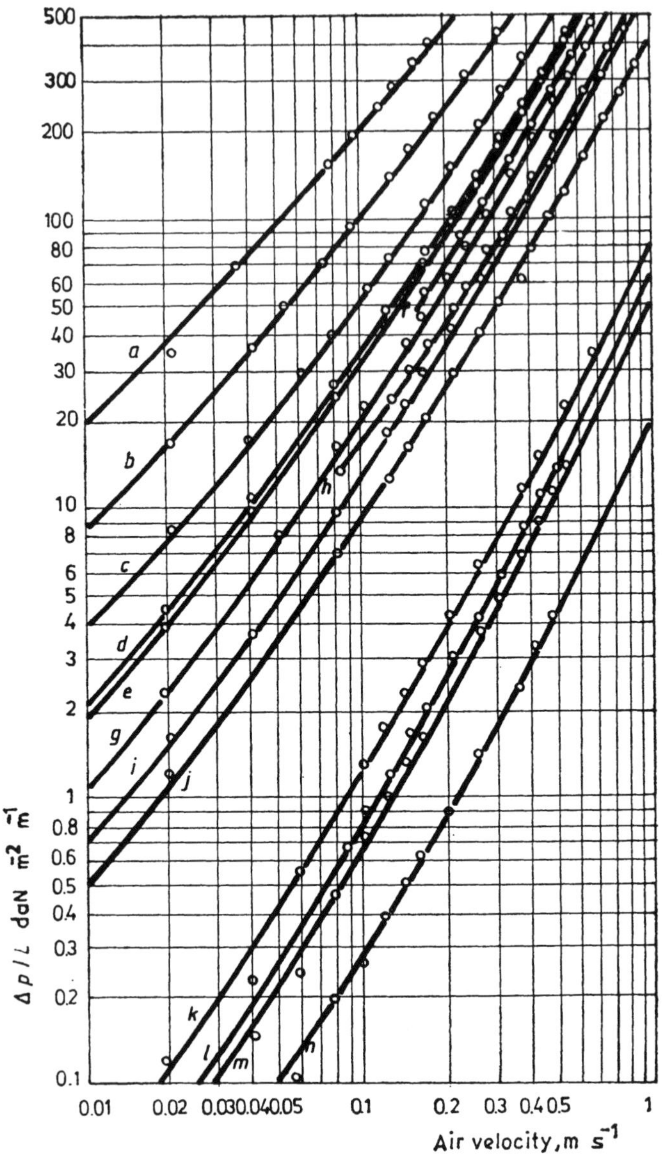

Fig. 211. Air resistance of crop and root tuberous products: *(a)* clover seed; *(b)* rape seed; *(c)* oats; *(d)* barley; *(e)* rye; *(f)* wheat; *(g)* maize; *(h)* vetch; *(i)* bean; *(j)* peas; *(k)* carrots; *(l)* potatoes; *(m)* sugar beet; *(n)* fodder beet

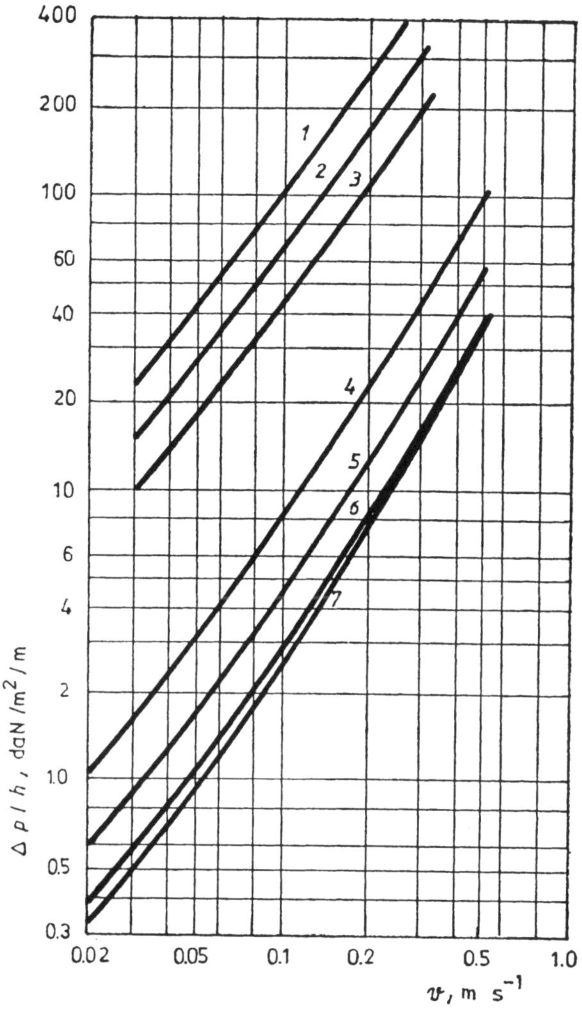

Fig. 212. Air resistance of various types of hay as a function of air velocity: *(1)* Meadow hay (moisture content 18%, $\gamma=140$ kg m^{-3}); *(2)* meadow hay (moisture content 17.5%, $\gamma=155$ kg m^{-3}); *(3)* meadow hay (moisture content 16.5%, $\gamma=170$ kg m^{-3}); *(4)* alfalfa (moisture content 23%, $\gamma=104$ kg m^{-3}); *(5)* alfalfa (moisture content 23%, $\gamma=84$ kg m^{-3}); *(6)* alfalfa (moisture content 23%, $\gamma=71$ kg m^{-3}); *(7)* alfalfa (moisture content 59%, $\gamma=95$ kg m^{-3})

where γ_{vw} is the volumetric weight. The exponent of velocity generally varies between 1.5 and 1.6. The value of the exponent m is 1.2–1.4 for straw species, 2.7–3.0 for alfalfa, and in the case of meadow-grass a value as high as $m=4.0$ may be found.

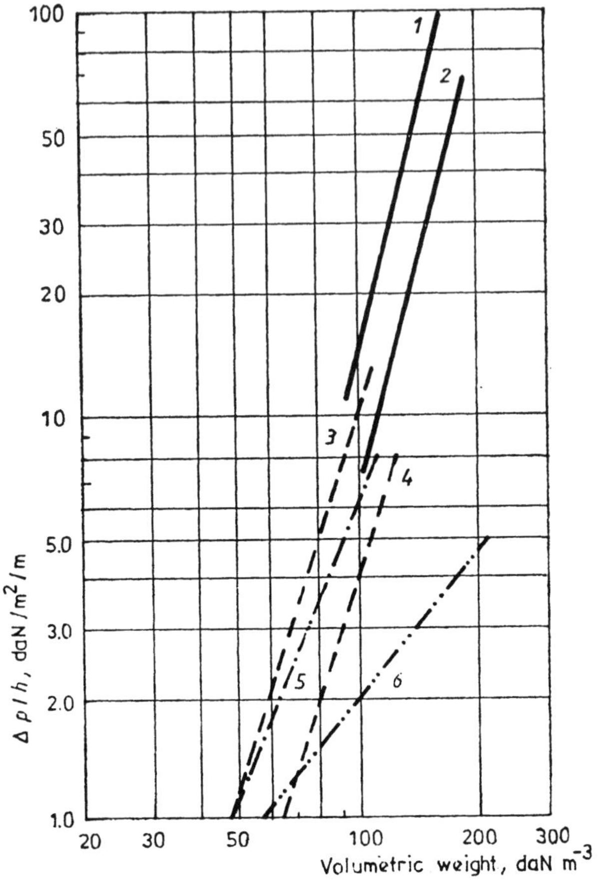

Fig. 213. Air resistance of forage materials as a function of volumetric weight. *(1)* Meadow hay (moisture content 17%, $v=0.1$ m s^{-1}); *(2)* meadow hay (moisture content 17%, $v=0.05$ m s^{-1}); *(3)* alfalfa (moisture content 12%, $v=0.1$ m s^{-1}); *(4)* alfalfa (moisture content 12%, $v=0.05$ m s^{-1}); *(5)* alfalfa (moisture content 23%, $v=0.1$ m s^{-1}); *(6)* rye straw (moisture content 20%, $v=0.1$ m s^{-1})

In ventilating hay or straw pellets the resistance depends greatly on the percentage of trimmings (fines) present. Figure 214 shows the specific aerodynamic resistance of hay pellets as a function of air velocity, for various degrees of cleanness. In the presence of 20% trimmings the specific resistance may increase as much as tenfold over the value for clean pellets.

Cereal piles are frequently aerated by introducing air through perforated ducts. Figure 215 shows the typical design of such a duct. The perforations take the form of round holes arranged in a honeycomb: the relative active surface area

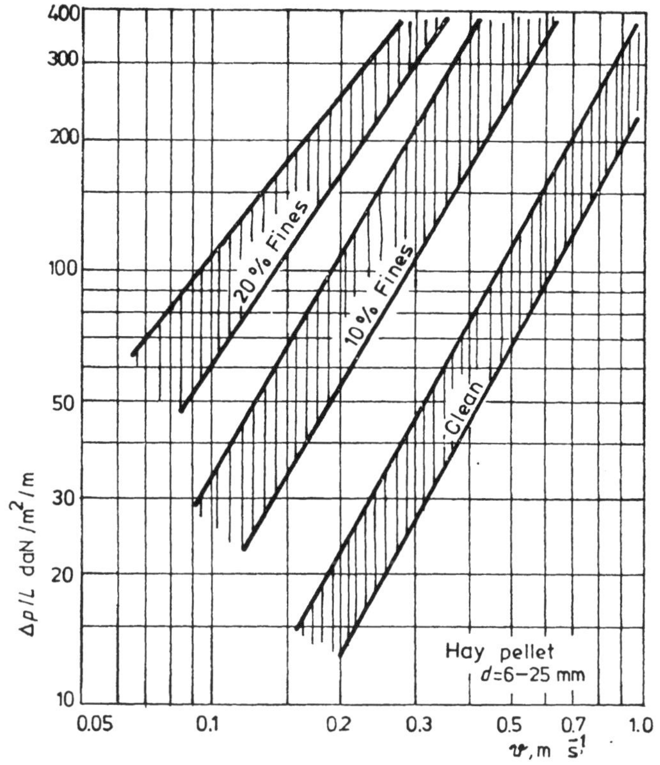

Fig. 214. Air resistance of a pile of hay pellets for various cleanness grades

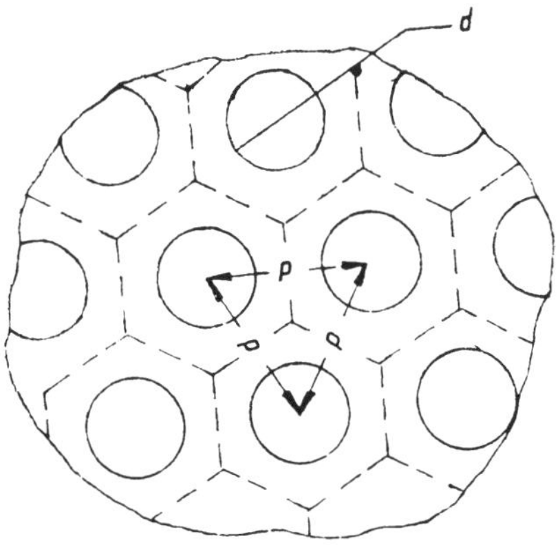

Fig. 215. Design of plate perforations in cereal aeration

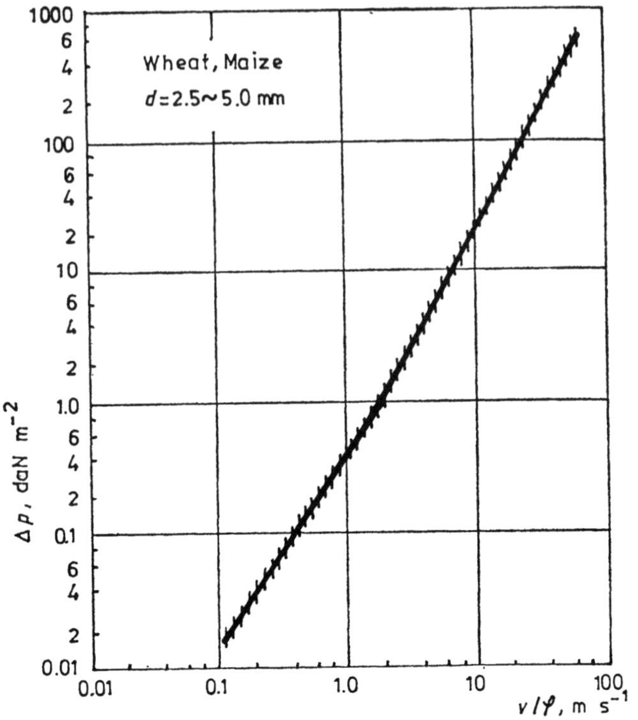

Fig. 216. Air resistance of perforated plates supporting grain

(ratio of the area of holes to the total surface area) is given by

$$\varphi = f/F = 0.907\,(d/p)^2$$

where d is the diameter of the holes, and p their center-to-center separation. The aerodynamic resistance of the perforated plate must be examined with the cereal placed on it, because a part of the cross-sectional area will be obstructed by the grains. Thus the resistance of the plate depends also on the species of cereal which is placed on it.

The aerodynamic resistance of a perforated plate may be examined by dimensional analysis [104] which supplies the general relationship

$$\Delta p/\varrho v^2 = f(Re, \varphi, \Phi)$$

where Δp is the pressure drop through the plate, ϱ the density of air, v the air velocity relative to the whole cross-sectional area, and Φ is a parameter characteristic of the cereal, φ is the relative opening of the perforated plate. On processing experimental results a simpler relationship is obtained, with the pressure drop Δp plotted as a function of v/φ (Fig. 216) [104]. The curve is valid, with a certain amount of scatter for wheat and maize with $d=2.5$—5.0 hole diameters. The value of φ varied during the investigations between 0.002 and 0.2. According to Fig. 216, the resistance of the perforated plate must also be taken into account when $\varphi<0.1$.

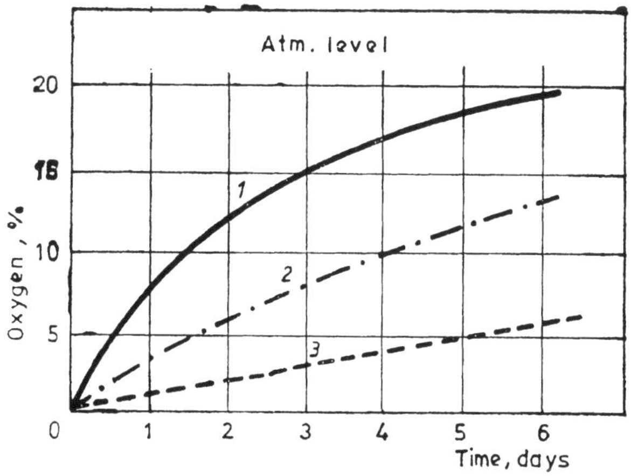

Fig. 217. Oxygen diffusion through concrete walls ($F/V=0.64\ dm^2\ dm^{-3}$). *(1)* Concrete wall 5 cm thick; *(2)* the same with latex paint coating; *(3)* the same with epoxy paint coating

An important requirement in the preservation of silage fodders is the prevention of all air exchange or oxygen diffusion through the walls of the silo. In the case of steel-plate silos the plate itself supplies perfect sealing, which must also be ensured at the junctions. Concrete silo walls are permeable, permitting diffusion of oxygen. The diffusion may be reduced greatly by painting the surfaces. The rate of diffusion of oxygen through a wall of thickness b may be described by the differential equation

$$dq/dt = D(F/C_1 b)(p_0 - p) \tag{150}$$

where dq/dt is the rate of oxygen flow, D the diffusion coefficient, F the cross-sectional area through which flow occurs, p and p_0 are the partial pressures of oxygen respectively in the chamber and in the ambient air, and C_1 is a constant. The quantity of oxygen in the chamber is proportional to the partial pressure, i.e., $dq=C_2 dp$, and so eqn. (150) may also be written in the form [103]

$$dp/dt = D(F/C_1 C_2 b)(p_0 - p)$$

integration of which gives

$$p = p_0(1-e^{-kt}) \tag{151}$$

where $k=DF/C_1 C_2 b$. Equation (151) describes the increase of the oxygen partial pressure (or concentration) as a function of time. Figure 217 shows the variation of oxygen concentration in a chamber 24.5 l in volume, one side of which is a concrete plate 25×62 cm and 5 cm thick [103]. The form of the curve corresponds to eqn. (151) and $k=0.37\ \mathrm{l\,h^{-1}}$ for the uppermost curve. Figure 217 shows that there is considerable oxygen diffusion through the concrete wall. This may be reduced greatly by painting the surface with an appropriate paint, but cannot be stopped completely. In silage towers oxygen is consumed very rapidly, and therefore the difference in partial oxygen pressure across the walls may be taken in practice as remaining identical to the partial pressure in the ambient air (155 mmHg). With this assumption, the permeability of a 5 cm thick concrete wall is $0.186\ \mathrm{m^3\,m^{-2}\,day^{-1}}$, while after coating with an epoxy paint it is $0.02\ \mathrm{m^3\,m^{-2}\,day^{-1}}$. The quantity of oxygen diffusing into a silo daily may be calculated easily, if the surface area of the silo is known. For increasing wall thicknesses the quantity diffused decreases according to eqn. (150).

14.4 Separation of foreign materials

The terminal velocities and resistance coefficients of individual materials may differ greatly, which offers the possibility of separating such materials from each other in an air stream. This method is used to separate threshing byproducts (chaff, husks, etc.) from grain, to separate grains of different specific weights, and to remove other impurities, etc. Figure 218 illustrates the mechanism of the

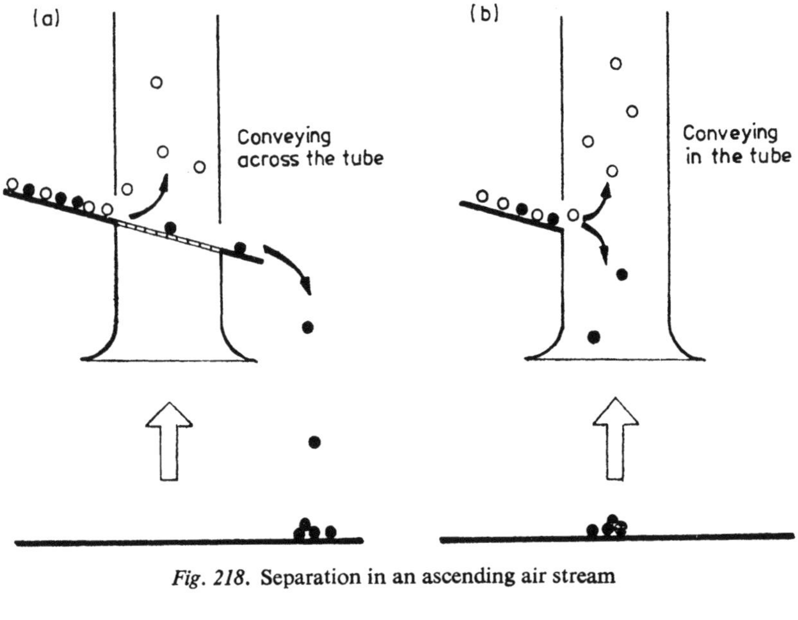

Fig. 218. Separation in an ascending air stream

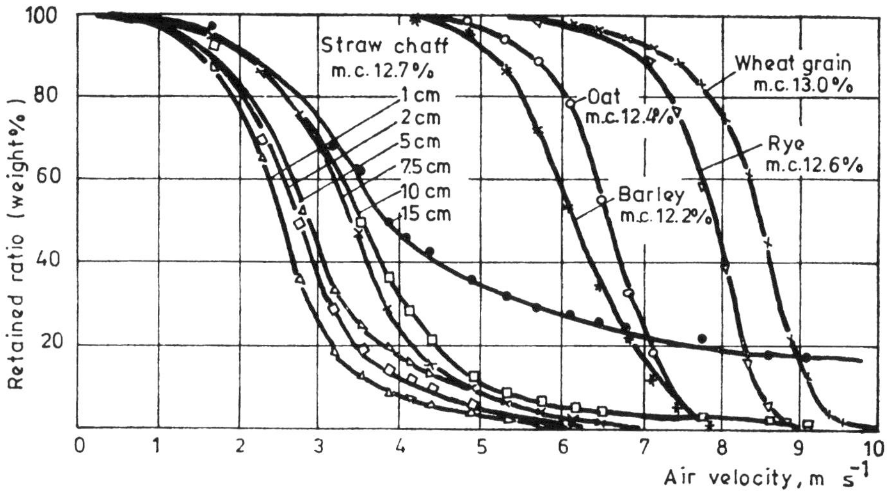

Fig. 219. Separation curves for granular and forage materials

separation in an ascending air stream. The air velocity must be selected so that it falls between the terminal velocities of the materials to be separated. The air stream will then entrain impurities of lower terminal velocity, while material of higher terminal velocity remains on the screen, or falls downwards. The terminal velocities of impurities and the individual grains of a product may differ only slightly. This means that a certain proportion of impurities will be removed at a lower air velocity, and a different proportion at a somewhat higher air velocity. The material may be best characterized in this respect by plotting a separation curve showing the retained ratio of the components as a function of air velocity (Fig. 219) [100]. The curve permits determination of the optimal air velocity and the degree of separation attainable. According to Fig. 219, wheat may be cleaned most effectively in an air stream of velocity 6 m s^{-1}. For oats, an air velocity of 5 m s^{-1} is best although contamination by longer straw haulms will not be affected.

The terminal velocity of a material is affected greatly by its moisture content, and so different air velocities are needed to separate the individual impurities, depending on their moisture content. Figure 220 shows separation curves for 5 cm long wheat straw and chaff of various moisture contents [100]. Attempts have also been made to separate potatoes, soil clods and stones in an air stream, but the separation curves overlap greatly (Fig. 221) [102], and so the method does not permit adequate separation in this case. Fruits shaken off a tree mechanically are frequently accompanied by leaves and smaller stems as impurities. These may be separated in an air stream.

Aerodynamic resistance also plays an important role in the operation of sep-

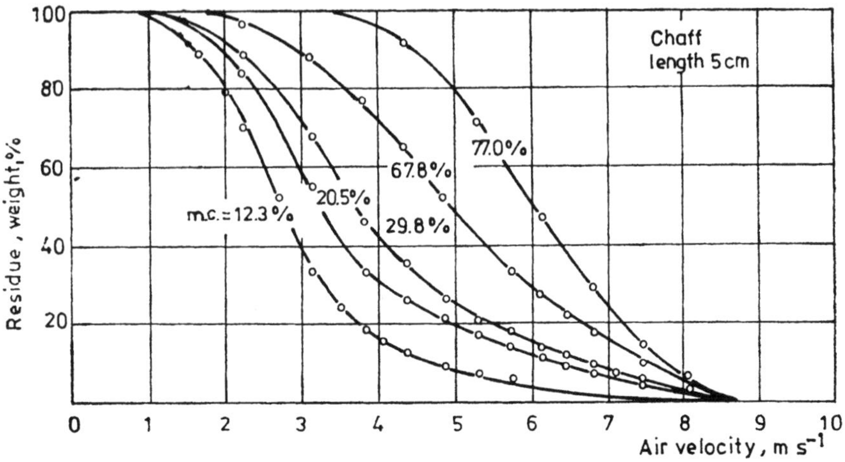

Fig. 220. Separation curves for wheat straw and chaff of various moisture contents

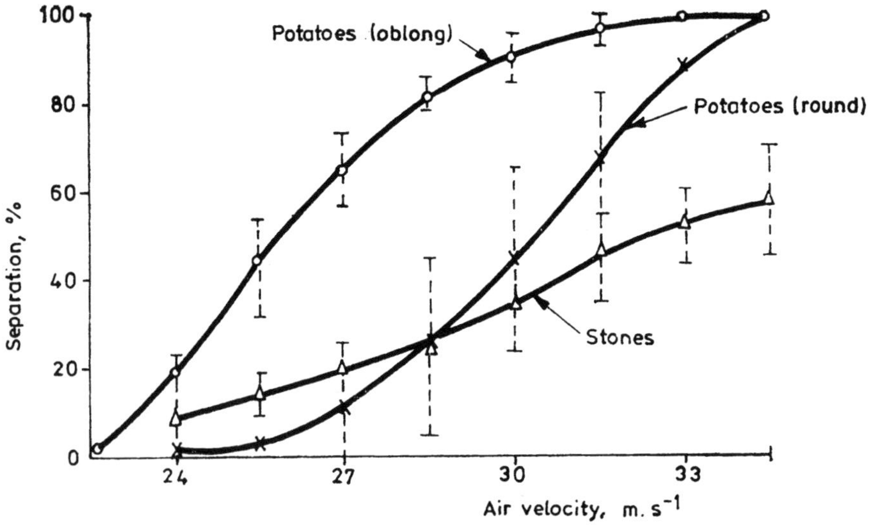

Fig. 221. Separation curves for potatoes and stones

arating cyclones. The principles of construction of such a cyclone may be seen in Fig. 222. Air entering tangentially creates a rotational movement in the cyclone, and then leaves it by ascending through the central tube. The motion of a particle in the rotational flow field is affected by the centrifugal force, by gravity and by aerodynamic resistance. The centrifugal force is given by

$$P_c = Gu_t^2/gr$$

where G is the weight of the particle, u_t its tangential velocity, and r the radius of its path. The aerodynamic resistance acting on the particle is, similarly to eqn. (146),

$$W = c_w F(\gamma/2g) v_r^2$$

where v_r is the radial velocity of the particle. If the two above forces are in equilibrium in the radial direction, the preceding equations may be equated to

$$G u_t^2 / gr = c_w F(\gamma/2g) v_r^2$$

from which the ratio of the tangential and radial velocities is

$$u_t/v_r = \sqrt{c_w F \gamma r / 2G} \tag{152}$$

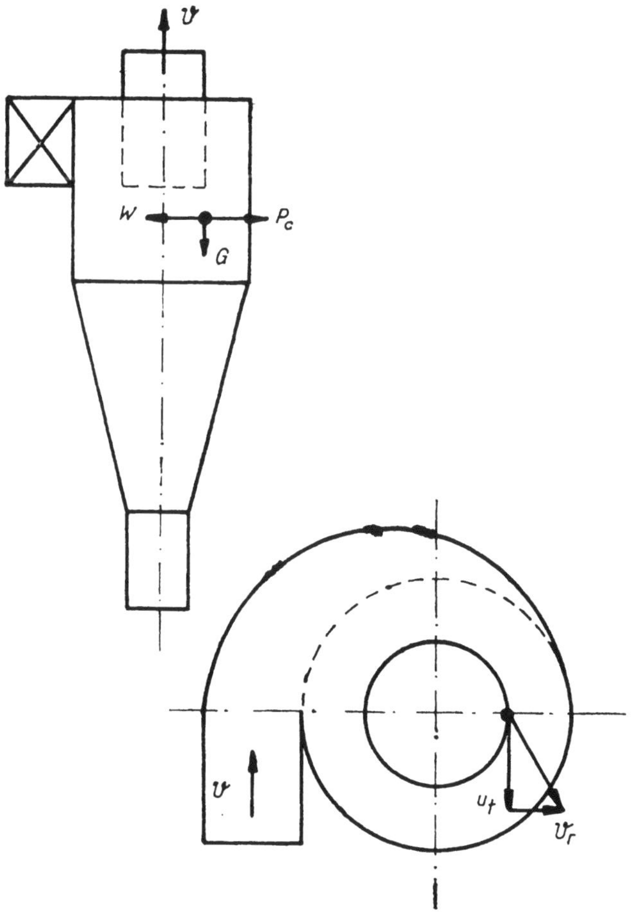

Fig. 222. Principles of construction of a cyclone

Evaluation of eqn. (152) on the basis of data for treshing products (grain, chaff, husks) shows that the centrifugal force greatly exceeds the aerodynamic resistance force. The cyclone thus separates all the solid particles together, and is unsuitable for separating the individual components from each other in this case [101].

14.5 Pneumatic conveying of agricultural materials

Pneumatic transport is frequently used in agriculture to convey granular materials. The force required for pneumatic transport is supplied by aerodynamic resistance: if this force exceeds the force required to displace the material, the latter starts to move.

Two basic types of pneumatic transport may be distinguished: flying and plug transport (Fig. 223) [108]. In the case of low particle concentrations the individual grains to all intents move freely (fly) in the tube and also have a velocity component normal to the main direction of motion, owing to collisions. In the case of plug transport the grain fills the tube and the material moves in common. A special case of plug transport is sectional plug transport, when large-sized materials (mail-tube cartridges, cereal sheafs, bales, etc.) individually fill the cross-section of the tube completely. The two transport modes differ basically. Gen-

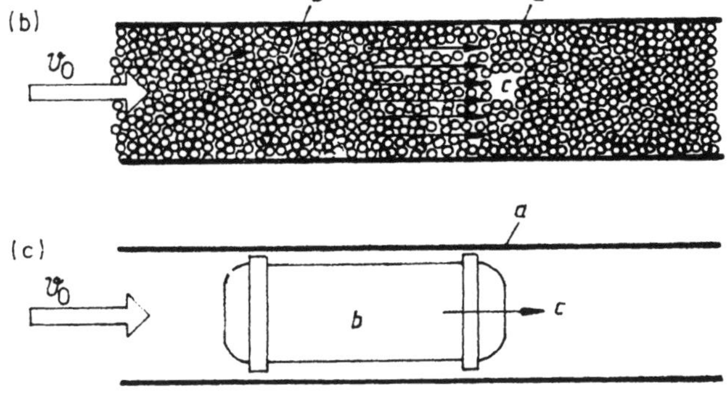

Fig. 223. Forms of pneumatic transport

erally an air velocity of 20–25 m s^{-1} is required for flying transport, while the figure for the plug transport is 0.5–2.0 m s^{-1}. For plug transport a high static pressure is necessary; for flying transport, a considerably lower static pressure.

14.5.1 State diagram for pneumatic transport

To understand better the phenomena encountered in pneumatic transport of various materials, it is convenient to study first the grain-air state diagram for transport vertically upwards in a tube. In the state diagram the pressure drop per 1 m tube length is measured on the vertical axis, and the mean velocity of the air flow (relative to the cross-sectional area of the empty tube) is plotted on the horizontal axis on a logarithmic scale, except for points in the immediate vicinity of the origin (Fig. 224) [108].

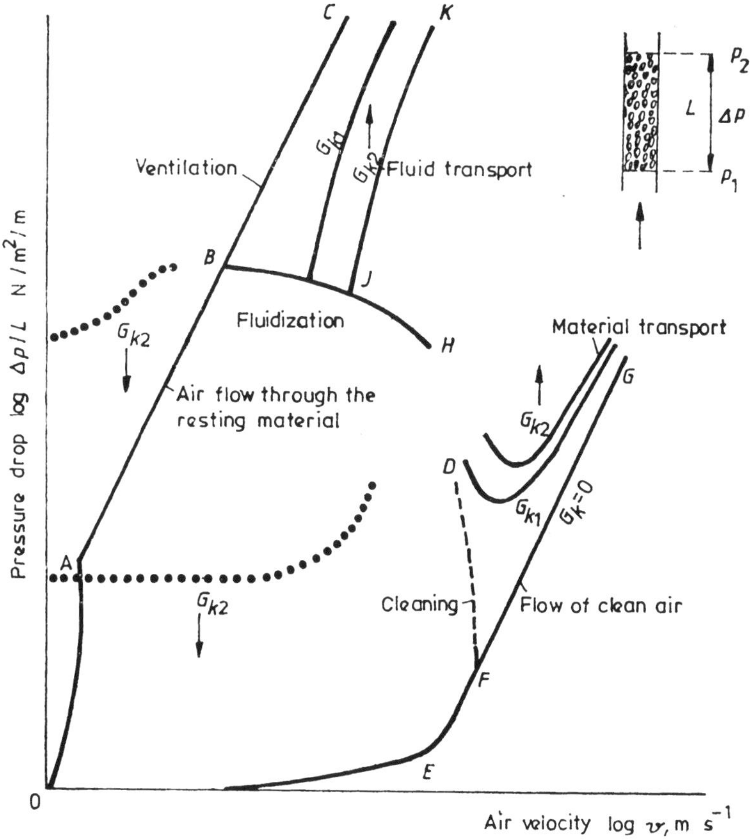

Fig. 224. State diagram for pneumatic transport

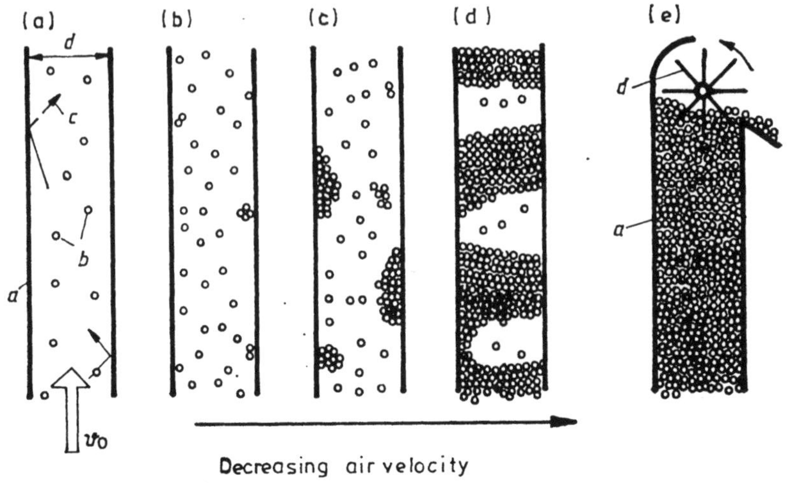

Fig. 225. Decrease in uniformity of transport with reduction of air velocity

The curve *E–G* represents the flow of air alone ($G_k = 0$) in the tube, and the curve *A–C* the aerodynamic resistance of the materials filling the tube completely, when in the stationary state. On charging the material to be transported the pressure drop increases, and therefore the curves take up positions above the curve *E–G*, depending on the concentration of material. When the air velocity decreases, the required pressure decreases for a while, and then increases again. The reason is that when the air velocity is decreased the uniformity of the flow is disturbed, bunching occurs in the tube, and the movement of both the air and the material transported becomes unsteady (Fig. 225). The line *F–D* represents equilibrium, with no transport. With further decrease in velocity the bunching increases and finally the tube is filled by the material. An equilibrium state may also be created here, without as yet any transport. This state, which is indicated by the line *B–H* in the state diagram, is termed the fluidized state. Above the line *B–H* (e.g., on the line *J–K*), plug transport is possible. Left from the line *F–D* and under the line *B–H* no transport is possible, the material moves downwards (see arrow).

The state diagram for horizontal transport is constructed similarly and the boundary curves *A–C* and *E–G* are identical. When the velocity decreases, similar phenomena appear in horizontal as in vertical transport: the development and appearance of the characteristic states are similar to those illustrated in Fig. 225.

14.5.2 Characteristic numbers and similarity laws

The shapes and sizes of individual particles of a material being transported may differ, and parameters characterizing shape and size must therefore be defined. The characteristic relationships for pneumatic transport are generally measured in model installations. The results are applicable to equipment of other dimensions only if the similarity criteria are known and the measurement results are plotted as a function of the similarity characteristics. The equivalent diameter of a particle (grain) may be calculated from the volume using the equation

$$d_e = \sqrt[3]{6V/\pi} \tag{153}$$

The shape coefficient expresses the actual surface area as compared to the surface area calculated using the equivalent diameter:

$$f = F/(d_e^2 \pi) \tag{154}$$

In the case of plug transport, when the tube is filled completely by material, the hydraulic diameter of the bulk material, calculated as the ratio of the pore volume between the grains to their surface area, is the characteristic parameter. Introducing the concept of bulk porosity,

$$\varepsilon = (V_t - \sum V)/V_t \tag{155}$$

the hydraulic diameter may be expressed as

$$d_h = 6(V_t - \sum V)/\sum F = [\varepsilon/f(1-\varepsilon)]\,d_e \tag{156}$$

where V_t is the total volume.

Geometrical similarity is ensured by the condition

$$D/d_e = \text{const}$$

where D is the diameter of the tube. In addition, the shapes of the tube and the material transported must also be similar.

Similarity of operational conditions is ensured by adopting the specific load, i.e.,

$$\gamma_m/\gamma_{vw} = \text{const}$$

where γ_m is the weight of material found in 1 m^3 tube volume during transport, and γ_{vw} the volumetric weight of the material.

Stationary flow for the transport medium is characterized by the Reynolds number (ratio of the forces of friction and inertia):

$$Re = v_0 D/\nu \tag{157}$$

where D is the tube diameter, v_0 the air velocity relative to the full cross-sectional area of the tube, and ν the kinematic viscosity of air. The ratio of inertial and gravitational forces gives the Froude number:

$$Fr = v_0/\sqrt{Dg} \tag{158}$$

In addition to the air velocity, the velocity c_0 and the terminal velocity of the particles v_{cr} also play an important role. The velocities c_0 and v_{cr} may also be used in the Froude number:

$$Fr^* = c_0/\sqrt{Dg} \tag{158a}$$

or

$$Fr' = v_{cr}/\sqrt{Dg} \tag{158b}$$

In the case of plug transport the tube is filled completely by material, therefore the hydraulic diameter must be used in expressing the Reynolds number:

$$Re = d_h v/\nu$$

where v is the relative velocity between the air and the material: its value is $v = v_0/\varepsilon$.

Taking d_h from eqn. (156), the Reynolds number may be written as

$$Re = [1/f(1-\varepsilon)]\, v_0 d_e/\nu$$

14.5.3 Distributions of air velocity and material in transport tubes

Material transported in horizontal tubes does not fill the space uniformly, but is found mainly in the lower part of the cross-section, owing to gravity. The air velocity profile is greatly modified by the unequal material distribution, as compared to the symmetrical velocity distribution prevailing during the transport of air alone. Figure 226 shows the distribution of air velocity in a tube 61 mm in diameter during transport of air and wheat at various air velocities [108]. The higher the concentration of transported material, the more asymmetrical the distribution of air velocity; this cannot be alleviated to any significant extent by increasing the air velocity.

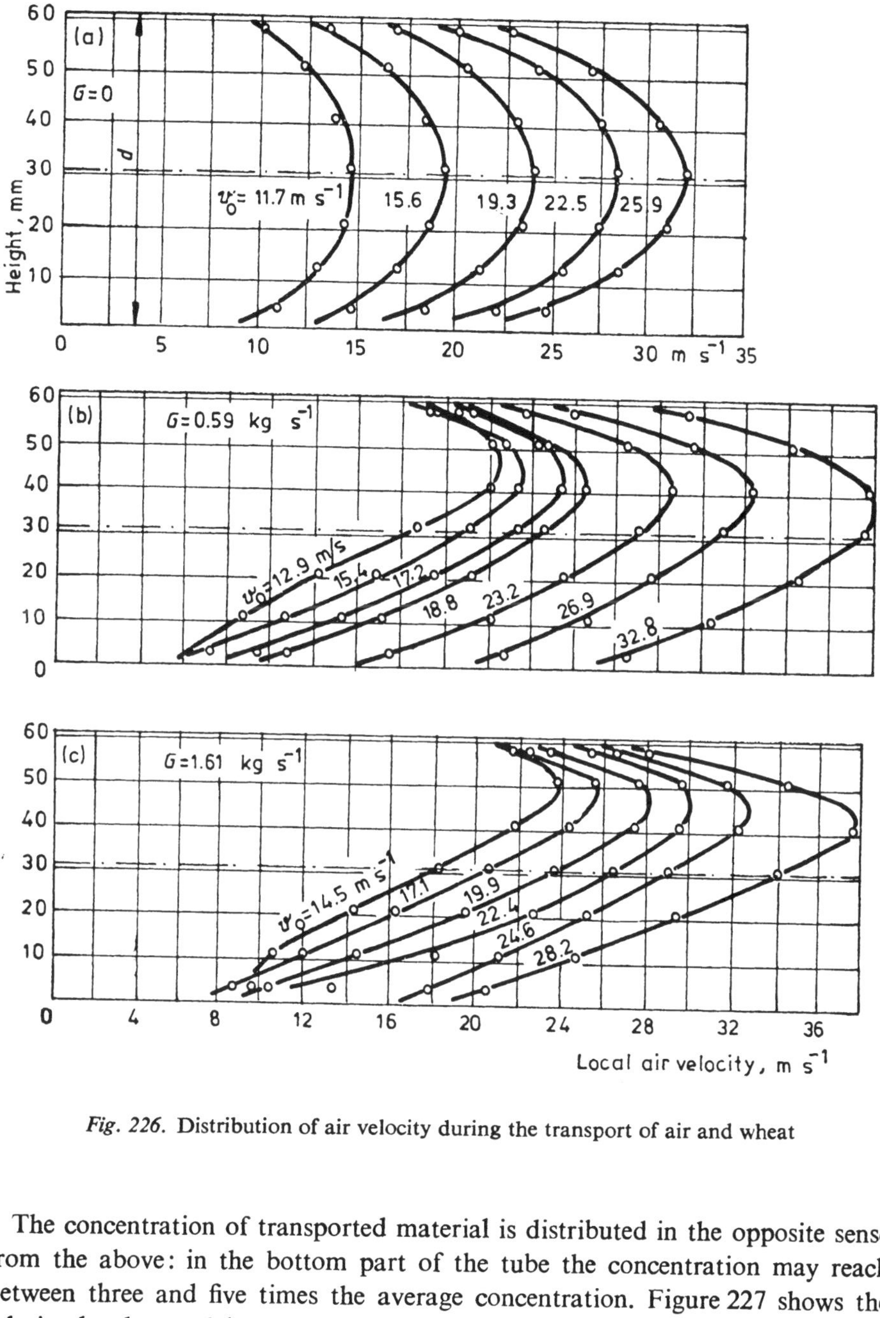

Fig. 226. Distribution of air velocity during the transport of air and wheat

The concentration of transported material is distributed in the opposite sense from the above: in the bottom part of the tube the concentration may reach between three and five times the average concentration. Figure 227 shows the relative local material concentration as a function of height in the tube cross-section during transport of wheat. The tube diameter is 61 mm, and two cases

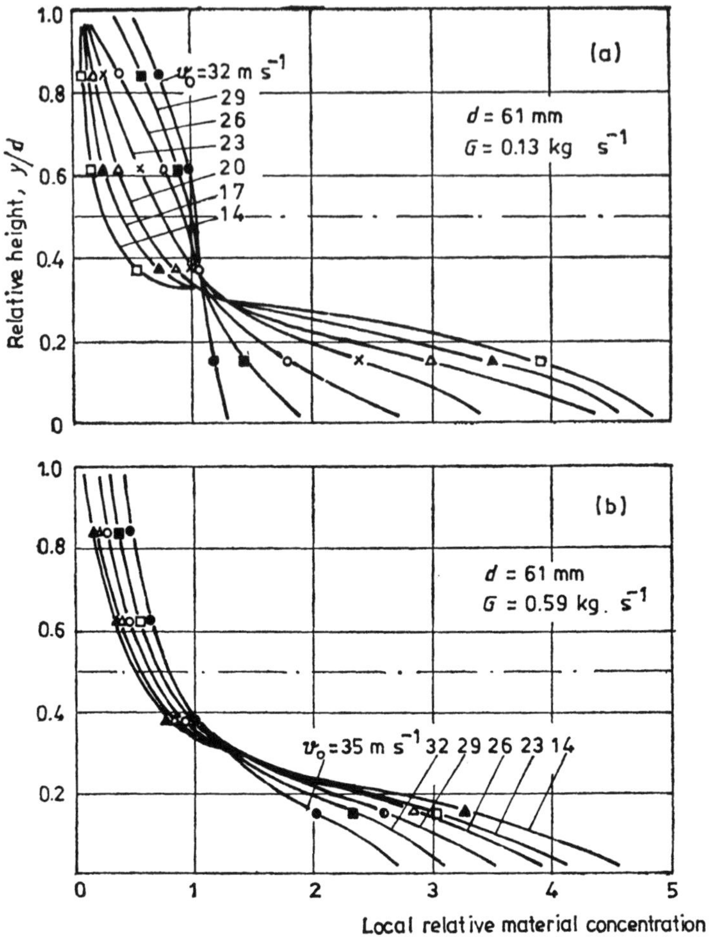

Fig. 227. Relative local material concentration as a function of height in transport tube during transport of wheat

are considered, with transport throughputs of 0.13 and 0.59 kg s^{-1}. The higher the transport throughput, the less the effect of the air velocity on the distribution of the material. From the above it may be seen that in a horizontal transport tube the material moves mostly in the lower part, and the air, at a higher velocity, in the upper part of the tube. Consequently the mean effective air velocity (v_0^*) acting on the material is lower than the mean air velocity v_0. The ratio v_0^*/v_0 is lower, the lower the air velocity v_0 and the higher the concentration of transported material (Fig. 228) [108].

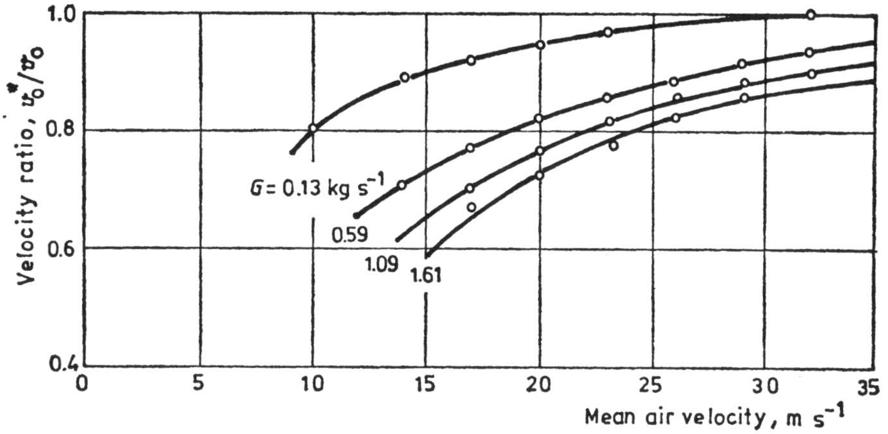

Fig. 228. Mean effective air velocity as a function of mean air velocity

14.5.4 Pressure loss in transport tubes

During pneumatic transport the pressure decreases along the transport tube, owing to various sources of resistance. The total resistance is composed of the following components:

(a) frictional resistance to air flow in the tube itself;

(b) resistance caused by the material being transported, comprising frictional, impact and lifting resistance (for a tube tilted at a given angle); and

(c) resistance to acceleration (i.e., inertia) within the starting section and at elbows.

The frictional resistance to air flow in a tube may be determined from the equation

$$\Delta p_l = \lambda_0 (L\gamma/D2g) v_0^2 \qquad (159)$$

where λ_0 is the friction coefficient of the tube, whose value varies between 0.013 and 0.02. In the case of long pipelines and greater pressure losses, expansion of the air must also be taken into account. The velocity of the gas increases as a result of expansion, and so the pressure loss will be greater [109]. The effect of expansion on pressure loss may be neglected up to $\Delta p_t = 1000$ daN m^{-2}.

The resistance caused by the transported material may also be given in the form of eqn. (159), i.e., in the form of resistance coefficients related to the velocity c_0 of the material. Thus the total resistance in steady-state operation may be calculated from the equation [108]

$$\Delta p_t = \lambda_0 (L\gamma/D2g) v_0^2 + (\lambda_1 + \lambda_2 + \lambda_3)(L\gamma_m/D2g) c_0^2 \qquad (160)$$

where λ_1, λ_2 and λ_3 are resistance coefficients characterizing respectively the friction acting on the material, collisions within the material, and lifting of the material ($\lambda_3=0$ for horizontal transport). Values of the coefficients λ_1, λ_2 and λ_3 may be plotted as functions of various variables [108]. In this way, transport systems can be dimensioned with reasonable accuracy. Theoretical calculation of the additional sources of resistance due to the material according to eqn. (160) is in principle not feasible, since the majority are not of a frictional nature. Consequently the expressions for λ_1, λ_2 and λ_3 are complicated and fail to indicate clearly the factors on which the individual resistances depend.

Pápai [109, 113] assumed that the additional resistance originates mainly from impact, i.e., equilibrium of the particles may be expressed by the equation

$$c_w F(\gamma/2g) w^2 = \xi(m_1 c_0^2/2)$$

where w is the relative velocity and m_1 the mass of the particles, and ξ is an impact coefficient characteristic of the transported material. Its value for dry wheat transported by air is 0.08–0.09. From the equilibrium equation,

$$w = \sqrt{B} c_0 = v_0 - c_0 \tag{161}$$

or

$$c_0 = [1/(1+\sqrt{B})] v_0 = (1-s) v_0$$

where

$$B = \xi G_1/\gamma F c_w$$

According to eqn. (161) the velocity of a given transported material is proportional to the air velocity, i.e., the slip s in transport is constant. The additional resistance due to transport in a 1 m long pipeline is

$$\Delta p_2/L = k_p(Q_m v_0/gF) \tag{162}$$

where Q_m is the material flow rate, F the cross-section of the tube, and $k_p = \xi(1-s)/2$; its value for dry wheat transported by air is 0.023–0.026, while the slip of the material flow is 38–40%.

During pneumatic transport of materials at higher concentrations and lower air velocities, the greater part of the material moves by sliding along the lower surfaces of the tube, and so the resistance in this case is due mainly to friction. The equilibrium equation in this case is

$$c_w F(\gamma/2g) w^2 = kG_1$$

from which

$$w = \sqrt{2kG_1 g/\gamma F c_w} = \text{const}$$

The transport velocity of the material is $c_0 = v_0 - w = v_0 - \text{const}$, i.e., the material lags behind the transporting air with a constant velocity difference. The pres-

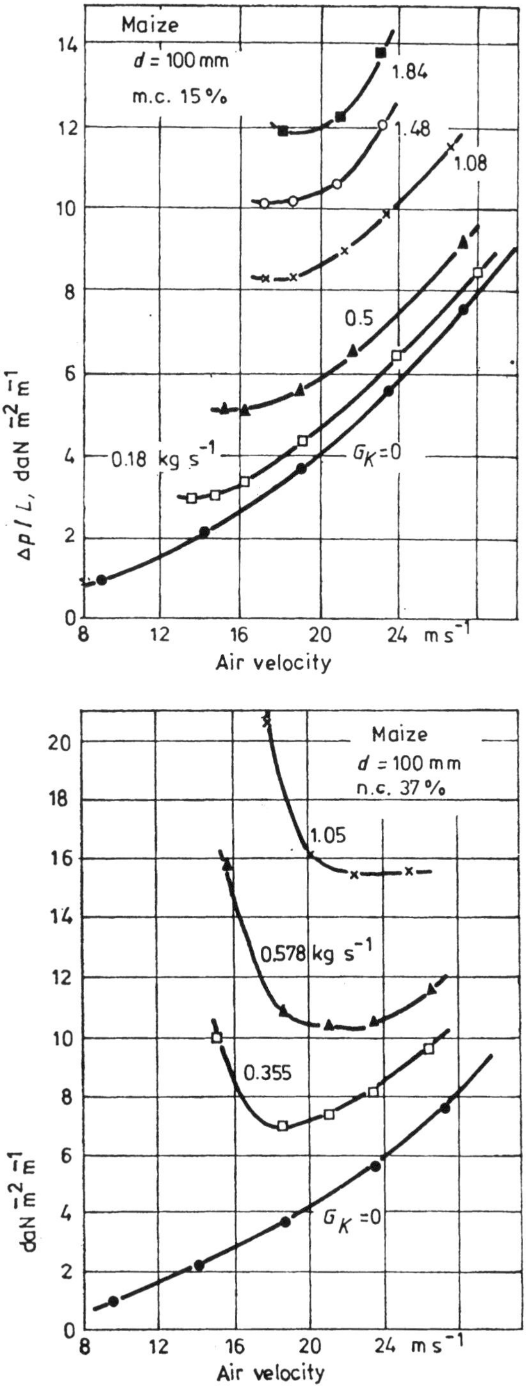

Fig. 229. Pressure drops required for transporting air-dried and wet maize as functions of air velocity

sure loss due to friction is

$$\Delta p_1/L = k(Q_m/c_0 F) \tag{163}$$

where the resistance coefficient k must be determined experimentally. A relative increase of this latter resistance component causes continual increase of the total additional resistance with decreasing air velocity, after reaching a minimum value. The moisture content of cereals has a large influence on the elasticity of the grains and on the friction coefficient between the grains and the tube wall, and so in transporting wetter cereals the additional resistance increases. Figure 229 shows the pressure loss as a function of the air velocity required to transport dry and wet maize (with 15 and 37% moisture contents) in a tube 100 mm in diameter [116].

During vertical tube transport the additional resistance has two components: the pressure loss spent in lifting the transported material, and the pressure loss spent in overcoming impact resistance. The pressure loss required to lift the transported material is

$$\Delta p_G/L = Q_m/c_0 F \tag{164}$$

while the resistance caused by impact, similarly to eqn. (162), is

$$\Delta p_2/L = k'_p(Q_m c_0/gF) \tag{165}$$

where $k'_p = 0.093\ \text{m}^{-1}$. Figure 230 illustrates the velocity conditions during vertical transport of wheat [109, 114]. As may be seen, the relative velocity of the grains ($w = v_0 - c_0$) increases with increasing air velocity.

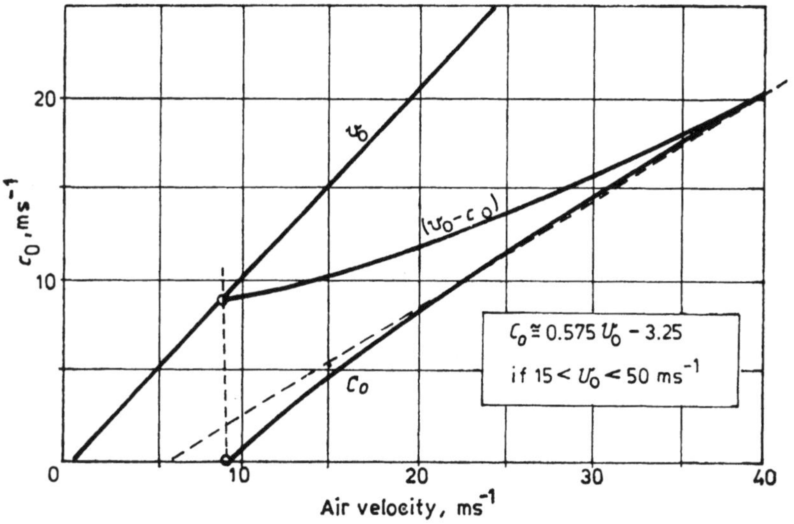

Fig. 230. Velocity relations in vertical material transport

14.5.5 Limiting velocity for pneumatic transport

The air velocity for transporting various materials should be selected so as to ensure safe and continuous forwarding of material in the tube. When the air velocity decreases, a proportion of the material will settle in the bottom of the tube and accumulate into bunches. Below a given air velocity transport becomes impossible [107, 108, 115].

In the case of air velocities that are too high, the material may be damaged and the energy efficiency of transport decreases. Therefore efforts must be made to effect transport at the lowest possible air velocity which permits stable, continuous transport.

Figure 231 presents limiting velocities for clogging in transporting maize and wheat. The values generally vary between 20 and 25 m s^{-1}. The greater the tube diameter, the higher the limit velocity for clogging [107, 108].

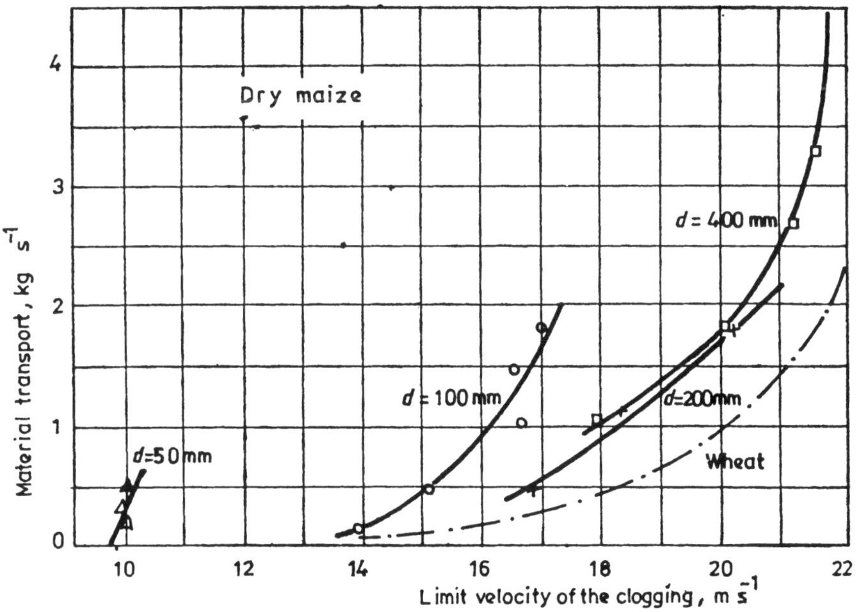

Fig. 231. Limiting air velocities for clogging during transport of maize and wheat

14.5.6 Unsteady states of motion

Pneumatic transporting equipment also contains tube sections in which the state of motion is not steady. Such sections may be those following a feeding device, or built-in elbows and subsequent sections. During charging the grains

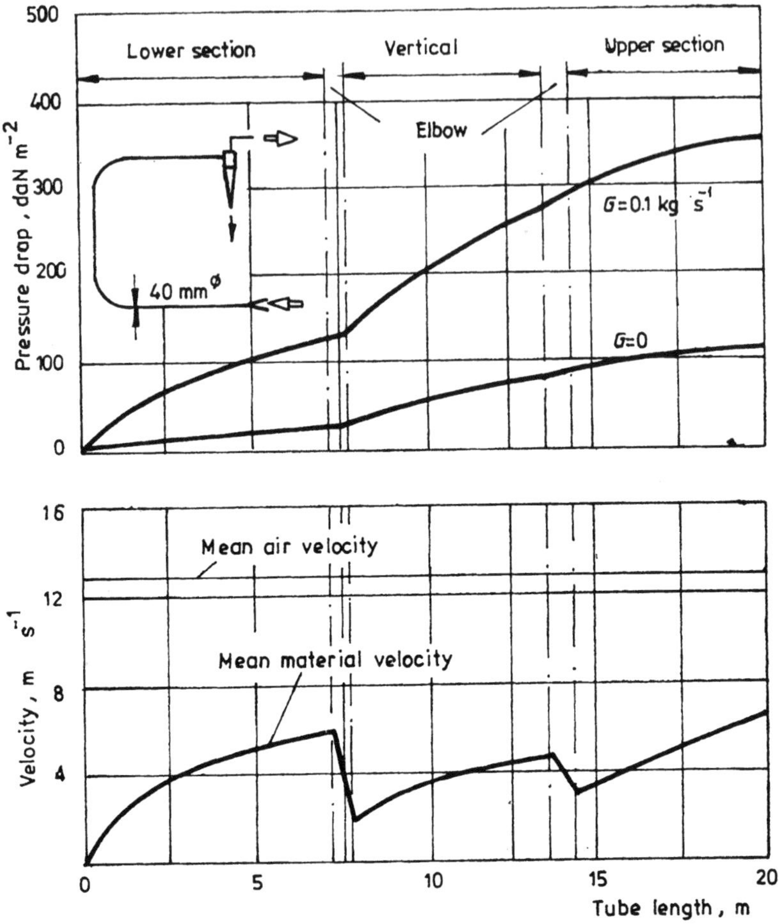

Fig. 232. Unstationary states of motion in a pipeline

must be accelerated practically from zero to the velocity corresponding to the operational state. This is the starting stage. In curved regions and elbows of a pipeline the velocity of the grains will decrease, followed by acceleration. This is illustrated clearly in Fig. 232, which shows the variation of the pressure loss and transport velocity over the length of such a tube.

Acceleration requires energy, and therefore losses due to acceleration must be taken into account in dimensioning. The pressure loss resulting from acceleration may be calculated on the basis of the momentum theorem. Accordingly,

$$F \Delta p_d = mc_0 - mc_1$$

where F is the cross-sectional area of the tube, m the mass of material charged per second, and c_1 the velocity of the material at the beginning of acceleration. During charging $c_1 \simeq 0$, therefore the additional pressure loss in the starting stage is

$$\Delta p_d = Q_m c_0 / Fg \tag{166}$$

where Q_m is the weight of material charged per second. The duration of acceleration and the distance covered by the grains during this time may be obtained by solving the equation of motion [109]. For wheat and maize, in the case of moderate concentrations for which the slip of material may be assumed to be 40%, the length of the starting section is 9.2 m, independent of air velocity.

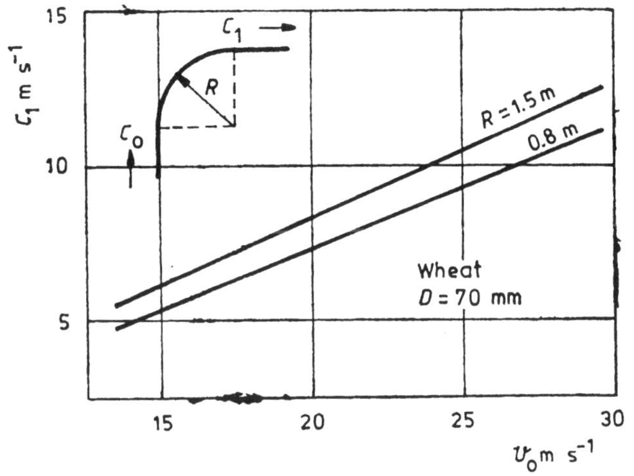

Fig. 233. Outlet velocity of wheat grains from a horizontal 90° elbow as a function of air velocity

Elbows may be installed in the horizontal or vertical plane. The decrease in transport velocity is less at a horizontal elbow, and so the total pressure loss is also lower. The radius of curvature of an elbow typically amounts to 4–6 times the tube diameter. Figure 233 shows the outlet velocity of wheat grains as a function of air velocity for a horizontal 90° elbow [110]. The curves are valid for relatively low material concentrations. The velocity loss at elbows installed in the vertical plane is higher, especially in the case of low inlet velocities. Figure 234 presents the outlet velocity as a function of inlet velocity for vertical elbows having various radii of curvature (the dashed curve refers to a horizontally disposed elbow). The additional pressure loss is created not so much in the elbow as in the straight section following it (Fig. 235) [110].

The total resistance is composed of the pressure loss arising in the transport of air alone, and the pressure loss due to the transport of material. For lower

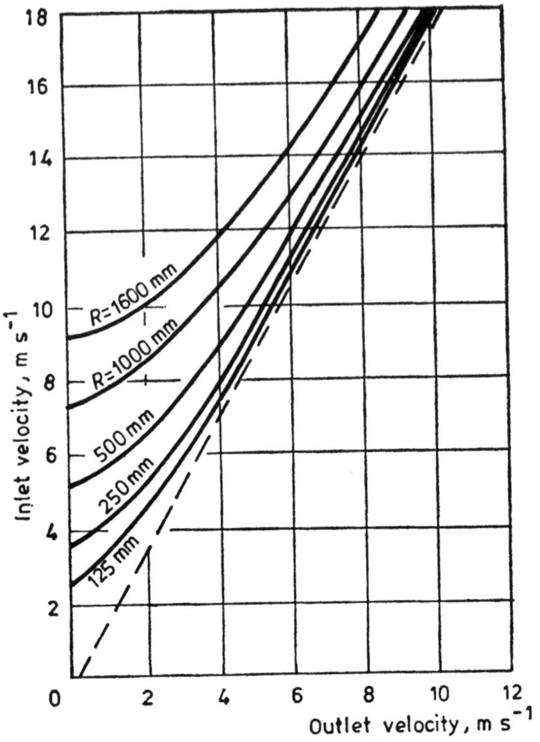

Fig. 234. Leaving velocity of grains as a function of entering velocity

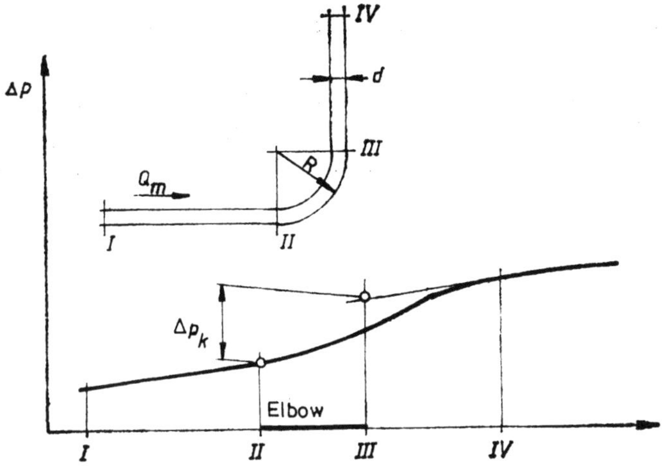

Fig. 235. Pressure loss at an elbow and in the subsequent tube length

material concentrations the pressure loss due to material transport increases linearly with the air velocity v_0 and material flow rate Q_m:

$$\Delta p_d / v_0 = kQ_m \tag{167}$$

where k is a constant which may be assigned a value of 8.5 for horizontal 90° elbows having $R/D = 6$–10.

14.5.7 Pneumatic transport of forage materials

The aerodynamic characteristics of forage materials are generally unknown, especially for piles with a random orientation of constituents and so transport equipment must be dimensioned on the basis of experimental data. During transport, forage materials again show an asymmetrical distribution in the transport tube: the material concentration is significantly higher in the lower part of the tube. The air velocity profile is correspondingly asymmetrical. As a result, the transport velocity is lower in the lower parts of the tube and higher in the upper parts. Thus, for example, during the transport of alfalfa chaff, 15 m s^{-1} transport velocity was found in the lower third and 20 m s^{-1} in the central third of the tube cross-section, for a mean air velocity of $v_0 = 26$ m s^{-1}.

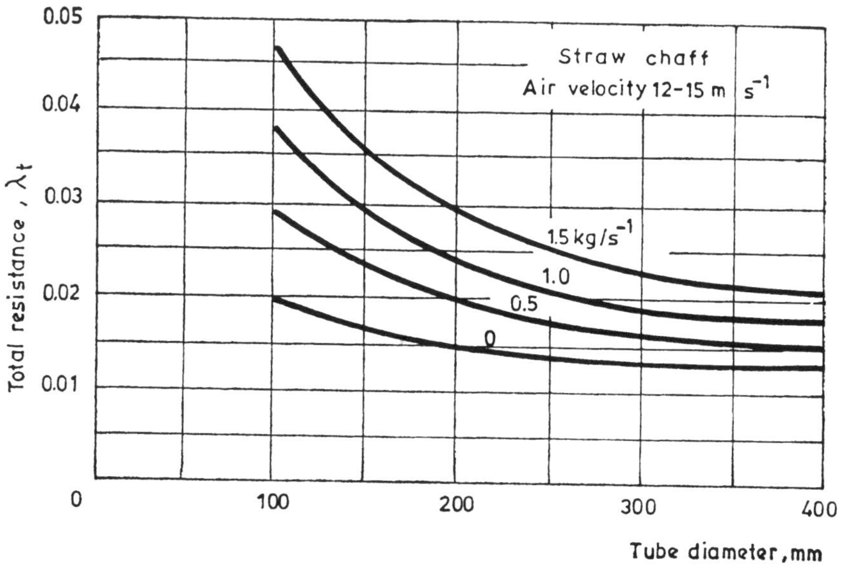

Fig. 236. Total resistance during transport of straw chaff as a function of tube diameter

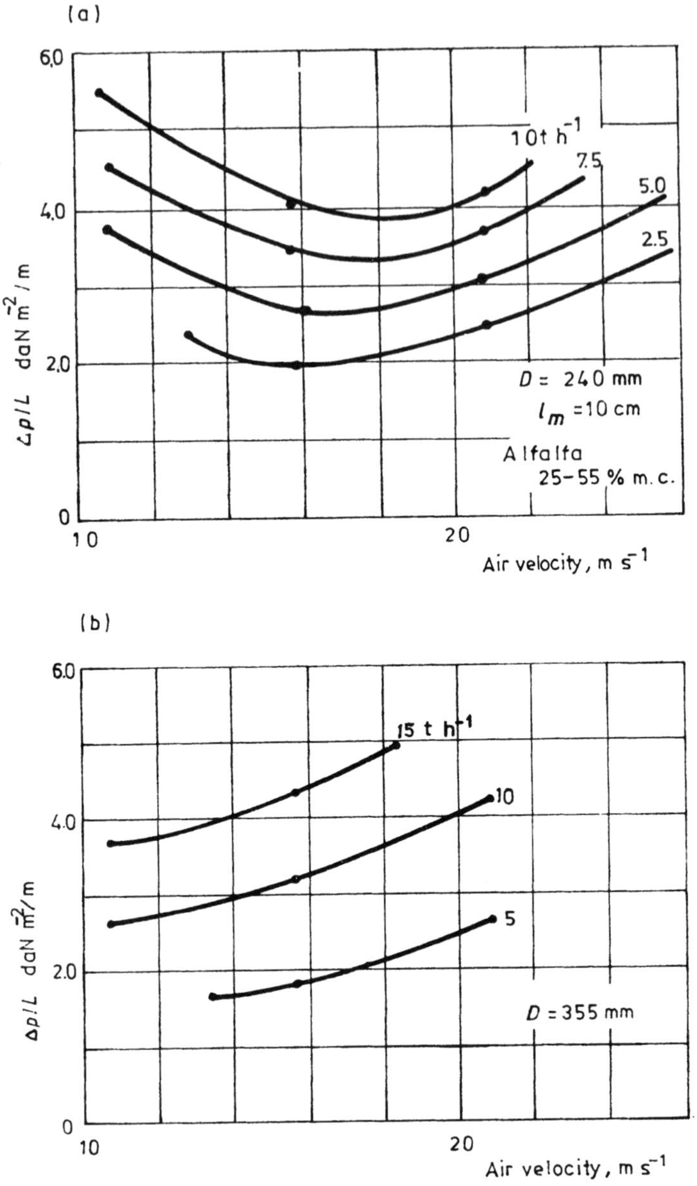

Fig. 237. Specific pressure loss required for transporting alfalfa chaff

The pressure loss in transport equipment may be calculated on the basis of the total resistance coefficient λ_t or the pressure loss per meter, with the addition of resistance in the starting section and any elbows. Knowing λ_t the pressure loss over a horizontal section may be found from

$$\Delta p_t = \lambda_t (L\gamma / D2g) v_0^2$$

Figure 236 shows the resistance coefficient for straw chaff as a function of tube diameter. The theoretical chaff length is 24 mm, the air velocity 15 m s^{-1} [107]. Figure 237 presents the specific pressure loss values required to transport alfalfa chaff, for two tube diameters [117]. By means of the data listed, pipelines of any arbitrary length may be dimensioned.

14.6 Fluid-bed conveying

A special case of pneumatic transport is that of fluid-bed conveying, in which bulk material is brought into the fluidized state by an appropriate air flow (see Fig. 244). Fluidized materials have properties closely similar to those of liquids, and may be transported either in pipelines or in open channels with a slightly downwards slope.

Figure 238 shows the principal elements of a fluid-bed transport device used in the emptying of cement transport vehicles and for transporting cement [112]. The bottom of the container is fabricated from air-permeable screen cloth, and the air at high pressure (2.9 bar) is introduced beneath it. The air passes through the cement layer, fluidizes it and undergoes a certain pressure loss. The remaining pressure entrains the loosened material along the tube. In long pipelines,

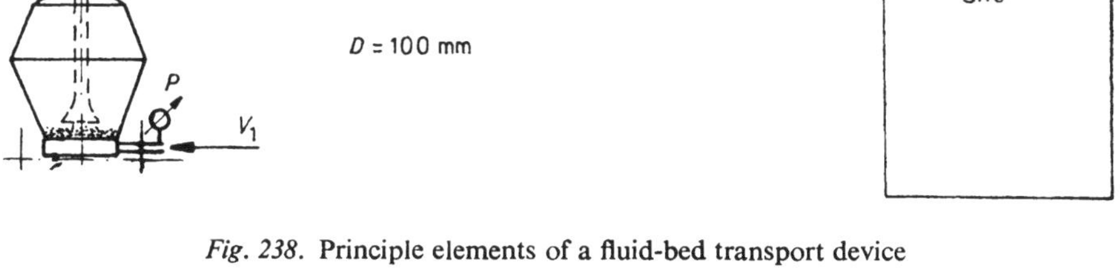

Fig. 238. Principle elements of a fluid-bed transport device

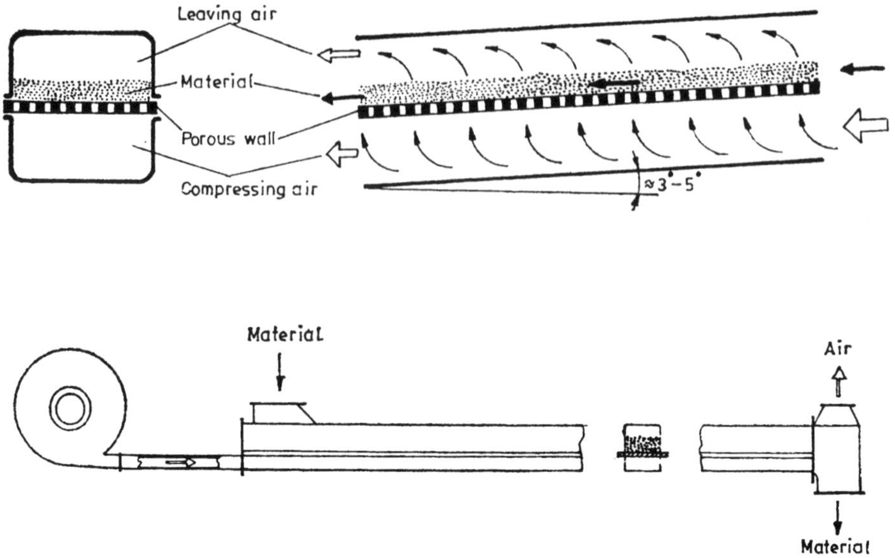

Fig. 239. Principles of construction of a fluidization channel

separation of the cement from the air is prevented by regular feeding of additional air.

Figure 239 illustrates the principles of fluid-bed transport in open channels [111]. The bottom of the channel is fabricated from porous material and has a slope of 3–5°. Air is introduced beneath the porous material at a pressure such that the velocity of the emerging flow is sufficient to fluidize the layer of material being transported. The fluidized material moves in the direction of the slope under the action of gravity.

During fluidized transport in pipelines, the total pressure loss results from the tube resistance and additional resistance due to material transport:

$$\Delta p = \Delta p_0 + \Delta p_m$$

The idle-run pressure loss (tube resistance) is low in relation to the total pressure loss, even in the case of longer lines, and so may be disregarded in preliminary calculations. The pressure loss required to transport a material is then given by the sum of the pressure required to maintain the fluidized state, and the pressures spent in overcoming friction and impact resistance for the material transported.

To maintain the fluidized state the weight of material in the fluidizing layer must be supported, i.e.,

$$\Delta p_G = G/F = lQ_m/c_0 F$$

where G is the weight of material in the fluidized layer, F the cross-sectional area and l the length of the fluidizing tube, Q_m the material flow rate, and c_0 the mean velocity of the transported material. The pressure loss required to surmount the forces due to friction between the material and the tube wall, as well as those due to friction of mixing, is also proportional to the weight G, i.e.,

$$\Delta p_{fr} = k_1 G/F$$

The total pressure required to transport the material is then

$$\Delta p_m = (G+k_1 G)/F = (1+k_1)G/F = k_f(G/F)$$

or, utilizing the transport characteristics,

$$\Delta p_m = k_f(lQ_m/c_0 F) \tag{168}$$

Equation (168) is valid for both vertical and horizontal transport, but the coefficient k_f has different values in the two cases. The value of k_f during horizontal transport of cement has been obtained experimentally as 0.9–1.1, and during vertical transport as 1.8–2.2 [112], i.e., a considerably lower pressure loss is involved in horizontal transport. The reason is that in horizontal transport the weight of the material itself need not be compensated fully.

In the case of long pipelines the pressure loss is not proportional to the length of the pipeline, owing to expansion of the transporting gas. In this case both the air velocity and material velocity increase over the length of the line. Figure 240 shows the pressure loss in the system illustrated in Fig. 238 as a function of tube length, for various transport capacities [112].

Fluid bed transport is designed primarily for transporting fine-grained (floury) materials, since the quantity of transporting air required is then relatively low.

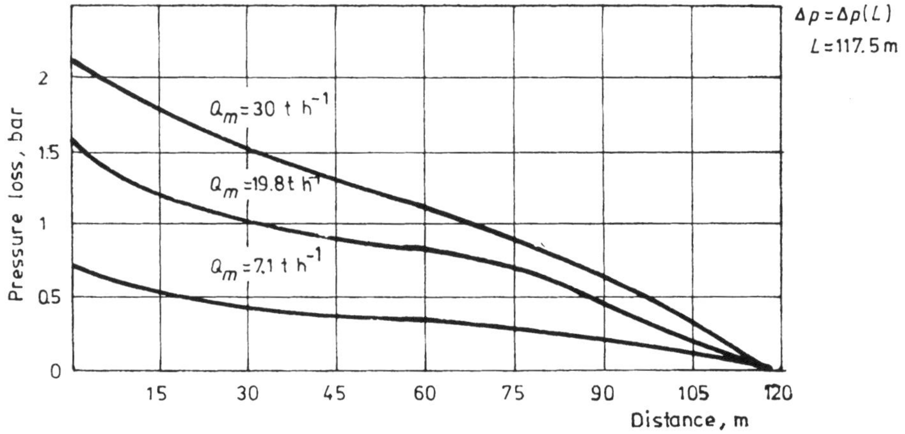

Fig. 240. Pressure loss during transport by fluidization as a function of tube length

14.7 Conveying by throwing

A special mode of transport of agricultural materials is conveying by throwing. Cereals may be lifted to small heights (4–6 m) by a throwing fan, but the most frequent application of the method is found in forage harvesters and silo-filling chaff cutters. The height of conveying may even attain 20–25 m during the filling of silos.

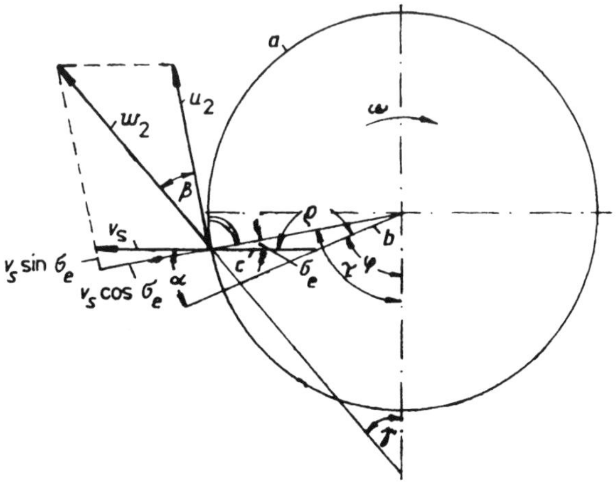

Fig. 241. Principles of construction of a throwing fan

Figure 241 shows the principles of construction of a throwing fan. Material is thrown from the end of an impeller, at a velocity agreeing closely with the circumferential velocity, into a transport tube. In the transport tube gravity and air resistance act on the mass moving upwards, and so the equation of motion in a stationary medium may be written in the form

$$m(dv/dt) = c_w F(\gamma/2g)v^2 + mg$$

At time $t=0$ the velocity of the mass equals the circumferential velocity, i.e., $v=u_2$, and so, after integration,

$$t = (v_{cr}/g)\text{arc tan}\,\{[(u_2-v)/v_{cr}]/[1+(u_2 v/v_{cr}^2)]\} \qquad (169)$$

where v_{cr} is the terminal velocity. In practice a certain air velocity is always present in the tube which promotes transport. This may be taken into account by introducing the relative velocity of the moving material; in this case the preceding equation may be written as

$$t = (v_{cr}/g)\text{ arc tan }\{[(u_2-v_a-v+v_a)/v_{cr}]/[1+(u_2-v_a)(v-v_a)/v_{cr}^2]\}$$

When the mass reaches the highest point of its path, $v=0$ and $t=t_{max}$: then

$$t_{max} = (v_{cr}/g) \arctan \{(u_2/v_{cr})/[1-(u_2-v_a)v_a/v_{cr}^2]\} \tag{170}$$

where v_a is the air velocity. The distance covered by the moving mass is obtained by integrating eqn. (169). Carrying out the integration, after substituting $v=dh/dt$, the maximum height reached by material starting at a velocity u_2 and moving in a stationary medium is

$$h_{max} = (v_{cr}^2/g) \ln [\cos (gt_{max}/v_{cr}) + (u_2/v_{cr}) \sin (gt_{max}/v_{cr})] \tag{171}$$

To take the air velocity v_a into account, u_2-v_a must be substituted for the velocity u_2; the highest point of the path is then

$$H_{max} = (v_{cr}/g) \ln \{\cos (gt_{max}/v_{cr}) + [(u_2-v_a)/v_{cr}] \sin (gt_{max}/v_{cr})\} + v_a t_{max}$$

Figure 242 shows the variation of the throwing height as a function of the

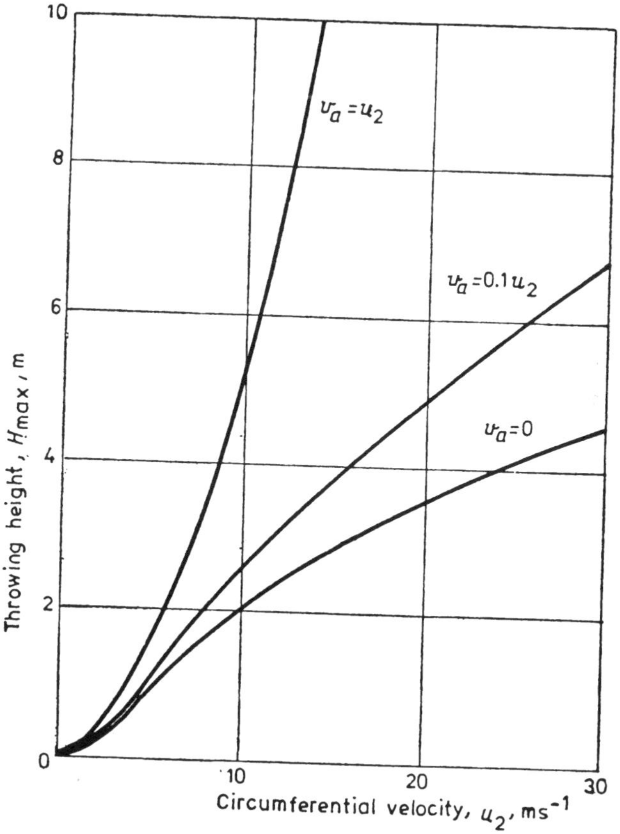

Fig. 242. Throwing height as a function of circumferential velocity

circumferential velocity, for various air velocities. As may be seen, the throwing height attainable may be increased greatly by increasing the air velocity [128].

A special case of conveying by throwing is the spreading of granulated fertilizers on soil by means of throwing discs. The particles leave the throwing disc at a velocity v_0, generally in the horizontal direction, or at some angle α, enclosed with the horizontal. The throwing disc is installed at a height h above the soil surface. The path of the particles, i.e., the throwing distance may be determined knowing the above data and including aerodynamic factors.

To determine the path of the particles the laws of inclined throw are utilized, with allowance for aerodynamic resistance. The instantaneous velocity v of a particle thrown at an angle α may be decomposed into horizontal and vertical components:

$$v = \sqrt{\dot{x}^2+\dot{y}^2}$$

The aerodynamic resistance may be calculated from eqn. (146),

$$W = c_w F(\gamma/2g)v^2,$$

and may also be decomposed into horizontal and vertical components:

$$W_x = W \cos \alpha$$

and

$$W_y = W \sin \alpha.$$

The equilibrium of forces in the x direction (with $\cos \alpha = \dot{x}/v$ taken into account) gives

$$m\ddot{x} = -W_x = -K_1 v_0^2(\dot{x}/v) = -K_1 \dot{x}\sqrt{\dot{x}^2+\dot{y}^2}$$

where $K_1 = c_w F(\gamma/2g)$. In the y direction gravity must also be taken into account, so

$$m\ddot{y} = -K_1 \dot{y}\sqrt{\dot{x}^2+\dot{y}^2} - mg$$

In the case of spherical particles the cross-sectional area F and mass m may be expressed in terms of the diameter, giving the differential equations

$$\ddot{x} = -K\dot{x}\sqrt{\dot{x}^2+\dot{y}^2}$$

and

$$\ddot{y} = -g - K\dot{y}\sqrt{\dot{x}^2+\dot{y}^2} \qquad (172)$$

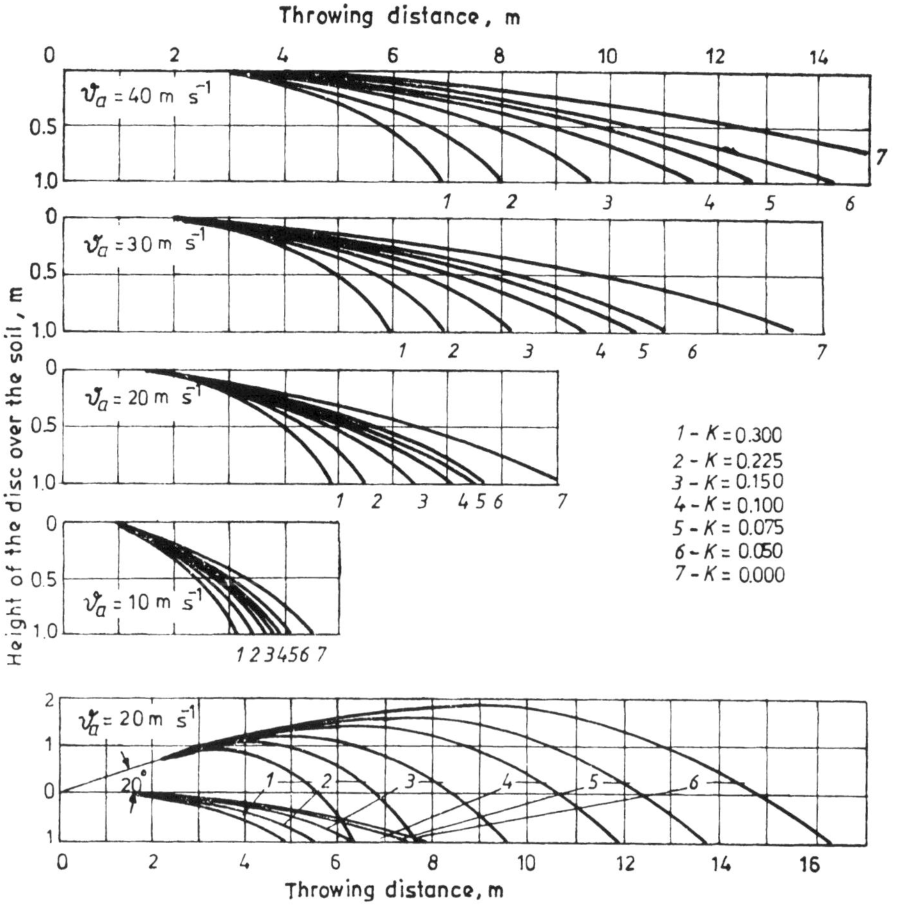

Fig. 243. Throwing distances for various *K* values

where $K=3\gamma_a c_w/4\gamma_m d$, γ_a and γ_m being the specific weights respectively of air and of the particles, and d the latter's diameter. For granulated fertilizers $d=1$–5 mm, $\gamma_m=1.5\sim2.0\times10^3$ kg m^{-3} and $c_w=0.5$. With these values, K varies between 0.32 and 0.05.

Equations (172) may be solved by series development [154] or using an analog computer [153]. Computation results obtained using the latter method are displayed in Fig. 243, for various K values. As may be seen, the paths of particles of different sizes deviate significantly, and the distance of throw may be increased greatly by increasing the angle α. Therefore mixed or inhomogeneous fertilizer compositions cannot be spread evenly by throwing discs.

14.8 Pneumatic conveying of non-Newtonian materials

Wet fodders (dense swill), fresh animal manure and other viscous materials may also be transported pneumatically over relatively great distances. Transport occurs with the material in a dense stream, i.e., with the use of little air, but at high pressure. The material to be transported is characterized by its plastic viscosity η and the initial shear stress τ_0. Figure 244 shows these values for pig and cow manure as functions of moisture content [118].

An unpleasant property of wet, viscous materials is their inclination to adhere to surfaces, whereby they stick to tube walls during transport, decreasing the free cross-sectional area and even blocking the tube. The inclination to adhere is a function of the composition and moisture content of the material; values are given in Fig. 245 for pig and cow manure [118].

The resistance resulting from friction during transport in horizontal tubes is

$$\Delta p = \lambda (L\gamma/D2g) c_0^2 \tag{173}$$

where γ is the volumetric weight of the transported material, whose value for manure is about 1000 kg m^{-3}, and λ is a resistance coefficient, whose values according to experimental data [118] are, for laminar and turbulent flow respectively,

$$\lambda_1 = (9.3+255D)/Re^*$$

and

$$\lambda_2 = (0.735+7.3D)/Re^{*0.555}$$

The Reynolds number is expressed in this case as

$$Re^* = 1/[\eta/\gamma c_0 D + \tau_0/6c_0^2\gamma] \tag{174}$$

To the pressure loss determined from eqn. (173) must be added the pressure loss required for acceleration [see eqn. (166)], the resistance of pipe elbows, and,

Table 15

c_0 (m s^{-1})	Moisture content (%)	η (daN s m^{-2})	τ_0 (daN m^{-2})	Δp_t (bar)
4.96	90.0	0.019	0.44	2.08
4.56	88.4	0.027	1.10	2.32
6.00	88.9	0.023	0.80	2.99
5.71	88.5	0.026	1.02	2.93
7.00	88.5	0.026	1.02	3.53
5.94	88.7	0.025	0.91	3.15
5.80	90.5	0.016	0.32	2.32

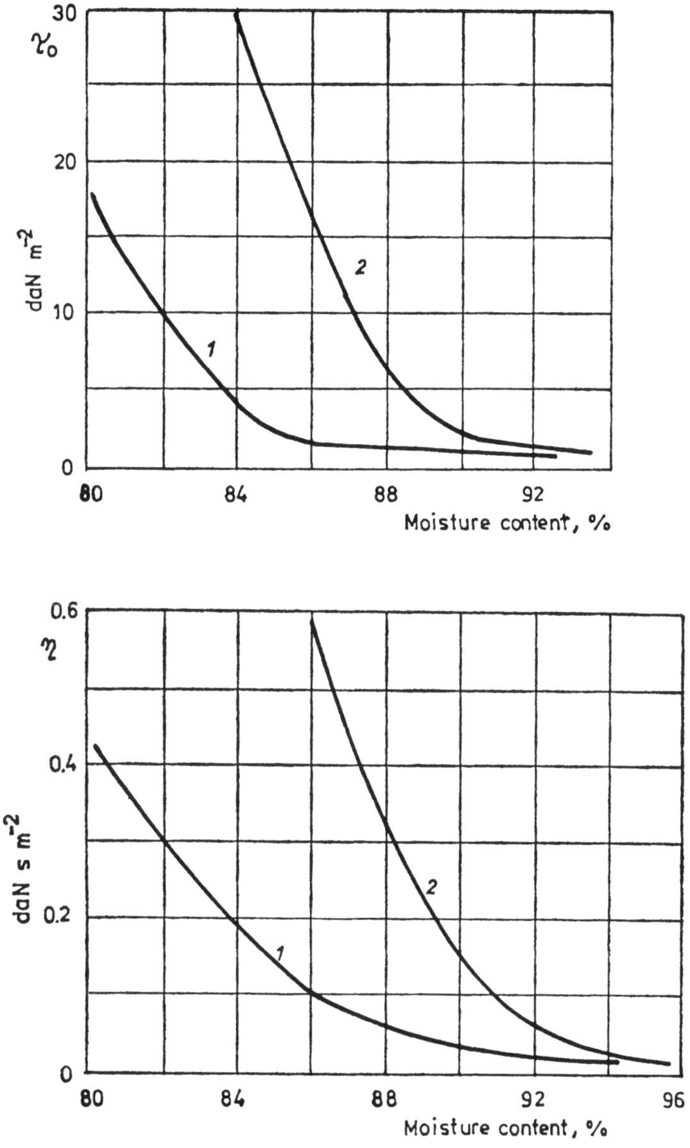

Fig. 244. Viscosity and initial shear stress of pig *(1)* and cow manure *(2)* as functions of moisture content

for vertical transport, the pressure required to lift the material. Table 15 shows the total measured pressure loss during the transport of pig manure over a distance of 100 m in a tube of diameter $D = 125$ mm under various conditions [118].

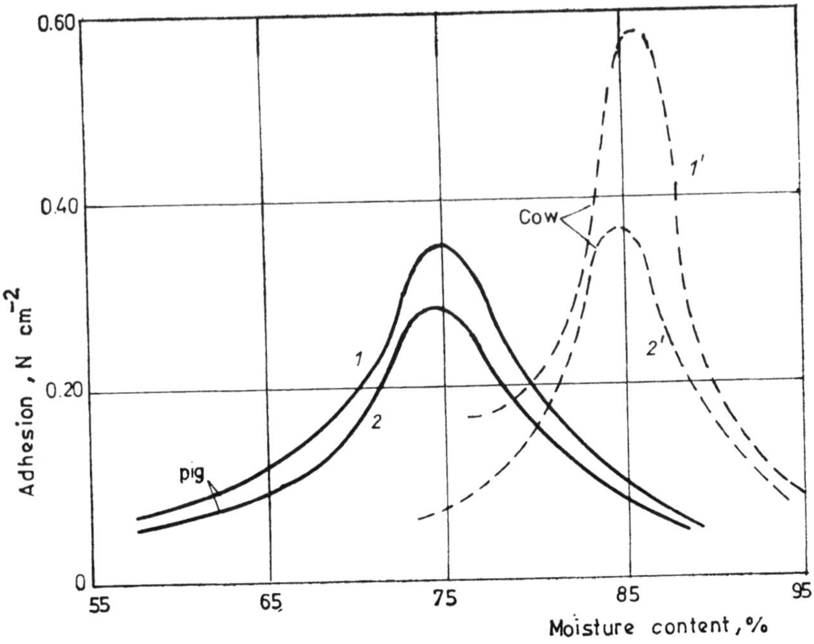

Fig. 245. Adhesion of pig and cow manure as a function of moisture content. *(1, 1')* Asbestos; *(2, 2')* steel

14.9 Flow in perforated ducts

Perforated ducts are frequently used in ventilating granular crop products, other agricultural materials and stables.

The perforation may be uniform along the duct, but then the distribution of air will not be uniform over the length. The perforation may be corrected, knowing the deviation, to obtain a uniform air distribution.

Perforated ducts may be operated under suction or under pressure, as shown in Fig. 246. The duct problem may be characterized by the distance x measured from the duct's closed end, its total length L, its equivalent diameter D ($D=4F/P$, where P is the perimeter of the duct) and the percentage area of perforations in relation to the total surface area.

The variation of the pressure head along the tube may be expressed by the differential equation [119]

$$dh/dx = -k(d/dx)(v^2/2g) \pm \lambda v^2/2gD \tag{175}$$

where λ is the friction coefficient of the duct, and k the reciprocal of the square

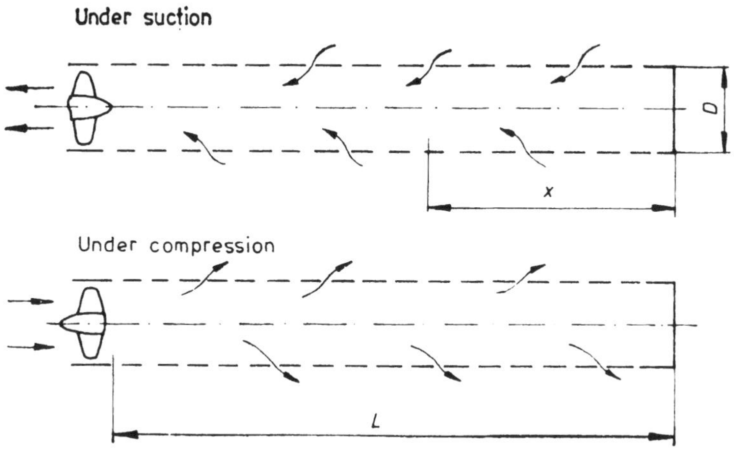

Fig. 246. Perforated tube operated by suction or compression

of the velocity coefficient φ $(v=\varphi\sqrt{2gh})$. The sign of the second term on the right-hand side is negative for a suction and positive for a pressurized system. The velocity v in eqn. (175) may also be expressed in terms of the transport quantity Q by means of the simple relationship $v=Q/F$. Then

$$dh/dx = -k(d/dx)(Q^2/2F^2g) \pm \lambda Q^2/2DF^2g$$

or

$$dh/dx = -(2kQ/2F^2g)dQ/dx \pm \lambda Q^2/2DF^2g \tag{175a}$$

If the air is distributed uniformly, and equal quantity q ($m^3\,s^{-1}\,m^{-1}$) of air enters, or leaves every 1 m length of the duct, and so the rate of flow at a distance x may be written in the simple form

$$Q = qx$$

Substitution of this latter into eqn. (175a) yields, after integration, the equation

$$p-p_0 = (8\gamma/g)(q/k)^2(x/D)^2[-k \pm \lambda x/3D] \tag{176}$$

where p_0 is the static pressure at the closed end of the duct. Figure 247 illustrates eqn. (176) graphically. It may be seen that the pressure differences $p-p_0$ are much greater in the suction system than in pressurized systems, and so it is advisable in general to employ the latter.

In practice uniform distribution of air may be ensured by means of non-uniform perforation over the length of the tube, according to the pressure variation determined by eqn. (176). For uniform perforation the quantity of air entering

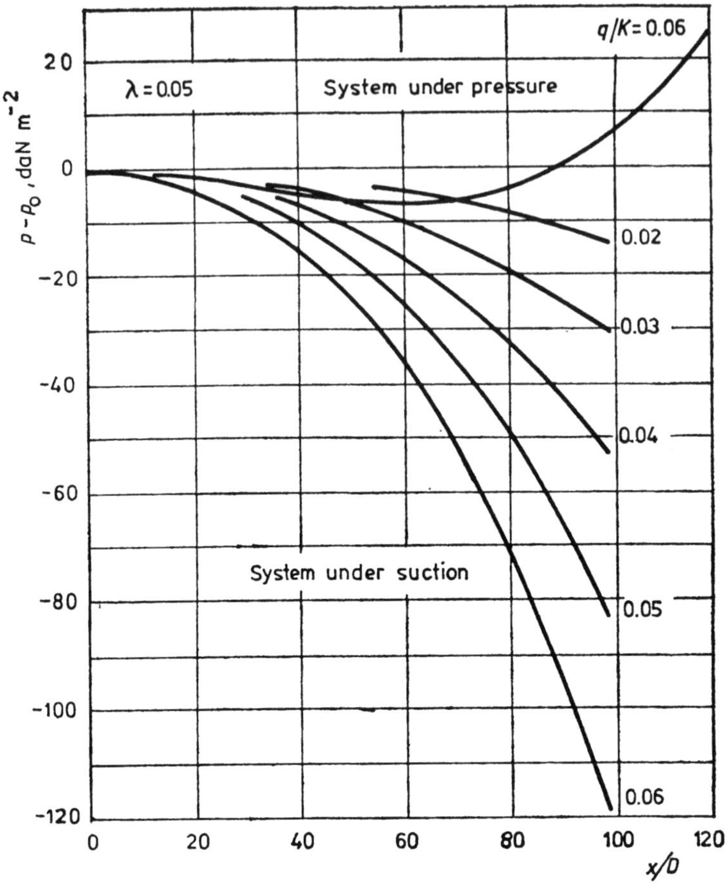

Fig. 247. Pressure distribution as a function of relative tube length

or leaving will not be distributed uniformly along the tube, owing to the varying pressure loss. In this case the quantity of air entering or leaving at an arbitrary point along the tube is

$$dQ/dx = \mu f_0 \sqrt{2gh}$$

where μf_0 is the effective perforation ($m^2\,m^{-1}$) per 1 m tube length. From this equation,

$$dh/dx = (1/\mu^2 f_0^2 g)(dQ/dx)d^2Q/dx^2$$

Equating the right-hand side of this equation with the right-hand side of eqn. (175a) yields

$$(2F^2/\mu^2 f_0^2)(d^2 Q/dx^2) = -2Qk \pm [\lambda Q^2/D(dQ/dx)]$$

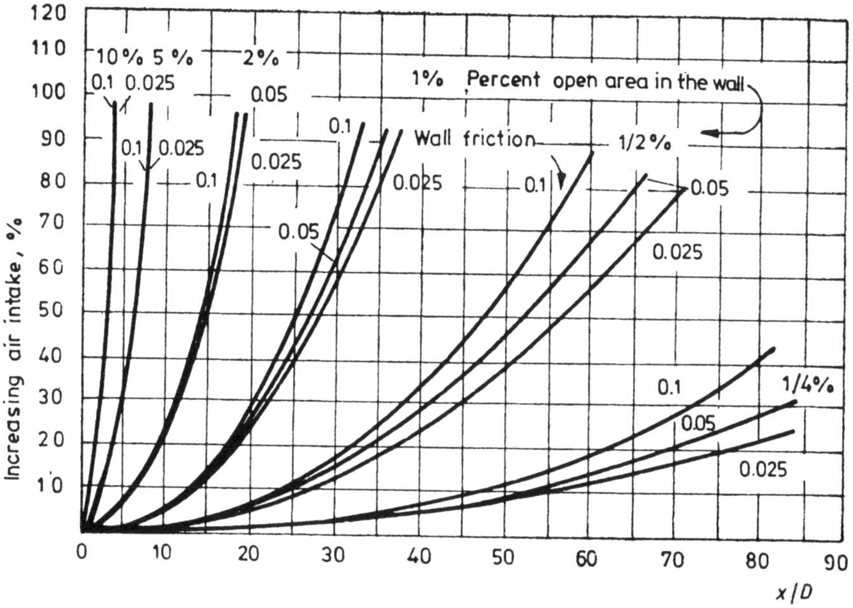

Fig. 248. Percentage deviation of quantity of air entering as a function of relative tube length

The perforation f_0 may be expressed as a percentage of the surface area of 1 m of tube ($f_0=(\%)P$), while the cross-sectional area of the tube may be written in terms of the equivalent diameter ($F=PD/4$). The final form of the differential equation with these substitutions is

$$\left(D^2/8\mu^2(\%)^2\right)d^2Q/dx^2 = -2Qk\pm[\lambda Q^2/D(dQ/dx)] \tag{177}$$

This differential equation may be solved conveniently using an analog computer [120]. Figure 248 illustrates the percentage deviation of the quantity of air entering the tube as a function of relative tube length for various perforations. It may be seen from the figure that the distribution is more uniform, the shorter the relative tube length and the smaller the perforated surface area.

In a pressurized system the static pressure variation originating from the variation in velocity will mitigate the variation in pressure caused by frictional losses, and so the curve for the air distribution may have a different form (Fig. 249) [120]. The distribution curve is again influenced decisively by the degree of perforation in this case, and in addition by the value of q_0, relative to the closed end, for small air quantities. The values of q_0 given beside the curves in Fig. 249 indicate the limiting values, above which the distribution is unchanged. The figure shows the effect of q_0 values below the limiting value on the distribution of air in the case of 2% perforation. It may be established on the basis of Figs. 247

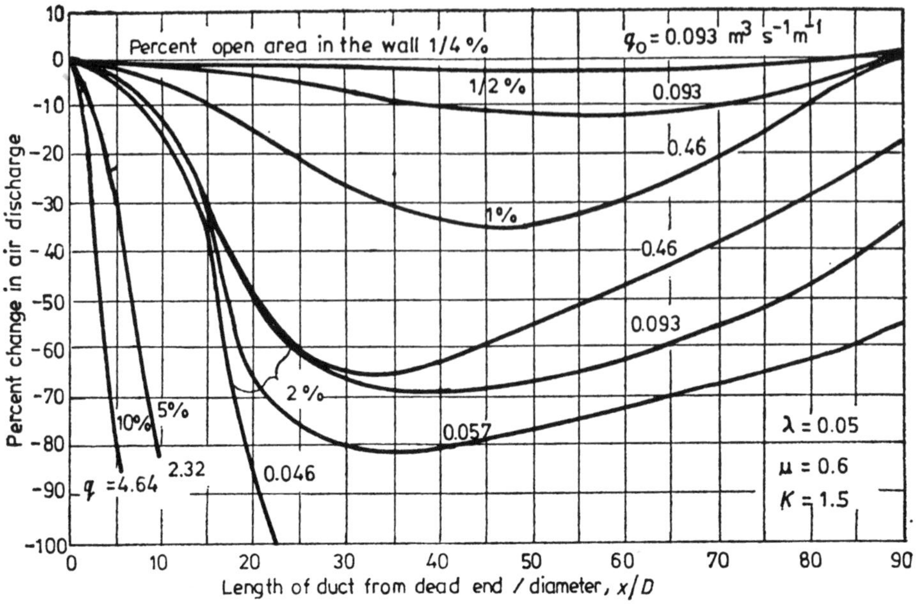

Fig. 249. Distribution curves for air in a pressurized system

and 248 that the same system will give a more even distribution under pressure than under suction.

A special application of perforated ducts is met with in the ventilation towers with a central distribution of air, the schematic design of which is shown in Fig. 250. Both the outer wall of the tower and the internal distributing tube are perforated, and the cereal or hay to be ventilated is packed between them. The ventilating air is forced into the central tube and passes radially through the crop. At the upper end of the tube a plug may be moved in correspondence with the level of the crop so as to close a length corresponding to the radial dimensions of the crop. In this way satisfactory air distribution is ensured in the upper layers.

The relationship between the air flow rate (Q_0) and the pressure loss may be determined with the assumption of radial air flow in the crop. The radial velocity along an arbitrary radius r is

$$v_r = Q_0/2r\pi H$$

where H is the free height of the central perforated tube. The expression for the pressure loss in differential form is

$$dp/dr = k''v_r^n$$

Fig. 250. Ventilation system with central air distribution

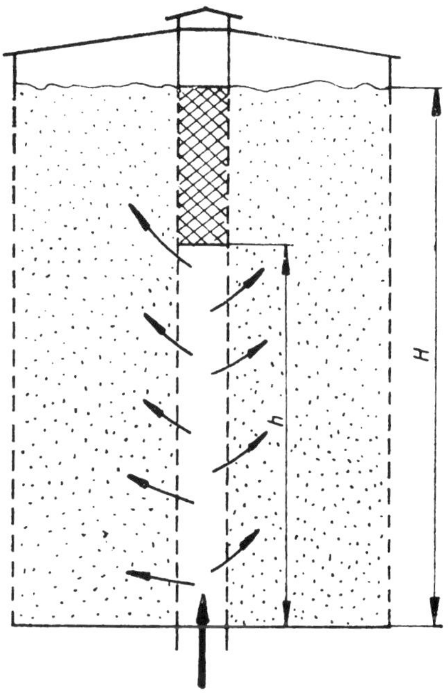

Substitution of v_r from above, followed by integration, yields

$$dp = k''(Q_0/2\pi H)^n \int_{r_0}^{R} (1/r^n)dr$$

$$\Delta p = k''(Q_0/2\pi H)^n[1/(n-1)][1/r_0^{n-1} - 1/R^{n-1}] \quad (178)$$

The value of the exponent n generally varies between 1.4 and 1.6, depending on the crop and air velocity. For maize, the values $k''=840$ and $n=1.5$ (for m, daN, s units) may be used.

14.10 Ventilation of bales and stacks

A frequent task in agriculture is the ventilation of bales, stacks of bales, or loose forage material, for drying or recooling. In this case a complex multi-dimensional flow develops, owing to the irregular shapes and frequent inhomogenities involved, whose calculation is a complicated task. The problem may be solved generally by the methods of finite differences [105] or finite elements [106].

In the calculations it is assumed on the basis of experimental results demonstrated earlier (Section 14.3) that the specific pressure loss may be approximated by an equation of the form

$$\Delta p/h = C\gamma_{vw}^{m} v^{n}$$

or

$$\Delta p/h = B\gamma_{vw}^{m} Q^{n}$$

In this case, two-dimensional flow may be described by the nonlinear partial differential equation [122]

$$[(\partial p/\partial x)^2+(\partial p/\partial y)^2](\partial^2 p/\partial x^2+\partial^2 p/\partial y^2)-2m[(\partial p/\partial x)^2(\partial^2 p/\partial x^2)+$$
$$+2(\partial p/\partial x)(\partial p/\partial y)(\partial^2 p/\partial x^2)+(\partial p/\partial y)^2\partial^2 p/\partial y^2]=0$$

where

$$m=(1-1/n)/2$$

The above differential equation may be solved by the method of finite differences. The cross-sectional area investigated is decomposed into square elements of equal size (Fig. 251). The partial derivatives $\partial p/\partial x$ and $\partial p/\partial y$ are expressed in terms of the nodal-point values (with approximation by Taylor series) as

$$\partial p/\partial x = (p_{1,0}-p_{0,-1})/2h$$

$$\partial p/\partial y = (p_{1,0}-p_{-1,0})/2h$$

$$\partial^2 p/\partial x^2 = (p_{0,1}-2p_{0,0}+p_{0,-1})/h^2$$

and so on. By substituting the above into the differential equation the individual pressure values at the nodal-points may be determined by the method of succes-

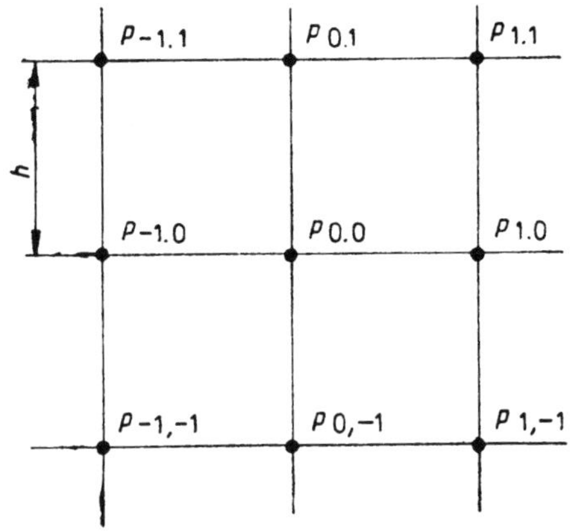

Fig. 251. Division of space into elements in application of the finite difference method

sive approximation. The expression for the pressure $p_{0,0}$ from the finite-difference equation is

$$p_{0,0} = [1/4(1-m)][a_3+a_4-m(2a_2^2a_3+a_2a_1(a_5-a_6)+2a_1^2a_4)/(a_2^2+a_1^2)]$$

where

$$a_1 = p_{0,1}-p_{0,-1} \quad a_4 = p_{0,1}+p_{0,-1}$$

$$a_2 = p_{1,0}-p_{-1,0} \quad a_5 = p_{1,1}-p_{1,-1}$$

$$a_3 = p_{1,0}+p_{-1,0} \quad a_6 = p_{-1,1}-p_{-1,-1}$$

The above equation must be written for every individual point and the calculation is continued by the method of successive approximation until the predicted pressures at the individual points stop changing significantly in subsequent calculation sequences. The pressure gradient is obtained from

$$\text{grad}\, p = \sqrt{(\partial p/\partial x)^2+(\partial p/\partial y)^2} = C\gamma_{vw}^m v^n$$

from which the magnitude of the velocity vector v may be calculated. The direction of the velocity vector, relative to the positive x axis, is

$$\tan \psi = (\partial p/\partial y)/(\partial p/\partial x)$$

In several cases the method of the finite differences may be applied only with difficulty, or with neglect of substantial terms. Thus the approximation for bodies having complex curved surfaces is difficult using square elements. Taking the inhomogeneities frequently found in bulk materials into account is also difficult. In these cases it is better to use the finite-element method (see Section 11.6), which permits allowance for inhomogeneities and approximation for more complex geometrical forms.

One method of ventilating or drying large bales (120×120×240 cm) is to place them tightly beside each other over a ventilation channel whose upper side is open. The width of the channel is selected according to the dimensions of the bales so as to ensure the most uniform air distribution possible within them. The flow may be regarded (apart from at the ends of the channel) as two-dimensional to good approximation. Figure 252 shows the calculation results obtained using the finite-element method [106]. The figure shows lines of constant pressure and the velocity distribution of the air leaving the surface of a bale. The dashed lines illustrate the subdivision into elements. By applying eight nodal-points, a sufficiently accurate result is obtained with a relatively small number of elements.

Bales and loose forage materials may be stacked and the stacks then ventilated. In the interest of uniform air distribution, the ventilation channel and the cross-

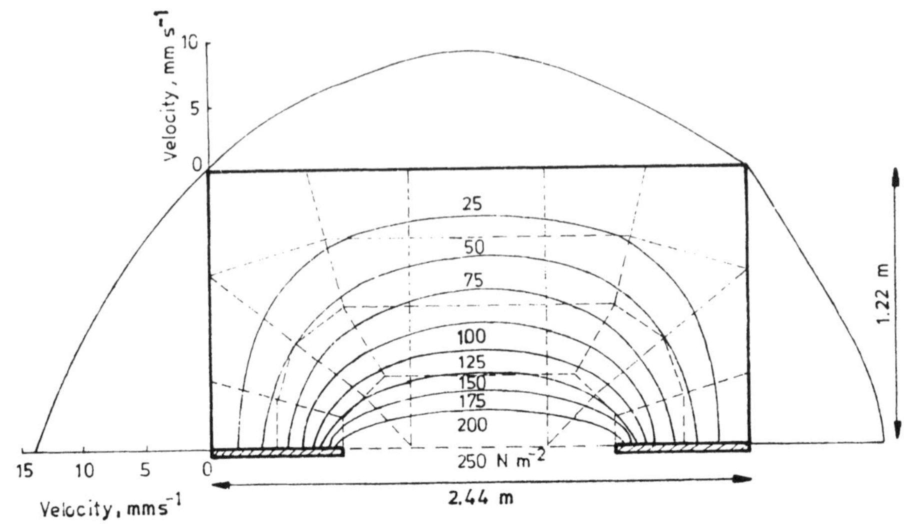

Fig. 252. Isobaric lines and distribution of outlet velocity during ventilation of a large bale (finite-element method)

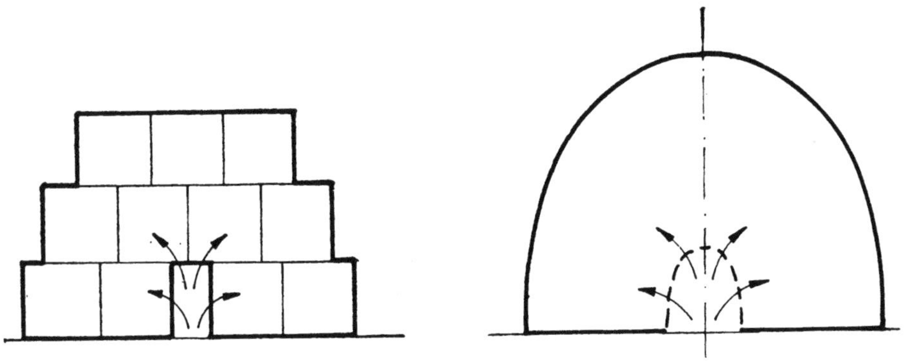

Fig. 253. Typical forms of ventilated stacks

section of the stack must be chosen so as to ensure closely similar aerodynamic resistance in every direction. Figure 253 illustrates typical stack forms. The dimensions of the main channel must be selected so as to minimize the pressure loss along it. Even after ensuring uniform aerodynamic resistance, settling of a stack, which is nonuniform, depending on the height and the presence of gaps between the bales, may cause problems. These effects may be taken into account by the finite-element method.

For approximate calculations eqn. (178) may be used, with consideration of the fact that here only semicircular cross-sections are involved:

$$\Delta p = k''(Q_0/\pi H)^n[1/(n-1)][1/r_0^{n-1} - 1/R^{n-1}]$$

The mean radius values must be substituted for r_0 and R. In the case of stacks built of bales, the air consumption will exceed that calculated for solid bales by about 20%, owing to the gaps between the bales.

14.11 Non-Newtonian flow in tubes

Non-Newtonian materials (manure, liquid fodders, fruit purees, etc.), are frequently transported in pipelines. For dimensioning such pipelines the viscoplastic characteristics and certain flow properties of the material must be known.

The flow behavior of plastic and quasiplastic materials are described respectively by the equations

$$dv/dy = (\tau - \tau_0)/\eta$$

and

$$dv/dy = (\tau - \tau_0)^n/\eta'$$

where η is the plastic viscosity, η' the apparent viscosity, and τ_0 the initial shear stress. According to observations, dilute materials of high moisture content (about 90%) behave as plastic materials, while materials of higher density show quasiplastic properties.

The flow here is also characterized by the Reynolds number, in the form

$$Re = 1/[(\eta/\gamma dv) + (\tau_0/6\gamma v^2)] \qquad (179)$$

Figure 254 illustrates the resistance to tube flow for fodders (swill) with two moisture-content values, as a function of flow velocity, for various tube diameters [123]. The curves contain break points, where the flow changes from laminar to turbulent. The critical Re value, which indicates the type of flow, may be calculated from the velocity corresponding to the break point. Processing of experimental data shows that the critical Re value for non-Newtonian liquids is not independent of the tube diameter. In the case of the fodder materials for which data are plotted in Fig. 254, the critical Re values decrease significantly as the tube diameter increases.

If the specific tube resistance is known, similarly to Fig. 254, dimensioning of a pipeline presents no difficulty. In the opposite case, dimensioning may be performed on the basis of the values of η and τ_0 for the material to be transported. The characteristics η and τ_0 are generally determined by means of a rotational viscosimeter. According to experience, viscosimeters fail to give reliable results

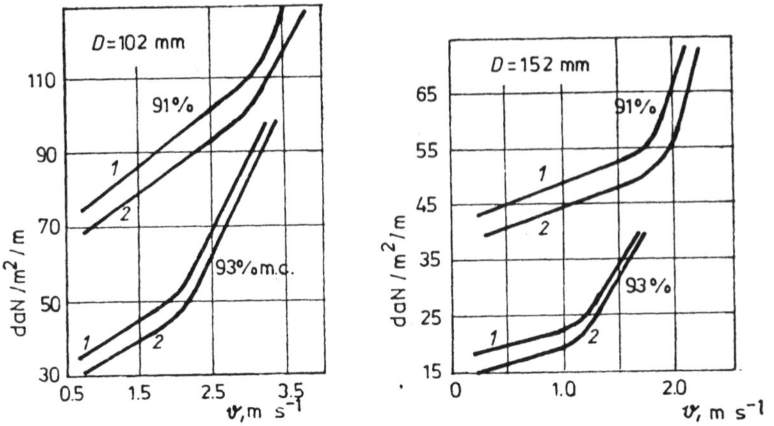

Fig. 254. Tube resistance of liquid fodder as a function of flow velocity. *(1)* Swill of factory spirit; *(2)* swill of beet slices

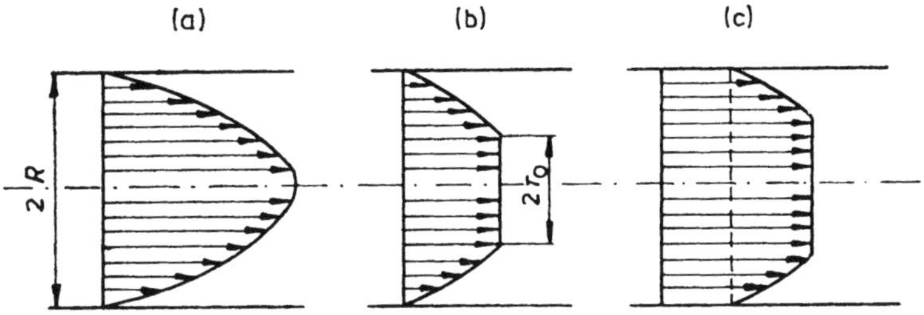

Fig. 255. Typical velocity profiles in tube flow

in numerous cases, so it is advisable to determine viscosity characteristics on the basis of measurements of the actual flow in the tube considered. The pressure loss Δp occurring in a tube of length L is spent in overcoming the stress τ arising in the individual layers, i.e.,

$$\Delta p r^2 \pi = \tau 2 r \pi L$$

from which

$$\tau = \Delta p r / 2L \tag{180}$$

Equation (180) is the basic equation of flow for a tube (and for a capillary viscosimeter).

During flow in a tube various velocity profiles may develop, depending on the characteristics of the medium. Figure 255 shows typical velocity profiles. During laminar flow, the distribution of the velocity for a Newtonian liquid is parabolic. In the case of plastic flow (for a Bingham body), the velocity distribution is parabolic close to the wall, but in the central part of the tube the velocity is constant (Fig. 255(b)). During the flow of dense, highly viscous materials, slipping occurs along the wall, as shown in Fig. 255(c). In the case of plastic flow, the stresses τ, arising along the wall and τ_0 on the radius r_0 may be expressed by means of eqn. (180) in the form

$$\tau_w = \Delta pR/2L$$

$$\tau_0 = \Delta pr_0/2L$$

Division of the two equations yields

$$r_0/R = \tau_0/\tau_w \quad (181)$$

The quantity of material flowing through the tube may be calculated from the velocity distribution as

$$Q = \int_0^R v(r) 2\pi r \, dr$$

from which the following expression may be obtained [125]:

$$Q = (\Delta pR^4\pi/8\eta L)[1-4r_0/3R+(1/3)(r_0/R)^4] \quad (182)$$

or, with eqn. (181) taken into account,

$$Q = (\Delta pR^4\pi/8\eta L)[1-4\tau_0/3\tau_w+(1/3)(\tau_0/\tau_w)^4] \quad (182a)$$

If the pressure Δp is such that the stress τ_w is sufficiently high in relation to the initial shear stress τ_0, then the last term of eqn. (182a) may be neglected and the equation may be written in the form

$$\tau_w = (4/3)\tau_0+\eta(4Q/R^3\pi) \quad (183)$$

Equation (183) may be represented using an appropriate system of coordinates by a straight line, as illustrated in Fig. 256. Consequently it is sufficient to measure

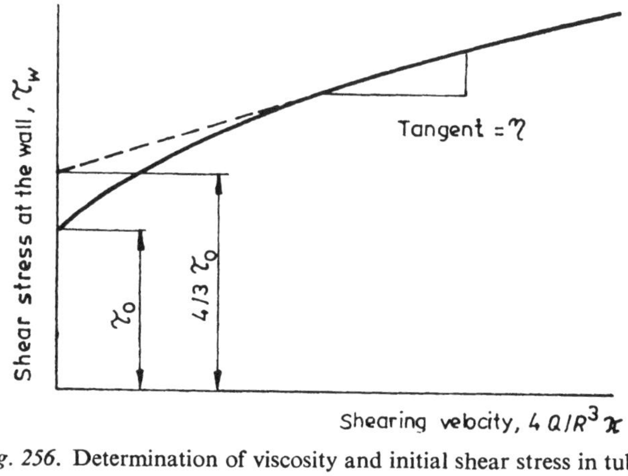

Fig. 256. Determination of viscosity and initial shear stress in tube flow

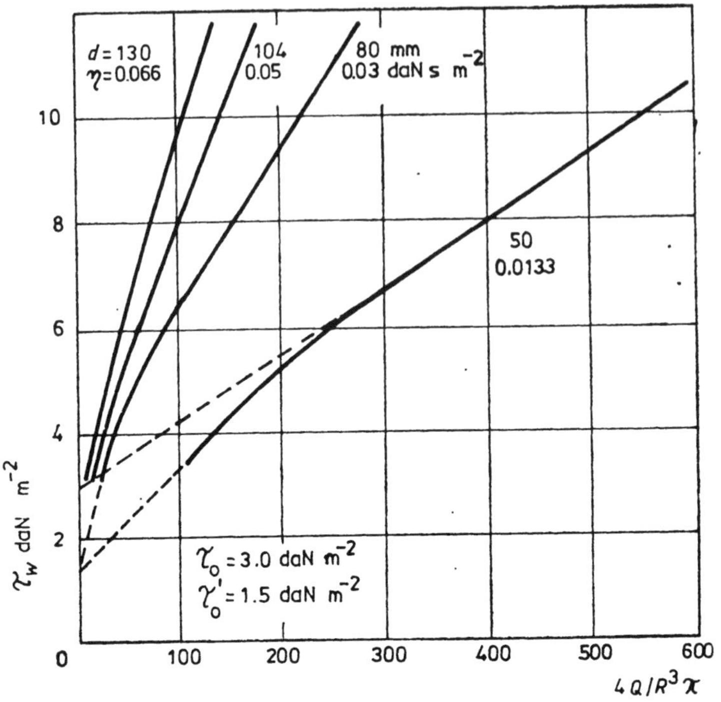

Fig. 257. Rheological curves for various tube diameters

the flow quantity Q for various pressure losses Δp in the pipeline and to represent the data according to Fig. 256. The slope of the straight line obtained gives the plastic viscosity, while the value of τ_0 is given by 3/4 of the intercept on the vertical axis.

Practical measurements have shown that the viscosity η depends on the tube diameter in numerous cases. Figure 257 shows rheological curves for a fodder mixture with 74.5% moisture content for various tube diameters [124]. Above 100 mm tube diameter the deviation is not significant, but for smaller diameters great deviations are obtained. The reason is that the material slides along the wall differently, depending on the tube diameter: the smaller the tube diameter, the greater the sliding.

The rate of flow for a material which slides along the wall may be calculated similarly to eqn. (182a), completed by a term allowing for the slippage:

$$Q/R^3\pi = v_s(\tau_w)/R + (\Delta pR/8\eta L)[1 - 4\tau_0/3\tau_w + (1/3)(\tau_0/\tau_w)^4] \tag{184}$$

where $v_s(\tau_w)$ is the sliding velocity along the wall, which is a function of the shear stress τ_w. The left-hand side of this equation may be expressed in terms of

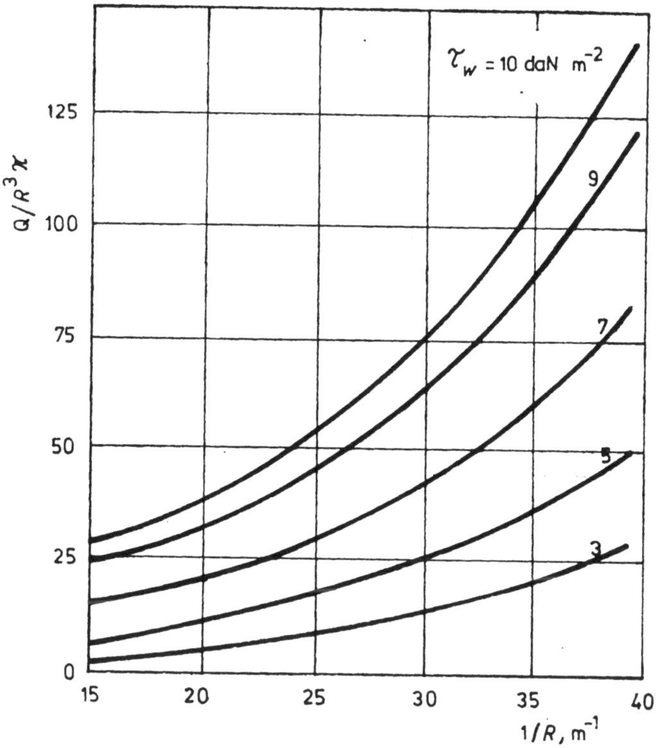

Fig. 258. Plotting method for determining sliding velocity along wall

an average flow velocity:

$$Q/R^3\pi = u/R$$

Again neglecting the last term in eqn. (184), as before the pressure loss per 1 m length may be expressed as

$$\Delta p/L = (8\eta/R)[(1/R)(u-v_s)+\tau_0/3\eta] \qquad (185)$$

The sliding velocity along the wall (v_s) must be determined experimentally. For this purpose the quantity $Q/R^3\pi$ is plotted as a function of $1/R$ for various shear stresses τ_w, as shown in Fig. 258 [124]. The curves may be described in a given case with sufficient accuracy by an equation of the form

$$Q/R^3\pi = A(1/R)^3+B$$

On comparing this with eqn. (184), the following equality may be established:

$$A(\tau_w)/R^3 = v_s(\tau_w)/R$$

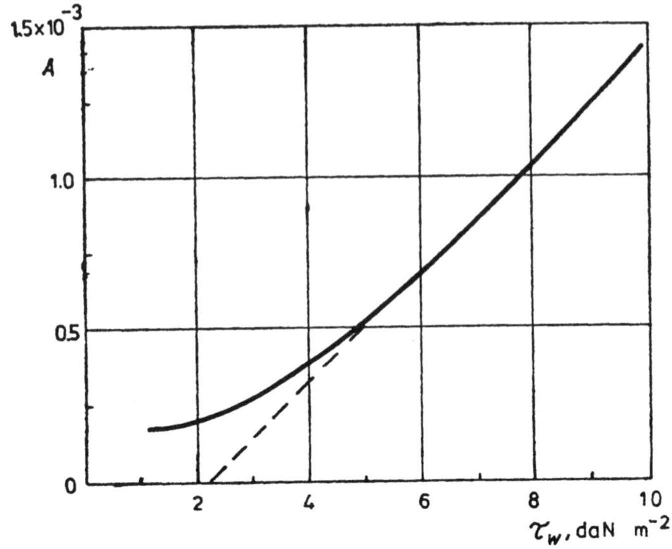

Fig. 259. Plot of $A(\tau_w)$ for tube flow

from which

$$v_s(\tau_w) = A(\tau_w)/R^2$$

The relationship $A(\tau_w)$ is illustrated in Fig. 259, based on the data of Fig. 257. The major part of the curve is straight, so

$$A(\tau_w) = a(\tau_w - \tau_{w0})$$

where τ_{w0} is the intercept of the straight line with the horizontal axis.

The tube resistance may also be determined as friction resistance on the basis of eqn. (179), if the tube friction coefficient λ is known. In the case of laminar flow, λ may be expressed as

$$\lambda = 64/Re$$

where the Reynolds number must be calculated on the basis of eqn. (179). According to investigations, the value of λ calculated in this way agrees well with experimental values in cases where no slippage occurs along the wall (Fig. 260). If the material shows sliding along the wall, its resistance coefficient decreases and the above equation must be multiplied by a correction factor of less than unity:

$$\lambda = K(64/Re)$$

The extent of sliding along the wall for a given material is a function of the tube diameter, and so the value of K also depends on the tube diameter, as may be seen from Fig. 261 [125].

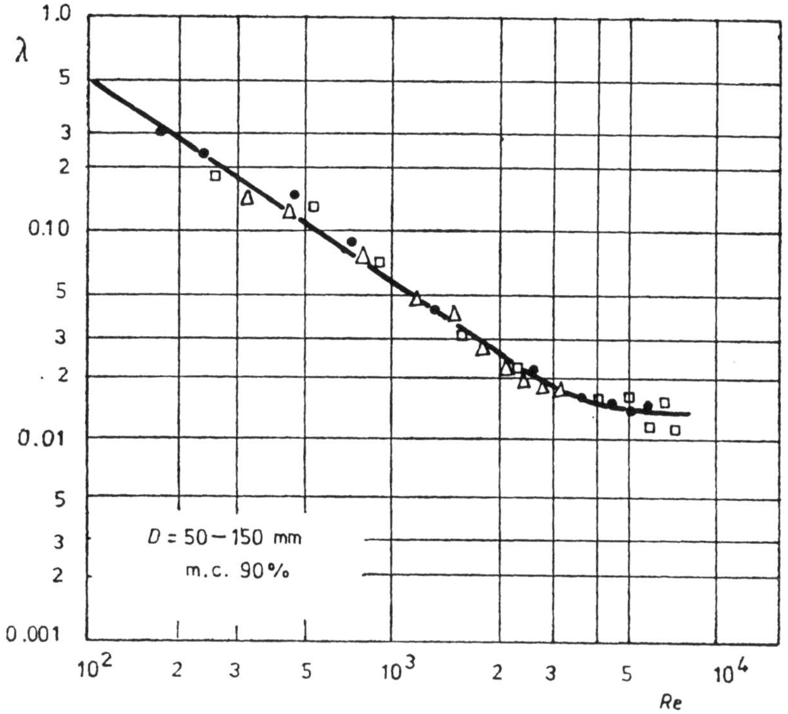

Fig. 260. Tube friction as a function of Reynolds number

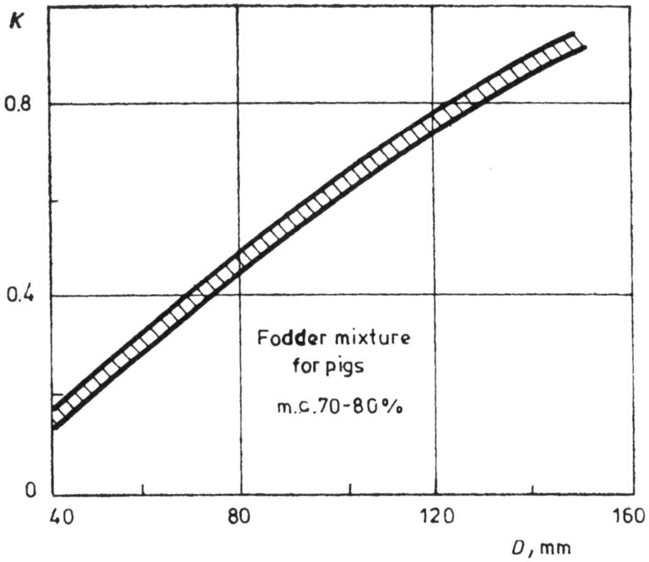

Fig. 261. Correction factor allowing for sliding along wall

In cattle sheds, sloped collection channels are frequently used to lead manure passing through a grid floor by gravity into a collecting basin. For correct operation of such a channel the slope of its base must be selected correctly as a function of the mechanical properties of the manure. The moisture content of fresh manure is 89–94%, depending on the feeding mode, and the corresponding values of η and τ_0 may be read from Fig. 245. The slope required for hydraulic flow of a plastic medium may be calculated from the relationship

$$i = \Delta p/\gamma L = 32\eta v/\gamma d^2 + 16\tau_0/3\gamma d$$

where d is the equivalent tube diameter (quadruple cross-sectional area divided by the wetted perimeter) expressed as

$$d = 4\delta b/(2\delta + b)$$

where δ is the layer thickness, and b the channel width. The velocity of flow in the collecting channel is very low, 20–40 m day^{-1}. Its value may be calculated from the number of animals (n) and the quantity of manure (q) produced by each, as

$$v = 1.16 \times 10^{-5} nq/\gamma\delta b$$

Since the velocity of flow is very low, the value of τ_0 must be selected carefully. If the initial section of the curve of τ_w versus $4Q/R^3\pi$ (Fig. 256) descends greatly, then τ_0 corresponding to $v=0$ may be considerably lower than the value corresponding to higher velocities.

The layer thickness δ depends on the moisture content: the denser the manure, the thicker the layer must be to ensure flow. Figure 262 illustrates the variation

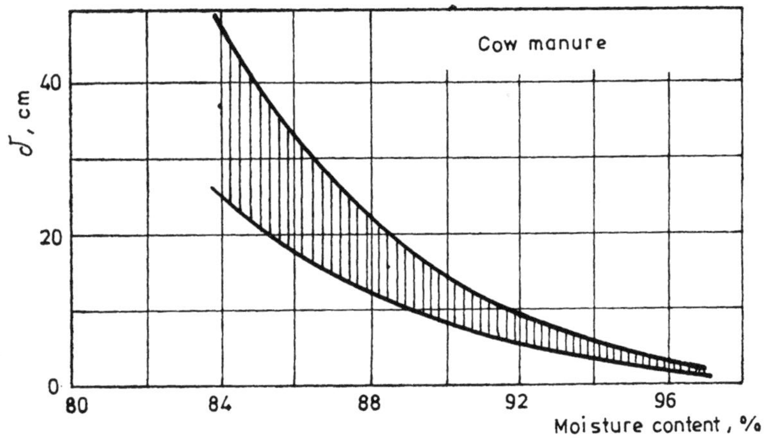

Fig. 262. Layer thickness of flowing manure as a function of moisture content

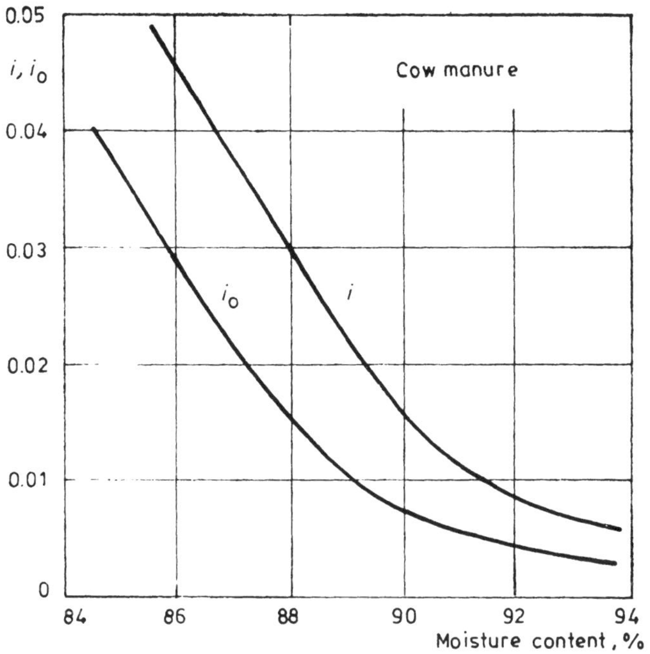

Fig. 263. Slope required for the flow of manure as a function of moisture content

of layer thickness as a function of moisture content. A thick layer of manure will flow at a somewhat smaller slope i_0 than the minimum layer thickness δ. Figure 263 shows the relevant experimental data, which may be utilized well in the design of collection systems [127].

Wet manure is inclined to stick, leading in certain cases to settling and blocking. The inclination to stick depends on the moisture content and the material of the wall, as seen from Fig. 246. The inclination to stick on steel is less than on asbestos cement. Sticking may be reduced greatly by correct selection of the moisture content.

14.12 Air resistance of fruit-tree crowns

In high-capacity spraying machines, an appropriate distribution of spray and an adequate penetration distance are obtained by means of an air jet. The penetration distance of an air jet and the distribution of velocity in it may be calculated on the basis of the general behavior of free jets. An air jet leaving an outlet aperture at a velocity v_0 gradually widens as a function of the distance covered, and at the same time its axial velocity decreases. The velocity v measured on the

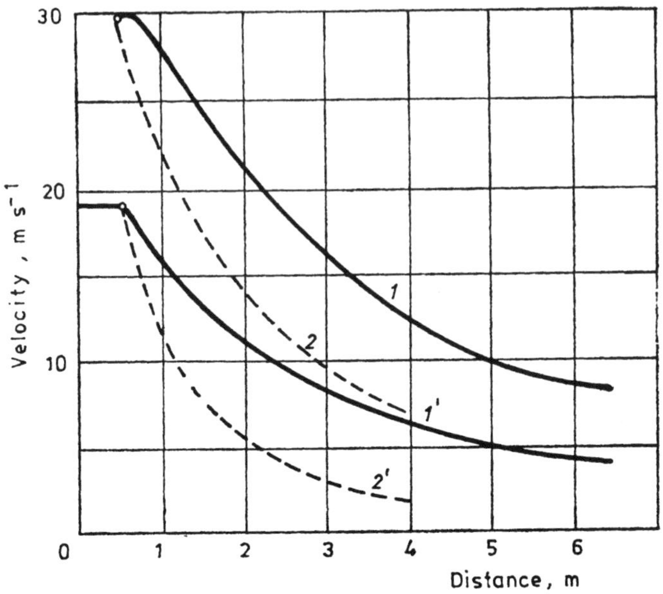

Fig. 264. Velocity in a free air jet as a function of distance, without hindrance (1) and in a tree crown (2)

axis of the jet may be calculated as a function of the distance covered by means of the equation

$$v_x/v_0 = 0.48/[(ax/d_e)+0.145]$$

where a is a constant depending on the cone angle of the jet, x the distance covered, and d_e the equivalent diameter of the outlet orifice. If the jet has to penetrate or traverse through the crowns of trees, the branches and leaves represent additional resistance and the axial velocity of the jet decreases to a greater extent (Fig. 264) [121]. The axial velocity may be calculated in this case from the relationship

$$v_{xk}/v_0 = 0.48/[(ax/d_e)+0.145+m(x-x_k)/d_e]$$

where x_k is the distance between the outlet orifice and the crown, and m is a constant depending on the crown (m=0.1–0.2). Processing of experimental data shows that it may be convenient to approximate the velocity loss for a free jet by functions of the form [121]

$$v_0^2/(v_0^2-v_x^2) = f(x/b_0)$$

or

$$v_0^2/(v_0^2-v_{xk}^2) = f(x/b_0)$$

Fig. 265. Velocity losses for a jet in air and in the crown of an apple tree

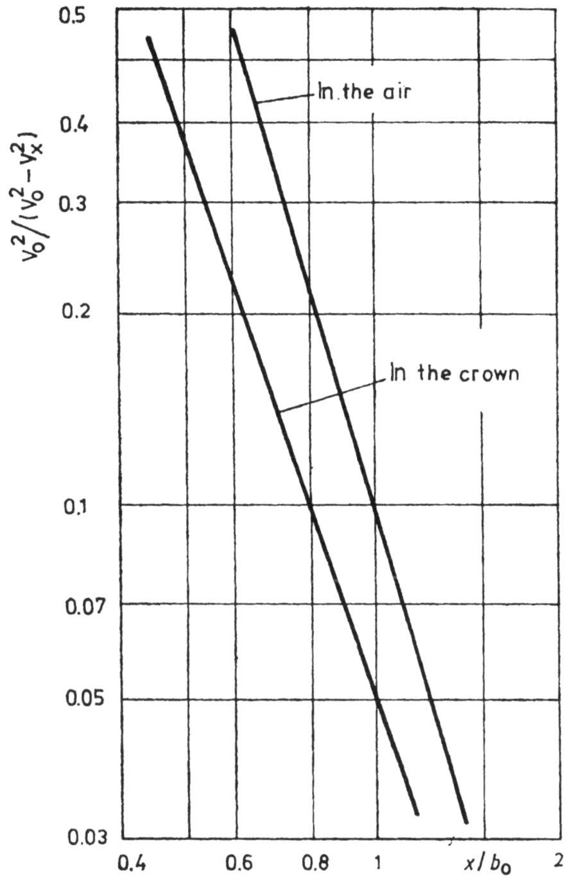

where b_0 is the width of the blowing orifice. Figure 265 shows the velocity losses for a free jet in air and in the crown of an apple tree. The distance from the blowing orifice is 0.5 m, and the diameter of the tree crowns is 6–8 m. The width of blowing orifices is generally 200–340 mm.

14.13 Hydrodynamic properties

Interest has recently increased in the hydraulic transport and treatment of fruits and vegetables (apples, cherries, tomatoes, potatoes, etc.). This is explained by the fact that hydraulic manipulating systems permit gentler treatment, with automatic washing and also cooling of the products [194]. In designing manipulation or transport systems utilizing liquid media, the hydrodynamic properties of the material being treated must be known. Since water is nearly always used as the liquid medium, the conditions governing the motion of the material treated must be determined in water.

The most important hydrodynamic properties are the settling or buoyant velocities, the drag coefficient, the orientations of the product in stationary and flowing water, the sinking depth into the medium on dropping the product, the angle of repose under the water surface, and the sliding and rolling angles of the product during submersion in water.

The gravitational force during uniform motion of a product in a stationary medium is compensated for by the buoyant force and the resistance of the medium, i.e.,

$$G = V\gamma_w + c_w F(\gamma_w/2g)v^2$$

By introducing the equivalent-sphere diameter the velocity of motion may be expressed as

$$v = [(4/3)d_e g(\gamma-\gamma_w)/\gamma_w c_w]^{1/2} \tag{186}$$

where γ and γ_w are the specific weights respectively of the product and the medium. The value of the drag coefficient c_w is 0.44 for a sphere in turbulent flow. The effective frontal surface area is greater for products whose shape deviates from spherical, and so the velocity of motion will be lower owing to the higher resistance of the medium. To take this effect into account it is usual to introduce a shape factor, as the ratio of the equivalent-sphere area to the maximum cross-sectional area:

$$\psi = F_e/F$$

The velocity of motion may then be calculated from the equation

$$v = [(4/3)d_e g(\gamma-\gamma_w)\psi/\gamma_w c_w]^{1/2} \tag{186a}$$

Agricultural products whose specific weight exceeds that of water move downwards at a settling velocity v, while products having a lesser specific weight move upwards when placed under water and the velocity v is the buoyant velocity.

It may be seen from eqn. (186) that the settling or buoyant velocity is determined primarily by the difference in specific weights and by the diameter if the shape does not deviate greatly from spherical. The specific weight of apples may be accepted as 0.8 g cm^{-3}, with the exception of very large fruits (Fig. 266), and so apples will float on water, a property which is utilized in both transport and classification. The buoyant velocity of apples in water is 0.5–0.6 m s^{-1}, depending on size. Calculations by means of eqn. (186a) supply similar values. If, during treatment, an apple moving upwards can impact against another apple

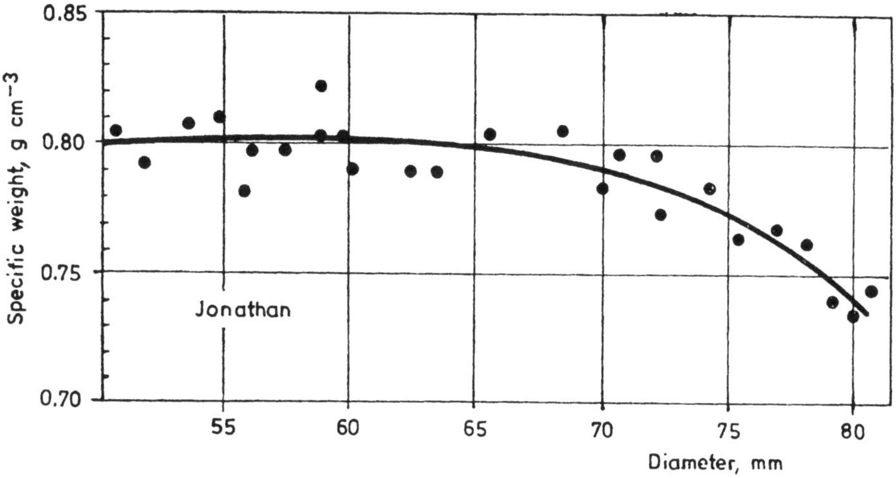

Fig. 266. Specific weigth of Jonathan apples as a function of diameter

or against the container wall, it is necessary to examine whether damage will be caused by impact at the given velocity.

In hydraulic classification of agricultural products, the orientations of individual products in a stationary or flowing medium are utilized. Generally, three orientations may be distinguished: with the stem up or down, or with a side uppermost. Table 16 gives data concerning the orientations of various apple species in water.

Table 16

Apple variety	Orientation		
	Stem up (%)	Stem down (%)	Side up (%)
Jonathan	95	4	1
Starking	66	28	6
Red Delicious	76	8	16
Golden Parmen	60	38	2

A relatively light product dropped into water will become submerged depending on its specific weight and the dropping height, and will then ascend to the surface. The depth of water selected must be greater than the expected maximum submersion if the product is to be prevented from impacting against the bottom of the container. The kinetic energy of a product falling into water is spent in overcoming the buoyant force and the resistance of the medium over the given depth of submersion. The differential equation of motion may be written as

$$m(dv/dt) = G - P_b - W$$

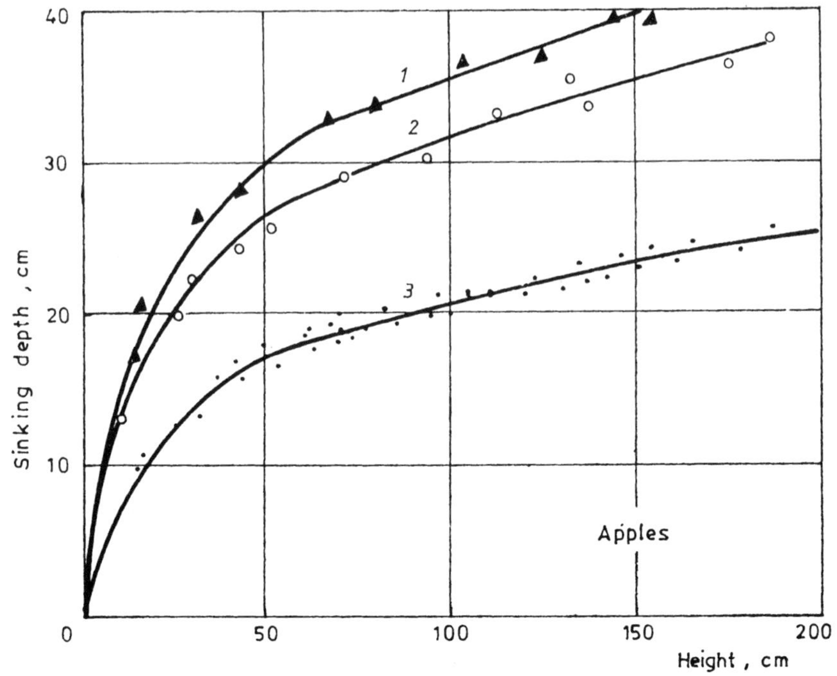

Fig. 267. Sinking of an apple dropped into water from various heights. *(1)* 60 mm diameter; *(2)* 52 mm diameter; *(3)* 60 mm diameter with air counterflow

where P_b is the buoyant and W the drag force. Substitution of the relevant expressions for the mass and the forces gives the differential equation

$$-dv/dt = g((\gamma_w/\gamma_{fr})-1)+(0.75c_w/d)(\gamma_w/\gamma_{fr})v^2$$

or, in shorter form,

$$-dv/dt = K_0+K_1v^2$$

Integration of this equation gives the sinking velocity of the body as a function of time:

$$v = \sqrt{K_0/K_1}[(\sqrt{K_1/K_0})v_0-\tan(t\sqrt{K_0K_1})]/[1+\sqrt{K_1/K_0})v_0\tan(t\sqrt{K_0K_1})]$$

The sinking velocity becomes zero at a given time t_0, which may be calculated as

$$t_0 = (1/\sqrt{K_0K_1})\ \mathrm{arc\,tan}\,(\sqrt{(K_1/K_0)}v_0)$$

The sinking depth is obtained by integrating the expression for velocity:

$$S = (1/2K_1)\ln[1+(K_1/K_0)v_0^2] \qquad (187)$$

Figure 267 shows the relationship between sinking depth and falling height obtained on the basis of experimental data. As may be seen, the sinking depth is always greater for apples of larger diameter.

The sinking depth may be decreased to about half its value by counterflow air circulation. Air flows upwards in the form of tiny bubbles in the water, and the downward motion of the apples is braked by the buoyant force of the bubbles.

The sinking depths given by eqn. (187) are slightly less than the measured values. This may be explained by the occurrence of complex phenomena during the impact of the body against the water surface, which cannot be taken into account by the present theory. Figure 267 illustrates that in the manipulation of apples a water depth of at least 50–60 cm is necessary. If the desired depth of water cannot be ensured, the bottom of the container must be coated with cushioning material.

Products emptied into water assume a single-layer disposition after a time, provided that the area of the water surface is sufficiently great. If it is insufficient, then a downwards-pointing pyramid is formed, whose angle is smaller, according to observations, than the natural angle of repose of the product. In the case of apples under water, angles of repose of 30–35° are observed.

15. FRICTION PROBLEMS

Friction plays an important and in numerous cases decisive role in all fields of agricultural mechanics. Friction is always present in some form during the movement of bodies, and it affects the force which has to be exerted. In silo bins and other storage structures the vertical load on the walls is determined by the friction coefficient. During pneumatic transport, especially for higher material concentrations, friction between the material and the wall is responsible for significant resistance. The elements of certain means of conveyance, such as screw conveyors, can be rated only if the friction coefficient is known. The behavior of bulk granular materials and cereals also depends greatly on the friction coefficient. Friction plays a smaller or greater role also during the cutting and pressing of agricultural products. The friction arising on the lateral faces of a knife or on the walls of a pressing canal increases the resistance, and in planning these devices a design must be found such that the relative value of frictional resistance is as low as possible. The winding of materials on rotating parts is also achieved as a result of friction.

These examples show clearly that friction plays a part in nearly all processes, and cannot generally be disregarded.

15.1 General laws of friction

If a body is placed on a plane surface (Fig. 268(a)), it can be displaced only by overcoming a frictional force. The force required to initiate displacement is termed the static frictional force. After the start of movement the frictional force generally decreases, and so movement can be maintained by a lower force. The force required to maintain movement is termed the kinetic frictional force.

The relationship between the force P required to start or maintain movement when a force G is acting in a direction normal to the contact surface may be expressed by Coulomb's law in the simple form

$$P = \mu G$$

where μ is the static or dynamic friction coefficient.

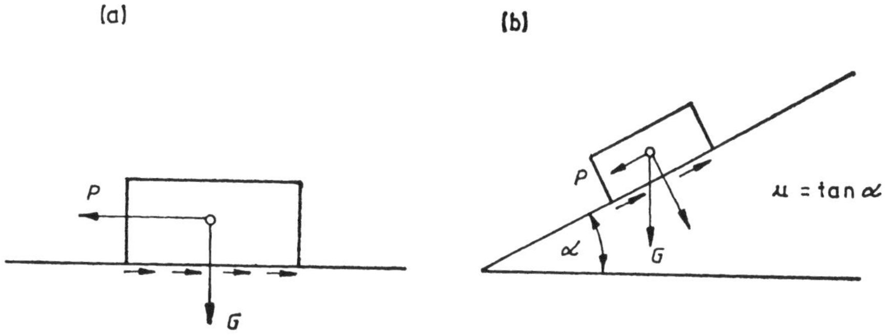

Fig. 268. Illustration of the friction process

The force required to start or maintain movement may also be produced by placing a body on a slope of angle α (Fig. 268(b)). In this case,

$$\mu = G \sin \alpha / G \cos \alpha = \tan \alpha$$

In formulating the general laws of friction it is assumed that the frictional force is

(a) proportional to the normal force;
(b) independent of the dimensions of the sliding surfaces;
(c) independent of the sliding velocity; and
(d) dependent on the nature of the contact surfaces.

These statements are not always valid in the case of biological materials, owing mainly to the fact that these materials may be deformed even by low surface pressures, and so their sliding surface varies. During the sliding of wet materials the phenomenon of adhesion frequently appears, and the friction characteristics may then be modified markedly. As a result of friction the temperature rises at the contact surface. Biological materials are sensitive to temperature variations, whereby the friction coefficient may also vary.

The above phenomena explain partly why the friction coefficient depends also on the friction path in the case of biological materials.

Formation of a liquid film on a contact surface depends on the wetting capacity of the liquid, understood as the ratio of the adhesive tension to the surface tension. The presence of adhesive forces also increases the frictional force. Simultaneously it also modifies the friction characteristics, since the adhesive force depends on the size of the surface area.

15.2 Friction coefficients of agricultural products

Friction coefficients may be measured experimentally using simple apparatus. Rotating discs or shafts are used to measure the dynamic friction coefficient. In most cases the static friction coefficient is measured on a slope of adjustable angle. The principal elements of the equipment used most frequently are shown in Figs. 269–271. The rotating-disc device shown in Fig. 269 may be used to measure

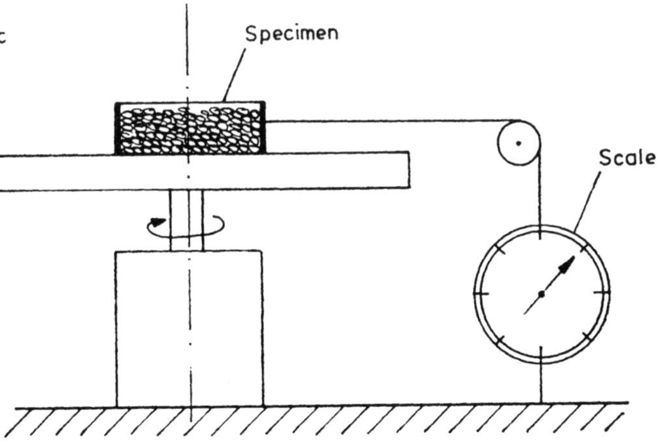

Fig. 269. Rotating-disc apparatus for measuring the friction coefficient

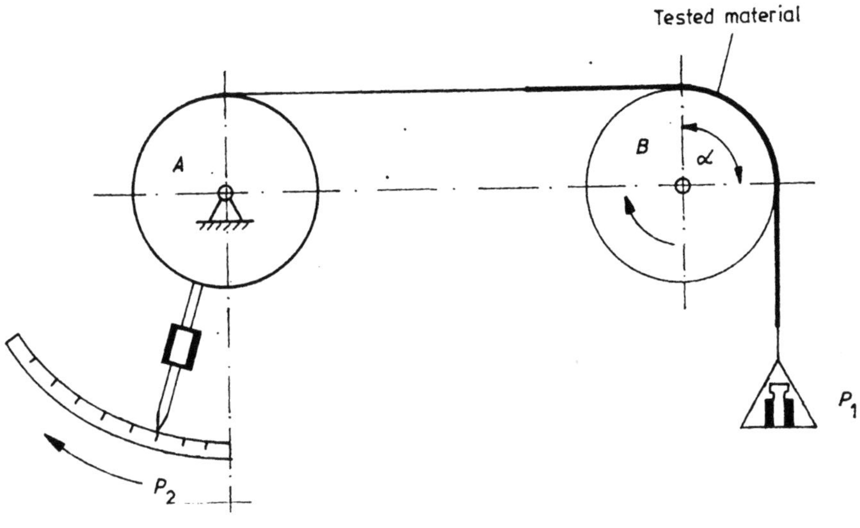

Fig. 270. Apparatus for measuring the friction coefficient of forage materials

friction coefficients of both bulk materials and solid masses. In measurements for bulk materials a light frame, not touching the rotating disc, is used. The friction coefficients of fibrous forage materials may be measured using the device shown in Fig. 270. The contact angle of the material with the rotary shaft is 90°, the required surface pressure is created by the force (weight) P_1, and P_2 is the retention force. The average surface pressure is given by

$$p_{av} = (P_2 - P_1)/\alpha r b$$

where b is the width of the fiber tested. Equilibrium of the fiber is described by the equation

$$P_2 = P_1 e^{\mu\alpha}$$

from which

$$\mu = (1/\alpha) \ln (P_2/P_1) \tag{188}$$

Using the device shown in Fig. 271, the static friction coefficient is given by the tangent of the angle at which the material starts to slide down the slope.

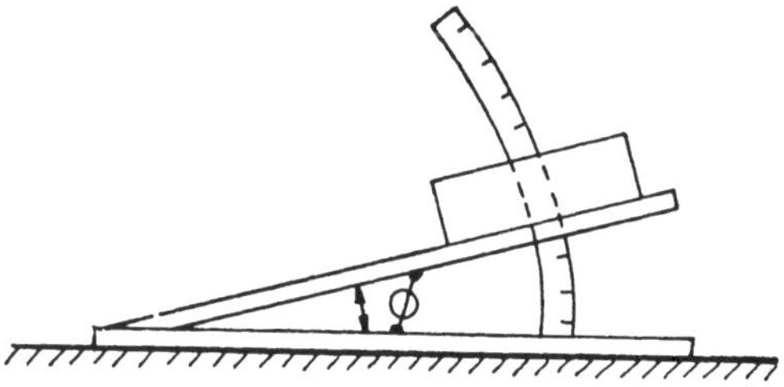

Fig. 271. Apparatus for measuring the static friction coefficient

The friction coefficients of agricultural products are influenced by numerous factors, the most important of which are the moisture content, the surface pressure, the sliding velocity, the nature of the surface and the characteristics of the environment.

The moisture content influences the friction coefficient markedly; generally, the friction coefficient increases with increasing moisture content. Figure 272 presents the dynamic friction coefficients of grass and alfalfa as functions of moisture content [130]. The friction coefficient is seen to be especially sensitive to moisture content in the range around 40%. The reason is that for moisture contents in excess of this value, water becomes available to wet the friction surfaces, thereby increasing the frictional force. A similar trend is also observed in the case of cereals (Fig. 273). Thus the power demands of all processing equipment involving friction (transport means, etc.) increase with rising moisture content. It frequently occurs that relatively dry material contains moisture impurities (e.g., straw

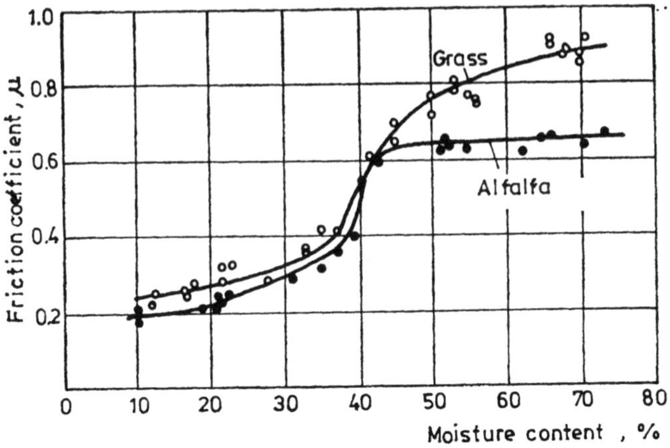

Fig. 272. Kinetic friction coefficient of grass and alfalfa as functions of moisture content

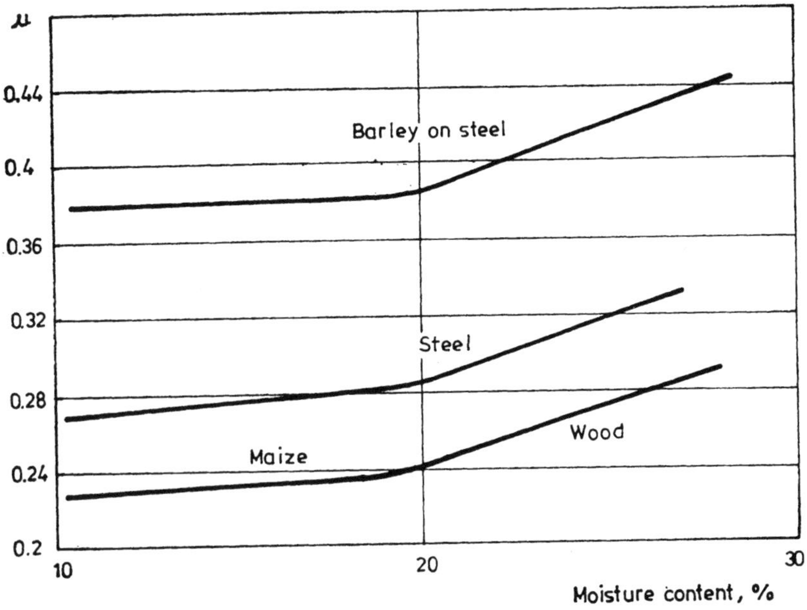

Fig. 273. Friction coefficient of crop products as functions of moisture content

on harvesting is contaminated by weeds). In such cases the friction coefficient increases, as shown in Table 17 [133].

The effect of surface pressure on the friction coefficient varies depending on the moisture content. For lower moisture contents, the effect of surface pressure

Table 17

Surface	Dynamic friction coefficient		Static friction coefficient	
	straw	straw with weeds	straw	Straw with weeds
Tarpaulin	0.39	0.44	0.51	0.63
Machined steel	0.35	0.46	—	—
Galvanized (zink-plated) steel	0.26	0.36	0.33	0.56

is not significant; in the case of cereals, for example, it is generally negligible. For materials of higher moisture content (silage, cuttings of root crops), the friction coefficient decreases with increasing surface pressure. Figure 274 shows the static friction coefficients of chopped grass and maize silage on blank steel plate as functions of surface pressure [131]. The friction coefficient of freshly cut root bulb crops depends to an even greater extent on surface pressure, since adhesion at the contact surface is more pronounced. Figure 275 shows the dynamic friction

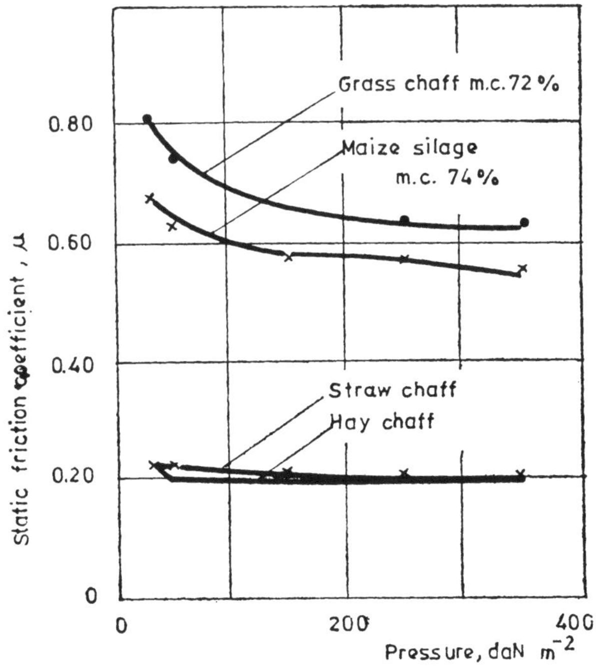

Fig. 274. Static friction coefficients of grass chaff and maize silage as functions of surface pressure

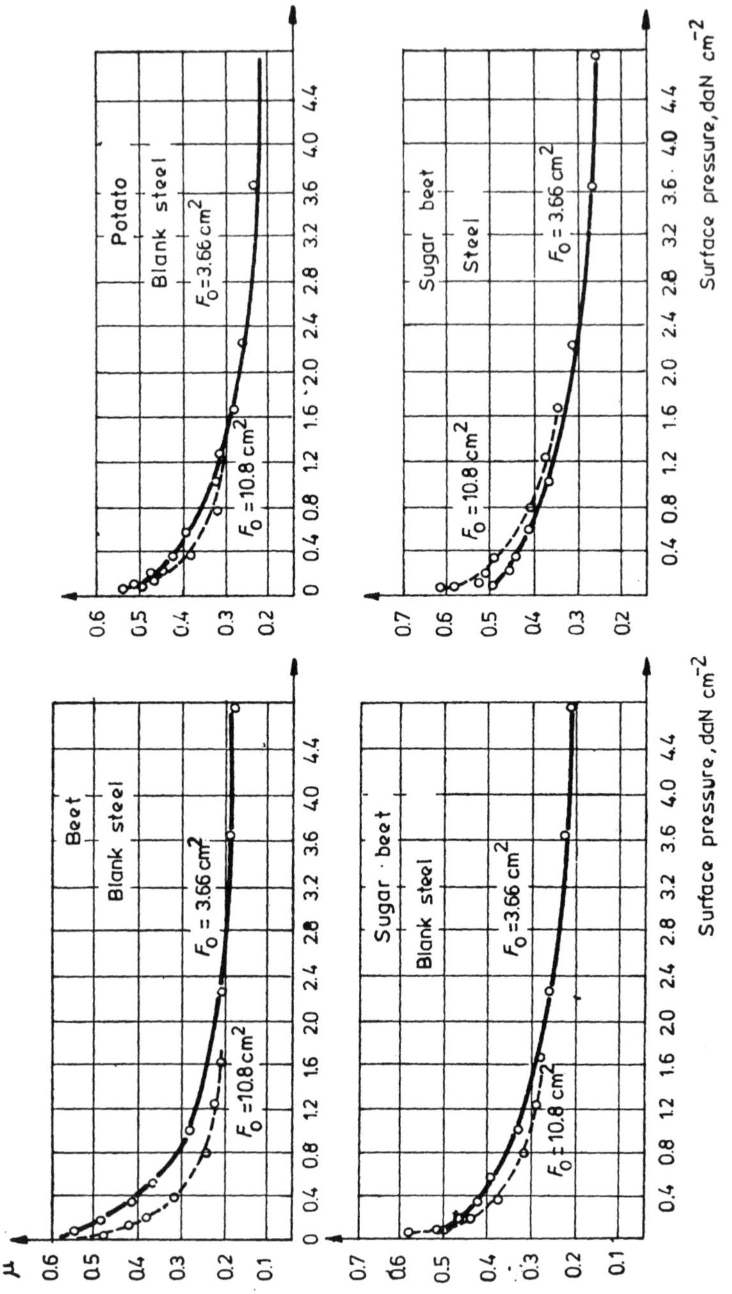

Fig. 275. Kinetic friction coefficient of freshly cut root tuberous products

coefficients of slices of fodder beet, sugar beet and potato on blank steel plate as functions of surface pressure [132].

The resultant frictional force obtained during the common action of friction and adhesion is given by the two-term expression

$$P = \mu^* G + aF = \mu G$$

where a is the adhesion coefficient, and F the contact surface area. On the basis of the data of Fig. 272 for grass and alfafa, the value of the adhesion coefficient is $a = 0.1\text{–}0.12$ daN cm^{-2}. It is of interest to note that similar adhesion-coefficient values are obtained for the friction of wet soils. The dynamic friction coefficient is generally not affected greatly by the sliding velocity. In the case of shelled maize a slight increase (of about 10%) in the friction coefficient is found in the velocity range 0–0.3 m s^{-1}, while at higher velocities the value remains constant. Essentially the same results are obtained for chopped silage.

The nature of the surface influences the value of the friction coefficient greatly. This applies to both the friction surface and the frictioning material. The surfaces encountered most frequently are steel (machined, polished or painted), wood and rubber, and in certain cases various plastics, etc. The friction coefficient is determined mainly by the smoothness of the surface and the material itself. The friction coefficient is not reduced by increasing the smoothness of the surface; in many cases it is increased owing to the better adhesion which results. Adhesion on painted surfaces is also excellent, and so the friction coefficient for such surfaces is high. Adhesion on roughly machined surfaces is poorer, and therefore a lower friction coefficient is generally obtained for these surfaces. The friction coefficient is also influenced by the nature of the surface of the frictioning substance. In particular, green forage materials have tiny hairs on their surfaces, which greatly increase the friction coefficient at the start of movement. If the friction path is longer, the hairs wear off and the friction coefficient decreases.

Plant stalks cannot be regarded as isotropic; their friction coefficient therefore depends on their orientation relative to the direction of movement. Figure 276 illustrates the friction coefficient of wheat straw on wheat straw as a function of moisture content, for various angles of orientation of the moving relative to the stationary layer. As may be seen, the friction coefficient is highest when the moving and stationary fibers are mutually parallel ($\beta = 0$). The friction coefficient is also affected substantially by the relative humidity of the ambient air. The surface of a crop may absorb moisture from or dissipate moisture into the air within a relatively short time, according to its equilibrium moisture curve, and the friction coefficient develops in correspondence with the instantaneous moisture content of the surface, independently of the internal moisture content. Figure 277 shows the kinetic friction coefficients of wheats with

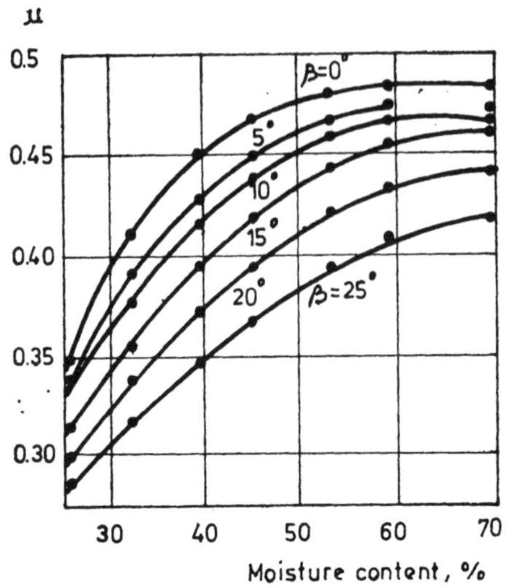

Fig. 276. Friction of straw on straw as a function of moisture content for various angles of orientation

various moisture contents as functions of the relative humidity of the ambient air [134]. Wheat with a lower moisture content absorbs water from the air at a lower rate, and so the friction coefficient increases to a lesser extent as the relative air humidity increases than in the case of wheat with a higher moisture content. The surface of cereal grains absorbs or dissipates moisture relatively rapidly. Figure 278 shows the variation of the friction coefficient over time for wheat grains with 18% moisture content placed into air of relative humidity 20%. In this case the wheat grains dissipate moisture to the ambient medium, their surface dries and their friction coefficient decreases over time. As may be seen, the vast majority of the reduction in this case occurs within 20 min [134].

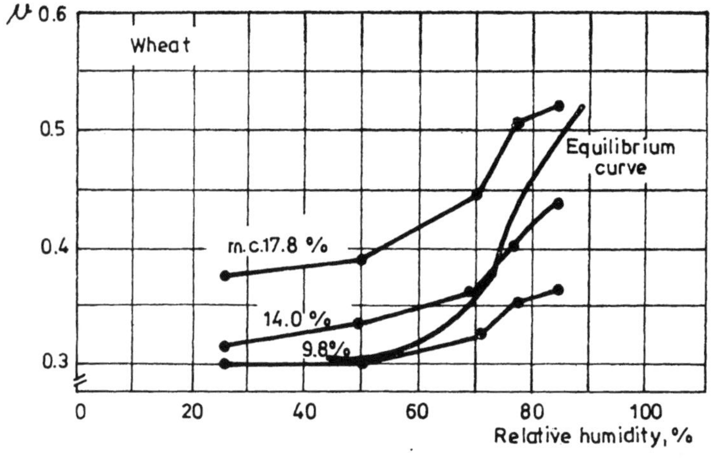

Fig. 277. Kinetic friction coefficient of wheat in air of various relative humidities

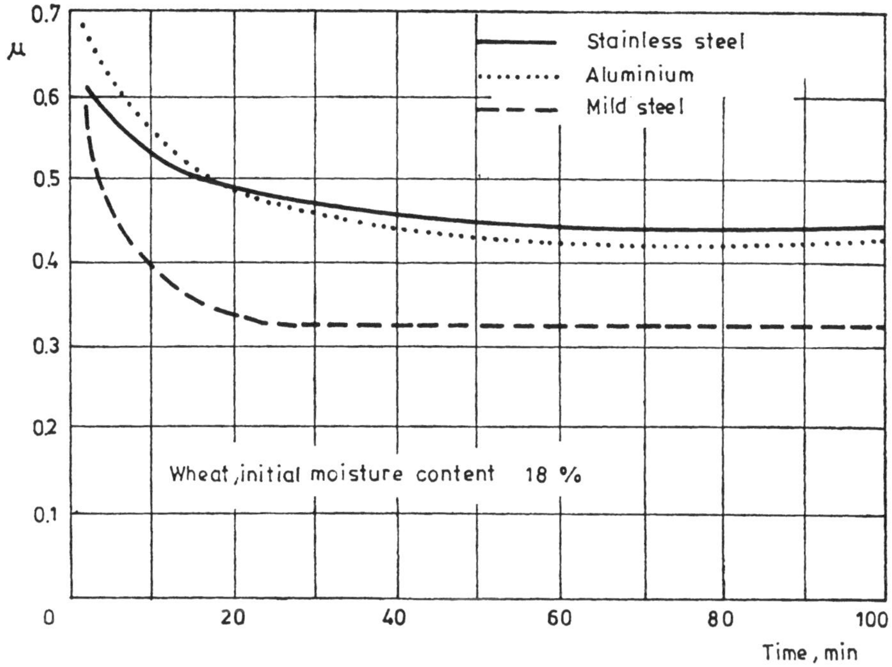

Fig. 278. Friction coefficient as a function of time for wheat drying in air (nonequilibrium state)

15.3 Rolling of agricultural products

Rolling plays an important part during the sorting and transport by gravity of many products, and during other technological processes. To select satisfactory slope angles, the rolling resistance of a product, namely, the static and dynamic rolling angles on various surfaces, must be known.

The rolling of agricultural products is a complex process, since the shape of the product is generally not regularly spherical and during rolling both the product and the surface may deform. Figure 279 shows the rolling of a rigid sphere on a deforming surface. On the basis of the figure,

$$P = cG/r$$

where c is the coefficient of rolling resistance. The rolling properties of agricultural products are studied on slopes. A deformable product resting without movement on a slope always lies on a part of its own surface which will be impressed relative to the original undeformed spherical surface. Therefore a relatively large slope angle is required to initiate rolling. For example, in the case of apples this angle is 13–18°. When rolling has started, a considerably smaller

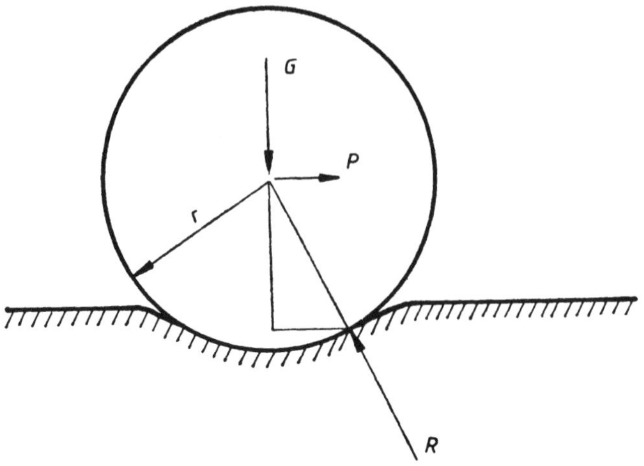

Fig. 279. Rolling of a rigid sphere on a deforming surface

slope angle suffices to maintain motion: for example, 2.5–4° in the case of apples Values of the rolling angles obtained for apples and tomatoes on various surfaces are given in Table 18.

Table 18

Surface	Apples		Tomatoes	
	Static	Dynamic	Static	Dynamic
Plywood	12–18°	2.5–4.5°	9–14°	3.6–4.8°
Steel plate	13–18°	2.4–4.0°	7–11°	3.6–4.8°
Rigid foam	13–18°	2.5–4.0°	11–13°	4.2–4.8°
Soft foam	11–16°	4.0–5.0°	11–13°	4.8–5.8°
Linen	12–16°	4.0–5.0°	13–14°	4.8–7.0°

15.4 Angle of internal friction and angle of natural repose

The most important friction characteristics of bulk materials (cereals and granular products) are the angle of internal friction and the natural angle of repose. The angle of internal friction is understood as the angle of friction appearing under the effect of friction between individual grains:

$$\mu_i = \tan \Phi_i$$

where μ_i is the internal friction coefficient, and Φ_i the angle of internal friction. The natural angle of repose (Φ_r) is understood as the angle enclosed with the

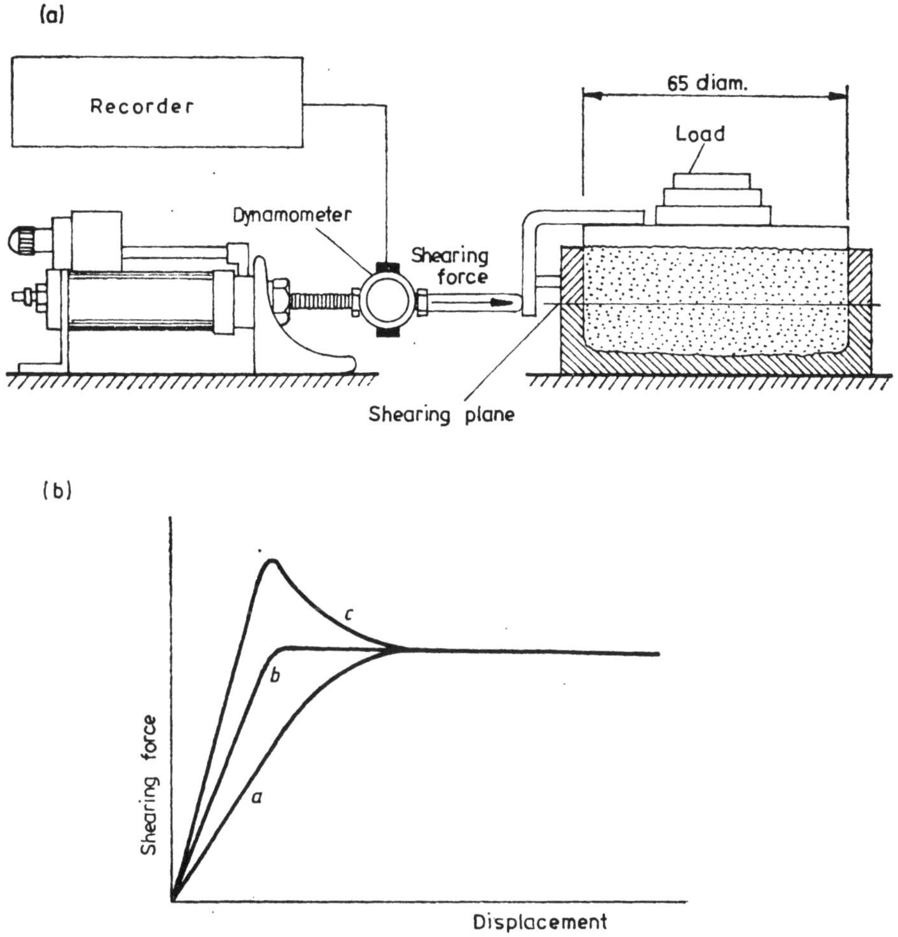

Fig. 280. Equipment for measuring the internal angle of friction and characteristic shearforce–displacement curves

horizontal, at which a piled bulk material will remain stationary. The angles Φ_i and Φ_r are not independent (apart from in extreme cases), but they are not identical.

The angle of internal friction is measured by means of a special shearing box (Fig. 280), in which bulk-material particles slide on each other along a plane. Loading of the material in the normal direction is achieved by weights stacked on each other. The upper ring, containing the material, is displaced during the measurements. The force required to overcome the stresses arising in the shearing plane is measured using a dynamometer cell and recorded as a function of displacement.

The force initially increases rapidly as a function of displacement, and after a certain displacement, remains constant. The shearing-force curve may show various courses under a given stress, depending on the state of consolidation of the material in the shearing box (Fig. 280(b)). Curve *(a)* is for material which was insufficiently preconsolidated, and so after the start of shearing it still continued to settle slowly. In the case of curve *(c)* the material was overconsolidated, and therefore at the start of shearing it loosened, i.e., its density decreased. In the case of curve *(b)* the material was consolidated properly, corresponding to the given normal stress.

Various granular materials may be compressed differently, depending on the shape, size and strength of the grains. Farinaceous and wet materials can generally be compressed to a greater extent, and so may easily be overconsolidated. Dry materials with high grain strength (e.g., dry sand) can be compressed in the main only elastically, and so they always assume the density corresponding to the normal stress, without any preparation.

During shear tests the maximum stress τ is plotted as a function of the normal stress σ. Various τ–σ relationships are obtained, depending on the behavior of the material; typical relationships are shown in Fig. 281. Figure 281(a) presents the relationship for ideal bulk material: the stress τ is proportional to σ and

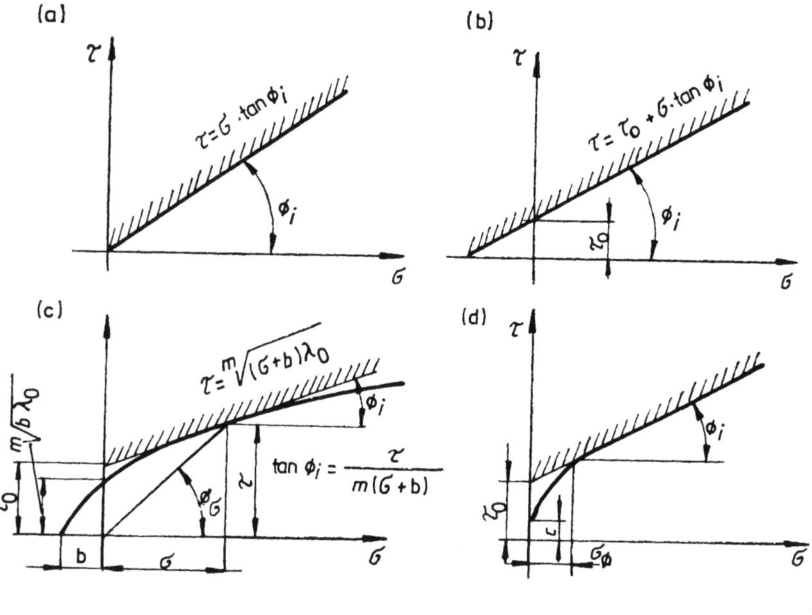

Fig. 281. Various τ–σ relationships

the straight line passes through the origin. The equation of the straight line is

$$\tau = \sigma \tan \Phi_i \tag{189}$$

where the angle of internal friction is given by the slope of the straight line. Figure 281(b) also shows a linear relationship between τ and σ, but this straight line does not pass through the origin. This means that a starting shear stress τ_0 appears even in the case of zero normal stress, as found in the cohesion of heavy soils (cohesive granular materials). The equation of the straight line is

$$\tau = \tau_0 + \sigma \tan \Phi_i \tag{190}$$

Figure 281(c) shows an exponential relationship:

$$\tau = \sqrt[m]{(\sigma + b)\lambda_0} \tag{191}$$

where m, b and λ_0 are constants. In this case different angles of internal friction correspond to each stress σ. For a given stress σ,

$$\tan \Phi_i = \tau / m(\sigma + b)$$

and the corresponding initial shear stress is

$$\tau_0 = mb \tan \Phi_i$$

If a straight line is drawn from the origin to a given point, then the angle of inclination of the line gives the effective angle of internal friction, characterizing the material at the given point as an ideal bulk material.

The relationship shown in Fig. 281(d) is linear beyond a given stress σ_Φ, and eqn. (190) is then applicable. In the interval $0 < \sigma < \sigma_\Phi$ the curve may be given by eqn. (191); the three constants appearing in this equation may be determined from the relations

$$c = \sqrt[m]{b\lambda_0},$$

$$\tau_0 + \sigma_\Phi \tan \Phi_i = \sqrt[m]{(\sigma_\Phi + b)\lambda_0}$$

$$\tan \Phi_i = (1/m)\sqrt[m]{\lambda_0/(\sigma_\Phi + b)^{m-1}}$$

For the curved region the angle of internal friction is variable; its value increases with decreasing σ.

In examining cereals behavior similar to that of Fig. 281(d) is generally obtained, so long as the stress σ does not exceed 0.2–0.4 daN cm^{-2}. For higher values of σ the curve varies similarly to Fig. 281(c), i.e., τ does not increase in proportion to the growth of σ.

The angle of internal friction of cereals depends on the shape, size and mois-

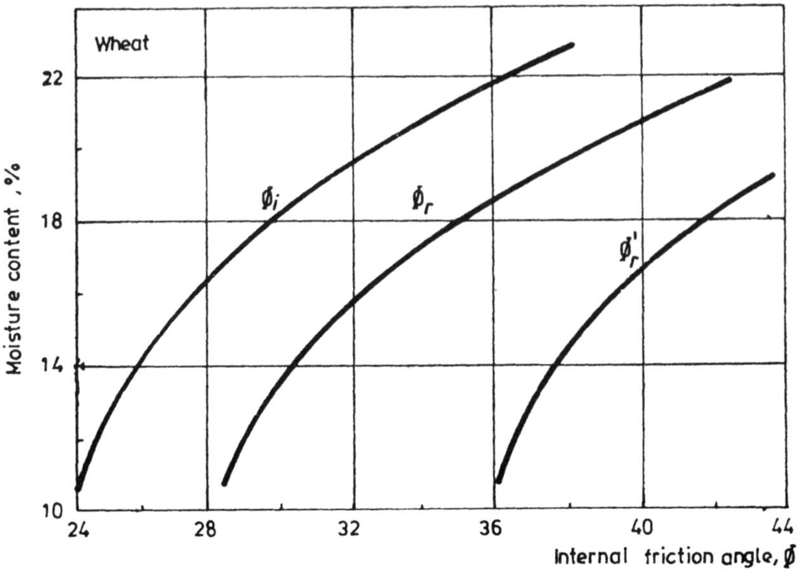

Fig. 282. Angle of internal friction and angle of repose of a wheat pile as functions of moisture content

ture content of the grains. The angle of internal friction and the initial shear stress τ_0 for granular bulk material are also influenced by consolidation or settling over longer periods. In particular, the value of τ_0 increases during settling. Figure 282 shows the angle of internal friction of wheat as a function of moisture content. With increasing moisture content the angle of internal friction increases significantly. The angle of internal friction is also increased by an increase in the bulk density (i.e., by compression) of a given material [135]. During investigations of sorghum seeds with 16% moisture content the value of Φ_i was found to increase from 24° to 28° on increasing the bulk density from 0.767 to 0.83.

Figure 283 presents the simpler devices used for measuring the natural angle of repose. Using the device shown in Fig. 283(a) the static angle of repose is determined as follows. The bulk material, whose surface is horizontal in the lower position, is placed into a frame, mounted on an articulated arm. Then the frame is tilted by means of the articulated arm until the grains on the surface start to move. The tilting angle gives the static natural angle of repose. The equipment shown in Fig. 283(b) is suitable for measuring the dynamic natural angle of repose. Bulk material placed in a box flows through an aperture in its base, whereby a funnel-shaped surface is formed in the box and a conical surface underneath. The angles of these surfaces enclosed with the horizontal give

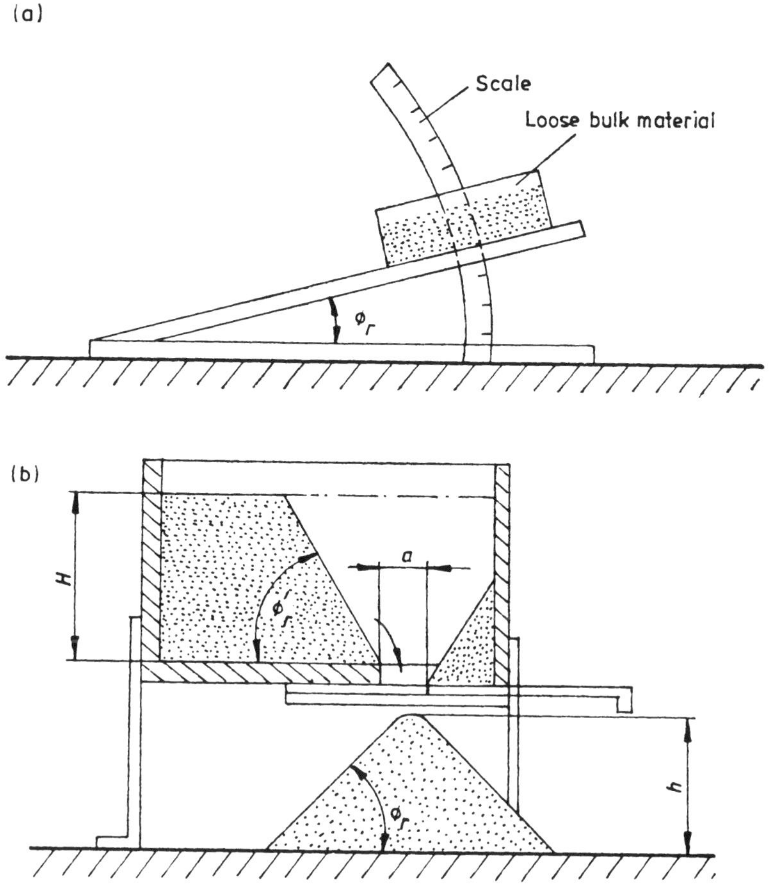

Fig. 283. Devices for measuring the angle of repose

the dynamic natural angle of repose. In the case of ideal bulk material, $\Phi_r' = \Phi_r$; for materials deviating from the ideal, $\Phi_r' > \Phi_r$ is generally found.

Figure 282 shows the variation of the angle of repose of wheat as a function of moisture content. Similar results have been obtained for maize, and so the values given here apply also to the latter.

15.5 State diagram for granular bulk materials

As has been seen earlier, the shearing force at a given normal stress depends on the extent of preliminary compression. A measure of the extent of compression is given by the porosity ε, and so it seems obvious to study the behavior of compressible materials in a τ–σ–ε system of space coordinates [136–138].

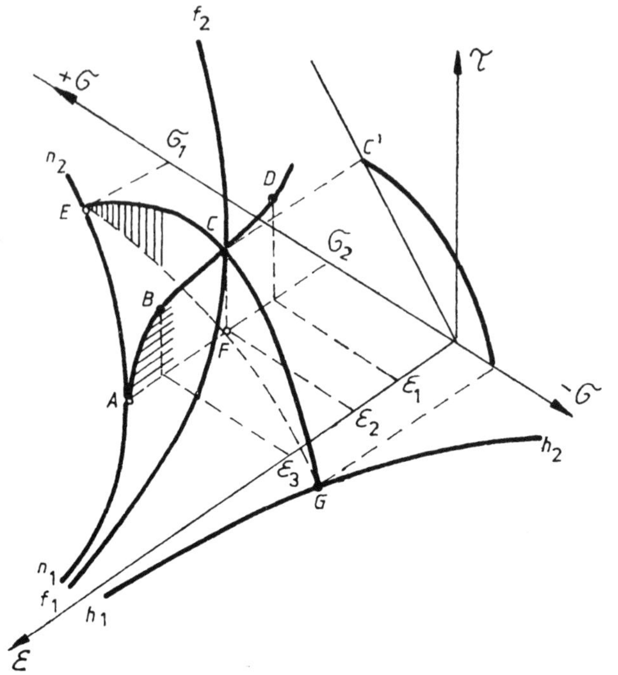

Fig. 284. State diagram for loose granular bulk materials

Figure 284 illustrates the state diagram characteristic of granular bulk materials. The curve n_1–n_2 lies in the σ–ε plane and describes the initial compression of the material. Material compressed by the stress σ_1 is unloaded along the curve section *EFG*. Cohesive materials can also take up a certain tensile stress, whose limiting value is shown by the curve h_1–h_2 [138].

The τ–σ–ε relationship corresponding to the stationary flow state is described by the three-dimensional curve f_1–f_2. Accordingly, the stationary flow state may be maintained under a normal stress σ_2 in the case of any preliminary consolidation by values of τ and ε corresponding to point *C*. In the case of a higher initial pore volume ratio (ε_3), the material first assumes the stress τ corresponding to point *B*, and then at the start of flow is compressed and becomes capable of assuming a higher stress τ. Thus the material again reaches the state corresponding to point *C*. For greater preliminary consolidation (ε_1), the material assumes the stress corresponding to point *D*; at the start of flow it loosens and again reaches the state corresponding to point *C*. However, in this latter case a higher stress τ is required to initiate flow than to maintain it.

The curves n_1–n_2, f_1–f_2 and h_1–h_2 supply a three-dimensional surface whose sections are the curves *ABCD* and *ECG*. For state characteristics below this surface, only elastic deformation occurs; state characteristics above this surface are not possible. Flow of material is possible for state characteristics found along the surface.

The projection of the curve f_1–f_2 on the τ–σ plane is approximately a straight line passing through the origin. The projection of the curve section CG found above this straight line supplies in a given case the stress τ required to initiate flow of a material consolidated by a normal stress σ_1 in the interval 0–σ_2.

Materials inclined to settling will consolidate over time to a low pore volume, i.e., they become overconsolidated relative to the given normal stress. A high stress τ is then required to initiate flow. Media resting in bins can flow out only if the stresses τ due to weight forces attain the critical values required to initiate flow.

15.6 Stress state of granular bulk materials

Granular bulk materials may be treated, with certain restrictions, as continuous solid materials. If the size of individual grains is small compared to the volume of the pile and the grains may be regarded as elastic (i.e., without the appearance of stresses σ and τ sufficient to cause plastic deformation), and provided that the mechanical properties are identical in every direction, then the stresses τ and σ caused by a load may be interpreted according to the relation [139–144]

$$\tau \leqq \pm\varphi(\sigma) \tag{192}$$

Characteristic forms of this function are shown in Fig. 281. The forces acting on a pile may be of two types: internal forces acting on each individual grains (weight and inertial forces), and external forces acting on the surface of the pile. The bulk material is in equilibrium when there is no displacement among individual grains.

To determine the plane stress state a triangular prism, on whose orthogonal sides the principal stresses σ_1 and σ_2 are acting, is sampled from the bulk body. On the third side, lying under the angle α_1, the stresses σ and τ which arise may be calculated from the equations [139]

$$\sigma = \sigma_1 \cos^2 \alpha_1 + \sigma_2 \sin^2 \alpha_1 \tag{193}$$

and

$$\tau = [(\sigma_1 - \sigma_2)/2] \sin 2\alpha_1 \tag{194}$$

The stresses σ and τ may also be determined graphically by means of Mohr's circle. Knowing the principal stresses σ_1 and σ_2, a circle may be drawn in the τ–σ system of coordinates. The straight line drawn at an angle α_1 from point 2,

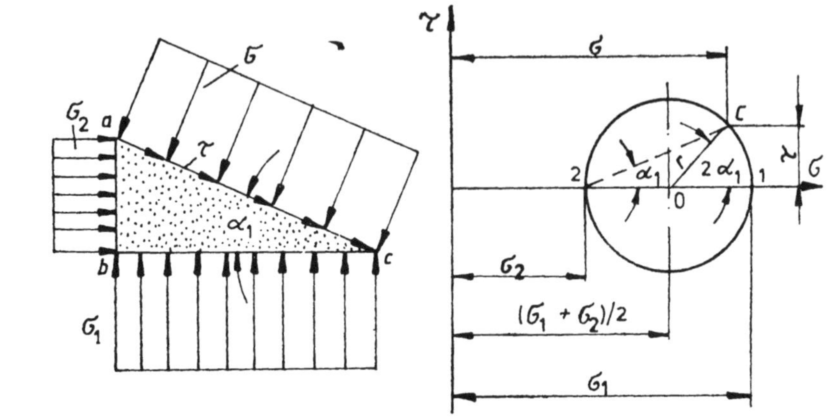

Fig. 285. Mohr circle for derivation of the stress state of a granular pile

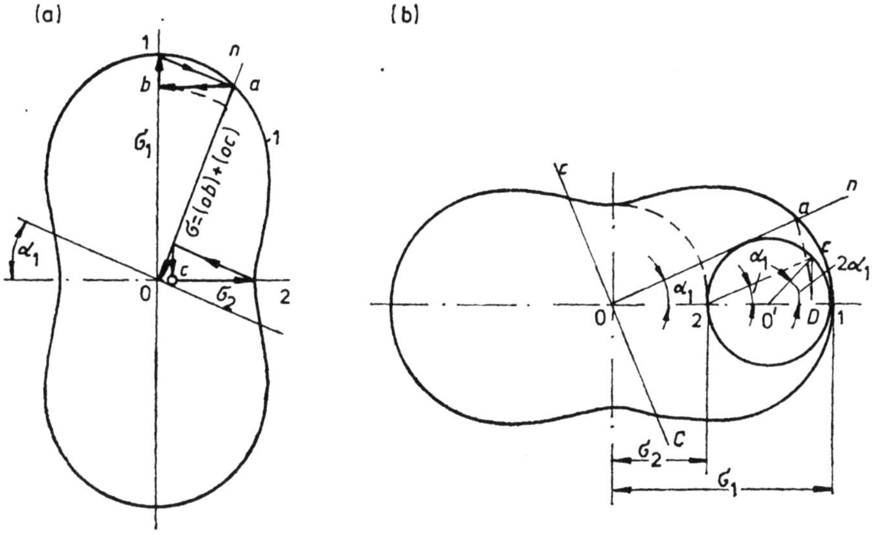

Fig. 286. Oval diagram of the normal stresses for a granular pile

or at an angle $2\alpha_1$ from the center, yields the point C as the intercept on the circumference, whose coordinates give the required stresses σ and τ (Fig. 285). The stress σ may also be determined from the oval diagram of normal stresses, as shown in Fig. 286. The principal stresses σ_1 and σ_2 are plotted perpendicularly to each other, the normal n is drawn on an arbitrary plane α_1, then a line perpendicular to the normal n is drawn from the endpoint of σ_1, and finally the point obtained is projected back to the σ_1 axis (point b).

The value of the distance $\overline{ob}$ obtained is $\sigma_1 \cos^2 \alpha_1$. By a similar construction, the distance $\overline{oc}$ obtained from the endpoint of σ_2 is $\sigma_2 \sin^2 \alpha_1$. In the sense of eqn. (193), the value of σ is obtained by measuring the distances $\overline{ob}$ and $\overline{oc}$ on the normal n. If the angle α_1 is varied between 0 and 360°, the vector of the stress σ describes a closed curve.

Figure 286(b) shows the relationship between the Mohr circle and the oval diagram. A line parallel to the normal n, drawn from point 2 yields the point C. The stress σ is given by the horizontal coordinate of C, so this must be translated by calipers to the normal n.

Bulk material, as a medium, will be in equilibrium if the relationship between the stresses τ and σ described by eqn. (192) prevails. Accordingly, the Mohr circle of stresses is always on the inside of the envelope curve $\pm\varphi(\sigma)$, or contacts it from the inside in extreme cases, as seen in Fig. 287. The case when the Mohr circle contacts the envelope curve is termed the equilibrium limit position, where $\tau = \pm\varphi(\sigma)$.

It is necessary to distinguish between the equilibrium states for a bulk medium and for a bulk body. Interpretation of the latter is necessary because bulk material is generally surrounded by walls on several sides, while its upper surface is free. On the free surface no stresses arise, while along the wall a stress given by

$$\tau \leqq \mu_1 \sigma$$

arises, where μ_1 is the friction coefficient between the bulk material and the wall.

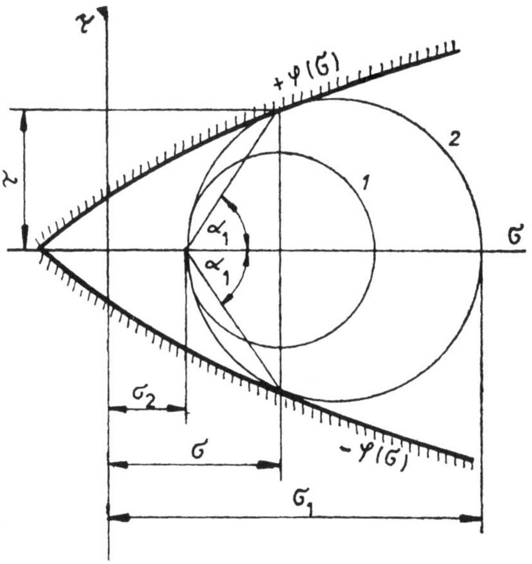

Fig. 287. Relationship between Mohr circle and envelope curve

On the basis of the above, there may be cases where the medium is in equilibrium (i.e., the grains do not move in relation to each other) but the whole body slides along the limiting walls (e.g., in a sloped canal).

When the equilibrium state for the bulk medium ceases to exist, the individual grains slide over each other along elemental planes, enclosing an angle α_1 with the plane of the local principal stress (Fig. 287). The elemental slip planes supply the slip surface.

15.7 Pressure distribution in bins

A considerable number of bulk agricultural products (cereals, fodder, etc.) are stored in variously designed storage spaces. The most frequent and modern method is the use of bins (silos of circular cross-section), permitting high-quality storage and mechanized loading and unloading.

Bulk material exerts a vertical pressure on the bottom of a bin and a lateral pressure on the vertical walls, which are also loaded by vertical forces due to friction against the walls. These loads must be determined for correct dimensioning.

15.7.1 Lateral pressure coefficient

At any arbitrary point in a bulk material a horizontal pressure, whose value depends on the friction properties of the material and the direction of the principal stress, is created by the effect of the locally acting vertical pressure.

The ratio of the horizontal pressure σ_h or p_h by the vertical pressure σ_v or p_v is termed the lateral pressure coefficient:

$$k = \sigma_h/\sigma_v \tag{195}$$

The vertical and horizontal pressures enclose an angle of 90°, and so σ_v and σ_h are found at the opposite ends of a diameter of the Mohr circle (Fig. 288). The lateral pressure coefficient is minimum when the principal stress σ_1 acts vertically in this case,

$$k_{\min} = \sigma_2/\sigma_1 \tag{195a}$$

The lateral pressure coefficient is maximum when the vertical pressure is the smallest and the horizontal pressure the greatest principal stress, i.e.,

$$k_{\max} = \sigma_1/\sigma_2$$

The latter case may be demonstrated by pushing a piston arranged in a bin wall inwards after filling.

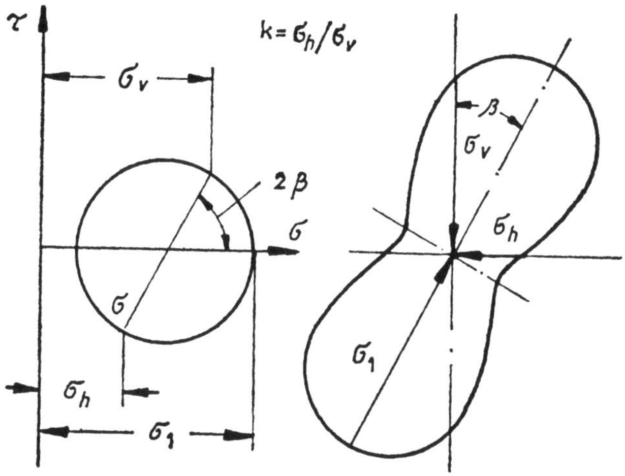

Fig. 288. Representation of vertical and horizontal pressures on the Mohr circle and in the oval diagram

(b)

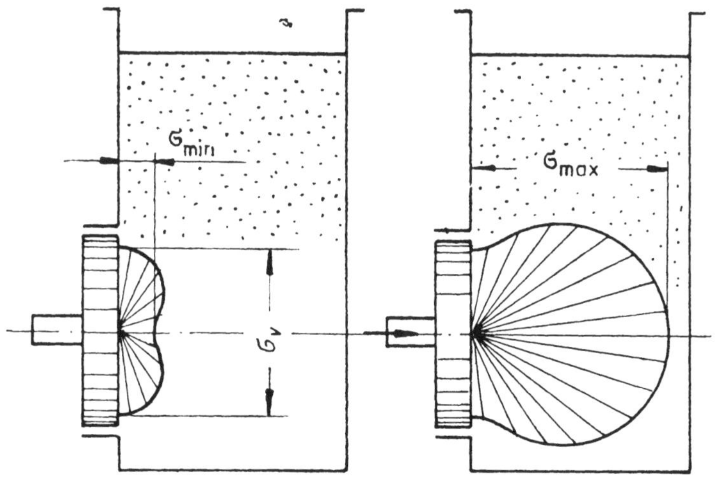

Fig. 289. Development of active and passive pressures in a granular pile

Figure 289 illustrates the pressure conditions corresponding to various directions of the principal stress. The left-hand figure (a) shows active pressure of the medium against a stationary piston; the right-hand figure (b) shows the passive pressure acting on a piston moving inwards. The lateral pressure coefficient may

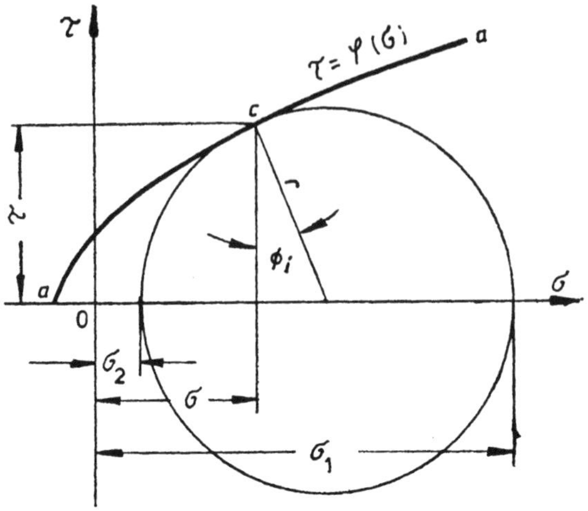

Fig. 290. Derivation of the lateral pressure coefficient for a granular pile

be calculated by making use of Fig. 289 in the following way. On the basis of the figure the following equalities may be written

$$(\sigma_v+\sigma_h)/2 = (\sigma_1+\sigma_2)/2$$

and

$$(\sigma_v-\sigma_h)/2 = [(\sigma_1-\sigma_2)/2] \cos 2\beta$$

from which

$$(\sigma_v-\sigma_h)/(\sigma_v+\sigma_h) = [(\sigma_1-\sigma_2)/(\sigma_1+\sigma_2)] \cos 2\beta$$

Now, taking eqn. (195) and (195a) into account, the lateral pressure coefficient is obtained as

$$k = (\tan^2 \beta+k_{\min})/(1+k_{\min} \tan^2 \beta) \tag{196}$$

where β is the angle enclosed by the principal stress with the vertical. The minimum value of the lateral pressure coefficient $k_{\min}$ may also be determined in a general form, independent of the form of the relationship $\tau=\varphi(\sigma)$ [139]. On the basis of Fig. 290, the radius of the Mohr circle, in terms of the angle of inclination Φ_i of the tangent at point C, may be written as

$$r = \sqrt{\tau^2+(\tau \tan \Phi_i)^2} = \tau\sqrt{1+(d\tau/d\sigma)^2}$$

On the other hand,

$$r = (\sigma_1-\sigma_2)/2 = \sigma_1(1-k_{\min})/2$$

The two above equations may be equated to give

$$k_{\min} = (\sigma_1-2r)/\sigma_1 = 1-(2\tau/\sigma_1)\sqrt{1+(d\tau/d\sigma)^2} \tag{197}$$

The stress τ is a function of σ; therefore the relationship between the values of σ and σ_1 must be determined:

$$\sigma = \sigma_1 - (r + \tau \tan \Phi_i)$$

or, by substituting the expression for r,

$$\sigma = \sigma_1 - [\tau \sqrt{1+(d\tau/d\sigma)^2} + \tau(d\tau/d\sigma)]$$

Now the general relationship $\tau = \varphi(\sigma)$ may be written in the form

$$\tau = \varphi\{\sigma_1 - \tau[\sqrt{1+(d\tau/d\sigma)^2} + d\tau/d\sigma]\} \qquad (198)$$

Using eqns. (197) and (198), the value of k_{min} may be determined for various functions $\tau = \varphi(\sigma)$. For ideal bulk material, according to eqn. (189),

$$\tau = \mu_i \sigma$$

and

$$d\tau/d\sigma = \mu_i$$

Substitution into eqn. (198) yields

$$\tau = \mu_i\{\sigma_1 - \tau[\sqrt{1+\mu_i^2} + \mu_i]\}$$

from which

$$\tau = \mu_i \sigma_1/(1 + \mu_i \sqrt{1+\mu_i^2} + \mu_i^2)$$

Substitution of this expression for τ into eqn. (197) gives

$$k_{min} = (\sqrt{1+\mu_i^2} - \mu_i)/(\sqrt{1+\mu_i^2} + \mu_i) = (1 - \sin \Phi_i)/(1 + \sin \Phi_i) \qquad (199)$$

It may be seen from eqn. (199) that for ideal bulk material the lateral pressure coefficient is independent of stress and depends only on the angle of internal friction.

In the case of the relationship given by eqn. (190),

$$\tau = \tau_0 + \mu_i \sigma$$

and

$$d\tau/d\sigma = \mu_i$$

By a similar calculation method as before, the following relationship is now obtained for k_{min}:

$$k_{min} = 1 - 2[(\tau_0/\sigma_1) + \mu_i]/[\mu_i + \sqrt{1+\mu_i^2}] \qquad (200)$$

It follows from this relationship that k_{min} now depends also on the principal stress σ_1, in contrast to the case for an ideal medium. However, for increasing σ_1 the value of k_{min} given by eqn. (200) gradually approximates the lateral pressure

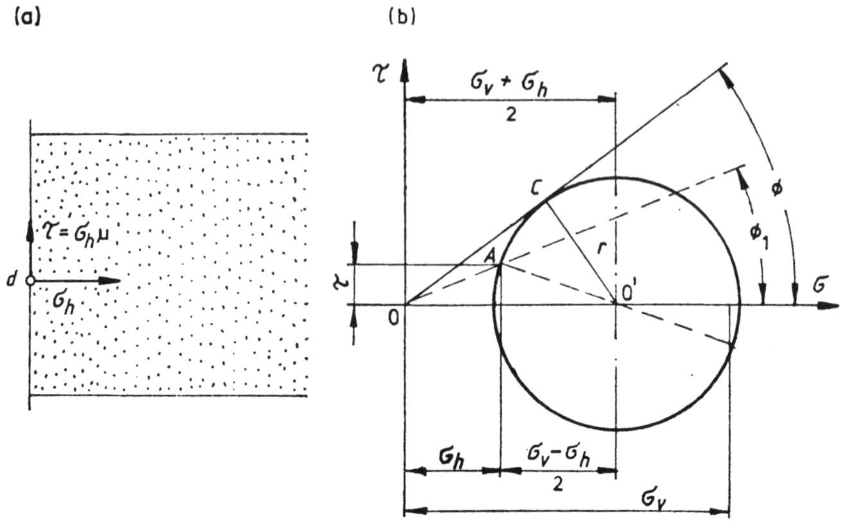

Fig. 291. Equilibrium of a granular medium along a wall

coefficient for an ideal medium, while for decreasing σ_1 the value of k_{min} also decreases, and there exists a σ_1 value at which $k_{min}=0$. This condition prevails obviously in the case of a σ_1 value for which the second term in eqn. (200) equals 1.

The lateral pressure coefficient along a vertical wall deviates from the cases discussed above, because shear stresses appear at the wall due to settling of the material. Taking the friction coefficient μ into account, the shear stress is given by $\tau=\sigma_h\mu$. The equilibrium conditions are shown in Fig. 291. If there were no friction at the wall, then k would be given by $k=\sigma_2/\sigma_1$, and the direction of σ_1 would be vertical. A shear stress τ can arise if the principal stress σ_1 encloses an angle ψ with the wall. The greater the angle of friction, the closer will point A come to point C on the Mohr circle, and the higher will be the value of the lateral pressure coefficient.

The lateral pressure coefficient along the wall may be calculated for an ideal bulk medium from the relationship

$$k = 1/[1+2\mu_i^2+2\sqrt{(1+\mu_i^2)(\mu_i^2-\mu^2)}] \tag{201}$$

In the case of $\mu_i=\mu$, the above expression simplifies to

$$k = 1/(1+2\mu_i^2) \tag{201a}$$

It follows from eqns. (201) and (201a) that high lateral pressure coefficients are obtained when μ_i is relatively low and the wall friction factor is nearly identical to the angle of internal friction. For example, during storage of maize both the internal and wall friction angles may be assumed to be about 25° and then $k=0.697$, which agrees well with values measured for silo towers of large diameter (15 m).

15.7.2 Calculation for flat bins

A bin is termed flat if its diameter is greater than its height. Similar conditions will also predominate during storage in any building when the height of the bulk material is less than the horizontal dimensions. Figure 292 illustrates the distribution of pressures acting on the bin.

The oval diagram of normal stresses is in a vertical position on the axis of the bin, while in the vicinity of the wall it encloses an angle ψ with the wall under the effect of the shear stresses τ arising there. The pressure at the base of the bin corresponds to the hydrostatic pressure at the center, i.e., $\sigma_0=h\gamma$. This pressure decreases towards the edges, since the shear stress τ arising at the walls supports a part of the weight.

The lateral pressure varies linearly with the height and its value may be calculated on the basis of the Rankine equation (earth pressure theory) as

$$\sigma_h = h\gamma \tan^2 (45-\Phi_i/2) \tag{202}$$

where the height h must be measured from the free upper surface. A similar

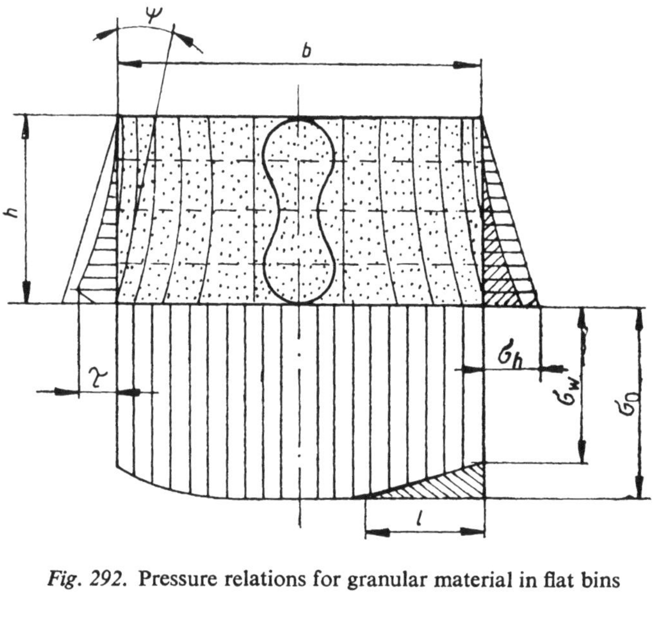

Fig. 292. Pressure relations for granular material in flat bins

result is also obtained by applying eqn. (199), when the horizontal pressure is calculated on the basis of the lateral pressure coefficient.

The shear stress τ varies in a nearly parabolic way with height, and decreases rapidly in the immediate vicinity of the base. This may be explained by the sudden decrease of settling found near the base, which prevents the development of shear stress. The maximum value of τ may be calculated from the relationship

$$\tau = \mu k \sigma_0$$

The base pressure near the wall (σ_w) may be determined from the Mohr circle of stresses (Fig. 285), or from the analytical relationship

$$\sigma_w = k\sigma_0[1+2\mu_i^2+2\sqrt{(1+\mu_i^2)(\mu_i^2-\mu^2)}] \tag{203}$$

The hatched area in Fig. 292 indicating the reduction of the base pressure corresponds to the area under the stress curve, i.e.,

$$(\sigma_0-\sigma_w)l/2 = \mu k \sigma_0 h/3$$

15.7.3 Calculation for high bins

In high bins, whose height significantly exceeds their diameter, significant frictional forces arise on the walls, whereby the base pressure becomes everywhere substantially lower than the hydrostatic pressure. Figure 293 illustrates the pressure conditions in a high bin.

The oval diagram of normal stresses is vertical on the axis of the bin; close to the wall, it encloses an angle ψ with the wall. The pressure on the base is maximum on the axis of the bin and decreases towards the edges. The distribution of lateral pressure as a function of height may be calculated on the basis of the following considerations. An element of thickness Δh is marked out at a height h in the bin. At equilibrium, the forces acting on this element may be written as [146]

$$F\sigma_v+\gamma F dh = F(\sigma_v+d\sigma_v)+\tau K dh$$

where F is the cross-sectional area and K the circumference of the bin. The stress on the wall is given by

$$\tau = \mu\sigma_h$$

and

$$k = \sigma_h/\sigma_v$$

where σ_v is the average vertical pressure. The initial equation may be solved to give values of σ_h, σ_v and τ as a function of height h using the two preceding rela-

Fig. 293. Pressure relations for granular material in high bins

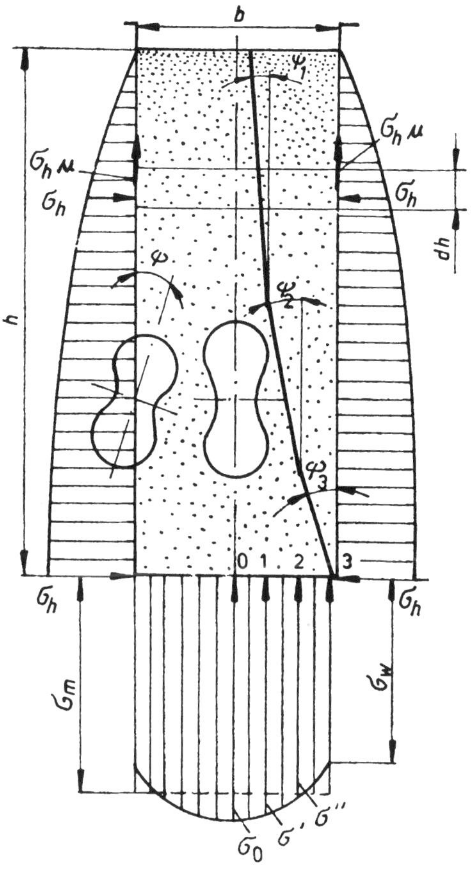

tionships: i.e.,

$$\sigma_h = (\gamma F/\mu K)(1 - e^{-\mu kKh/F}) \tag{204}$$

$$\sigma_v = (\gamma F/\mu kK)(1 - e^{-\mu kKh/F}) \tag{205}$$

and

$$\tau = (\gamma F/K)(1 - e^{-\mu kKh/F}) \tag{206}$$

According to eqn. (204), with increasing height the horizontal pressure tends to a maximum value

$$\sigma_{h\,\max} = \gamma F/\mu K \tag{206a}$$

The vertical pressure on the base in the vicinity of the wall may be calculated, similarly to eqn. (203), as

$$\sigma_w = \sigma_h[1 + 2\mu_i^2 + 2\sqrt{(1+\mu_i^2)(\mu_i^2 - \mu^2)}]$$

With the assumption that the horizontal pressure is nearly constant along the radius near the base, the maximum base pressure may be calculated as

$$\sigma_0 = \sigma_h/k_{min}$$

The above relationships apply to the equilibrium state immediately after filling. A deficiency of the theory is that it disregards settling of the medium over the course of time. During settling not only does the volumetric weight increase, but the internal friction angle and initial shear stress τ_0 also vary. Only few numerical values of the latter are available. Nevertheless, the equilibrium state immediately after filling may be calculated relatively well using the above theory.

The settling of a granular pile may be determined if its mechanical properties (modulus of elasticity, Poisson's ratio, viscoelastic characteristics) are known. The increase in the volumetric weight may be calculated from the relationship

$$\gamma = \gamma_0/\{1-(\sigma_v/E)[1-(2\nu^2/(1-\nu))] \qquad (207)$$

where E is the apparent modulus of elasticity, and ν the Poisson's ratio. Values of E for crop products can be taken from Fig. 170 as a function of vertical pressure, with creep taken into account [67]. A value of 0.35 may be substituted for the Poisson's ratio. With a decrease of vertical pressure σ_v the value of E also decreases. According to calculations, the volumetric weight of wheat increases by 3% and that of maize by about 2% under a vertical pressure of 1 bar in a silo tower. However, for certain bulk materials the compaction caused by settling is considerably greater (e.g., in the case of fodder mixtures and milling products). Figure 294 shows the variation of the volumetric weight of mixed fodder in a bin 5 m high [149]. The curve may be described by the empirical equation

$$\gamma = \gamma_0 + Ap_v^n$$

where A and n are constants. When deriving the pressures acting on the walls, Gutjar [147] allowed for settling of material by means of eqn. (207), modifying eqn. (205) in the following way:

$$\sigma_v = (1-e^{B\gamma_0 h-\mu kKh/F})/((\mu kK/\gamma_0 F-B)) \qquad (205a)$$

with

$$\sigma_h = k\sigma_v$$

where $B=(1/E)[1-2\nu^2/(1-\nu)]$. Application of eqn. (205a) is aggravated in certain cases by the absence of reliable E and ν values.

High bins (silos) are generally emptied by gravity through an orifice arranged in the center of the base. Measurements show that increase in horizontal wall pressure found during discharge in relation to the pressures obtained at filling. This phenomenon is not yet completely characterized in all its details, nor is a

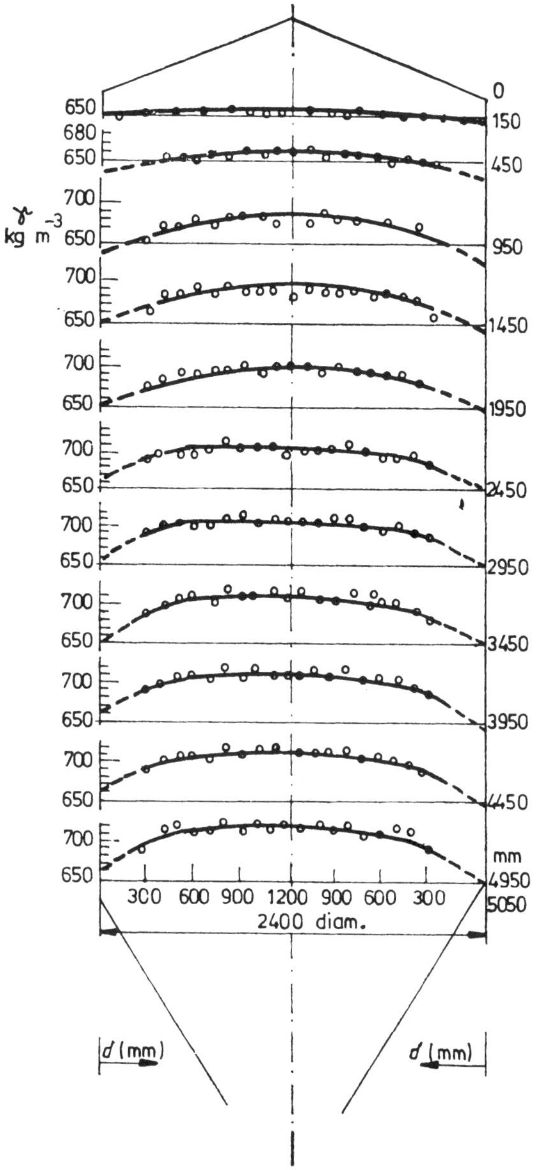

Fig. 294. Distribution of volumetric weight in a bin as a result of settling

reliable calculation method available. According to the measurements the additional wall pressure during discharge is the greater, the higher the tower in relation to its diameter. If the height approximates the diameter, the additional wall pressure becomes negligible.

During discharge the vertical pressure along the axis of the bin decreases,

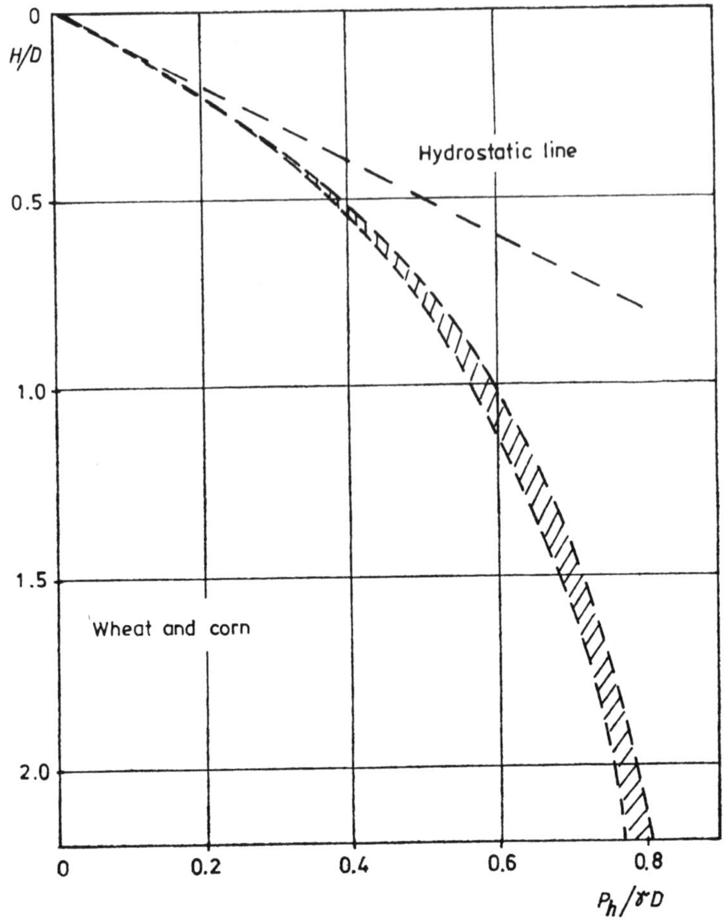

Fig. 295. Experimental data for lateral pressure distribution in silos containing wheat or corn

especially in the vicinity of the base, and so the equilibrium state of the pile moves from the limit position (where the Mohr circle contacts the envelope curve) into an intermediate equilibrium state, whereby the value of k increases and tends to the limiting value given by eqn. (201a) [148].

Experimental data for the lateral pressure distribution in large silos after consolidation are given in Fig. 295. The generalized pressure distribution can be approximated by the equation

$$p_h/\gamma D = C(H/D)^{0.5}$$

where D is the diameter and H the height of the silo. For wheat and maize the constant C has values of 0.57–0.60.

Silage of high moisture content (maize, alfalfa, etc.) is freqently stored in silo towers. Silage is a bulk material chopped to a mean size of 2–3 cm, and its characterization is much more difficult than in the case of cereals. The sizes and shapes scatter greatly, the deformation of individual components in storage is considerable, and the mechanical properties vary as a result of biological processes taking place during storage. Thus silo towers must be dimensioned mainly on the basis of available measurement results. Figure 296 shows the horizontal pres-

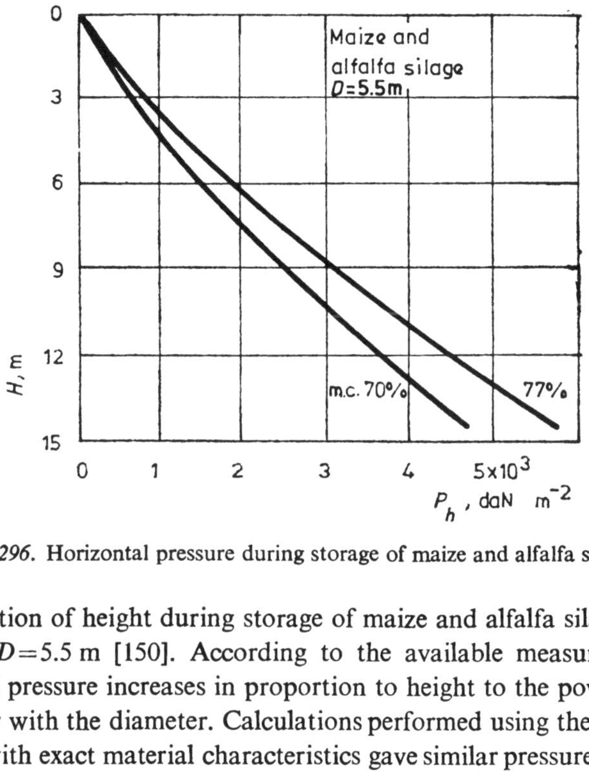

Fig. 296. Horizontal pressure during storage of maize and alfalfa silage

sure as a function of height during storage of maize and alfalfa silage in a tower of diameter $D=5.5$ m [150]. According to the available measurement results the horizontal pressure increases in proportion to height to the power of 1.5 and nearly linearly with the diameter. Calculations performed using the finite-element method and with exact material characteristics gave similar pressure distributions, as shown in Fig. 296 [82].

15.8 Flow of granular materials from an orifice

The flow of granular material from circular and square orifices plays an important role in numerous technological processes (discharge of silos, flow from grain bins into dosage elements, etc.). The outflow cannot be calculated on the basis of generally known hydrodynamic relationships, since the flow properties of granular materials differ fundamentally from those of liquids.

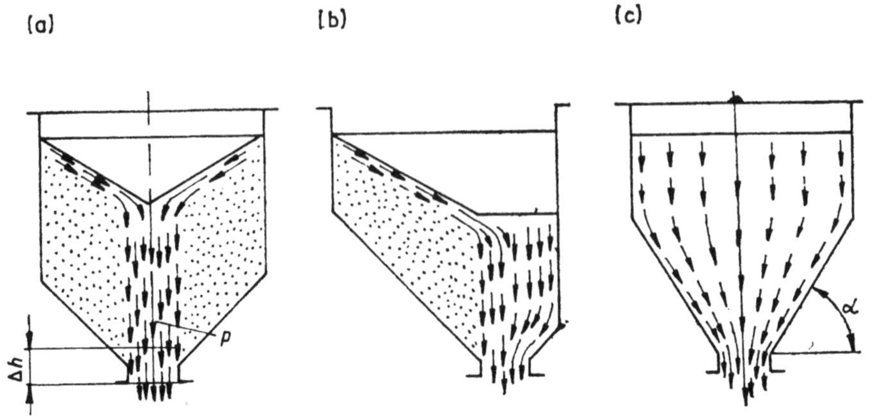

Fig. 297. Characteristic modes of outflow forms from bins

Figure 297 shows typical modes of flow from a bin. When the base plane encloses a relatively small angle with the horizontal, so-called funnel-type outflow occurs (Fig. 297(a)). The material moves only in the column located above the orifice and flows into this column from the upper layers. Another typical form of outflow is obtained when the angle of inclination of the base to the horizontal is large (Fig. 297(c)). In this case the whole volume moves and accordingly inflow occurs from the whole bin. Figure 297(b) shows the intermediate situation, where the outflow is asymmetrical and partly of the funnel type and partly volumetric (mass flow). Which mode of outflow occurs depends primarily on the hopper slope angle and the friction coefficient of the wall. The relationships are illustrated in Fig. 298.

The cohesion of certain granular products is zero, or very low, and so they flow easily from bins under the action of gravity. The flow of compactable and adhesive materials (flours, fertilizers, etc.) implies numerous problems, and correct geometrical design of the outlet orifices in bins for such materials is achieved even today by semiempirical methods. Jenike [136, 137] has, however, developed an approximate method for characterizing the flow properties of a compactable material whose essence is as follows. The maximum shear stress is determined as a function of normal stress using a shearing box, applying a given consolidation stress σ_1. By applying several consolidation stresses σ_1 the straight line of the stationary yield locus and the curves of the yield loci corresponding to individual σ_1 values may be plotted (Fig. 299). It may be seen from the Mohr circles drawn in the figure that for a consolidation stress σ_1 the development of stationary flow will be ensured by a normal stress σ corresponding to the center of the Mohr circle, without the occurrence of any peak value of shear stress. For lower normal stresses the material is overcompacted, and so a higher stress τ is needed to

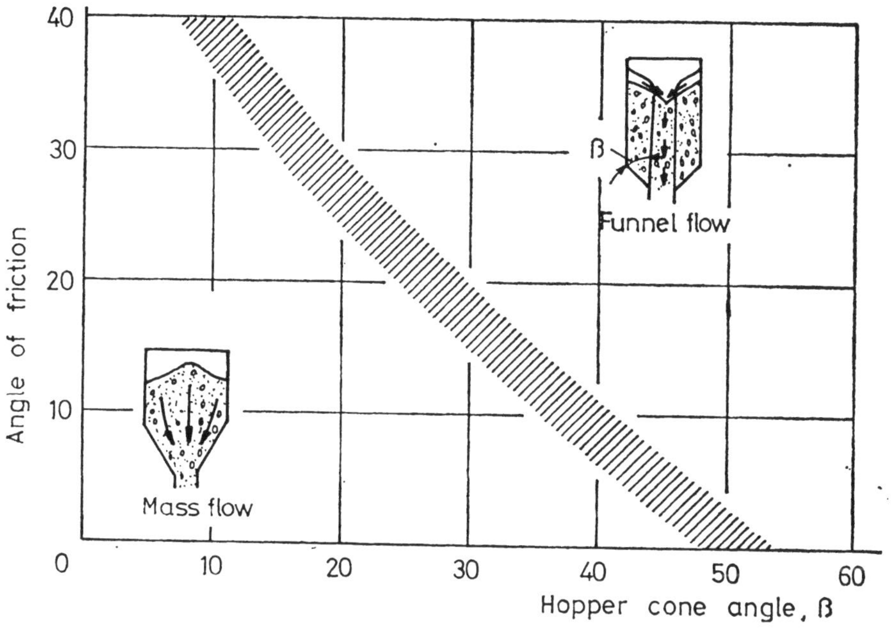

Fig. 298. Areas of funnel-type and mass outflow as functions of the friction and slope angle

initiate flow; the curves of the yield loci lie above the straight line. Mohr circles also may be constructed through the origin ($\sigma_2 = 0$) for the curves of the yield loci, and the main stress σ_c obtained in this way is characteristic of the strength of a cohesive material. For cohesionless materials $\sigma_c = 0$, i.e., the curve passes through the origin. The ratio of the stresses σ_c and σ_1 is characteristic of the flow properties of bulk granular materials, and the relationship obtained is termed the characteristic flow function (F).

Flow from bins is usually disturbed by the phenomenon of arching, when material remains in the form of an arch and leans against the wall. Arching remains stable if the strength σ_c of the material at the point of support exceeds the compression stress. For a bin with an outlet orifice of a given geometry, it is possible to determine a flow function F^* experimentally, as the ratio of the compressile stress σ^* arising at the regions of support of the initial arching, to the principal stress σ_1. The flow function F^* is a function of the internal angle of friction Φ_i, the angle of wall friction, and the angle of the conical base. The function F^* for an actual construction may be represented by a straight line passing through the origin (Fig. 299).

The outflow conditions for granular material may be determined by plotting the functions F and F^* together. Outflow is possible where the characteristic func-

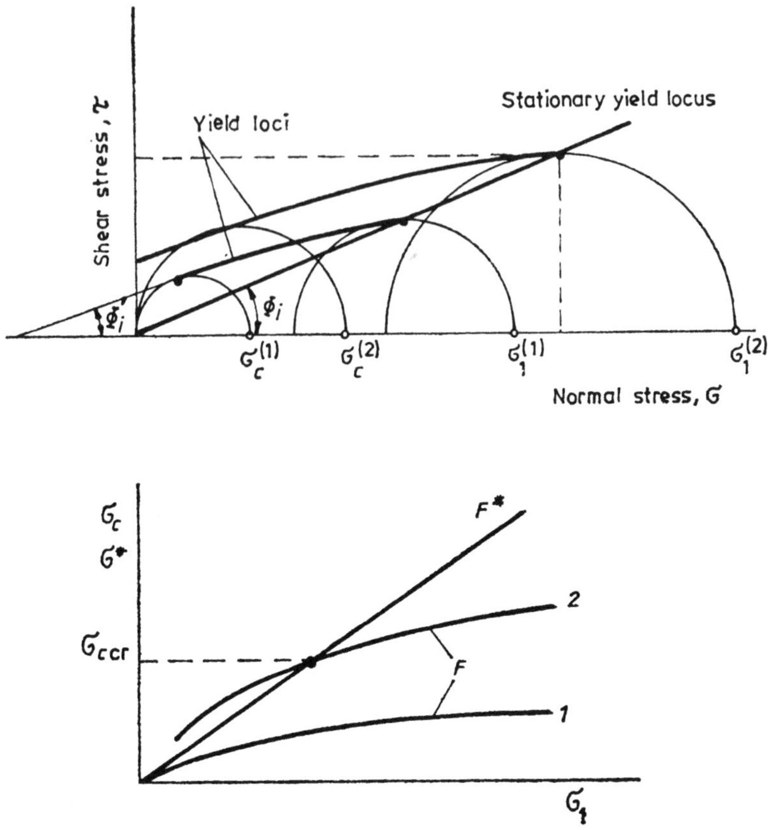

Fig. 299. Determination of the flow properties of granular piles; the dimensioning of outlet openings (Jenike process)

tion F passes under the function F^*. In this case,

$$\sigma_c/\sigma_1 < \sigma^*/\sigma_1$$

or

$$\sigma_c < \sigma^*$$

If the F curve lies everywhere beneath the F^* curve, outflow is possible whatever the stress state. At the intersection of the two curves, the condition $\sigma_c < \sigma^*$ is fulfilled only for values exceeding the critical material strength $\sigma_{c\,cr}$. In the limiting case, where $\sigma_{c\,cr} = \sigma^*$, the minimum width of the outlet opening is

$$B_{min} = H(\alpha)\sigma_{c\,cr}/\gamma$$

where γ is the volumetric weight of the granular material. The theoretical value

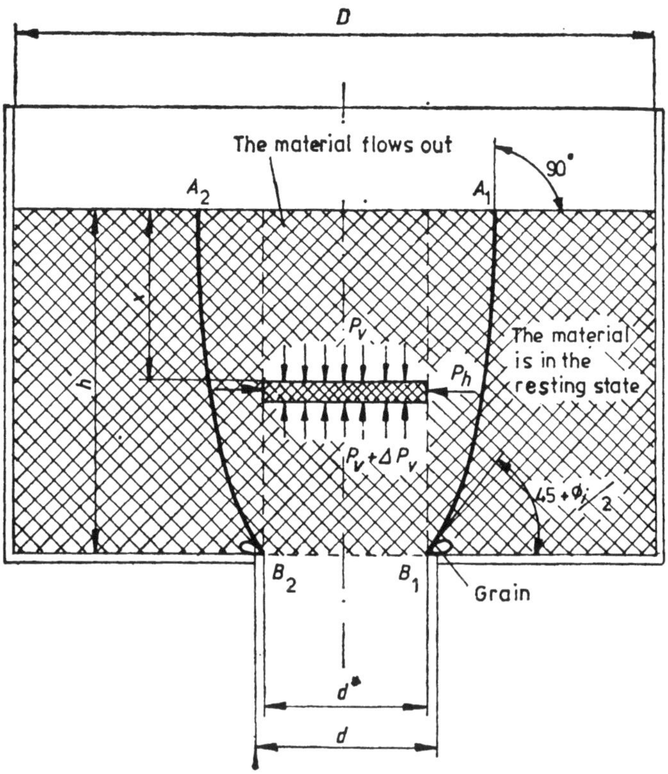

Fig. 300. Equilibrium of granular material over an outflow opening

of the factor $H(\alpha)$ for conical outflow orifices is 2.0; for slot outflow openings it is 1.0. These values increase slightly with increasing aperture angle [138].

Granular materials may be classified according to the ratio σ_1/σ_c as follows:

$\sigma_1/\sigma_c < 2$, highly cohesive materials, whose outflow from a bin can be ensured only by auxiliary equipment;

$4 > \sigma_1/\sigma_c > 2$, cohesive materials;

$10 > \sigma_1/\sigma_c > 4$, easy-flowing materials; and

$\sigma_1/\sigma_c > 10$, free-flowing materials.

Material flows out under the effect of the vertical pressure prevailing at an opening. Figure 300 shows the equilibrium conditions for the column above the opening [145]. The effective cross-sectional area of outflow is smaller than the geometrical cross-sectional area, because grains may protrude along the outlet edge such that the cross-sectional area of the opening is reduced. The boundary surface of the column which starts moving above the opening experiences friction against the material surrounding it, and so the vertical pressure depends on the height, as in high storage towers. When the height of the column of material above the outlet opening reaches five times its diameter, the pressure becomes independent of height, and its value is, similarly to eqn. (205),

$$\sigma_v = d^*\gamma/4\mu_i k$$

According to the general equation for outflow, the quantity emerging in unit time is

$$G = \mu_0 \gamma F^* \sqrt{2g(\sigma_v/\gamma)}$$

Substitution of the preceding expression for σ_v, and taking into account that $d^* = d - d'$, gives

$$G = \mu_0 \gamma (\pi/4) \sqrt{(g/2\mu_i k)} (d - d')^{2.5} \tag{208}$$

The value of d' for wheat is 6 mm, i.e., 1.5 times the mean grain diameter. The coefficient of discharge (μ_0) is 0.46 for circular orifices. On reduction of the opening the flow stops at a critical diameter, because the bulk material remains in an arch over the opening, owing to wedging of the grains. The minimum orifice diameter for outflow of wheat is about 15 mm. From eqn. (208) it may be calculated that an opening 9 cm in diameter is necessary for the outflow of wheat at 10 t h^{-1} and 21 cm in diameter for 100 t h^{-1}.

The vertical pressure over a quadrangular outflow opening is

$$\sigma_v = \gamma F^*/\mu_i k K = \gamma a^* b^*/\mu_i k 2(a^* + b^*)$$

where a^* and b^* are the effective length and width of the orifice ($a^* = a - d'$ and $b^* = b - d'$). With the above taken into account, the quantity discharged is

$$G = \mu_0 \gamma \sqrt{(g/\mu_i k)} (a^* b^*)^{1.5}/(a^* + b^*)^{0.5} \tag{208a}$$

where the coefficient of discharge $\mu_0 = 0.5$ for $a/b \geqq 1$; for increasing side ratios μ_0 decreases slightly, and for $a/b \geqq 4$ the value $\mu_0 = 0.42$ may be used. Intermediate values may be obtained by interpolation [145].

15.9 Flow of granular materials in chutes

Bulk materials are frequently transported over small horizontal distances by gravitational chutes. The chute is essentially an inclined trough or tube, in which material moves, under the effect of its own weight, against frictional forces arising on the walls.

Chutes may be straight, bent in a circular arc, or a combination of the two (Fig. 301). Straight chutes are used most frequently, since production and the

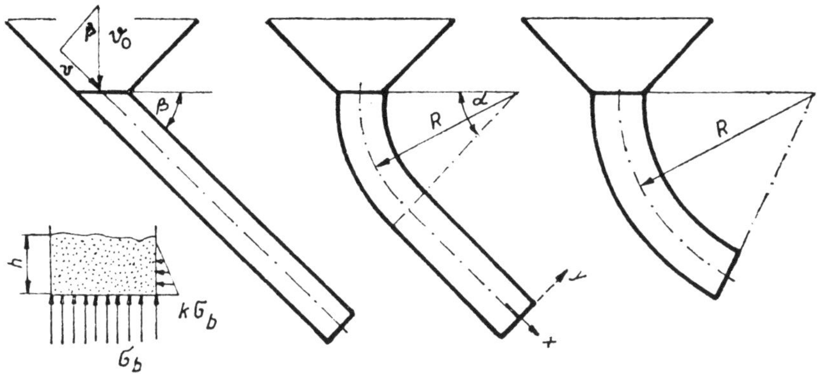

Fig. 301. Types of transport chutes

requirements to be met are simple. To dimension a chute a relationship must be found between the quantity to be transported, the thickness of material flowing, the slope of the chute, and its cross-sectional area.

The motion of material in a chute tilted at an angle β may be described by the equations

$$(G/b)(\sin\beta + a_x/g) - \sum \mu\sigma = 0$$
$$(G/b)(\cos\beta + a_y/g) - \sum \sigma = 0 \qquad (209)$$

where a_x and a_y are the accelerations of the material in the x and y directions. The material quantity per unit length is

$$G = bh\gamma$$

where b is the width of the chute, h the height of the material in it, and γ the volumetric weight of the material.

In the case of straight chutes $a_y = 0$, while the acceleration in the x direction may be calculated from the inlet and outlet velocities:

$$a = (v^2 - v_0^2)/2l$$

where l is the length of the chute.

When $v > v_0$, the vector of accelerations points into the direction of movement, while inertial forces act against movement, so that

$$a_x = -a = (v_0^2 - v^2)/2l$$

The normal stress (pressure) on the base of the chute is given in the general case by

$$\sigma_b = k\gamma(\cos\beta + a_y/g) \qquad (210)$$

and for straight chutes, for which $a_y=0$, by

$$\sigma_b = h\gamma \cos \beta \tag{210a}$$

With the assumption that the pressure on the side walls decreases linearly on advancing from the maximum value $k\sigma_b$ upwards, the sum of the frictional forces per unit length is

$$\sum \mu\sigma b = \sigma_b \mu b + k\sigma_b \mu h = \sigma_b \mu b(1+kh/b)$$

or

$$\sum \mu\sigma = \sigma_b \mu_e$$

where the expression for the equivalent friction coefficient is

$$\mu_e = \mu(1+kh/b) \tag{211}$$

On the basis of the above considerations, eqn. (209) may be written for a straight chute in the form

$$\sin \beta + (v_0^2 - v^2)/2gl - \mu_e \cos \beta = 0$$

from which

$$v = \sqrt{v_0^2 + 2gl(\sin \beta - \mu_e \cos \beta)} \tag{212}$$

If v_0 is the vertical inlet flow velocity, then the effective inlet velocity in the direction of the tube is $v_0 \sin \beta$ (Fig. 301).

Knowing of the velocity v, the depth of the flowing material, at every point of the chute, may be determined as

$$h = Q/b\sqrt{v_0^2 \sin^2 \beta + 2gl\,(\sin \beta - \mu_e \cos \beta)} \tag{213}$$

where Q is the volume flow rate. From eqn. (212) the smallest angle of inclination β at which flow is still maintained $(v=v_0)$ may be written as

$$\tan \beta_{cr} = \mu_e = \mu(1+kh/b) \tag{214}$$

The limiting value of β for wheat and maize (in steel plate chutes) is 24–28°, depending on moisture content. This value may be significantly higher in the case of farinaceous materials; for example, values in the range 36–42° are obtained for wheat flour. To ensure stable operation, values higher than the critical must be employed.

In circular arc-shaped chutes the material at first accelerates and then decelerates with the gradual decrease in inclination of the individual arc elements. Accordingly, the thickness of flowing material at first decreases and then increases (Fig. 302). If the circular arc extends beyond a given angle α, the material fills the whole cross-sectional area and the flow decelerates greatly, finally ceasing completely. To a first approximation, the value of the limiting angle α_f is

$$\alpha_f = 90° - \beta_{cr} = 90° - \arctan \mu_e \tag{215}$$

i.e., α_f is the angle such that the slope of the lower tube end is β_{cr} (see eqn. (214)). In chuting wheat and maize, $\alpha_f = 53°$ may be accepted as a mean value.

If a circular arc is to be connected to a straight section, it is advisable to locate the connection at or near the point where the velocity is maximum (α_m). This point may be determined from the differential equation of motion, written purposely with moving coordinates $\alpha(t)$ in the form [151]

$$\ddot{\alpha} - \mu_e(\dot{\alpha})^2 + (g/R)(\mu_e \sin\alpha - \cos\alpha) = 0 \tag{216}$$

The equivalent friction coefficient μ_e does not remain constant during variation of the layer thickness (see eqn. (211)), but its variation is generally slight, and so it is convenient to perform the calculations using a constant mean value, which greatly facilitates the solution of eqn. (216). To solve the equation, the velocity of flow may be calculated from the relation [151]

$$v = \sqrt{[2gR/(4\mu_e^2+1)][\sin\alpha(1-2\mu_e^2)+3\mu_e\cos\alpha]+e^{-2\alpha\mu_e}[v_0^2-6\mu_e Rg/(4\mu_e^2+1)]} \tag{217}$$

where R is the radius of curvature of the tube and g the acceleration due to gravity.

The angle α_m corresponding to the maximum velocity is influenced by the radius of curvature R, the initial velocity v_0, and the friction coefficient μ_e. Figure 303 shows the effects of these factors on α_m. Since v_0 is rarely higher than 1.0–1.2 m s^{-1} in practice, α_m is between 40° and 50°. This means that for cereals it is advisable to join a straight section to a circular arc at an angle $\alpha = 45$–$50°$, which will ensure stable and rapid flow in the chute.

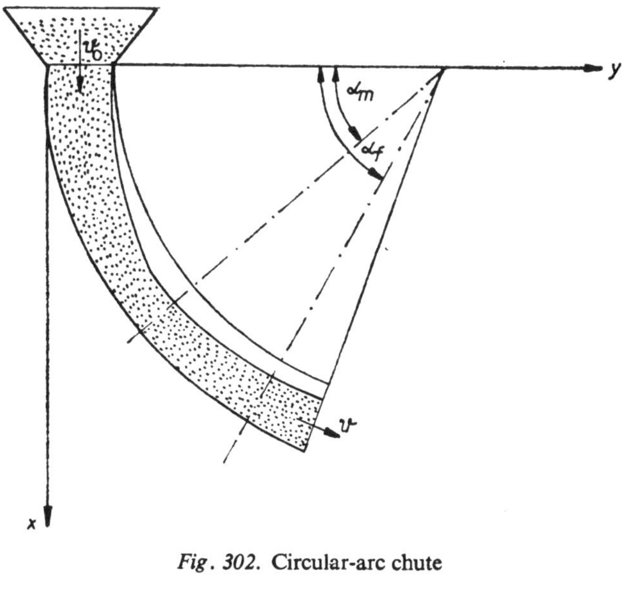

Fig. 302. Circular-arc chute

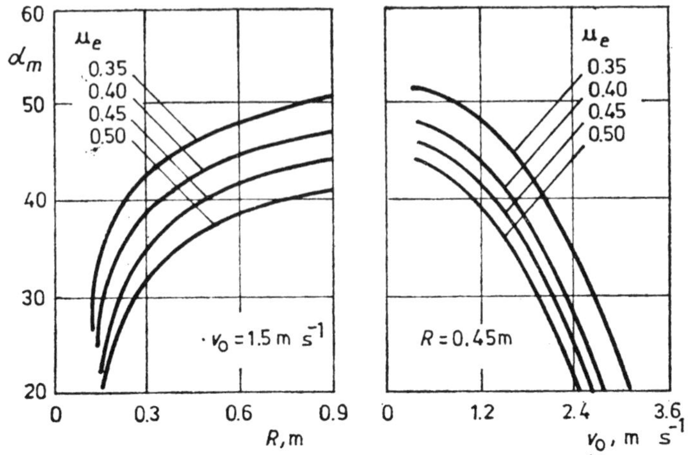

Fig. 303. Position of maximum flow velocity in a circular-arc chute

The thickness of the flowing material may be calculated from the continuity equation as

$$F_m v = \text{const}$$

where F_m is the cross-sectional area of the flowing material. In the case of a quadrangular cross-section,

$$h/h_0 = v_0/v$$

The value of h may be calculated knowing of the initial values h_0 and v_0 and after calculating the value of v from equation (212) or (217).

15.10 Further friction problems

During the design and operation of agricultural machines an immense number of friction problems are encountered. Hardly any machine exists whose operation involves no friction. Not all friction problems can be discussed here, but some of the more important are mentioned briefly in the following.

A part of the power consumption of disc- and drum-shaped rotating parts of machines is due to frictional losses. For rapidly rotating parts frictional losses result primarily from air resistance (as in the case of the rotating parts of hammer mills, chaff choppers or threshing drums).

The air resistance of a rotating disc may be calculated from the relationship [152]

$$N_s = 1.925 \times 10^{-5} k (\gamma/g) R^5 n^3 \tag{218}$$

where k is a constant for a particular disc, R the radius, n the rate of revolution, and γ the specific weight of air.

For rotating drums, the power consumption due to friction is

$$N_s = 4.82 \times 10^{-5} k' (\gamma/g) R^4 n^3 \tag{219}$$

As may be seen, the power consumed in overcoming friction increases theoretically with the cube of the rate of revolution.

However, experimental data show that the exponent of the rate of revolution varies between 2.2–2.3 for a range of rotating parts. Figure 304 illustrates the power consumption resulting from friction as a function of circumferential velocity

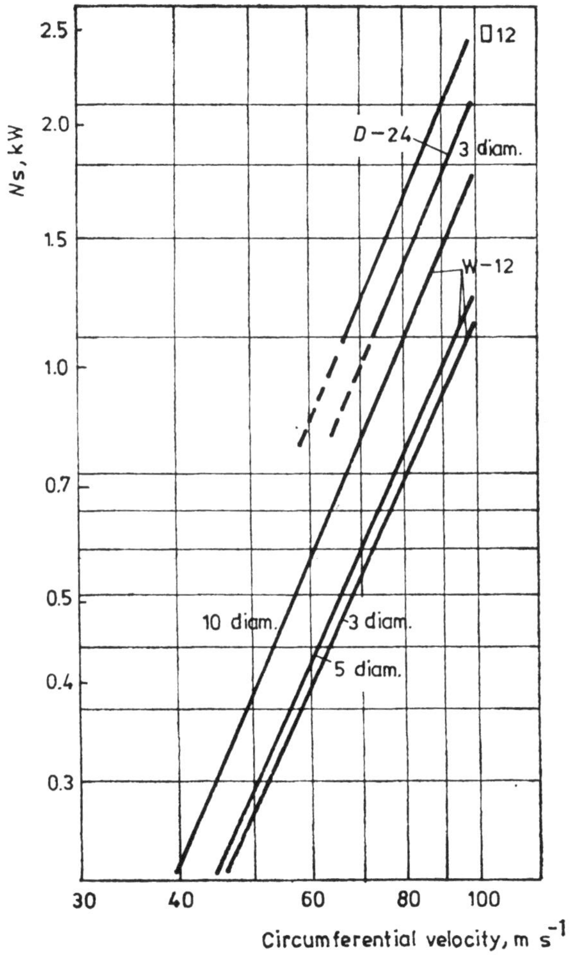

Fig. 304. Energy used in overcoming friction in hammer mills.

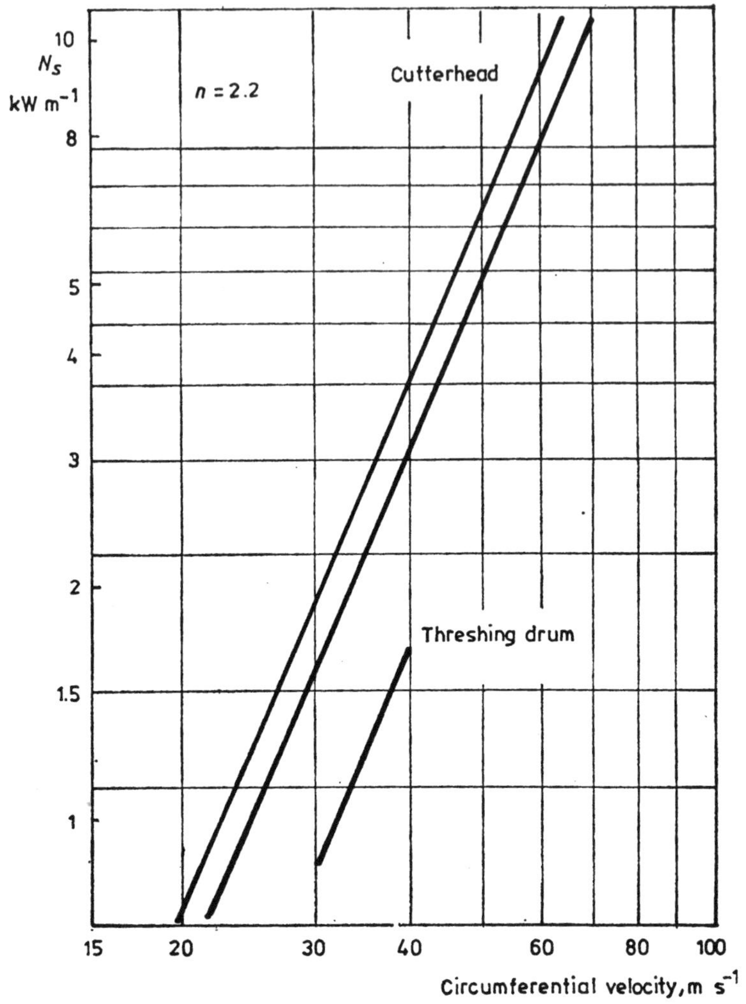

Fig. 305. Frictional energy consumption of cutter heads (forage harvesters) and threshing drums

for hammer mills. Figure 305 shows the corresponding relationship for self-propelled forage harvesters and threshing drums [152].

The air resistance of a hammer mill depends on the mesh size of the sieve surrounding the rotating part: the smaller the perforation, the less air transported by the rotating part and the lower the loss. The frictional power increases with increased width of the rotating part. The value of the constant k is 0.02–0.03

for smaller mills (with 12 hammers) and 0.018–0.026 for larger ones, depending on the rate of revolution.

For the rotating parts the Reynolds number may with certain assumptions be interpreted in the form

$$Re = Dv_c/\nu$$

where D is the diameter of the rotating part, v_c the circumferential velocity, and ν the kinematic viscosity of the medium. Usually, k values found experimentally are plotted as a function of the Re-number. Generalization in order to apply experimental k values to other machines must be performed with due circumspection, since the Reynolds number characterizes uniquely only the flow conditions for rotating parts of completely identical construction.

The operation of various throwing discs is also greatly influenced by friction. Throwing discs are used mainly to spread fertilizers and transport chaff, but other applications are also possible. The axis of rotation of a throwing disc may be vertical (for spreading fertilizers) or horizontal (for filling silos), the blades may be radial or curved forwards or backwards, and in certain cases may be prepared with a special profile. Figure 306 shows the forces acting on particles moving on a vertical-axis throwing disc. The forces acting on a particle of mass m are the centrifugal force ($mr\omega^2$), the Coriolis force ($2m\omega dr/dt$), the

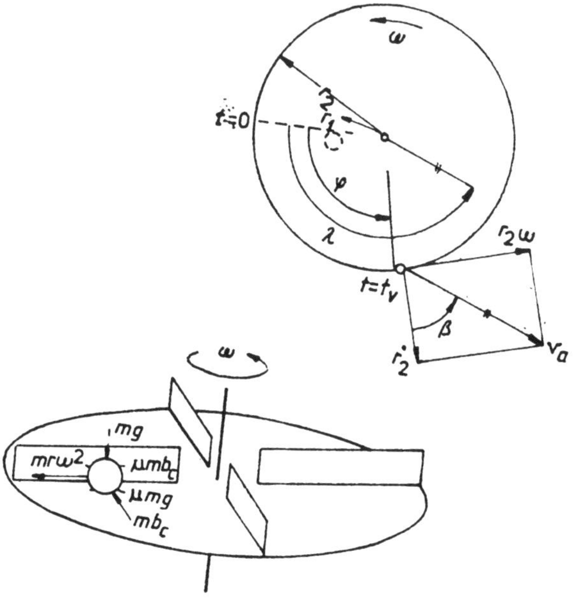

Fig. 306. Forces acting on moving particles on a throwing disc

gravitational force (mg) and the frictional force ($\mu mg+\mu 2m\omega dr/dt$). The differential equation for the particle's motion is accordingly

$$d^2r/dt^2+2\mu\omega(dr/dt)-r\omega^2 = -\mu g \tag{220}$$

In most practical cases the gravitational force is negligible compared to the centrifugal force, and so the above inhomogeneous equation may be replaced by the homogeneous second-order differential equation

$$d^2r/dt^2+2\mu\omega(dr/dt)-r\omega^2 = 0 \tag{220a}$$

The general solution of this equation may be written as

$$r = \mu g/\omega^2+C_2[(\lambda_2/\lambda_1)e^{\lambda_1 t}+e^{-\lambda_2 t}] \tag{221}$$

$$dr/dt = C_2\lambda_2\omega(e^{\lambda_1 t}-e^{-\lambda_2 t}) \tag{222}$$

where

$$\lambda_1 = \omega(\sqrt{\mu^2+1}-\mu)$$

and

$$\lambda_2 = \omega(\sqrt{\mu^2+1}+\mu)$$

On the basis of the initial conditions ($v_r=0$ on the radius r_1, at time $t=0$),

$$C_2 = [r_1-\mu(g/\omega^2)]/[1+(\lambda_2/\lambda_1)]$$

From eqns. (221) and (222) the radial velocity corresponding to a given external radius r_2 may be determined, and the absolute velocity v_a may be found from the circumferential velocity. As the particle moves from the starting radius r_1 to the external radius r_2, the throwing disc rotates through an angle φ, and the absolute velocity of the particle encloses an angle β with the radius. For higher rates of revolution ($n > 300$ *RPM*), the angle φ depends only on the radii r_1 and r_2 and on the friction coefficient μ; it no longer depends on the rate of revolution. Figure 307 shows the radial velocity of a particle leaving at radius r_2 for various r_1 values, for two different friction coefficients [153]. Figure 308 illustrates the paths of particles starting from various radii r_1, for two different friction coefficients. The value of the angle φ for any arbitrary exit radius r_2 may be read from the figure. The most important characteristics may be determined rapidly using the above two figures. The angle β is given by

$$\tan\beta = r_2\omega/\dot{r}_2$$

and the absolute exit velocity is

$$v_a = r_2\omega/\sin\beta$$

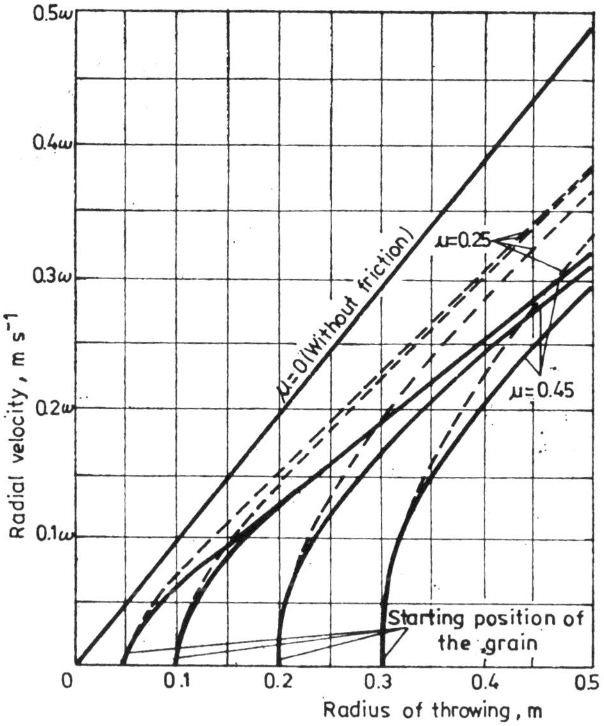

Fig. 307. Radial velocity of leaving particles in the case of various starting radii on a throwing disc

For example, suppose that $r_1=0.1$ m, $r_2=0.25$ m, $n=400$ rpm and $\mu=0.3$ for spreading a fertilizer. From Fig. 308, $\varphi=105°$. From Fig. 307, the radial velocity required to calculate the angle β is 0.18ω. With these values,

$$\tan\beta = r_2\omega/0.18\omega = 0.25/0.18 = 1.39$$

and

$$\beta = 54°$$

The absolute exit velocity is

$$v_a = 0.25 \times 42/0.81 = 13.0 \text{ m s}^{-1}$$

A frequent problem is to determine the state of motion of materials on oscillating sieve surfaces. The sieve surface encloses an angle β with the horizontal and is suspended on plate springs of length l (Fig. 309). Oscillatory motion is generated by a crank drive mechanism, and the direction of displacement encloses an angle β with the plane of the sieve. In practice, $\sigma \simeq \beta$ in the majority of cases. The crank radius is short in relation to the length of the connecting rod, and so

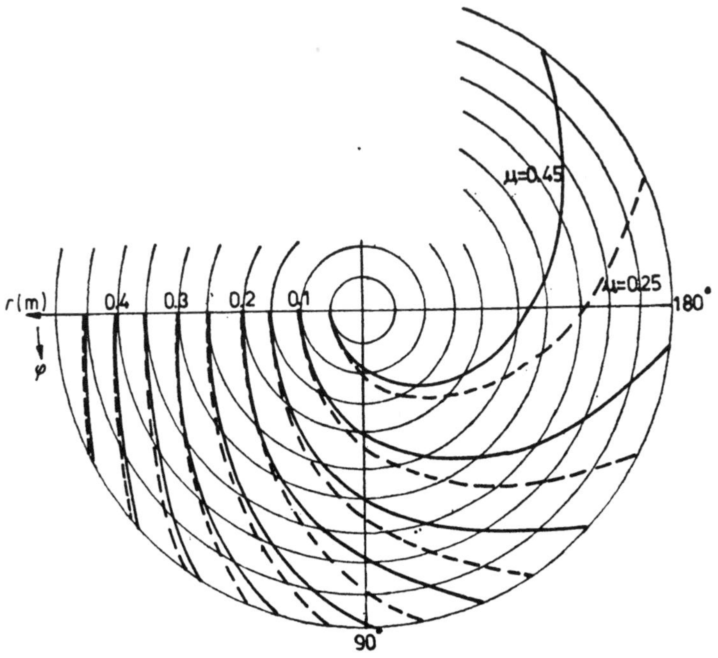

Fig. 308. Pathways of particles with various starting radii on a throwing disc

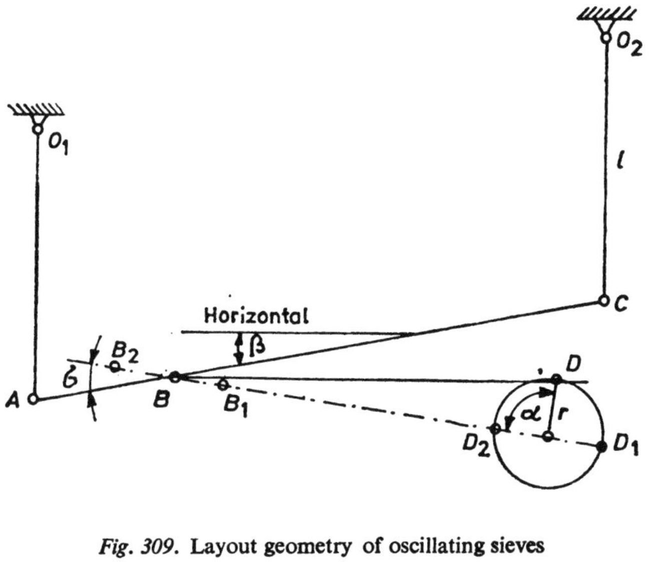

Fig. 309. Layout geometry of oscillating sieves

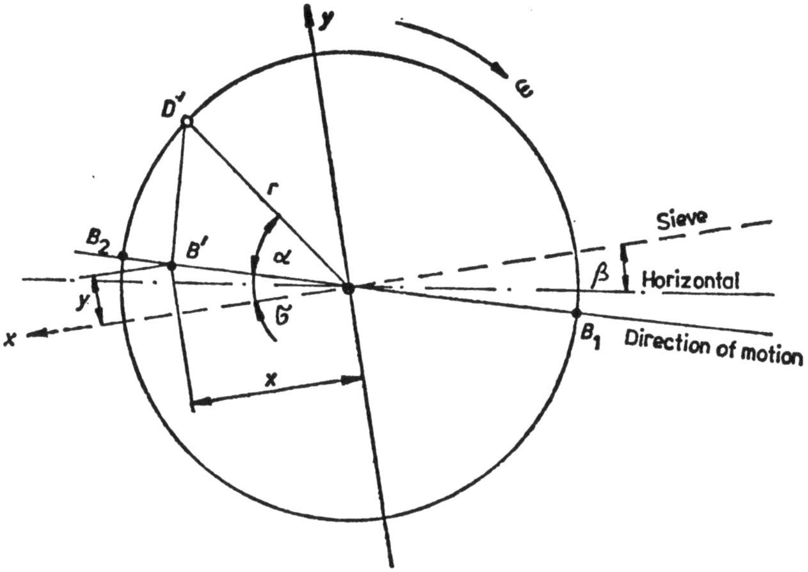

Fig. 310. Determination of motion for an oscillating sieve

harmonic motion results as the projected motion of the uniformly rotating crankshaft. The coordinates of a projection point B' corresponding to an arbitrary angular displacement α (Fig. 310) are

$$x = r \cos\sigma \cos\alpha$$

$$y = r \sin\sigma \sin\alpha$$

The velocity of the point B' is

$$v_x = -r\omega \cos\sigma \sin\alpha$$

$$v_y = -r\omega \sin\sigma \sin\alpha$$

while the acceleration is

$$a_x = -r\omega^2 \cos\sigma \cos\alpha$$

$$a_y = -r\omega^2 \sin\sigma \cos\alpha$$

The inertial force acting on a particle of mass m moving with the sieve consists of the components

$$I_x = -ma_x = mr\omega^2 \cos\sigma \cos\alpha$$

$$I_y = -ma_y = mr\omega^2 \sin\sigma \cos\alpha$$

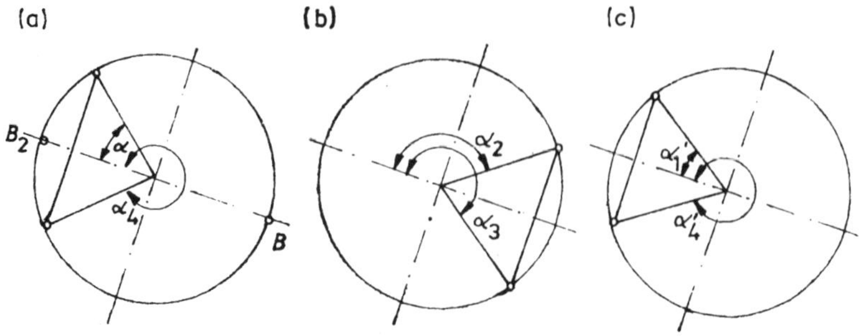

Fig. 311. Zones of characteristic states of motion on a sieve

In quadrants I and IV of full rotation, the inertial force is positive and the particle moves downwards, if

$$I_x + mg \sin \beta > \mu(mg \cos \beta - I_y)$$

where $\mu = \tan \varphi$ is the friction coefficient. Substitution of the expressions for I_x and I_y into the above equation yields the condition

$$r\omega^2 \cos \alpha > g\,(\sin(\varphi - \beta)/\cos(\sigma - \varphi)) \tag{223}$$

The angular interval $\alpha_1 - \alpha_4$ where slip occurs may be determined from eqn. (223) (Fig. 311(a)). When the crankshaft is in quadrants *II* and *III*, the force of inertia I_x acts to move the particle upwards. Displacement is possible when

$$I_x - mg \sin \beta > \mu(mg \cos \beta + I_y)$$

Taking negative values of I_x and I_y into account, the following condition is obtained:

$$-r\omega^2 \cos \alpha > g\,(\sin(\varphi + \beta)/\cos(\sigma + \varphi)) \tag{224}$$

from which the angular interval $\alpha_2 - \alpha_3$ in which the particle moves upwards may be determined (Fig. 311(b)). In quadrants *I* and *IV* the particle may be torn (lifted) from the surface by the force of inertia I_y, if $I_y > mg \cos \beta$. This condition yields the inequality

$$r\omega^2 \cos \alpha > g\,(\cos \beta/\sin \sigma) \tag{225}$$

from which the angular interval $\alpha_1' - \alpha_4'$ over which the particle is torn off the surface may be determined (Fig. 311(c)). Accordingly, the start of the particle's displacement upwards occurs at an angle α_2 and downwards at an angle α_4.

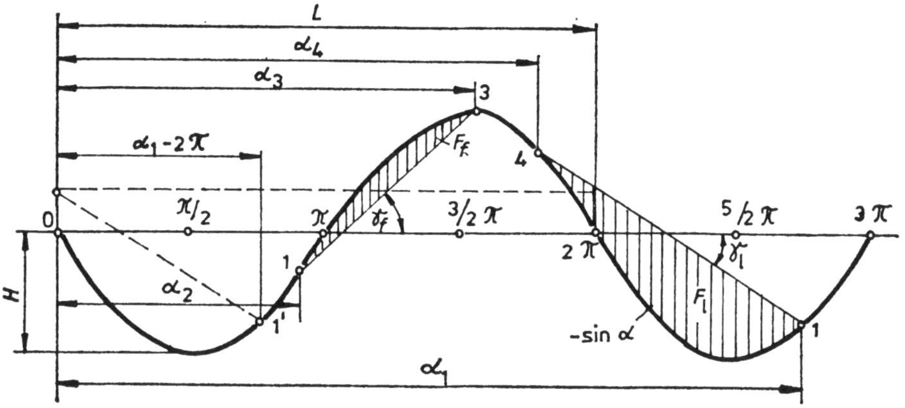

Fig. 312. Graphical representation of states of motion on a sieve

The angular interval over which slip occurs depends on the ratio of the frictional force to the inertial force. The angles α_3 and α_1 marking the limits of displacement may be calculated from the equations [162]

$$-\sin\alpha_4-[g\sin(\varphi-\beta)/r\omega^2\cos(\sigma-\varphi)](\alpha_1-\alpha_4)=-\sin\alpha_1$$

and (226)

$$-\sin\alpha_2+[g\sin(\varphi+\beta)/r\omega^2\cos(\sigma+\varphi)](\alpha_3-\alpha_2)=-\sin\alpha_3$$

Figure 312 shows a graphical solution of eqn. (226). Since the second terms of the equations are linear in α, the solutions are given by the intersection of the sine curve and the straight lines drawn respectively at the angular positions α_2 and α_4. The hatched area between the sine curve and the straight line is proportional to the displacement of the particle downwards or upwards: i.e.,

$$S_1=[\cos(\sigma-\varphi)/\cos\varphi]rF_1m_\alpha m_y$$

and (227)

$$S_2=-[\cos(\sigma+\varphi)/\cos\varphi]rF_2m_\alpha m_y$$

where m_α and m_y scale in the directions of angular displacement and of the vertical axis. The mean velocity of the material's movement downwards or upwards is

$$w_1=S_1n/60$$

$$w_2=S_2n/60$$

where n is the rate of revolution of the crankshaft. Movement of the material on the sieve depends essentially on four factors: the angle of friction, the angle of inclination of the sieve, the direction of sieve motion (σ), and the kinematic

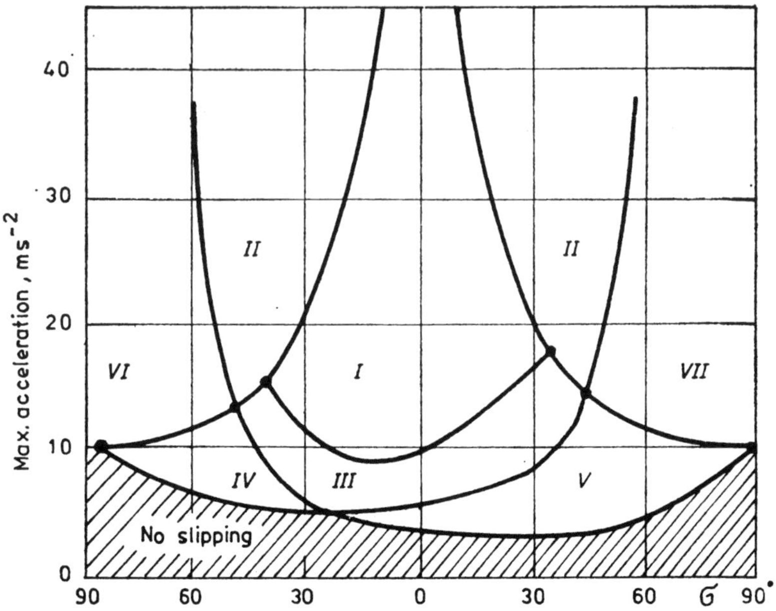

Fig. 313. State diagram for motion on a sieve. *(I)* Sliding down and up (stationary motion); *(II)* sliding down and up, throwing with stopping; *(III)* sliding down and up, with stopping; *(IV)* sliding up, with stopping; *(V)* sliding down, with stopping; *(VI)* sliding up, throwing with stopping; *(VII)* sliding down, throwing with stopping

characteristics ($r\omega^2$). If values of β and φ are given, a state diagram illustrating the appearance of individual typical states of motion may be plotted using an $r\omega^2$–σ system of coordinates. Such a state diagram is shown in Fig. 313 for the values $\beta=5°$ and $\varphi=25°$ [162]. From the point of view of maximum sieve throughput it is advisable to select steady-state motion such that the material slides continuously without stopping.

16. WAFERING AND PRESSING OF AGRICULTURAL MATERIALS

During agricultural technological processes certain materials must be compressed in order to reduce their volume (as in the case of forage materials), to obtain a defined shape which facilitates handling (granulation of farinaceous materials), or to obtain juice (e.g., grape pressing). To calculate compaction processes and determine their power requirements, the physicomechanical properties of compacted materials and the factors influencing the mechanism of compaction must be known [155–161].

16.1 General relationships for pressing processes

The general relationships for pressing may be studied most simply by means of a pressing cylinder. A given quantity of material is placed without prior compaction into the pressing cylinder and compressed slowly by a piston (Fig. 314). The volumetric weight of the material increases with the decrease in volume. The pressure to be exerted by the piston increases progressively with increase in volumetric weight, and therefore it is advisable to plot the pressure on a logarithmic system of coordinates. The volumetric weight of a bulk material can be reduced until all the air is pressed out from among it. The volumetric weight attainable in this way corresponds to the specific weight of the material for a given moisture content.

The compressibility of forage materials depends mainly on the plant species and moisture content, but the length and orientation of the strands also have a certain effect. These variables also influence greatly the initial volumetric weight γ_0 of the material.

Figures 315 and 316 show pressure–volumetric weight relationships for various hay species. The moisture content is seen to have the greatest effect [155, 157, 159]. The effect of moisture content may be demonstrated more clearly by plotting the volumetric weight as a function of moisture content, with pressure as a parameter. The volumetric weight may be related to the wet material or to its dry content. Figures 317 and 318 show the corresponding plots for meadow grass [158].

Exponential and parabolic relationships are used to describe the pressure–volumetric weight relationship. The exponential relationship is used in the form

$$p = K(e^{a(\gamma-\gamma_0)}-1) \tag{228}$$

where K and a are material-dependent constants. For forage materials with 10–20% moisture content, $K=5.2\times10^{-3}$ and $a=5\times10^{-2}$ if the initial volumetric weight γ_0 is less than 80 kg m^{-3} [164]. Parabolic relationships are generally used in the forms

$$p = C(\gamma-\gamma_0)^m \tag{229}$$

or

$$p = C(\gamma^m-\gamma_0^m) \tag{230}$$

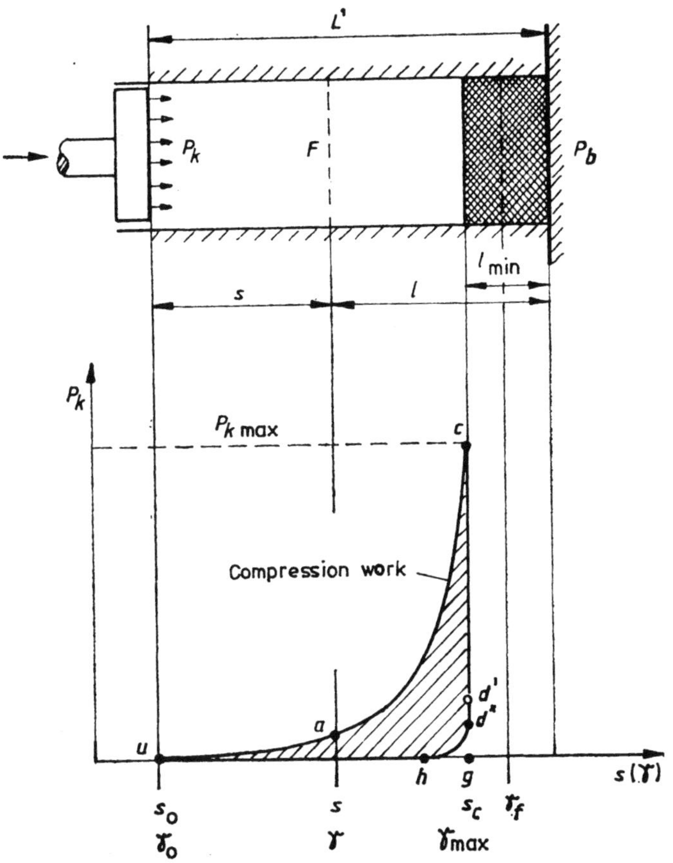

Fig. 314. Compaction of material under a piston in a pressing cylinder

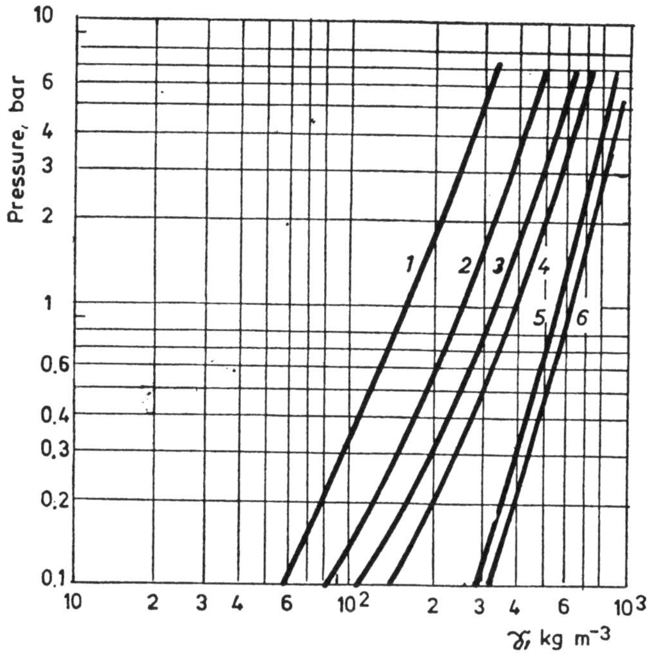

Fig. 315. Compaction curves for forage materials in the pressure range encountered in balers. *(1)* Hay, moisture content 8%; *(2)* hay, moisture content 16%; *(3)* hay, moisture content 23%; *(4)* hay, moisture content 36%; *(5)* grass, moisture content 64%; *(6)* clover, moisture content 82%

where C and m are material-dependent constants. For the pressure range encountered in balers, a relationship also allowing for the moisture content of the material and the velocity of compression has been developed [166]:

$$p = [a(1-U)^2 + b](1+0.5v)\gamma^c(1+0.5U) \tag{231}$$

where the volumetric weight γ must be substituted in units of t m^{-3} and the values of the constants a, b and c may be taken from Table 19.

Table 19

Product	a	b	c	Validity limits
Meadow grass	80	5	2.2	U=0.1–0.8
Alfalfa	60	10	1.9	U=0.1–0.8
Wheat straw	180	10	2.3	
Barley straw	180	20	2.0	U=0.08–0.22
Rye straw	180	30	2.0	

The velocity of compression may be characterized by the reduced piston velocity according to the expression

$$v = v_p/l^*(1-U)$$

where l^* is the length, for given dimensions of the pressing space, in which the quantity of the material charged attains the density $\gamma = 0.05 \text{ t m}^{-3}$.

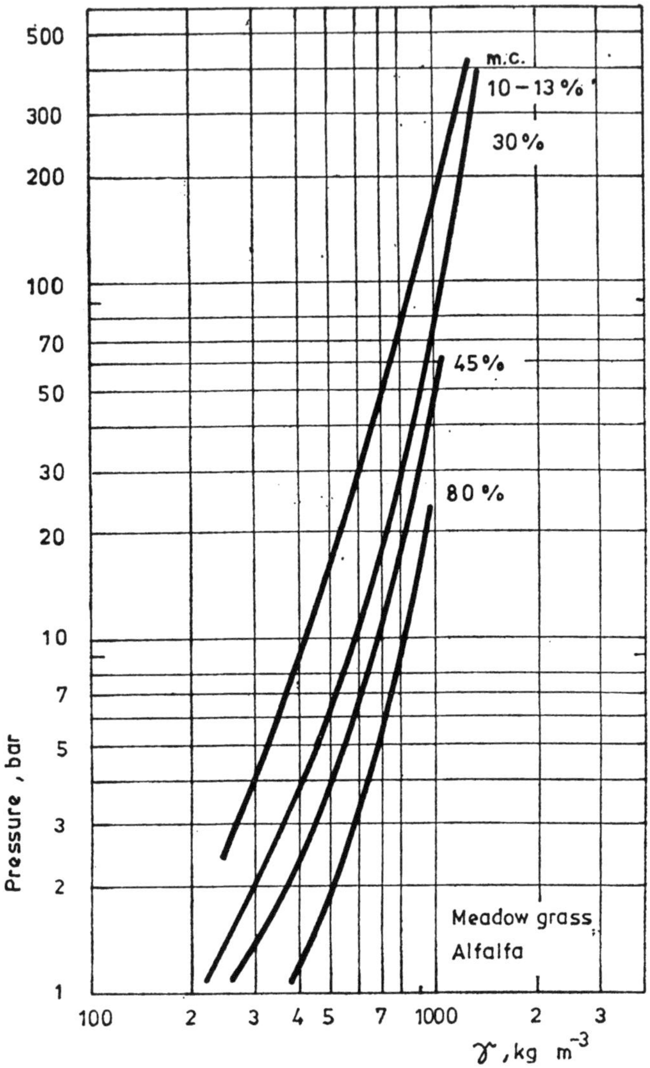

Fig. 316. Compressibilities of meadow grass and alfalfa with various moisture contents

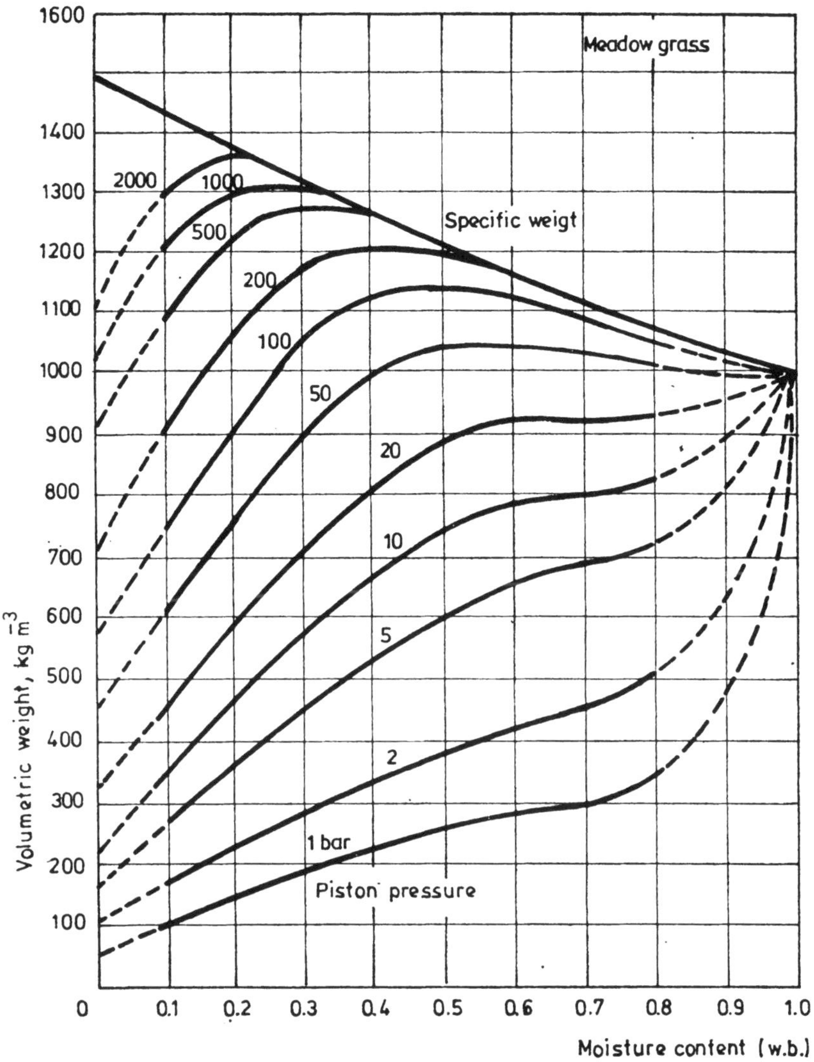

Fig. 317. Relationship between volumetric weight and moisture content for various pressures (wet basis)

In pistontype balers material is recompressed during subsequent cycles from the recovered state to that corresponding to the maximum pressure, as new material arrives in the pressing space. The exponent *m* (in eqn. (229)) for recompressed material increases with the number of cycles, as shown in Fig. 319 [166].

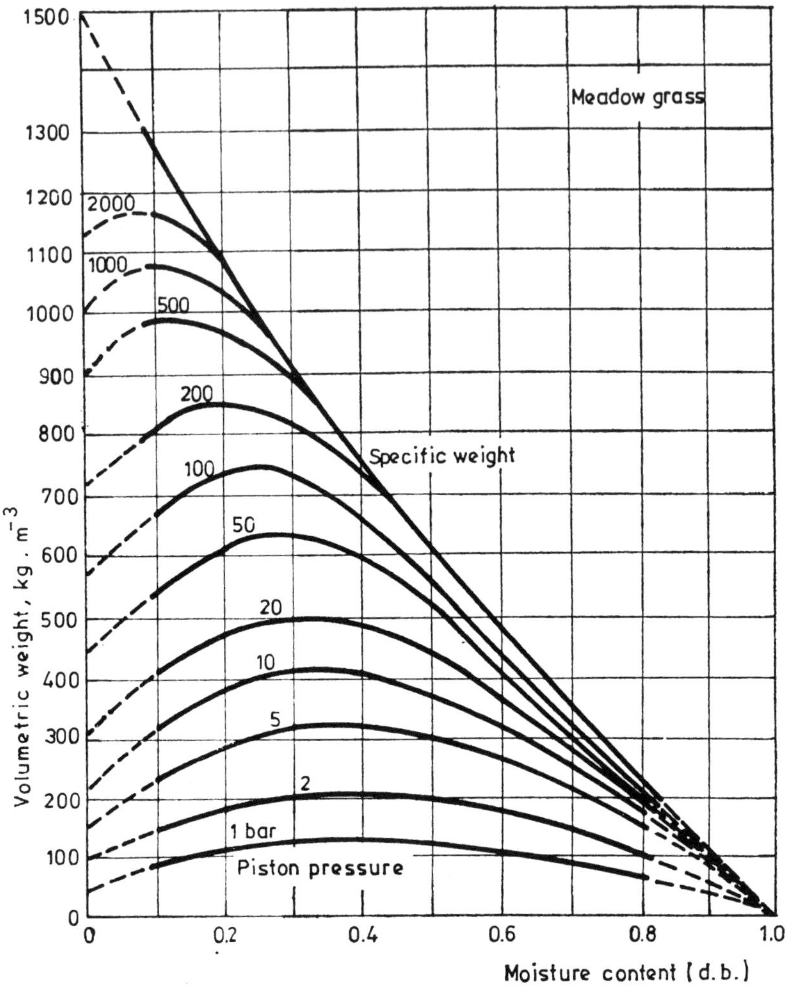

Fig. 318. Relationship between volumetric weight and moisture content for various pressures (dry basis)

Agricultural materials which are compressed are almost without exception viscoelastic [163], and so the density attainable at a given pressure depends on the velocity of compression and also on the subsequent period for which the material is kept compressed. Viscoelastic properties may be taken into account, as has been seen, using various models comprising a spring and a dashpot element. The model used most frequently is the three-element model, with compression at a velocity v_0 or with a sinusoidally alternating velocity (see Fig. 92). The behavior

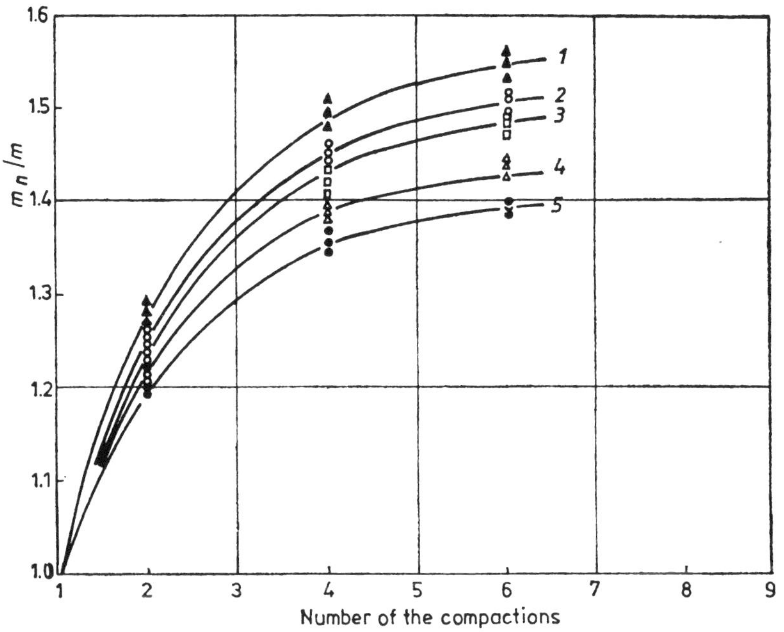

Fig. 319. Effect of the number of repeated compactions on compaction regularity. *(1)* Barley; *(2)* alfalfa; *(3)* rye; *(4)* oat; *(5)* meadow grass

of the three-element model for various loading modes is described by eqns. (42)–(50a). Application of the above rheological models is aggravated by the circumstance that the modulus of elasticity of many agricultural materials varies greatly during compression, even over narrow pressure ranges, and so the quantities E_1, E_2 and η are not constant.

In rapid compression the air found among the particles of forage materials also plays a role: some of this air is compressed, while a part escapes past the compressing piston. This effect may be taken into account by connecting a Maxwell model, i.e., a serially connected spring and dashpot element, in series with the existing three-element model. The elastic element corresponds to the compression, the dashpot to the escaping outflow. The resulting five-element model may be solved in closed form both for motion at constant velocity v_0 and also for motion according to a crankshaft drive [167], but application is much more difficult than in the case of three-element model.

The measure $\sigma_\infty/\sigma_{max}$ of relaxation for forage materials (Fig. 320) is independent, to a first approximation, of the value of σ_{max}, and varies over a relatively narrow interval for various materials: generally, $\sigma_\infty/\sigma_{max}=0.63$–$0.67$. The relaxation time T is a function of the maximum pressure, the type of material,

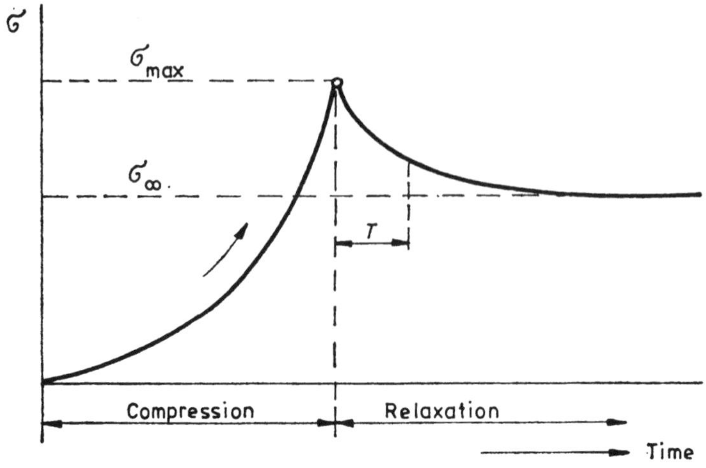

Fig. 320. Relaxation of compacted material

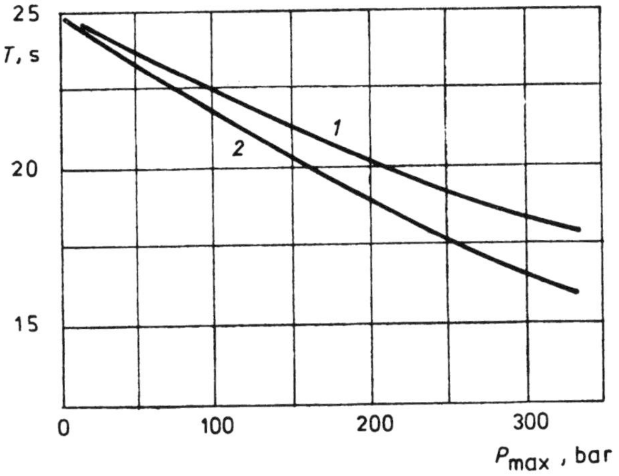

Fig. 321. Relaxation times of wheat straw and alfalfa as functions of pressure. *(1)* Wheat straw; *(2)* alfalfa

and the moisture content. Figure 321 shows the relaxation times for wheat straw (12% moisture content) and alfalfa (29% moisture content) as functions of the maximum pressure [164]. As may be seen, there is no significant difference between the relaxation times of the two materials.

As mentioned above, basic problem in applying rheological equations is that the quantities E_1, E_2 and η appearing in them are not constant but depend on the volumetric weight. Both E_1 and E_2 increase with increasing volumetric weight.

Therefore, these factors must be expressed as functions of the volumetric weight, or of the related values of displacement or time. These relationships may be established experimentally for concrete cases. The following relationship exists between the momentary volumetric weight and the strain:

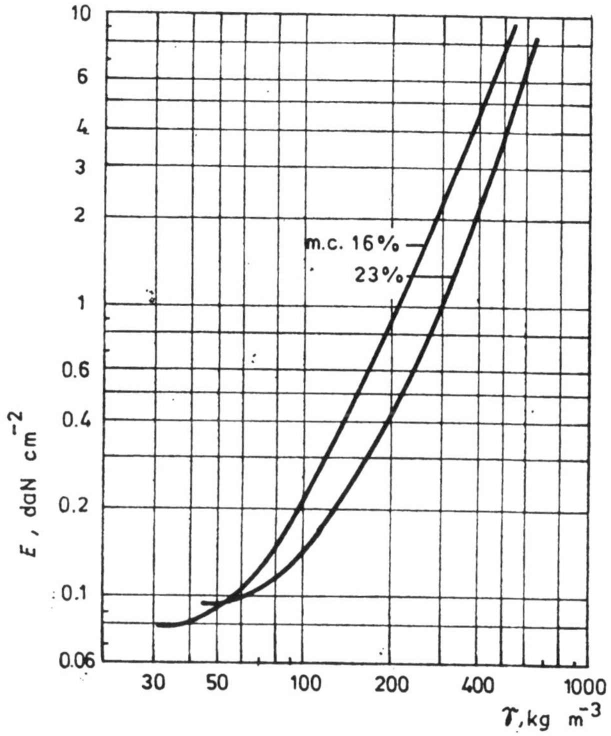

Fig. 322. Variation of the modulus of elasticity during compaction of wheat straw

$$\gamma = \gamma_0/(1-\varepsilon) \tag{232}$$

or

$$\varepsilon = 1-(\gamma_0/\gamma) \tag{233}$$

Substitution of γ from eqn. (232) into eqn. (229) yields

$$p = C\gamma_0^m[\varepsilon/(1-\varepsilon)]^m \tag{234}$$

For the pressure range encountered in balers, $m \simeq 2$ for forage materials, and comparison of eqn. (234) with the well-known relationship $\sigma = E\varepsilon$ yields E as

$$E = C\gamma_0^2\varepsilon/(1-\varepsilon)^2 \tag{235}$$

However, processing of experimental data (Fig. 322) shows that in the case of $\varepsilon=0$ (i.e., at the start of compression), $E=E_0$ and not zero, as would follow from the above expression. Therefore, the modulus of elasticity of forage materials may be approximated using the relationship

$$E = E_0 + C\gamma_0^2\varepsilon/(1-\varepsilon)^2 \tag{235a}$$

The initial modulus of elasticity E_0 is a function mainly of the initial volumetric weight γ_0, but is also influenced to a certain extent by the moisture content. According to the curves shown in Fig. 322, for meadow hay with two different moisture contents, $E_0=0.08$ daN cm^{-2} when $\gamma_0=30$ kg m^{-3}, and $E_0=$ $=0.1$ daN cm^{-2} when $\gamma_0=45$ kg m^{-3}. In the case of 16% moisture content, $C=3.0\times10^{-5}$. With increasing moisture content the value of C decreases; thus for 23% moisture content, $C=2\times10^{-5}$. Equation (235a) is valid until the volumetric weight of the material reaches its specific weight, i.e., until the material becomes virtually incompressible. The functions of the moduli of elasticity E_1 and E_2 appearing in the models agree in form with eqn. (235a); thus it is only necessary to select E_{01}, E_{02} and the constants C_1 and C_2 correctly on the basis of experimental data. The relaxation time T may also be obtained from experimental results.

16.2 Energy requirements of pressing

The energy requirements of pressing by balers or pelleting machines comprise the net pressing work and the work spent in pushing. The net pressing work may be determined most simply by means of a pressing cylinder (Fig. 314). A piston moving from position s_0 compresses the material so that its volumetric weight γ_0 increases to the value γ_{max}, while the pressure acting on the piston gradually increases. The work spent in pressing is given by the area under the pressure curve:

$$A_c = F\int p\,ds$$

or, related to weight units,

$$A_c/G = (F/G)\int p\,ds \tag{236}$$

Alternatively, since $G=FL\gamma_0$, the expression

$$\gamma/\gamma_0 = L/(L-s)$$

or

$$-ds = L\gamma_0 d(1/\gamma)$$

yields the equation

$$A_c/G = \int_{\gamma_0}^{\gamma} p d(1/\gamma) \tag{236a}$$

i.e., if the compression curve is plotted using a system of p–$1/\gamma$ coordinates, the area under the curve gives the specific net compression work.

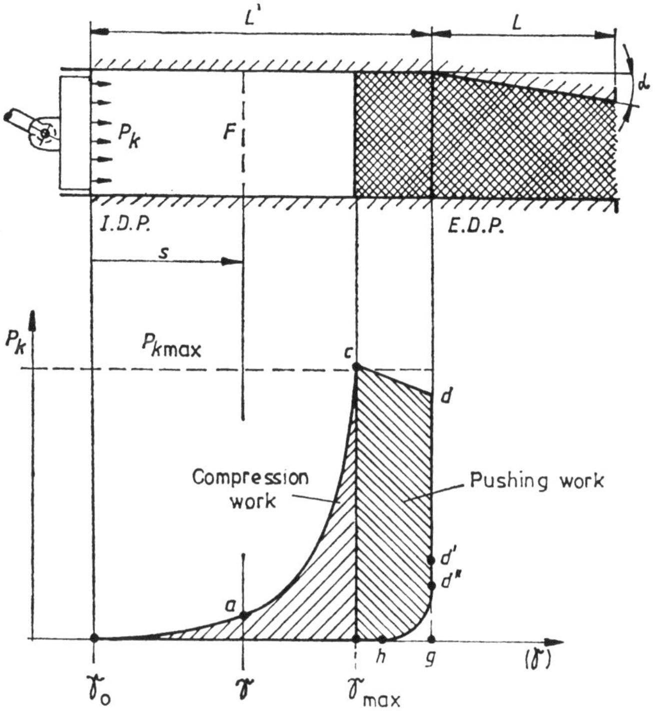

Fig. 323. Compression and pushing work for compaction

In continuous pressing processes material is compressed not in a closed space but in a suitably designed channel. Countersupport of the material is ensured by narrowing of the channel and by wall friction. In this case the net compression work is accompanied by pushing work, which must be added to it, as shown in Fig. 323. The pushing work may be calculated on the basis of eqn. (236a), with the assumptions that the pressure $p = p_{max} = \text{const}$ (disregarding the slight variation over the section c–d), and that $1/\gamma = 1/\gamma_{max} = \text{const}$. For these values,

$$A_s/G = p_{max}/\gamma_{max} \tag{237}$$

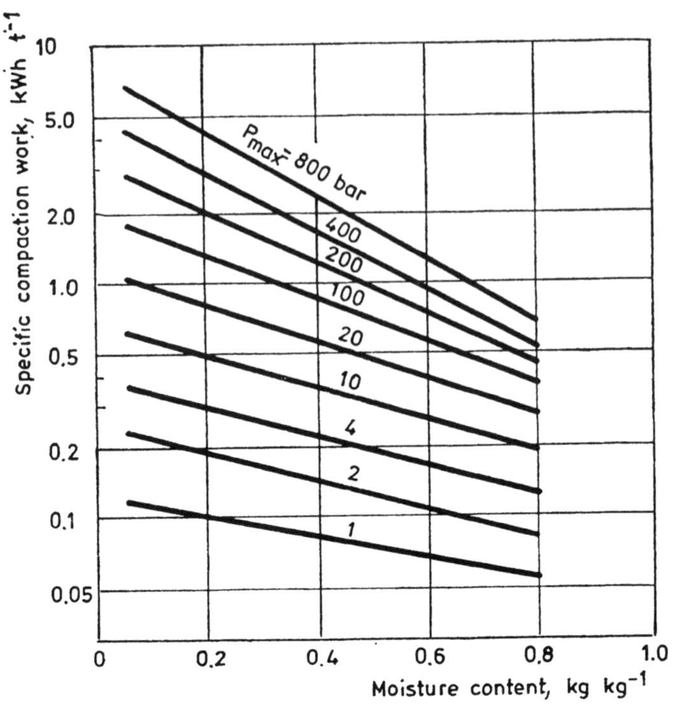

Fig. 324. Specific compaction work as a function of moisture content for meadow grass and alfalfa

and the total specific work is

$$A/G = A_c/G + A_s/G = \int_{\gamma_0}^{\gamma} p d(1/\gamma) + p_{\max}/\gamma_{\max} \tag{238}$$

The specific work spent in pressing depends primarily on the pressure range and the moisture content. Figure 324 shows the specific compression work as a function of moisture content for various final pressures, for meadow grass and alfalfa [158]. Figure 325 presents the total specific work as a function of maximum volumetric weight for meadow grass with various moisture contents [155]. As may be seen, the energy requirements are influenced decisively by the moisture content.

The energy requirements are also affected by the velocity of compression. For increasing velocity the pressure required to attain a given volumetric weight generally increases, and together with it the energy requirements. The ratio of the compression and pushing work also depends crucially on the dimensions of the pressing channel. The compression work is independent of the channel dimensions, but the pushing or friction work decreases significantly as the cross-sectional area

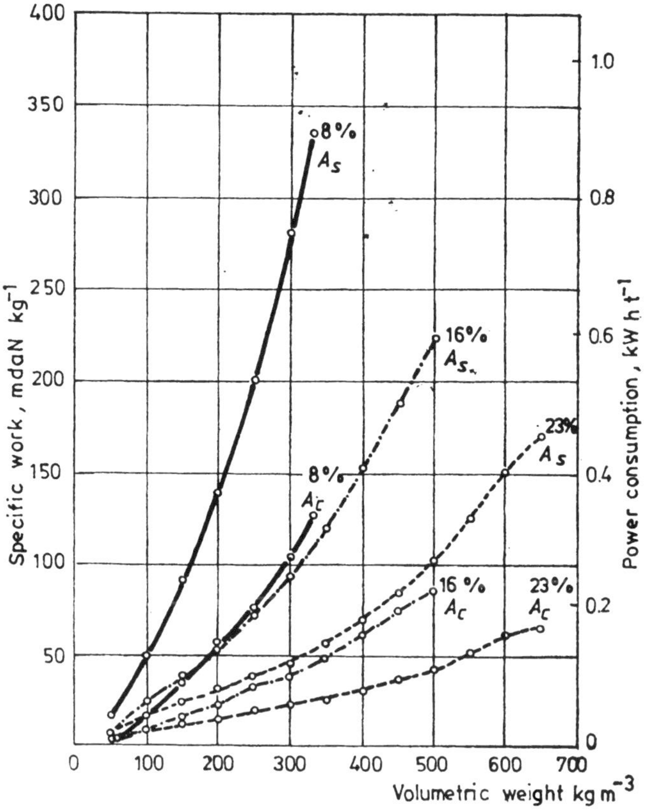

Fig. 325. Total specific work as a function of maximum volumetric weight for meadow grass with various moisture contents

of the channel increases. Figure 326 shows the variations of the compression and friction work required in pressing air-dry straw, as functions of the cross-sectional area of the pressing channel. The friction work is also slightly influenced by the weight of material charged during individual cycles [165]. Figure 327 illustrates the variation of the total compression work as a function of the quantity of material charged per cycle, for two channel dimensions and the conditions given in the preceding figure. As may be seen, from the point of view of conserving energy it would be advisable to use a channel having as great a cross-sectional area as possible. However, with increasing cross-sectional area the forces acting on the connecting rod and crankshaft increase considerably, necessitating more robust construction.

The power requirements of pelleting machines show a similar pattern. Production of small-diameter pellets is highly energy-demanding, owing to the high

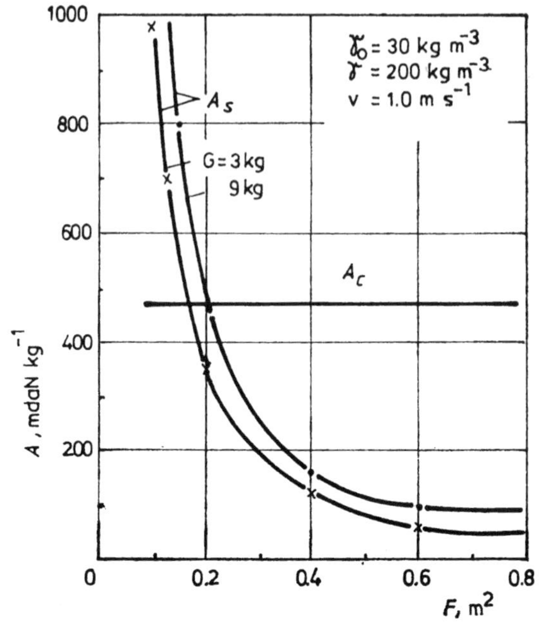

Fig. 326. Effect of cross-sectional area of compacting channel on compaction work

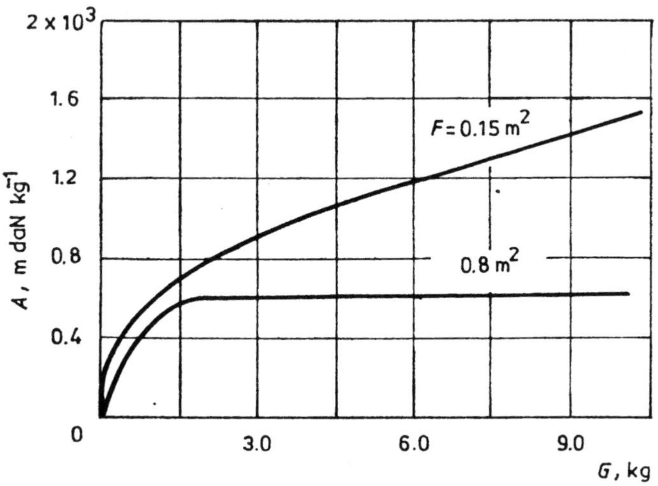

Fig. 327. Total compaction work as a function of quantity of material cyclically charged

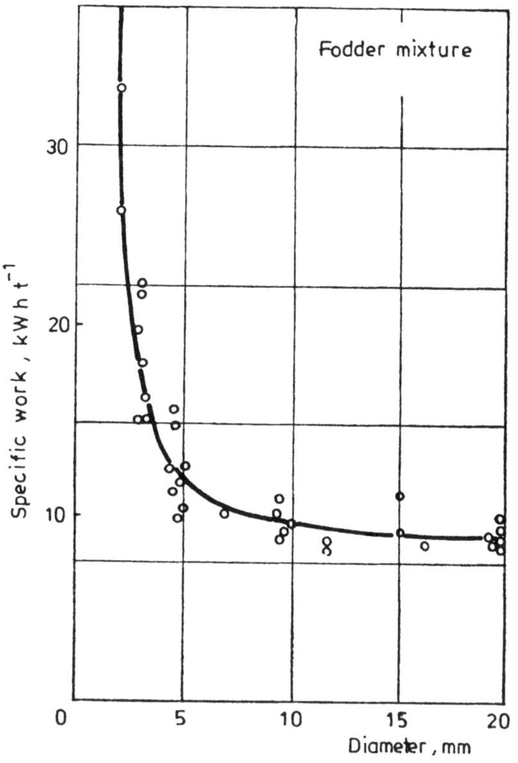

Fig. 328. Specific energy requirements of pelleting process as a function of pellet diameter for fodder flour

friction work. Figure 328 shows the power requirements for pelleting fodder flour, as a function of pellet diameter [168]. The curve may be described by the empirical equation

$$N/Q = d/0.09(d-1.5)$$

where Q is the throughput of the machine ($\mathrm{t\,h^{-1}}$), and d the diameter of the pellets (mm).

16.3 Rebound of material after pressing

Materials are compressed by a pressure p to a volumetric weight γ_{max} during pressing. After unloading the material rebounds according to its viscoelastic properties, i.e., its volumetric weight decreases to a value γ_k (Fig. 329(a)). The equation for the unloading curve may be given, similarly to eqn. (229), as

$$p = K_1(e^{a_1(\gamma-\gamma_k)}-1)$$

where K_1 and a_1 are constants depending on the type of material, its moisture

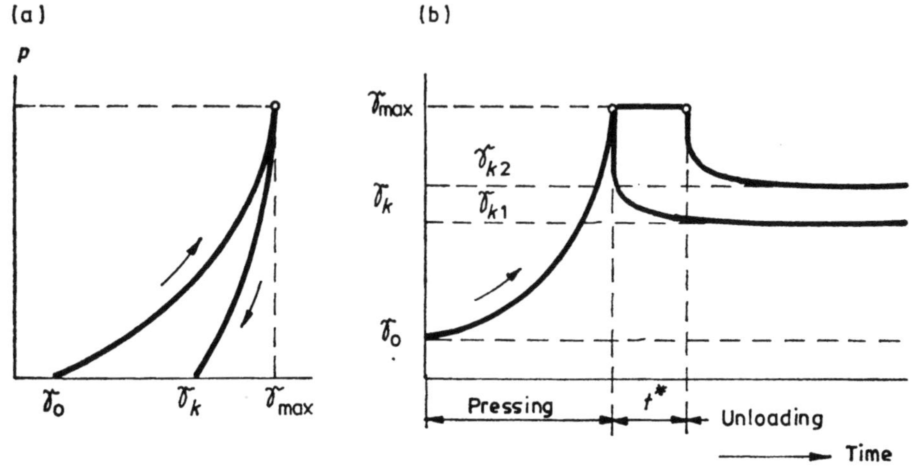

Fig. 329. Rebound of a material after compression

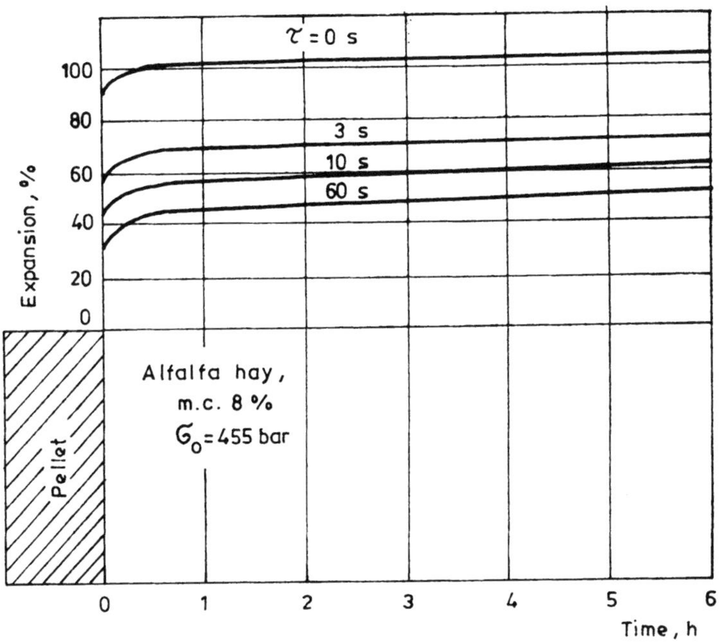

Fig. 330. Rebound of alfalfa wafers as a function of time

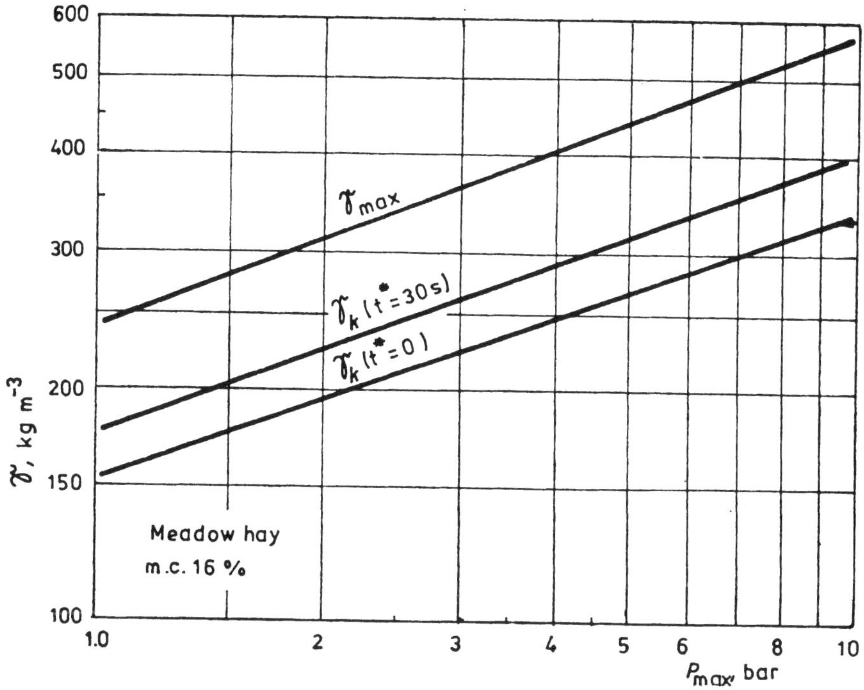

Fig. 331. Rebound of meadow hay in the pressure range encountered in balers

content and the period of time spent under the maximum pressure. Unloading may be performed either immediately after the value γ_{max} is attained, or after a certain time t^* has elapsed (Fig. 329(b)). The longer the time t^*, the less the extent of rebound of the material, and the higher the value of γ_k. The volumetric weight γ_k corresponding to the stable state is not attained immediately, even in the case of instantaneous unloading. Rebound of a material may continue for several days, naturally with very slight increments. However, a considerable proportion of the rebound usually occurs within a short time after unloading (Fig. 330) [173].

The decrease in volumetric weight during rebound may be calculated from the empirical relationship [164]

$$\gamma_k = a\gamma_0 + b\gamma_{max} + ct^*(\gamma_{max} - \gamma_0) \tag{239}$$

where the constants may be substituted by the values $a=0.4$, $b=0.57$ and $c=4.6\times10^{-3}$ for forage materials, for the pressure range encountered in balers. The value t^* must then be substituted in s units, and the range of validity for the constant c is $0 \leq t^* \leq 60$ s.

Figure 331 shows the rebound of meadow hay for the pressure range encoun-

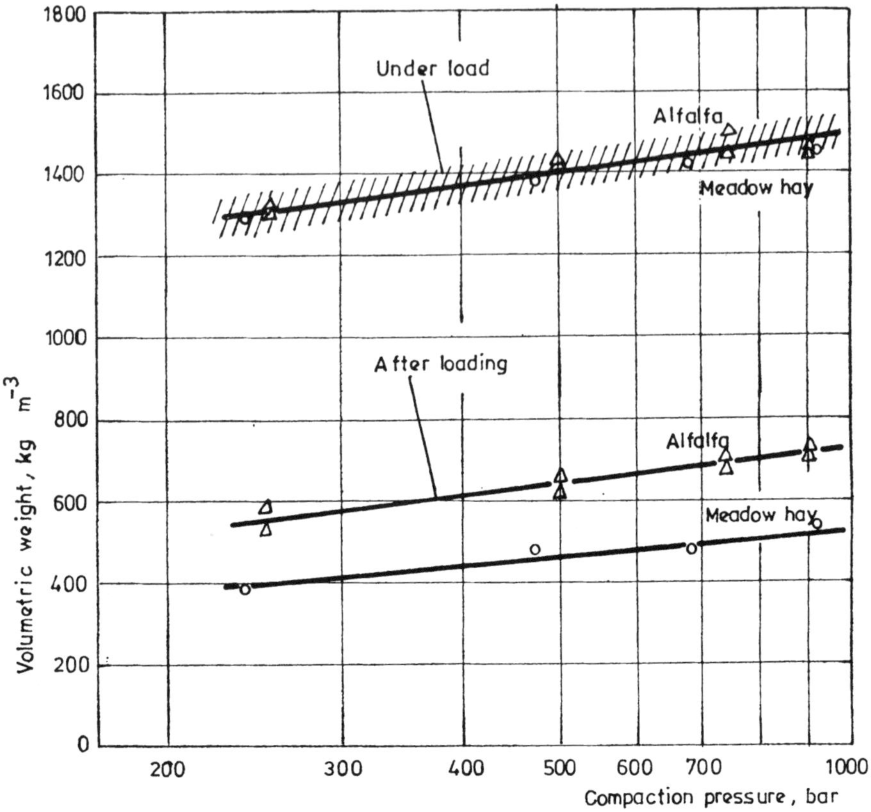

Fig. 332. Rebound of alfalfa pellets as a function of pressure

tered in balers, for immediate unloading and after maintaining the pressure for 30 s. It may be seen that the volumetric weight of the bales may be increased visibly by maintaining the pressure. Figure 332 illustrates the rebound of alfalfa and meadow grass with 17–20% moisture content for the pressure range encountered in pelleting machines, as a function of maximum pressure, after three days' storage [156].

16.4 Pressure distribution in the space before a compressing piston

During the compression of materials both in cylinders and channels, the pressure in the space before the piston is not uniform. Accordingly, the density (volumetric weight) of the material varies as a function of the distance from the piston. The reason for this is that the material is exposed to friction along the

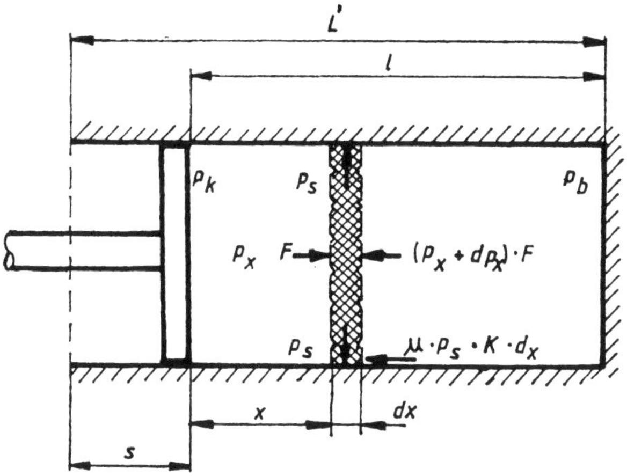

Fig. 333. Force relations in a pressing cylinder

walls surrounding the compression space, and the axial pressure arising in the material is reduced by the frictional force.

Figure 333 illustrates the force relations developing in a compression cylinder. At a distance x from the piston a pressure p_x acts, creating a transverse wall pressure p_s depending on the Poisson's ratio. Knowing the friction coefficient, the frictional force acting on an element of width dx may be calculated as

$$S = \mu p_s D\pi \, dx$$

In the case of unidirectional deformation, the relationship between the pressures p_x and p_s is

$$p_s = [\nu/(1-\nu)]\, p_x$$

and so equilibrium of the element dx may be expressed by the differential equation

$$dp_x F + \mu[\nu/(1-\nu)]\, p_x D\pi dx = 0$$

The solution of this equation is

$$p_x = p_k e^{-kx} \tag{240}$$

where

$$k = (4/D)\mu\nu/(1-\nu)$$

The pressure on the base of pressing cylinder is obtained by substituting the length l for x:

$$p_b = p_k e^{-kl}$$

Figure 334 shows schematically the pressure curves for various degrees of compression.

In order to apply eqn. (240), the Poisson's ratio and friction coefficient μ of the material being compressed must be known as functions of the pressure and moisture content. The Poisson's ratio of forage materials increases with their volumetric weight and with their moisture content, as shown in Fig. 335 [158]. For the pressure range encountered in balers $\nu=0.25$–0.35, while for pelleting machines, $\nu=0.35$–0.45. The friction coefficient is a function mainly of the moisture con-

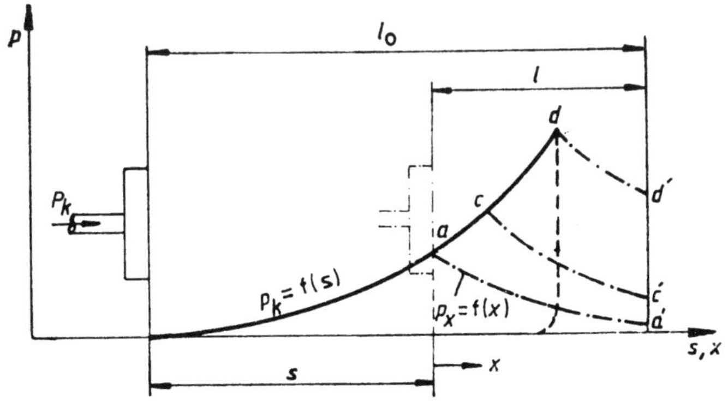

Fig. 334. Pressure distribution before a pressing piston

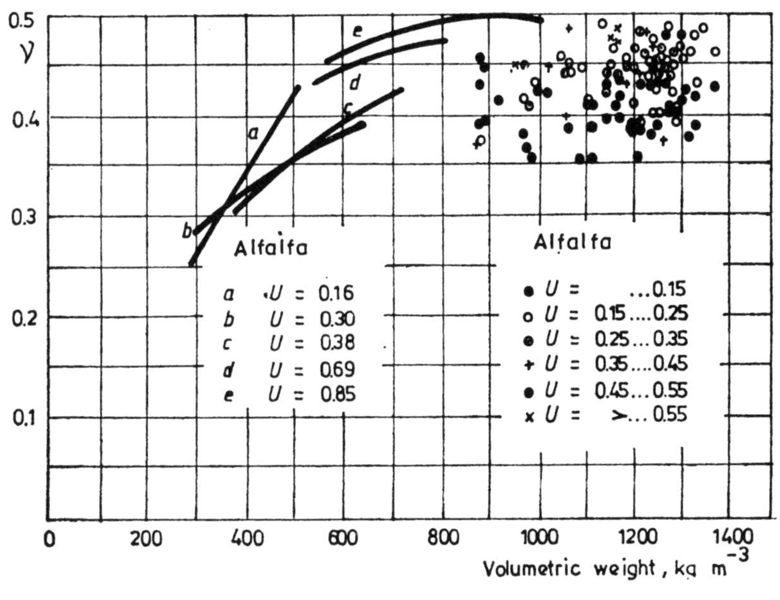

Fig. 335. Poisson's ratio of forage materials a as function of volumetric weight

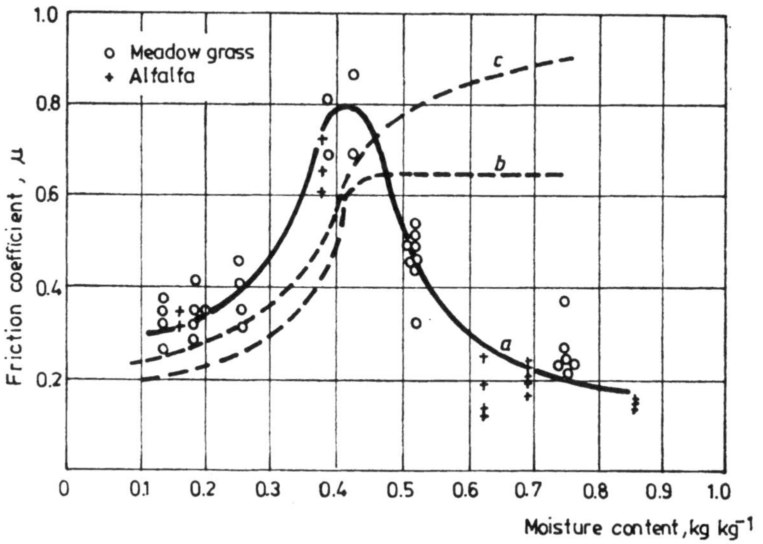

Fig. 336. Friction coefficient of forage materials as a function of moisture content for various pressure ranges. *(a)* 10–15 daN cm^{-2}; *(2)* 0.2–0.3 daN cm^{-2}; *(c)* 0.005 daN cm^{-2}

tent, but it is also influenced significantly by the pressure. The latter effect appears partly as a result of the deformation of the material under high pressures, whereby the contact surface area is modified, and partly as a result of water being pressed out of the material, which in this case serves to lubricate the sliding surface. As a consequence the friction coefficient decreases at high pressures for materials with higher moisture contents, as shown in Fig. 336 [33, 160]. At low pressures, high friction coefficients are obtained.

Calculation of the pressure distribution in the pressing channel of balers is much more complex. The material undergoes deformation in two directions, since the density of the bales is adjusted to the desired value by varying the angle of the upper plate. Figure 337 shows the form of the pressing channel and the forces acting on the material. The individual stress components in the case of a three-dimensional stress state may be written as:

$$\begin{aligned} p_x &= E(\varepsilon)\varepsilon_x+(p_y+p_z)\nu \\ p_y &= E(\varepsilon)\varepsilon_y+(p_x+p_z)\nu \\ p_z &= E(\varepsilon)\varepsilon_z+(p_y+p_x)\nu \end{aligned} \tag{241}$$

In the present case $\varepsilon_z=0$, and so the first term on the right-hand side of the third equation may be omitted. The modulus of elasticity may be expressed according to eqn. (235a) as

$$E(\varepsilon) = E_0+C\gamma_0^2(\varepsilon/(1-\varepsilon)^2)$$

In the case of two-dimensional deformation, ε may be calculated from the relative reduction in volume:

$$\varepsilon = \Delta V/V = \varepsilon_x + \varepsilon_y - \varepsilon_x \varepsilon_y \tag{242}$$

From eqn. (241), p_x may be derived as

$$p_x = [(1-\nu)^2/(1-2\nu)(1-\nu^2)]E(\varepsilon)\{\varepsilon_x + [\nu/(1-\nu)]\varepsilon_y\}$$

According to this model, eqn. (234) may be written in the form:

$$p_x = C\gamma_0^2[\varepsilon_{xy}/(1-\varepsilon_{xy})]^2$$

where $\varepsilon_{xy} = \varepsilon_x + [\nu/(1-\nu)]\varepsilon_y$. From the above equation,

$$\varepsilon_x = \sqrt{p_x/C\gamma_0^2}/(1+\sqrt{p_x/C\gamma_0^2}) - [\nu/(1-\nu)]\varepsilon_y \tag{243}$$

The strain in the y direction may be expressed simply as

$$\varepsilon_y = \alpha x/a \tag{244}$$

On the basis of Fig. 337, the equilibrium of an element of width dx may be

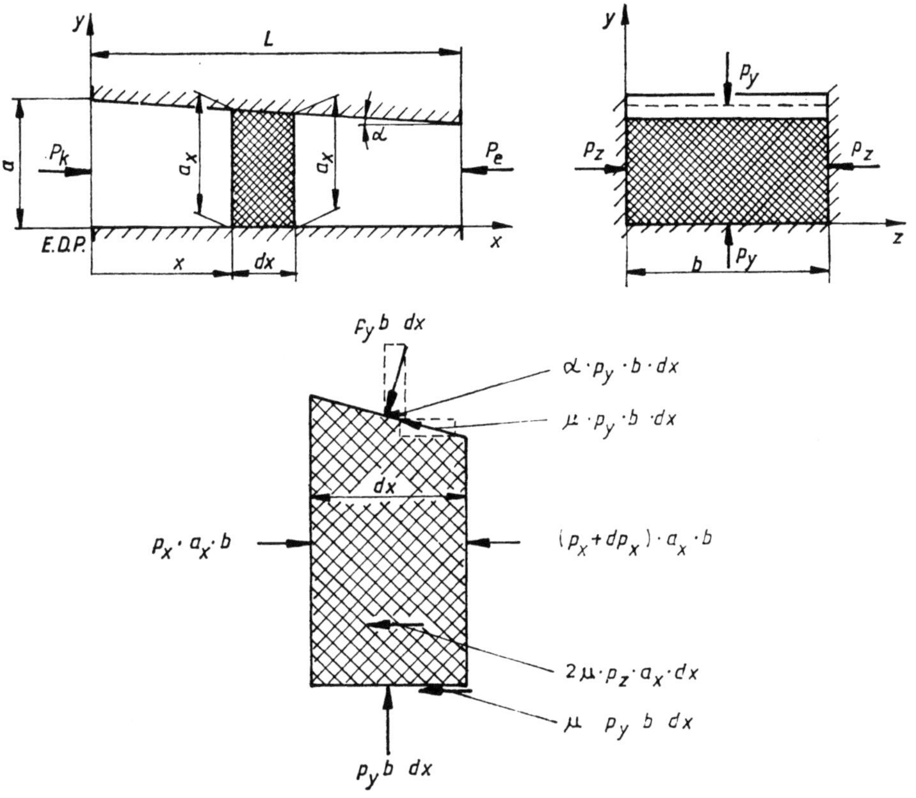

Fig. 337. Pressure relations in a pressing channel

written in the form

$$-p_x a_x b+(p_x+dp_x)a_x b+\alpha p_y bdx+2\mu p_y bdx+2\mu p_z a_x dx = 0$$

Introduction of the relation $a_x=a-\alpha x$ yields the differential equation

$$dp_x/dx+[(\alpha+2\mu)/(a-\alpha x)+2\mu/b][\nu/(1-\nu)]p_x+$$
$$+E(\varepsilon)(\alpha/a)[(\alpha+2\mu)/(a-\alpha x)(1-\nu^2)+(2\mu/b)\nu/(1-\nu^2)]x = 0 \qquad (245)$$

while the pressure component in the y direction is

$$p_y = [1/(1-\nu^2)]E(\varepsilon)\varepsilon_y+[\nu/(1-\nu)]p_x \qquad (246)$$

The differential equation may be solved, with eqns. (235a), (242), (243) and (244) taken into account, only by means of computer, with the boundary condition that at the end of the pressing channel (at the point $x=L$), $p_x=p_e$.

Knowing p_x and the corresponding ε_x values, the p_y component may be calculated simply from eqn. (246). From eqn. (245) it may be seen that when $\alpha=0$ (in the case of parallel walls), the last term of the differential equation is zero and the solution of the remaining homogeneous equation agrees with the equation obtained earlier (eqn. (240)). The modulus of elasticity E does not appear in this equation, i.e., p_x is independent of ε_x, and so eqn. (245) may be simplified, although only with certain omissions. One simplification is that calculations are performed using the mean value of a_x instead of the variable quantity, thus

$$a_x \simeq a_m = a-\alpha(L/2)$$

while the modulus of elasticity E is used in the simplified form

$$E = C\gamma_0^2(\varepsilon_y/(1-\varepsilon_y))^2$$

In this case, eqn. (245) may be written as

$$dp_x/dx+Ap_x+B[x/(a-\alpha x)]^2 = 0 \qquad (245a)$$

where

$$A = [(\alpha+2\mu)/a_m+2\mu/b]\nu/(1-\nu)$$

and

$$B = C\gamma_0^2\alpha^2[(\alpha+2\mu)/a_m(1-\nu^2)+(2\mu/b)\nu/(1-\nu^2)]$$

The solution of eqn. (245a), with the boundary condition (i.e., $p_x=p_e$ at $x=L$) taken into account, may be written in the form

$$p_x = p_e e^{A(L-x)}+(B/A\alpha^2)(e^{A(L-x)}-1)+(Ba^2/\alpha^3)(e^{A(L-x)}/\xi_L-1/\xi)+$$
$$+e^{(A/\alpha)\xi}K[\int_{\xi=a}^{\xi_L}(e^{-(A/\alpha)\xi}/\xi)\,d\xi-\int_{\xi=a}^{\xi}(e^{-(A/\alpha)\xi}/\xi)d\xi] \qquad (247)$$

where
$$K = ABa^2/\alpha^4 + 2aB/\alpha^3$$
$$\xi = a - \alpha x$$
and
$$\xi_L = a - \alpha L$$

The pressure acting on the piston is obtained by substituting $x=0$: thus

$$p_0 = p_e e^{AL} + (B/A\alpha^2)(e^{AL}-1) + (Ba^2/\alpha^3)(e^{AL}/\xi_L - 1/a) + e^{(A/\alpha)a} K \int_{\xi=a}^{\xi_L} (e^{-(A/\alpha)\xi}/\xi)d\xi \tag{247a}$$

The values of the integrals

$$\int_x^\infty (e^{-mt}/t)dt = -Ei(-mx)$$

appearing in the above equations may be taken from integral tables. Since the upper limit of integration is smaller than the lower, the last term of eqn. (247) is always negative.

Figure 338 shows the results of calculations performed using eqn. (247), with the following initial data: dimensions of pressing channel, $a=40$ cm; $b=80$ cm; $L=100$ cm; $\alpha=0.175$ (10°); $\mu=0.4$; $\nu=0.35$; $C=3\times10^{-5}$; $\gamma_0=30$ kg m^{-3}; and $p_e=0.2$ daN cm^{-2}. According to the calculations, the pressure in the pressing

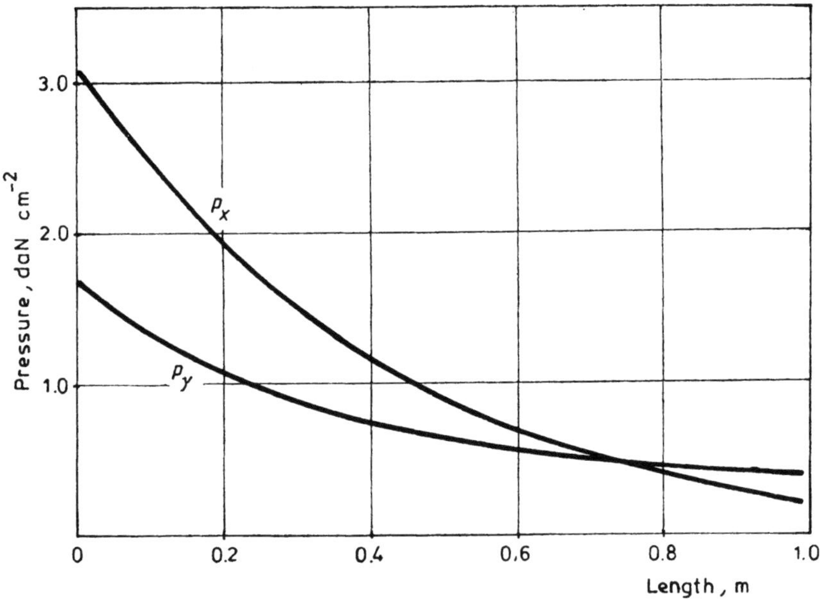

Fig. 338. Pressure in a pressing channel as a function of length

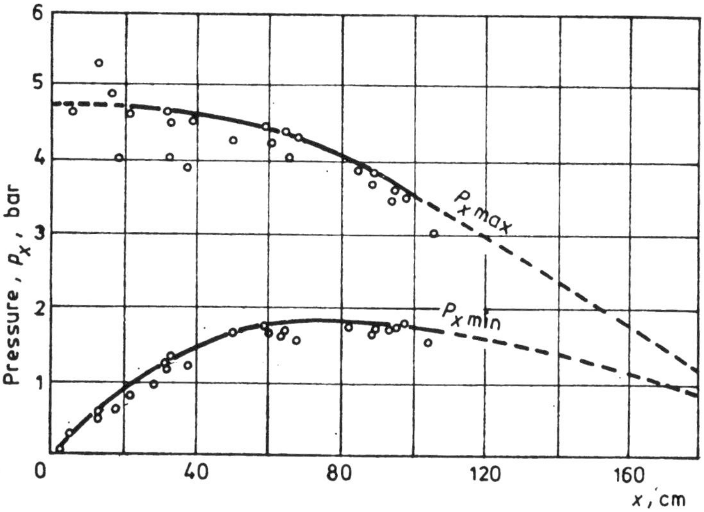

Fig. 339. Maximum and minimum pressures developing in a pressing channel as functions of distance from piston dead-point

channel decreases relatively rapidly in the x direction. The vertical pressure is created partly by the horizontal pressure and partly under the effect of variation of the cross-sectional area; therefore the p_y curve intersects the p_x curve.

The above calculation is valid for the case when compaction is brought about by a single stroke. However, in reality the material reaches the rear part of the pressing channel and is pressed out after several pressing cycles. During repeated cycles the material is compressed harder, and therefore the deformation in the y direction contributes a larger component p_y to the pressure. Figure 339 presents the pressure conditions developing in the pressing channel of a baler as a function of the distance measured from the dead-point position of the piston [156]. The figure shows the maximum pressure obtained at the dead-point position and the residual pressure after unloading. It may be seen that the rebound of material per cycle decreases with distance from the dead-point position. This is the result partly of the increasing number of cycles and partly of the fact that rebound is also impeded by wall friction.

16.5 Pressure conditions in pelleting machines

A basic method for reducing the volume of forage and granular-farinaceous fodders is pelleting. Handling (transport, storage, feeding) of pelleted material is simple, and in the case of fodder mixtures the risk of separation of the individual components is prevented. The pressure required for pelleting varies between 500 and 1500 bar. The diameter of fodder flour pellets is 5–15 mm and of forage fodder pellets 20–50 mm.

Fodder flours are pelleted mostly in die rings, whose layout is illustrated in Fig. 340. The farinaceous material charged is compressed by a suitably adjusted

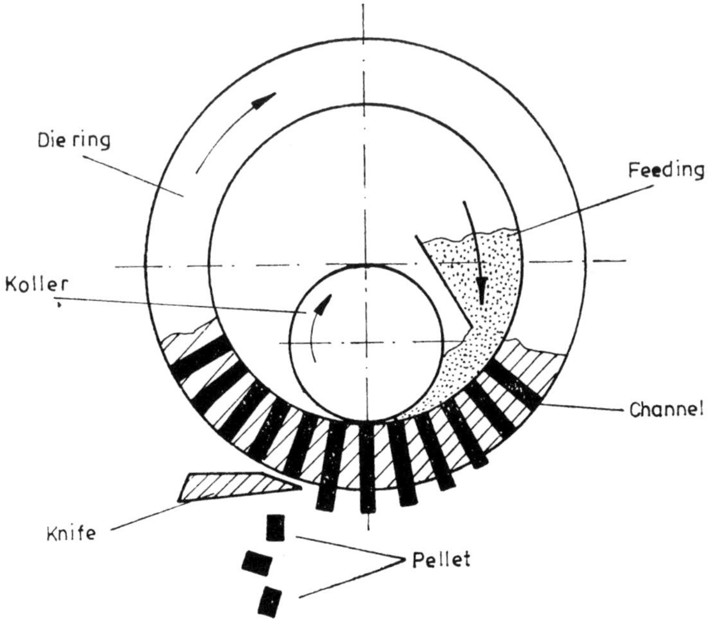

Fig. 340. Principles of pellet production

roller (similar to that of a chaser mill) and pressed into the boreholes of the ring. The compressed material emerging from the boreholes is cut by a knife on the outer side. Figure 341 shows the deformation conditions for the material under the roller [168]. The roller is set relative to the die ring with a gap y_r, and so a compressed layer (carpet) is formed on the running surface of the ring, whose thickness y_t exceeds the value y_r, due to rebound of the material. The layer thickness s of fresh material charged before the roller is gradually reduced as a result of compression by the latter. With increasing pressure the thickness y_t of the layer decreases gradually until it reaches the minimum value y_r. At point b the pressure attains its maximum value and the material is pressed by the roller into the extrusion channels. The pressure remains practically constant in the material during pushing and then decreases rapidly as the gap increases.

Figure 342 shows the distribution of pressure between the roller and die ring during the compression of fodder flour [168]. The width of the roller is 50 mm, the inner diameter of the die ring is 300 mm and the diameter of the boreholes is 4.5 mm. The compressibility of floury materials may be described by an equation of the form

$$p = e^{A\gamma+B} - D \tag{248}$$

while for straw-grit-molass mixtures the following equation is used:

$$p = a\gamma^b e^{c\gamma} \tag{249}$$

The constants appearing in the equations are determined experimentally. With regard to the fact that the volumetric weight increases proportionally to the decrease in layer thickness, i.e.,

$$\gamma/\gamma_0 = y_0/y$$

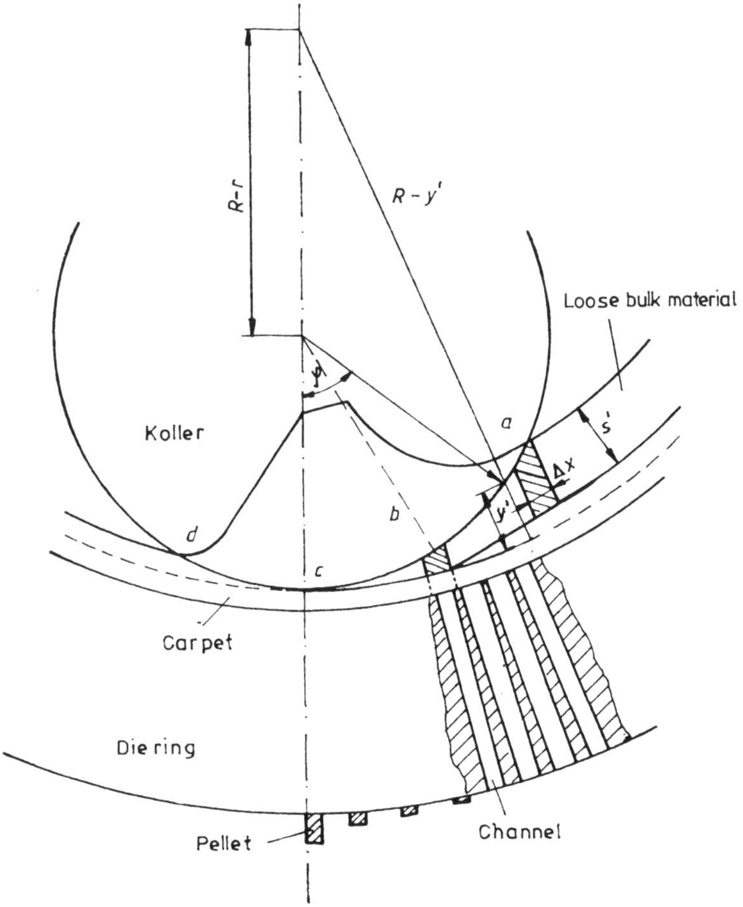

Fig. 341. Deformation of material under the roller of a pelleting machine

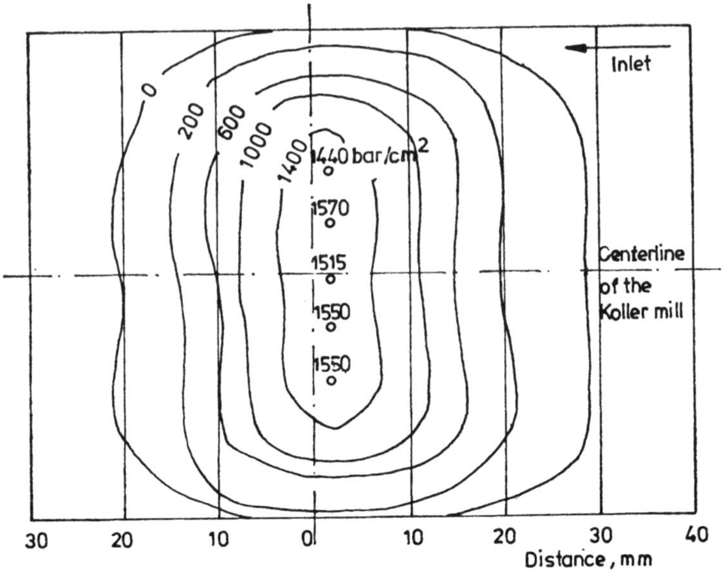

Fig. 342. Pressure distribution between the roller and die ring in pressing fodder flour

it is possible to write

$$y/y_0 = 1-\varepsilon = A\gamma_0/[\ln(p+D)-B]$$

For example, for the pelleting of fodder flour (γ_0=500 kg m^{-3}), the following relationship was found experimentally [168]:

$$y/y_0 = 5.05/[\ln(p+0.453)+5.842] \quad (250)$$

The thickness of the unloaded layer remaining after the roller has passed may be calculated as

$$y_t = y_r + p_{max}/C \quad (251)$$

where C is the volumetric compaction coefficient, whose value for fodder flour amounts to 30 000 daN cm^{-3}.

The gap y_r is generally selected in the range 0.4–0.8 mm, since the throughput decreases rapidly for wider gaps. The modulus of elasticity of fodder flours increases steeply with increasing pressure, as shown in Fig. 343.

Individual amounts of relative compaction have been marked on the curve. As may be seen, the material can be compressed to 40% of its original volume by a pressure of 1000 bar, whereas compression to 50% requires only 70 bar. The maximum pressure required for compression is determined primarily by the required durability of the pellets.

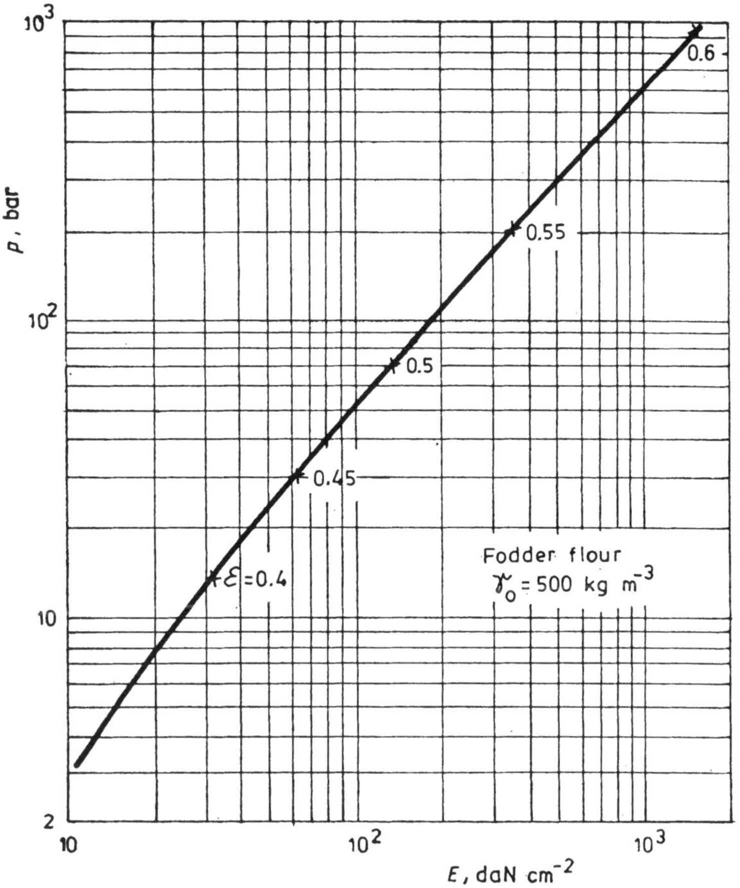

Fig. 343. Relationship between pressure and modulus of elasticity in pressing fodder flour

The throughput of a given die ring is determined by the operation of the roller and the thickness of the new layer charged. The greater the diameter of the roller, the thicker the layer which can be drawn under it. Any superfluous material is pushed before the roller; the throughput does not increase in this case. Figure 344 shows the throughput for one revolution as a function of the thickness of the layer charged for various roller diameters [168]. The curves become horizontal for layer thicknesses above a certain value, depending on the diameter of the roller, i.e., the throughput fails to increase further. The throughput is also influenced by the friction coefficient between the roller and the material. The higher the friction coefficient, the thicker the layer which can be drawn in by the roller, and the greater the throughput.

With increasing roller diameter the contact surface area increases and together

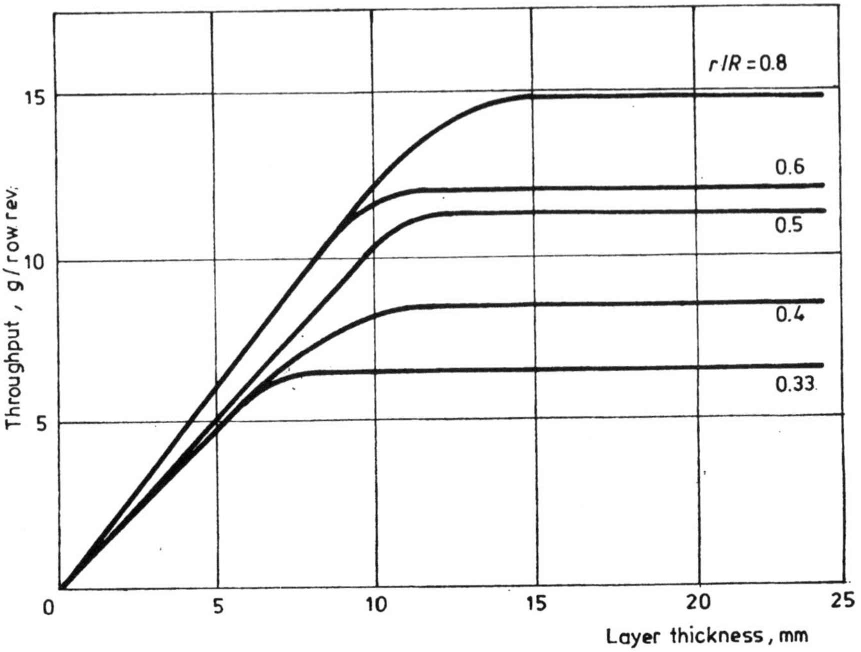

Fig. 344. Pellet throughput as a function of layer thickness

with it also the force acting on the roller and frame. According to investigations, the pelleting capacity per 1.0 t load on the roller is greatest when the relative roller dimension r/R is in the range 0.3–0.4; for r/R above 0.5, the pelleting capacity increases only slightly while the load on the roller is highly increased.

16.6 Effects of various parameters on the pelleting process

The density and durability of pellets fabricated under a given pressure depend on numerous factors, the most important of which are the material's structure, temperature, initial volumetric weight and moisture content, the velocity of pelleting and the duration for which pressure is maintained [169–172].

With increasing moisture content, many agricultural materials assume plastic properties, facilitating compression. Thus a given wet volumetric weight may be attained with a lower pressure if the moisture content is higher. However, a given volumetric weight relative to the dry content can generally be attained only by higher pressure if the moisture content is higher.

The compressibility increases also with heating for many materials, permitting the same volumetric weight to be attained by a lower pressure. This means that the specific work requirements of pelleting may be reduced by preheating the material. Figure 345 shows the variation of the specific work required to pellet straw and hay (pellet diameter 78 mm) as a function of temperature. In pressing farinaceous materials it has been found that the shear strength (characterizing the stability) is highest in relation to the pelleting pressure for the temperature range 60–80° C. This means that the energy consumption needed to attain greatest stability is also optimal in this temperature range.

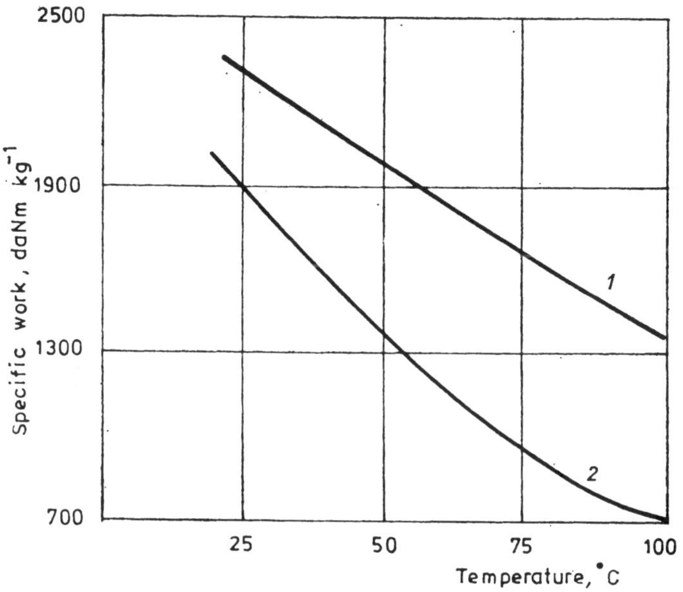

Fig. 345. Specific energy requirements for pelleting straw and hay as a function of temperature. *(1)* Straw, 12,6%; *(2)* hay, 9,6%

The duration for which pressure is maintained influences markedly the pressure and energy required to attain a given volumetric weight. Figure 346 shows the pressure required to attain a volumetric weight of 450 kg m^{-3}, relative to the dry content, as a function of moisture content, for various loading periods and temperatures [171]. As may be seen, the required pressure and also the energy requirements, are reduced considerably even by a loading period of 30 s.

On increasing the length of the boreholes the energy requirements increase and the pellets produced are also stronger. For given dimensions and a given number of revolutions, increase of throughput lowers the strength of the pellets [170].

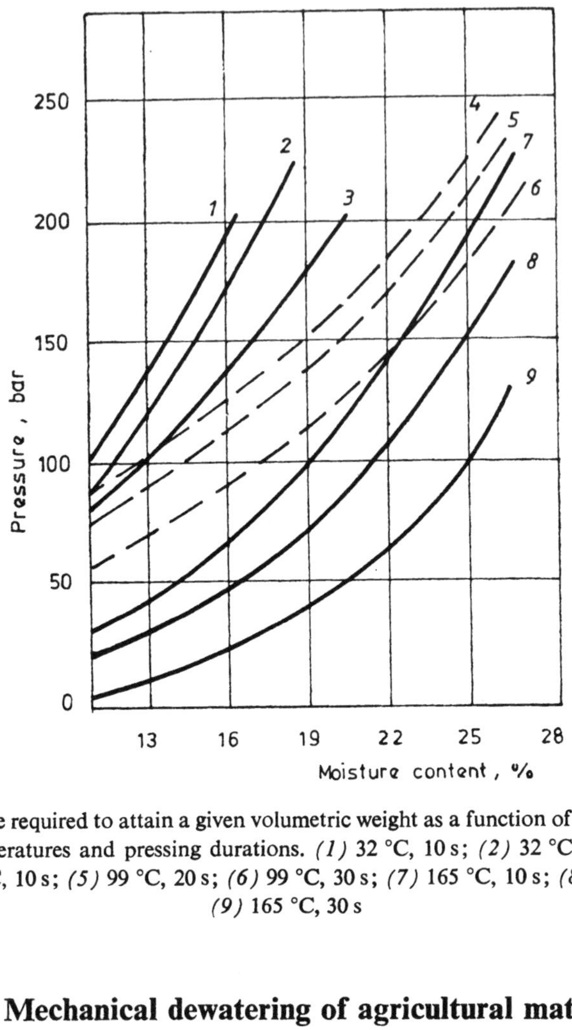

Fig. 346. Pressure required to attain a given volumetric weight as a function of moisture content, for various temperatures and pressing durations. *(1)* 32 °C, 10 s; *(2)* 32 °C, 20 s; *(3)* 32 °C, 30 s; *(4)* 99 °C, 10 s; *(5)* 99 °C, 20 s; *(6)* 99 °C, 30 s; *(7)* 165 °C, 10 s; *(8)* 165 °C, 20 s; *(9)* 165 °C, 30 s

16.7 Mechanical dewatering of agricultural materials

The classic method of obtaining juice mechanically is represented by the pressing of grapes; similar methods are also applied widely in the preserving industry to other fruits and products (e.g., tomatoes). Pressing has recently also been utilized for dewatering green fodders and preparing protein concentrations. In this context, it should be noted that the energy requirements of mechanical dewatering are about 1000 times less than those of thermal dewatering.

The mechanism of mechanical dewatering is as follows. Pressure applied to a plant body first breaks open the thin walls of the parenchymal cells, and the liquid contained in the latter migrates toward the free surface, in the direction of the pressure drop. Dewatering is greatest during this first stage. After the removal of a part of the juice from the material, the pressure is supported increasingly by the solid parts, whereby the pressure in the intermediate pores decreases and juice leaves the material by seepage at a decreasing velocity. Dewatering thus takes place over a certain time, and the period for which pressure is maintained plays an important role in determining the amount of juice extracted.

Flow created in porous materials may be described by Darcy's law in the simple form

$$v = k(dp/dx)$$

where k is the coefficient of permeability, and dp/dx the pressure gradient in the body. The permeability of biological materials shows complex relationships with various factors. The most important of these are the dimensions of the pores and the viscosity of the juice, but these factors do not remain constant during dewatering. The dimensions and shape of the pores vary as a result of deformation and creep of the material, and the viscosity is a function of temperature and of the concentration of the solid particles. The dry material content of the juice extracted also depends on numerous factors, such as the pressure and temperature, the duration of pressing, the thickness of the layer pressed, the extent to which the material is destroyed, etc.

Because of the complexities mentioned above, semiempirical and empirical methods must be used. The results are less capable of generalization, but are well utilizable in practice.

To describe the filtration resistance of compressed material, Körmendy [192] suggested the relationship

$$dq/dt = (F/\alpha\mu)(p_1 - p_2)/(q_i/F)$$

where dq/dt is the mass flow velocity over surface area F, α the specific resistance and μ the dynamic viscosity, $p_1 - p_2$ is the pressure drop in the layer, and q_i the initial (total) quantity of liquid to be pressed out.

Given appropriate experimental data, the specific resistivity to dewatering may be calculated by means of the above equation. For example, according to measured data the specific resistivity on compressing Jonathan apples varies according to the empirical relationship [192]

$$\log(\alpha\mu) = 5.13\times10^{-4}(100y)^{1.98} + 7.56 \text{ (dyn s cm}^2\text{ g}^{-2}\text{)}$$

where $y = q/q_i$ is the relative yield of juice.

The empirical relationship shows clearly that the specific resistivity increases in proportion to the square of relative yield of juice.

In describing drying processes the equilibrium moisture content of a material as a function of relative air humidity plays an important role. The equilibrium moisture content as a function of pressure (similarly to the pF number for soils) may also be defined in the case of dewatering by pressing. In practice, the equilibrium moisture content relative to a 30 min pressing period is generally adopted [174]. From the appropriate curves the quantity of water which may be removed by a given pressure in the case of a given initial moisture content may be evaluated simply. Figure 347 shows juice yield curves for apple pulp and the variation of the equilibrium moisture content as a function of pressing pressure [183]. The variation of juice yield over time may be calculated from the equation

$$q(t) = q_\infty(1-ae^{-t/T}) \tag{252}$$

where a and T are constants, and q_∞ is the maximum juice yield obtainable under the given pressure. Juice pressed from agricultural materials contains both dissolved and solid dry material, and therefore the dry content of the initial material also decreases during pressing (see Fig. 347(a)).

The variation of the equilibrium moisture content as a function of pressure may be described by an equation of the form:

$$U_e = U_0(1-ap^n)$$

where a and n are constants. For the curves given in Fig. 347(b) $n=0.3$, $a_1=0.088$ and $a_2=0.1676$ ($t=\infty$). The decrease of moisture content may also be described, similarly to eqn. (20), by the empirical equation [176]

$$(X-X_e)/(X_0-X_e) = e^{-ap^bH^ct^d}$$

where p is the pressure, H the layer thickness, t is time, and a, b, c and d are constants. The exponent d of time usually varies between 0.3 and 0.35, meaning that in the initial period of pressurization dewatering is rapid and then decreases greatly.

The quantity of liquid removable from freshly mowed alfalfa has been studied using pressing cylinder 110 mm in diameter [175]. Figure 348(a) shows the dewatering curves for pressures of 10 and 120 bar. It may be seen that the stage of rapid dewatering is complete within 2 min, and subsequently relatively little liquid may be obtained. Figure 348(b) shows the variation of the quantity of water removable as a function of pressure maintained for 2 min. For pressures greater than 35–40 bar the extent of dewatering increases only slightly, and so increasing the pressure further is unfavorable from the point of view of energy consumption.

The energy requirements of dewatering depend greatly on the extent of liquid removal, since at the beginning of the process little energy is needed, and then increasingly more. The average energy demands of must pressing lie in the range

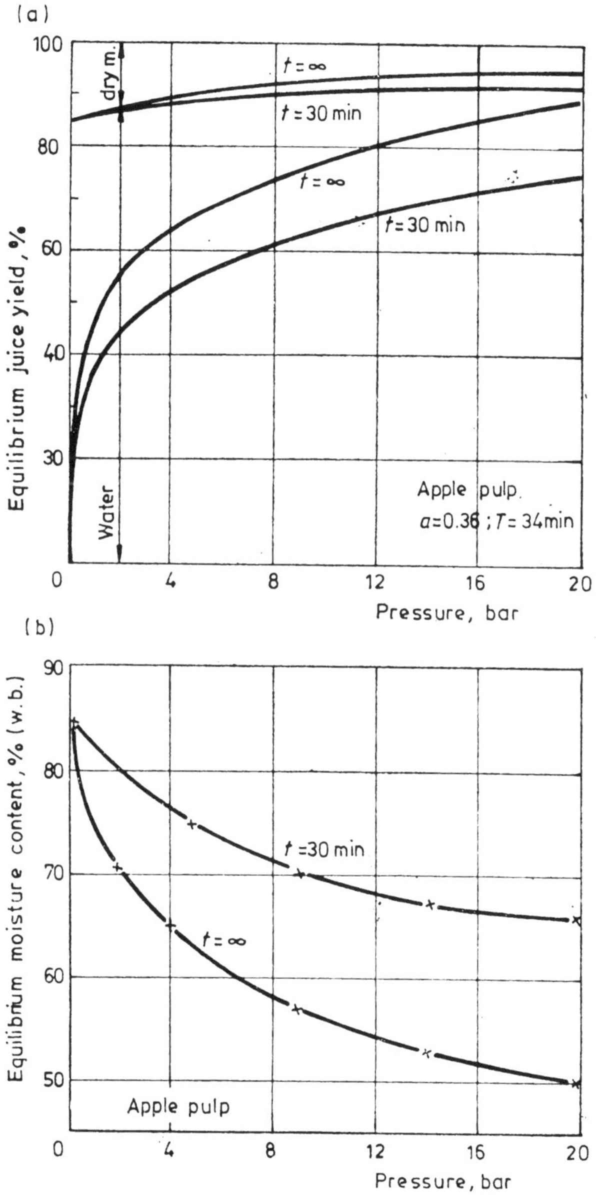

Fig. 347. Equilibrium juice yield for apple pulp and equilibrium moisture content in relation to pressing pressure

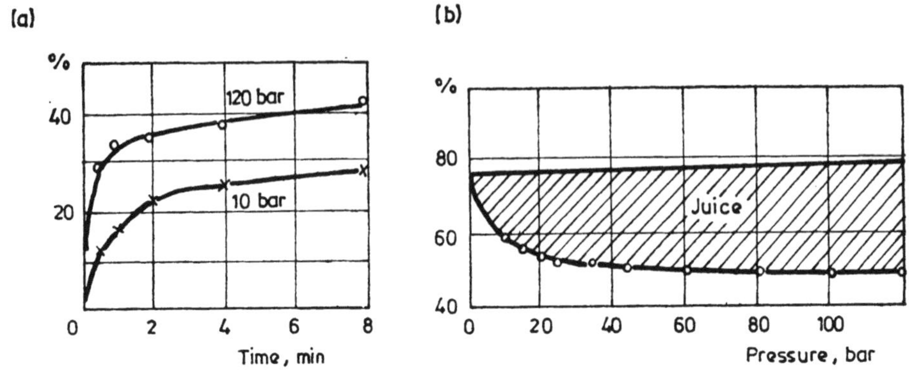

Fig. 348. Dewatering curves for alfalfa

0.24–0.30 kW h hl^{-1} [177]. For dewatering of alfalfa with 76% moisture content the energy needed to extract 1 kg of moisture after the removal of about half the total liquid content is 150–200 daN m, while the work required for 1 t of green material is 0.2–0.25 kW h t^{-1}. The energy requirements of mechanical dewatering may be reduced considerably by prior destruction of the texture of the material being pressed. For example, during mechanical dewatering of alfalfa it is advisable to combine cutting with crushing, which facilitates all types of water removal.

17. CUTTING OF AGRICULTURAL MATERIALS

Cutting of agricultural materials is one of the most frequent operations, and is applied nearly always during harvesting, in the separation and subsequent comminution of plant components. Cutting (with significant energy consumption) is also the main operation in fodder preparation. Other processing operations also frequently require cutting.

During cutting a cutting edge (knife) penetrates into a material, overcoming its strength and thereby separating it. During cutting various deformations occur in the material, depending on the form of the cutting edge and the kinematics of the process. Thus it is not possible to speak generally of the cutting resistance of a material except in relation to a given shape of cutting edge and given kinematics of cutting.

17.1 Cutting methods

Figure 349 shows the most frequent forms of cutting. The first figure (a) shows a process involving countermoving blades, where both sets of blades participate in cutting. The second figure (b) illustrates cutting by means of a resting and a moving blade, where the material is supported by the resting blade. The third figure (c) shows the cutting of thin layers (e.g., beet cutting), where the stress distribution around the cutting edge is significantly distorted by the free surface found close to the cutting plane. The material may be fixed more or less rigidly. The fourth (d) figure presents the recently widespread method of free cutting, where one end of a relatively long stalk is fixed and countersupport is ensured by the moment of inertia of the stalk. In this case the velocity of the cutting edge must be high (20–40 m s^{-1}).

Products may be cut individually or in bundles depending on the type of material and the technological process. An example of the cutting of bundles is provided by the chaff-chopper, which cuts bundle compressed into a quadrangular cross-section.

Material is first compressed and deformed under a cutting edge, depending

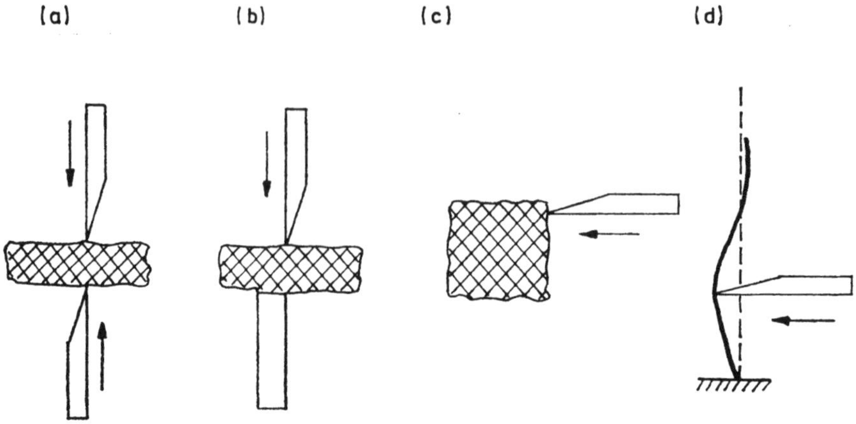

Fig. 349. Various forms of cutting

on the latter's shape and velocity; cutting begins only subsequently. The cutting edge may move normally to the material or at a certain angle. In the latter case, the cutting edge will be displaced during cutting in the direction of the surface cut.

Figure 350 shows the kinematic relations obtained during cutting [179]. A given point on a cutting edge rotating around a point O has a circumferential velocity v which may be decomposed into normal and tangential components. The angle λ enclosed by the vectors, termed the angle of slide, is given by

$$\tan \lambda = v_t/v_n = c/\sqrt{r^2-c^2} \tag{253}$$

The value $\varepsilon = \tan \lambda$ is termed the sliding coefficient. The peripheral force P acting at point A is the sum of two components:

$$P = P_1 + P_2 = N \cos \lambda + T \sin \lambda$$

The force N acting normally to the cutting edge is the product of the specific cutting resistance per cm and the cutting length:

$$N = pl$$

while the tangential force is

$$T = \mu N$$

where μ is the friction coefficient. Using these relations, the peripheral force may be written in the form

$$P = pl(\cos \lambda + \mu \sin \lambda) \tag{254}$$

As may be seen from Fig. 350, the resultant force and the displacement vector do not coincide in a straight line ($\lambda \neq \varrho$). During sliding, the edge angle β of the cutting edge is smaller in the direction of motion: the reduced edge angle may be calculated from the equation

$$\tan \beta_1 = \tan \beta \cos \lambda$$

The relative reduction of the cutting angle may be expressed by the factor

$$K = (\beta - \beta_1)/\beta$$

This factor is correlated with the relative variation of the normal force (Fig. 351). The latter is given by

$$\psi = (N_0 - N_1)/N_0$$

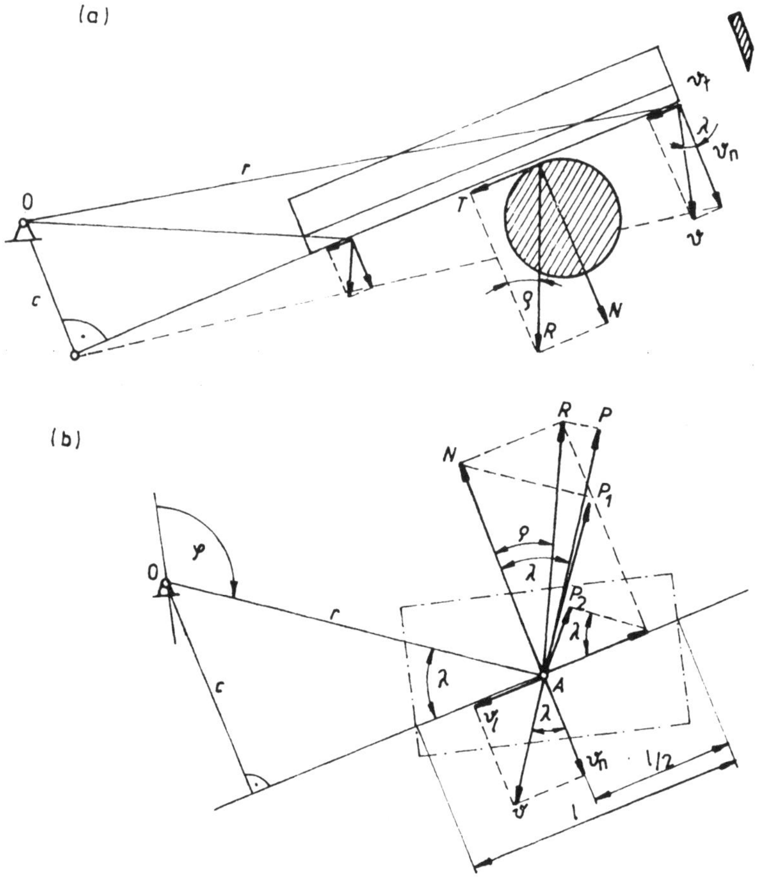

Fig. 350. Kinematic relations in cutting

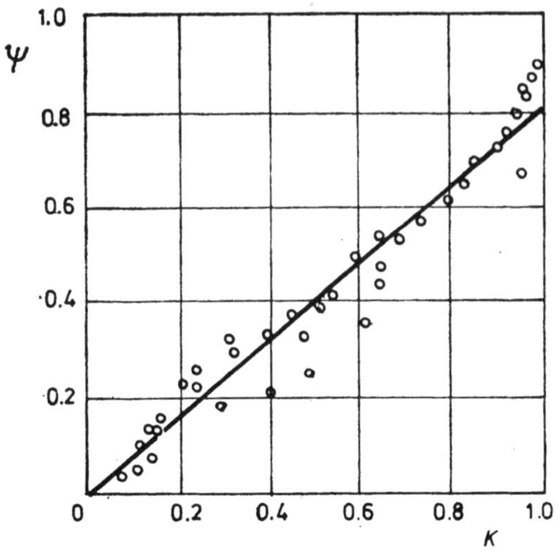

Fig. 351. Relationship between reduction of edge angle and reduction of cutting force

where N_0 is the normal force for displacement in the direction normal to the cutting edge, and N_1 the normal force during sliding. As may be seen, the normal force may be reduced greatly by cutting with a large amount of sliding.

17.2 Deformations caused by cutting

Cutting knives are generally sharpened on one side, with a sharpening angle β and an edge thickness δ. The penetration of a knife into a material causes deformation of the latter and various forces are acting on the surface of the knife, as illustrated schematically in Fig. 352. The normal force acting on the inclined face of the knife is the sum of the horizontal and vertical force components:

$$N = P_v \sin \beta + P_h \cos \beta$$

while the tangential force arising on it is

$$T_2 = \mu N = N \tan \varphi$$

where $\mu = \tan \varphi$ is the friction coefficient. On the vertical side of the knife, the tangential force is

$$T_1 = \mu P_h$$

The vertical component of the tangential force T_2 is

$$T_2' = T_2 \cos \beta = \mu[(1/2) P_v \sin 2\beta + P_h \cos^2 \beta]$$

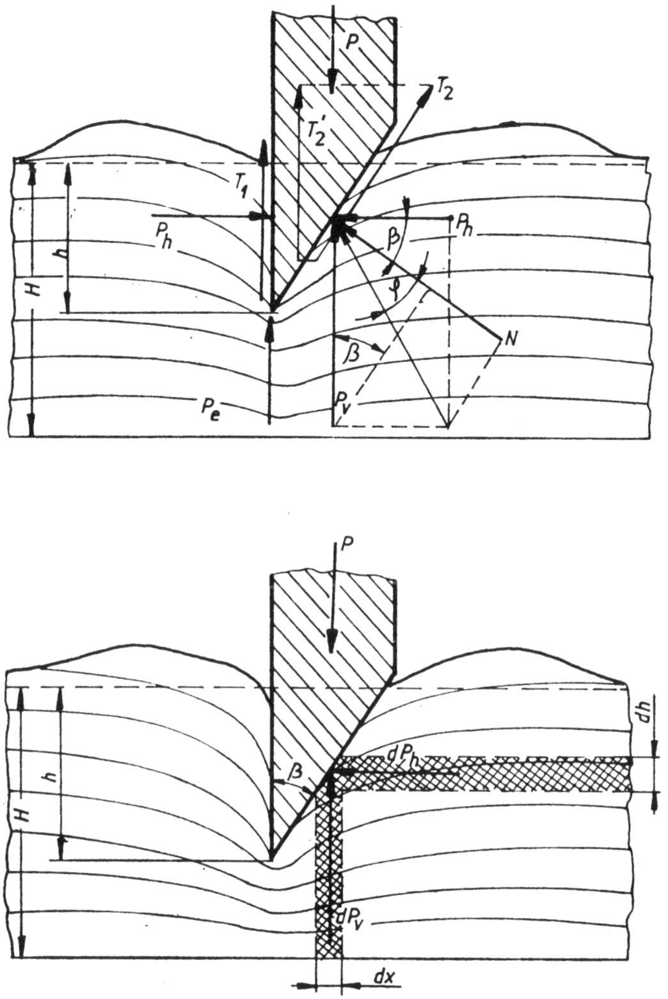

Fig. 352. Force relations for a knife penetrating into a material

At the instant of cutting, a force

$$P_e = F\sigma_B = \delta l \sigma_B$$

appears on the knife edge, where δ is the thickness of the edge, σ_B the yield strength of the material under the edge. Equilibrium of the vertical forces may be expressed by the equation

$$P = P_e + P_v + T_1 + T_2' \tag{255}$$

The forces P_v and P_h may be determined approximately as follows. Assume that Hooke's law is applicable to the material or

$$\varepsilon = h_x/H = \sigma/E$$

where E is the mean modulus of deformation relative to a given loading interval. The elemental force acting on a surface of width dx and of unit length is

$$dP_v = \varepsilon E dx = \varepsilon E \tan\beta \, dh$$

Substitution of the preceding expression for ε yields

$$P_v = (E/H)\tan\beta \int_0^h h_x dh = (E/2H)h^2 \tan\beta \tag{256}$$

The horizontal force increment is

$$dP_h = \nu\varepsilon E dh$$

where ν is the Poisson's ratio. Then, similarly to the above derivation,

$$P_h = (\nu E/H)\int_0^h h_x dh = (\nu E/2H)h^2 \tag{257}$$

The Poisson's ratio is not high in the direction of the stalk, especially in the case of short chaff. For maize silage values of ν=0.1–0.15 may be used. The equilibrium equation (eqn. (255)) may be written for unit length, with eqns. (256), (257) and the expression for T_2' taken into account, in the form

$$P = \delta\sigma_B + (E/2H)h^2[\tan\beta + \mu\sin^2\beta + \nu(\mu + \cos^2\beta)] \tag{258}$$

The edge thickness of the knife is generally δ=50–150 μm. The stress σ_B for maize stalks is 250–500 daN cm^{-2}. Thus a force of 2.5–5.0 daN falls on every cm length of cutting edge. The value of E is 3.5–4.0 daN cm^{-2}. The first term of eqn. (258) gives the useful cutting force; the second term expresses the force used to overcome other sources of resistance. The second term depends on the square of the preliminary compaction h lasting until the beginning of cutting proper, whose value varies linearly with the layer thickness H. Consequently, the additional resistance increases rapidly with increasing layer thickness, and the efficiency of cutting is lowered.

Figure 353 shows the deformation of root products during cutting into thin layers. Penetration of the knife into the material first causes compaction, until a given pressure is reached at which rupture occurs. In advancing, the knife compacts successive regions of material until rupture occurs again. The ruptured surface typically shows conchoidal form. The distance between individual ruptures is a function of the cutting thickness and the angle of the edge.

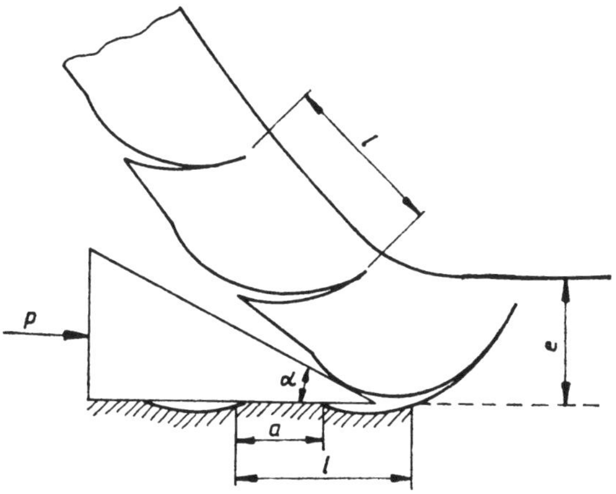

Fig. 353. Separation of thin layers of root products (slicing)

17.3 Energy requirements of cutting

As has been seen, two stages may be distinguished in the cutting process: the first stage involves preliminary compaction of the material until a pressure is reached at which the material under the edge yields, while the second stage concerns the motion of the edge in the material (cutting). These two stages may be observed clearly from static cutting diagrams. Figure 354 shows the static cutting diagram for a bundle of maize stalks. The material is compressed up to a height h until the cutting resistance is overcome, and the energy required is given by the area A_c under the curve. The energy requirements of effective cutting are given by A_v. The total work is thus

$$A = A_c + A_v$$

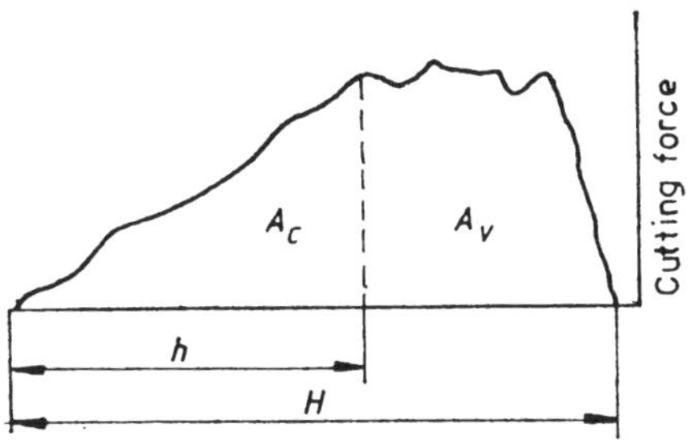

Fig. 354. Static cutting diagram for maize-stalk bundles

while the proportion of useful cutting work is

$$\eta_v = A_v/(A_c+A_v) \tag{259}$$

The specific energy requirements of cutting may be obtained by dividing the total work by the cross-sectional area of cutting:

$$A_f = A/F$$

Figure 355 shows the variation of the static cutting characteristics as a function of the layer thickness [66]. With increasing preliminary compaction height the proportion of useful cutting work is lowered, and the specific energy consumption increases.

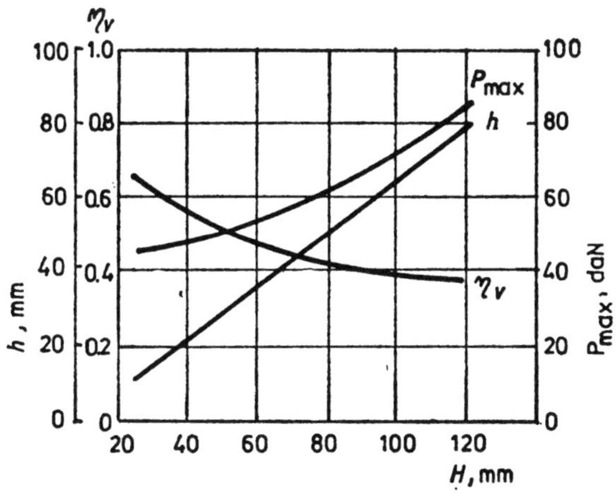

Fig. 355. Static cutting characteristics as a function of layer thickness

In practice, cutting is not a static but a dynamic process. With increasing cutting velocity, preliminary compaction decreases as a result of the material's inertia and plastic behavior, whereby the energy requirements of cutting are lowered. The latter effect operates as follows. Owing to the low velocity of deformation of plastic materials, impact loading by a rapidly moving knife edge propagates only slowly in the material, i.e., is concentrated in the material found around the edge. Consequently, cutting may be effected with lower energy consumption.

Figure 356 shows the most important characteristics of cutting as functions of cutting velocity for green maize stalks with 63% moisture content [66]. The cut cross-sectional area amounts to 70 cm^2, the thickness of the knife is 4 mm, its sharpening angle is 25°, and the edge thickness is 80 μm. It may be seen that

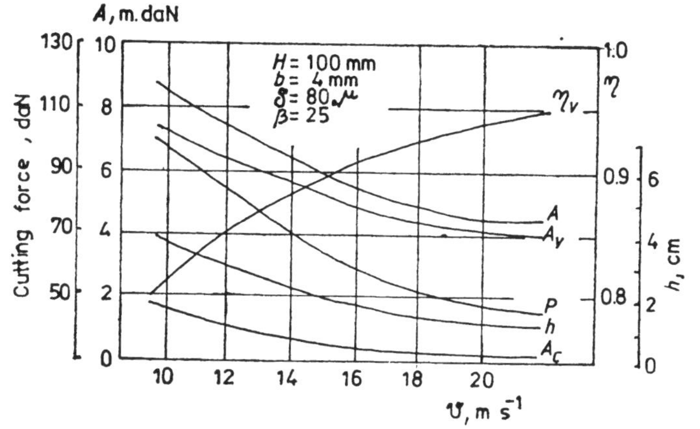

Fig. 356. Effect of cutting velocity on cutting characteristics of green maize stalks

with increasing cutting velocity the energy requirements decrease considerably, and the proportion of useful cutting work reaches 95%.

Numerous factors in addition to the above influence the energy requirements of cutting, some of which are related to the mechanical properties of the material, while others depend on the geometry and adjustment of the cutting edge and on the kinematic conditions. The mechanical properties depend on the type of material, its stage of growth and moisture content, the location of the cutting (close to the root or higher), etc.

Figure 357 illustrates the energy required to cut alfalfa stalks as a function of their thickness [178]. By dividing the various values by the corresponding cross-

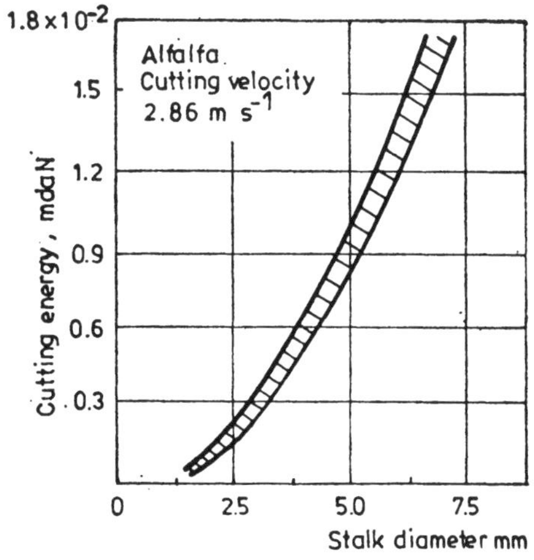

Fig. 357. Energy requirements for alfalfa as a function of stalk diameter

sectional areas, the result is obtained that the specific cutting resistance increases slightly with the diameter (0.041–0.046 daN m cm^{-2}).

The cutting resistance of younger plants may be significantly lower than that of older plants (by as much as one-half). This is a result mainly of corresponding variations in the texture, primarily in the proportions of fibrous and ligneous material. The thickness and texture of plant stalks also vary as functions of height, and so the cutting resistance also depends on the location of the cut: it is highest close to the soil, and decreases going upwards.

The thickness of the cutting edge influences the cutting resistance in various ways. The cutting force is practically constant for thicknesses up to 70–80 μm but for greater thicknesses it increases significantly (Fig. 358) [178]. Consequently, it is not advisable to use knife edges that are too thin, since they represent no improvement in terms of energy consumption while they wear rapidly and deform easily. A badly worn, thickened edge consumes much surplus energy. With increasing knife thickness the additional deformation increases, whereby the energy spent in cutting increases. Figure 359 shows the energy requirements of cutting for maize stalks, as a function of knife thickness. The increase is not very great once the knives become thicker [66].

The angle of sharpening (bevel angle) is one of the most important parameters, not only in terms of energy consumption but also as concerns the life of the knife. Figure 360 shows the maximum cutting force as a function of the sharpening angle [178]. According to measurements, the energy requirements increase rapidly for angles above 30°, and so in practice the angle is chosen between

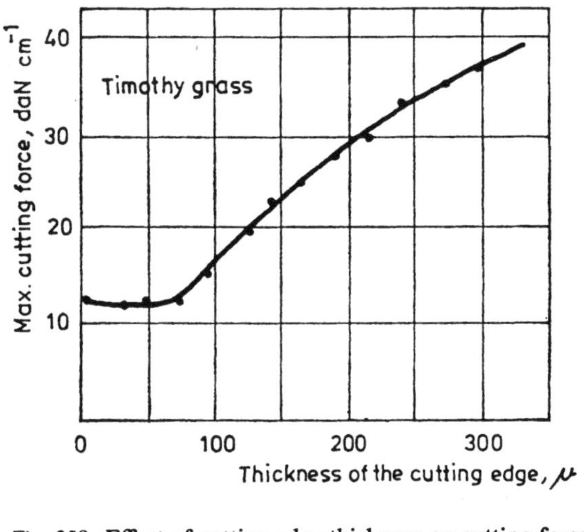

Fig. 358. Effect of cutting edge thickness on cutting force

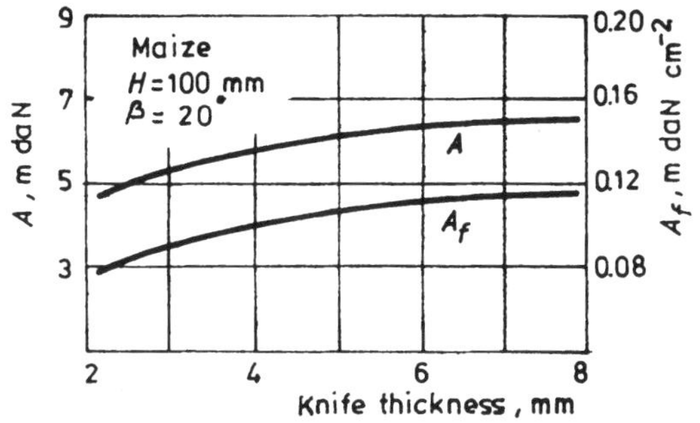

Fig. 359. Energy requirements for cutting maize stalks as a function of knife thickness

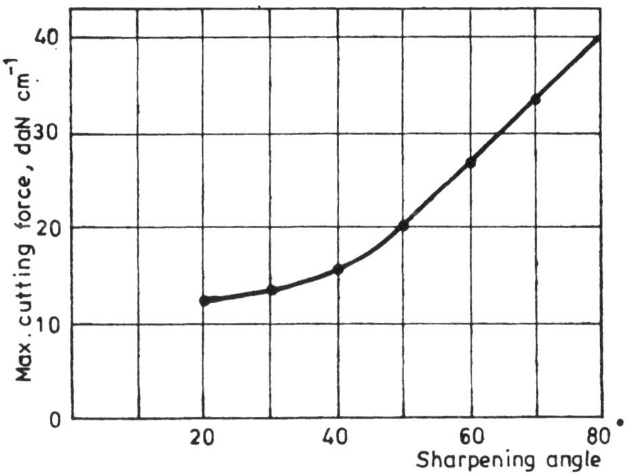

Fig. 360. Effect of sharpening angle on maximum cutting force

20 and 30°. The use of angles less than 20° is not advised, since the edge will then be deformed easily and thereby the life of the knife reduced.

Knives sharpened on one side are generally mounted at a positioning angle γ in order to reduce friction on the side face. In this case, the additional deformation is determined by $\gamma+\beta$, or the angle $\varphi=90°-(\gamma+\beta)$. It is easily admissible that the smaller the angle φ, the greater the additional deformation and the higher the energy requirements. Figure 361 shows the specific energy requirements as a function of the angle φ. In practice, φ must not exceed 60° [66].

A certain clearance is always present between a knife and a counterblade which may influence the quality and energy requirements of cutting. If the clearance is large, stalks will be bent rather than cut by the knife, meaning that further additional deformation occurs. This effect is more pronounced, the blunter the knife. Therefore keeping the clearance at a low value (a few tenths of 1 mm) is an important requirement.

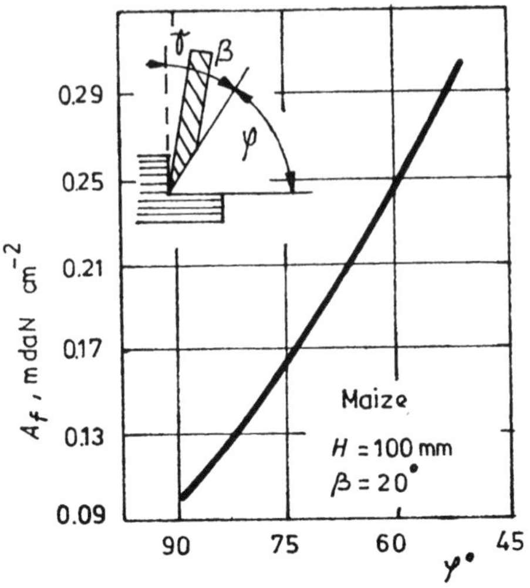

Fig. 361. Relationship between positioning angle of knife and specific energy requirements

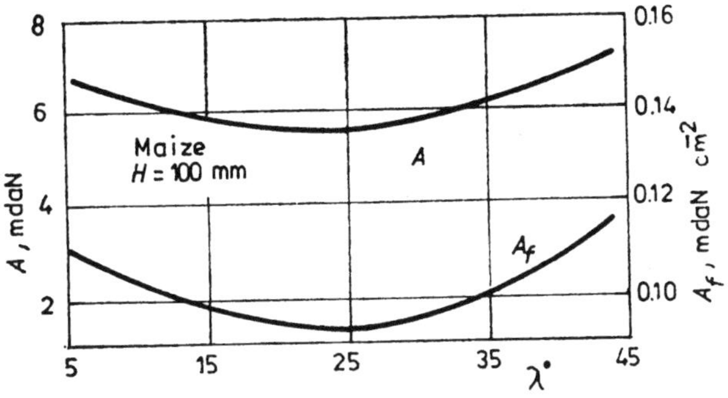

Fig. 362. Effect of sliding angle on specific energy requirements

It was seen earlier that during a sliding cut the normal force acting on a knife may be reduced significantly by increasing the angle λ. However, this does not imply that from the point of view of energy consumption the angle and thus the sliding should be as great as possible, since when sliding increases, sliding losses also increase, and so an optimal angle λ is found, generally between 20° and 25° (Fig. 362) [66].

17.4 Free cutting

Rotary grass mowers contain no counterblade, and so the process is one of free cutting. Free cutting is possible only if a reaction force corresponding to the maximum cutting force arises in the material. The components of this reaction force are as follows:

(a) the mass inertia (resistance to acceleration) of the stalks being cut; and

(b) static reaction forces due to bending of the stalk and angular displacement of the cross-sectional area cut.

Figure 363 shows the deformation of a stalk and the forces acting on it. The differential equation of motion may be written in the form

$$\Delta m(d^2x/dt^2) = S(\delta) - P_h(x) - P_h'(x) \qquad (260)$$

where Δm is the equivalent mass of the accelerated parts of the stalk, $S(\delta)$ the cutting force, $P_h(x)$ the horizontal force originating from bending of the stalk, and $P_h'(x)$ the horizontal component of the tension force R originating from the motion along a circular arc. The horizontal force due to bending is given by

$$P_h(x) = (3EI/h^3)x$$

where E is the modulus of deformation of the stalk under bending, I the moment of inertia of the cross-sectional area, and h the height of the cutting

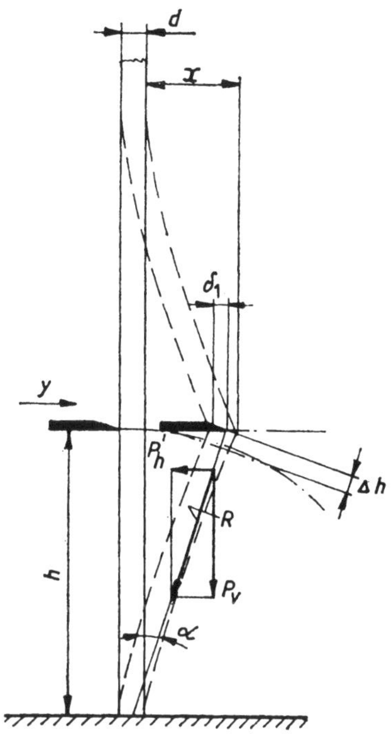

Fig. 363. Free cutting of standing forage materials

The value of P_h' may be calculated approximately as follows. The force required for elongation of the stalk by Δh is

$$R = FE(\Delta h/h)$$

The horizontal component of this force is

$$P_h' = \left(1/h - 1/\sqrt{h^2+x^2}\right) FEx$$

The equivalent mass of the accelerated parts may be calculated approximately by dividing the stalk into elements of mass Δm_i. If the acceleration of the elemental parts are known, then the kinetic energy of the stalk is

$$E = (1/2) \sum_{i=1}^{n} \Delta m_i \dot{x}_i^2$$

If an equivalent mass Δm moving at velocity x_0 is assumed to be present at the point of deflection, then, from equality of the energies,

$$\Delta m = \sum_{i=1}^{n} \Delta m_i \dot{x}_i^2/\dot{x}_0^2$$

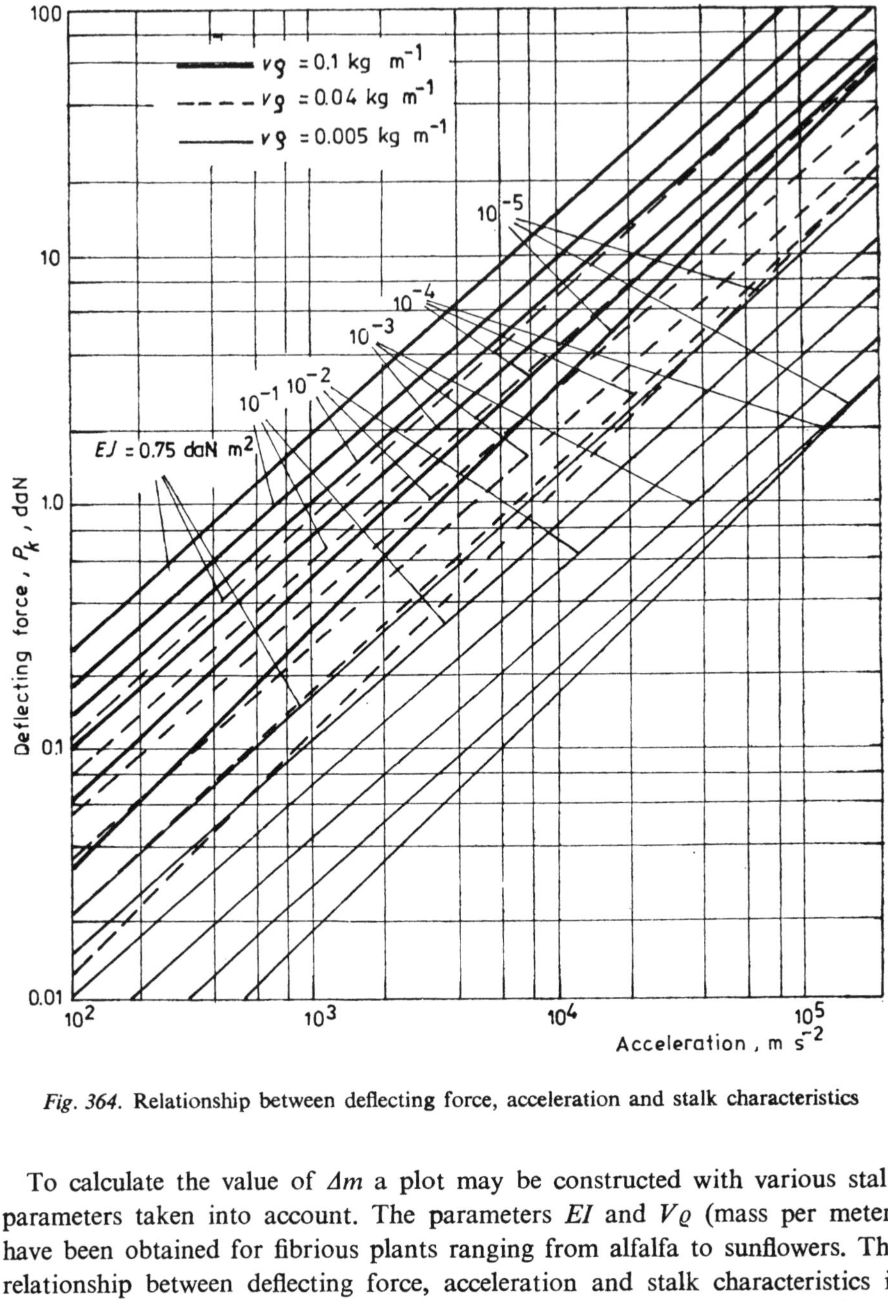

Fig. 364. Relationship between deflecting force, acceleration and stalk characteristics

To calculate the value of Δm a plot may be constructed with various stalk parameters taken into account. The parameters *EI* and $V\varrho$ (mass per meter) have been obtained for fibrious plants ranging from alfalfa to sunflowers. The relationship between deflecting force, acceleration and stalk characteristics is illustrated in Fig. 364, where the deflecting force measured on the vertical axis equals the maximum cutting force [180]. The cutting force is generally known from experiment, and so the required acceleration may be read from the diagram. Knowing the acceleration, Δm may be obtained from the relation $\Delta m = S/a$.

To calculate the critical velocity of the knife, the following expression, obtained by the method of dimensional analysis [180], may be applied:

$$\dot{y}_{cr} = v_{cr} = C\sqrt{dS_{max}/\Delta m} \tag{261}$$

where d is the diameter of the stalk (m), S_{max} the maximum cutting force (daN), Δm the equivalent mass (kg), and C is a constant, whose value may be accepted between 3.13 and 4.43. The relevant characteristics for a few plants are listed in Table 20 [180].

Table 20

Plant	Moisture content (%)	V_Q (kg m^{-1})	d (mm)	EI (daN m^2)
Sunflower (at flowering)	81	0.085	12	0.23
Alfalfa (after flowering)	54	0.050	3–6	0.009–0.004
Wheat (ripe for harvesting)	47	0.004	3.5	0.004

Using the data for wheat stems, with a maximum cutting force $S_{max} = 3$ daN, the acceleration at the deflection point as read from Fig. 364 is $a = 1.3 \times 10^5$ m s^{-2} and the equivalent mass is obtained as

$$\Delta m = 3g/1.3 \times 10^5 = 2.3 \times 10^{-4} \text{ kg}$$

Now, from eqn. (261), $v_{cr} = 21 \sim 30$ m s^{-1}, depending on the constant C. This value agrees well with the minimum cutting velocity found experimentally.

For plants with thick stalks (e.g., maize, sunflowers), significant horizontal and vertical forces may arise at the point of cutting as the knife lodges in the material. The vertical force also acts on the knife and increases the load on it. Figure 365 shows the forces appearing during cutting of a sunflower stalk, as functions of cutting velocity (knife thickness 3 mm; cutting angle 15°) [180]. For the given range of velocities the cutting force decreases only slightly, but the horizontal force appearing at the cutting point decreases considerably. Thus the difference between the cutting and horizontal forces increases with velocity, showing that countersupport is ensured increasingly by the mass inertia of the material. The vertical force also decreases with increase of the cutting velocity as the stalk deflection lessens, and may even assume a negative value. This may be explained by the fact that the upper part of the stalk is accelerated upwards by the knife, which is bevelled on its upper side, and the reaction force presses the blade, and with it the lower part of the stalk, downwards.

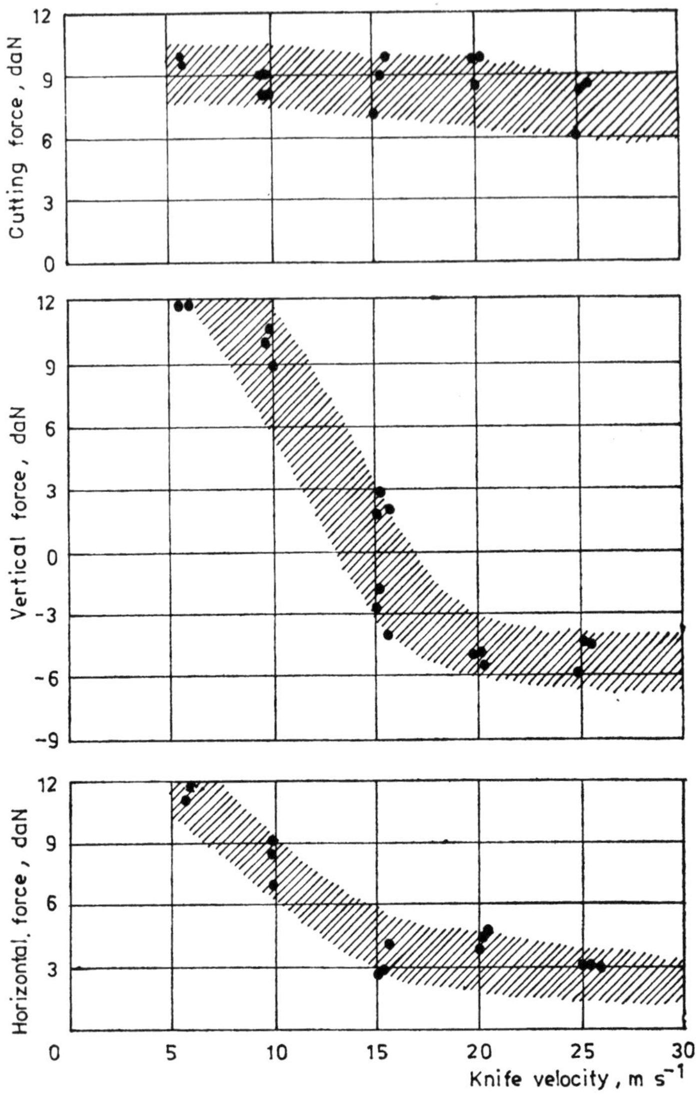

Fig. 365. Forces appearing in cutting sunflower stalks

According to experiment the quality of cutting also depends greatly on the knife velocity. For velocities close to or below the critical value, the stalk is broken and torn by the knife; the cut surface will not be smooth or perpendicular to the longitudinal axis. These phenomena are also observed at higher velocities if dulled knives are used; therefore cutting velocities applied in practice should exceed the critical value considerably.

17.5 Energy requirements of forage harvesters

The energy requirements of the various chaff cutters and silo-fodder harvesting machines differ significantly from those of net cutting. The reason is that in addition to the energy required to overcome the cutting resistance, the energy requirements of other auxiliary operations (transport, preliminary compaction) and sources of loss (ventilation and friction) must be included. These energy requirements depend variously on the load and the rate of operation, and so the measured total energy consumption follows no strictly generalizable laws. Nevertheless, a primary parameter influencing the energy consumption can be found.

Figure 366 shows the energy requirements of the cutter head in a self-propelled forage harvester as a function of throughput relative to the dry-material content

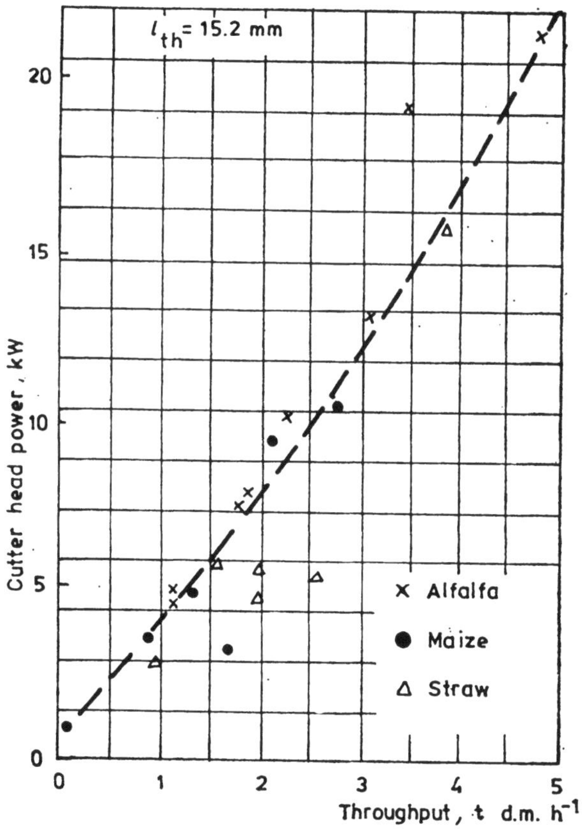

Fig. 366. Energy requirements of cutter head in a self-propelled forage harvester as a function of dry-material throughput (with 15.2 mm theoretical chaff length)

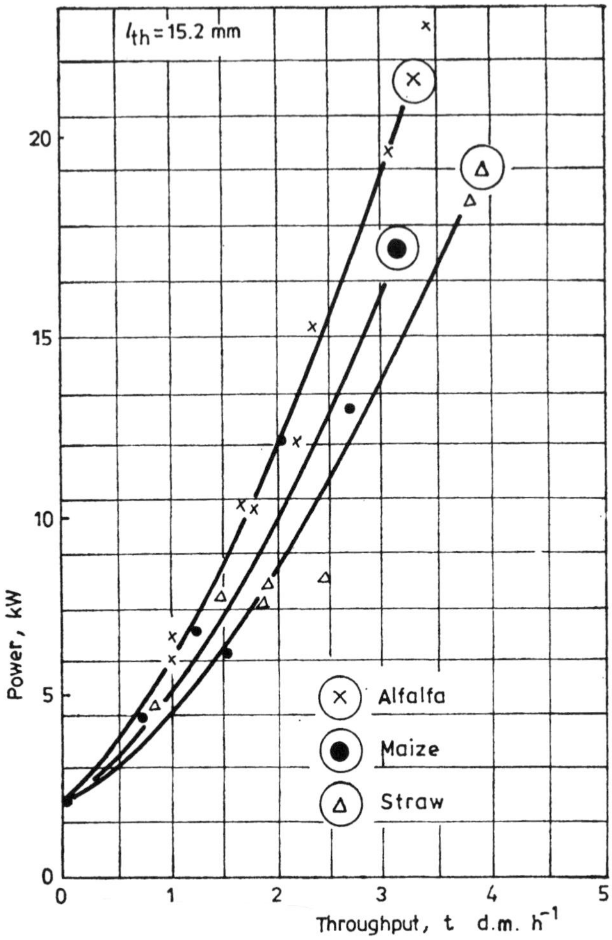

Fig. 367. Total energy requirements of a forage harvester as a function of throughput

[129]. Despite the scatter of the measurement points, it may be established that the energy required for cutting plants of various sorts and differing moisture contents varies similarly as a function of the dry content. The curves deviate from each other only to a slight extent and so the throughput in terms of dry material is of paramount and determining importance as concerns the total energy requirements (Fig. 367). The above data apply to a defined theoretical chaff length. The chaff length determines the number of cuts required for comminution of a given volume of material, and so the cutting energy is inversely proportional to the chaff length: the shorter the chaff, the more cuts necessary, and the higher the energy requirements of cutting.

Fig. 368. Idle-run and total energy requirements of a rotary grass mower as functions of peripheral velocity

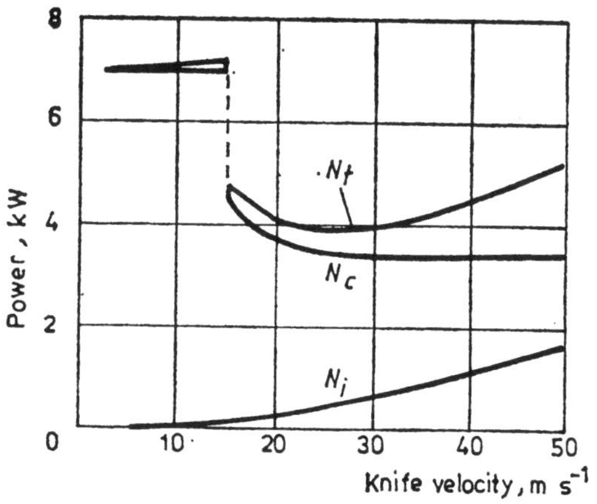

Figure 368 shows the idle-run and total energy requirements of a rotary grass mower as a function of peripheral velocity [180]. The idle run energy increases nearly as the square of the rate of revolution, because in this case not only air-resistance losses but also other friction-type losses (in the drive-gear) are involved. The cutting energy decreases suddenly at a critical cutting velocity, and then remains practically constant for velocities above 30 m s^{-1}. The total energy required shows a minimum in the range 20–30 m s^{-1}.

Recently the application of the chaff cutting combined with crushing has become widespread. Chaff cutting comminutes the material, while crushing destroys the texture. The crushed material can be compacted better, and also desorbs water more easily.

The crushing process involves very complex deformation phenomena, reliable mechanical characterization of which has not yet been achieved. In self-propelled forage harvesters so-called "recutters" are often used, and in these, friction phenomena predominate. Consequently, their energy consumption is high. In practice, the specific energy requirements double and the throughput is reduced by one-half on the application of a recutter.

Crushing drums may also be designed so that crushing is achieved by a progress similar to shearing, whereby the energy requirements may be reduced considerably, as shown by measurements for such equipment developed for use with forage harvesters [181].

18. GRINDING (COMMINUTION) OF AGRICULTURAL MATERIALS

One of the most important operations in fodder preparation is the comminution of granular products and forage fodders. In the overwhelming majority of cases comminution is done by hammer mills; only wheat is ground by cylindrical comminution equipments (roller mills) in the milling industry.

In preparing fodder it is desirable to attain as homogeneous a grain composition as possible, without the formation of floury material or dust. If fodder is to be granulated after grinding, it is advisable to prepare a finer grit, since this can be compacted better and gives a granulate of higher strength. Mixing operations also require a homogeneous grain distribution, since separation of the individual components is less probable in this case.

Various animals require ground fodders of differing grain distributions; therefore various fractions are distinguished, under the following names:

(a) grit (fine, 90% of grain diameters less than 1.0 mm; medium fine, 90% of grain diameters between 1.0 and 2.0 mm; coarse, 90% of grain between 3 and 5 mm);
(b) middling (0.12–0.3 mm grains);
(c) flour (0.07–0.2 mm).

The processes taking place in a hammer mill are the result of complex loads. The material charged into a mill first impacts against the hammer, whereby it is broken into fairly large pieces and accelerated to a velocity similar to the peripheral value. The accelerated particles impact against the surface of a screen, whereby they are further comminuted. Particles rebounding from the screen impact anew against the hammer. A contribution to comminution is also made by rubbing in the clearance between the hammer and the screen, i.e., by friction and impact among the particles.

A general condition in comminution is that the impact load must exceed the dynamic breaking strength of the material. Both the load and the breaking strength depend on the mechanical properties of the material and in numerous cases also on the orientation of the particles during impact. Agricultural materials are all viscoelastic, and their degree of elasticity varies considerably depending on the stage of ripening and the moisture content. This explains why the comminution of agricultural materials cannot be calculated theoretically, and so it is necessary to rely mainly on experimental results. Nevertheless, analysis of experimental results permits fundamental regularities to be revealed in the comminution process. Systematic investigations in this field have been performed among others by Bölöni [184–188].

18.1 Mechanisms of comminution

The aim of comminution is to reduce the size and increase the specific surface area of particles. Material breaks when the local load transferred by impact exceeds the breaking strength of the material and the energy transmitted is sufficient to overcome cohesive forces at the new surfaces created. The maximum pressure on the contact surface for impact of a sphere at a velocity v on a plane surface is

$$p_{max} = 0.251[mv^2\pi^4/A^4R^3]^{1/5}$$

For the material to break $p_{max} \geqq \sigma_B$. Substitution of this relation yields the critical impact velocity for breaking as

$$v_{cr} = (1.56/\sqrt{\gamma})\sigma_B^{5/2}[(1-\nu^2)/E]^2(R/r_e)^{3/2} \qquad (262)$$

where R is the radius of curvature of the body at the point of impact, and r_e the equivalent radius of the body. In the case of a sphere, $R=r_e$. The above relationship shows the qualitative effects of the most important material and geometrical characteristics on the critical velocity. However, in examining the effect of the breaking stress the fact must also be taken into account that the modulus of elasticity E is not independent of σ_B. If for the sake of approximation it is assumed that σ_B is proportional to E, then the critical velocity increases with the square root of σ_B. In the case of irregular bodies the local curvature may also have an important effect, and so the critical velocity will also depend on the orientation of the body. The breaking stress of certain products (e.g., wheat and barley) also depends on the orientation: the material can support a higher load in the direction of the longitudinal axis than normal to it.

The energy required to reduce the size of a body with a given dimension x may be expressed in the general differential form:

$$dE/dx = -k/x^n \qquad (263)$$

where the constant k accounts for all the material characteristics. The exponent n

generally varies between 1 and 2. If $n=1$, then the equation due to Kick is obtained:

$$E = k_K \ln (x_0/x_1) \tag{264}$$

In the case of $n=1.5$, Bond's law is valid:

$$E = k_B(1/\sqrt{x_1}-1/\sqrt{x_0}) \tag{265}$$

while if $n=2$, the Rittinger's law prevails [182]:

$$E = k_R(1/x_1-1/x_0) \tag{266}$$

where x_0 is the initial grain size and x_1 that after comminution. Numerous experimental results [184–188] have shown that comminution of agricultural materials takes place according to the Rittinger equation. From eqn. (266) the energy required to comminute the unit volume (1 cm³) of a material may be determined. Assume that an amount of the work C (cm daN cm^{-2}) is required to cut a cube of volume 1 cm³. Smaller cubes are then obtained by three cuts. The required specific energy is

$$A = 3kC(1/x_1-1/x_0) \tag{267}$$

The surface area of the original cube is $6x_0^2$ and its weight is γx_0^3. The specific surface area (cm² kg^{-1}) with these values is:

$$f_0 = 6/\gamma x_0$$

Similarly, the specific surface area after comminution is

$$f_1 = 6/\gamma x_1$$

If x_0 and x_1 expressed from the latter equations are substituted into eqn. (267), the following relationship is obtained:

$$A = kC\gamma(f_1-f_0)/2 = kC\gamma \Delta f/2 \tag{268}$$

where Δf is the increase in specific surface area. This means that the energy required for to comminute 1 cm³ of material is proportional to the increment of specific surface area.

The volume of a material of weight G is $V=G/\gamma$ and the work required for its comminution is

$$A_1 = kC\Delta fG/2$$

while the energy required is:

$$N = A_1/\Delta t = (kCG)(\Delta f/\Delta t)/2 \tag{269}$$

where Δt is the time for which the material is retained in the mill. The quotient

$\Delta f/\Delta t$ expresses the mean rate of comminution, while the quotient $G/\Delta t$ gives the throughput Q per hour, and so eqn. (269) may also be written in the form

$$N = kC\Delta fQ/2 \tag{269a}$$

This equation may be developed further as follows:

$$N = (1/2)kCdF/dt = vdF/dt \tag{270}$$

where dF/dt is the increment of surface area per hour in the mill, and v the specific energy of comminution (kW h cm^{-2}, or cm daN cm^{-2}).

18.2 General relationships for hammer mills

In the stable operating state the quantity of material charged into the mill equals the quantity discharged. A certain quantity of material is always found in the grinding space, which is termed the "filling" (F). Thus if the quantities charged and discharged are plotted as functions of time, then after the steady state of operation has been obtained two parallel straight lines are found (Fig. 369). The vertical distance between the two straight lines gives the magnitude of the filling F, while the horizontal separation determines the grinding time t_g. The grinding capacity is defined by the simple relationship

$$Q = F/t_g \tag{271}$$

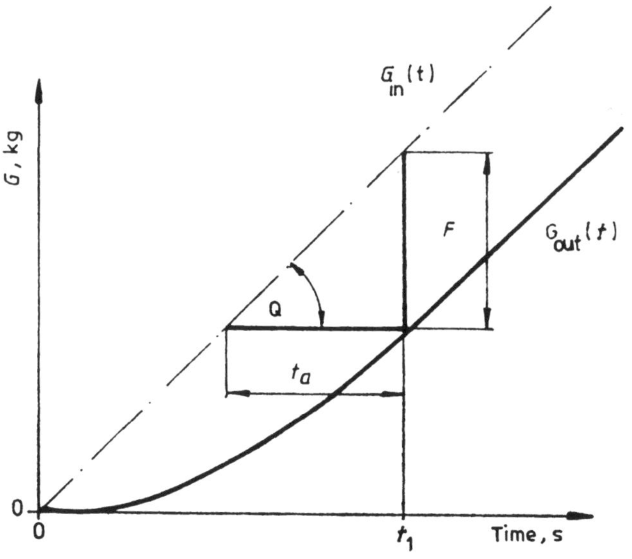

Fig. 369. Operation diagram for a hammer mill

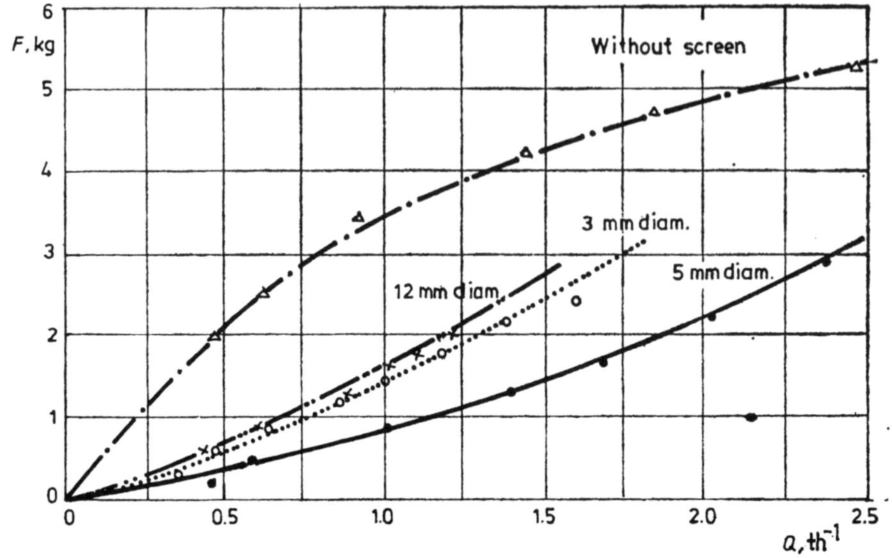

Fig. 370. Variation of hammer-mill filling for barley as a function of throughput

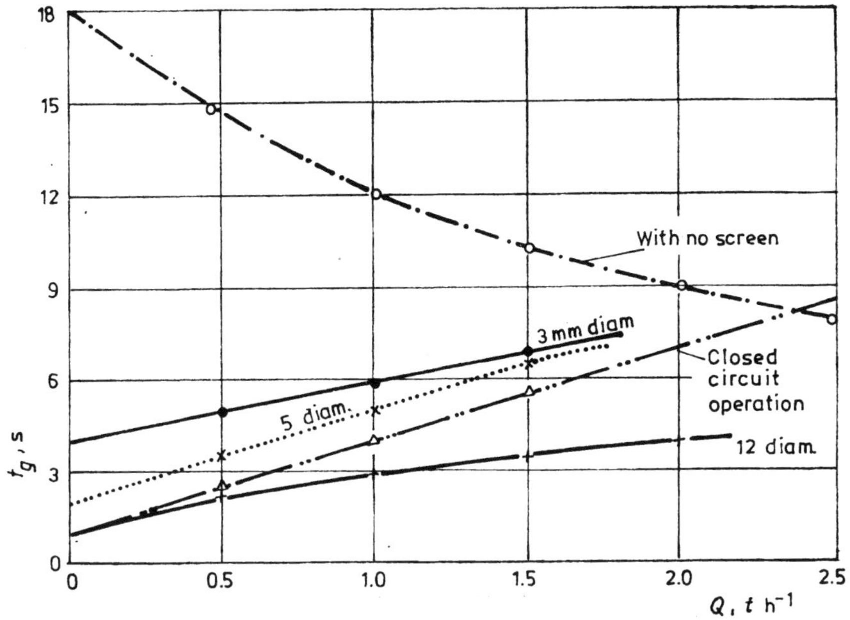

Fig. 371. Variation of hammer-mill grinding time for barley as a function of throughput

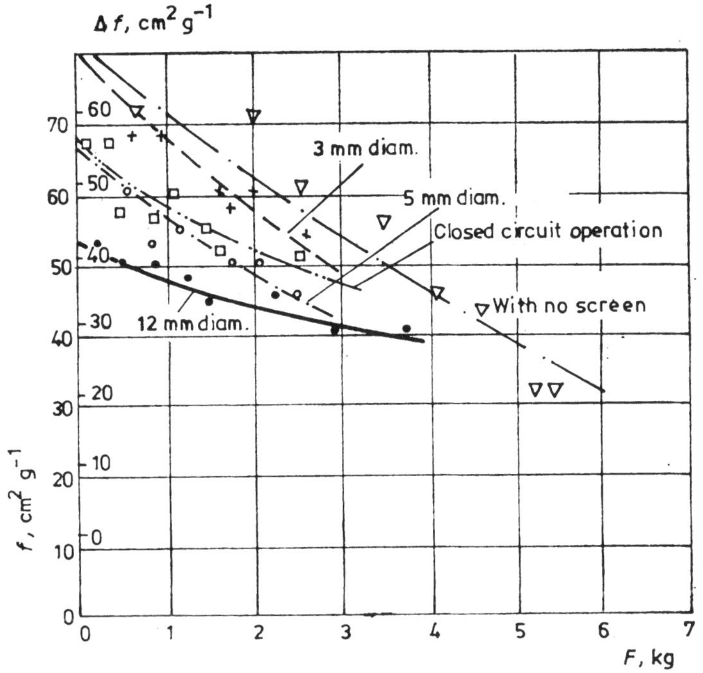

Fig. 372. Relationship between specific surface area of grit and magnitude of filling in a hammer mill

from which the filling may be determined as

$$F = Qt_g \tag{271a}$$

Figure 370 shows the measured filling for barley as a function of grit production capacity during operation of a D-24 hammer mill [188]. For a greater number of screen holes the filling in the mill is decreased while higher values are obtained for grinding without a screen. The tangent to the filling-grit-throughput curves gives the grinding time.

Figure 371 shows the grinding time as a function of grit throughput in the case of a D-24 hammer mill. It may be seen that with the application of screens the grinding time increases nearly linearly with throughput. In operating without a screen the inverse situation prevails and the grinding time is considerably longer. Thus, in the former case, with increasing grit throughput the grinding time, and still more the quantity of filling found in the mill, increases. The presence of a large quantity of material in the mill worsens the grinding conditions, and so the grit fineness deteriorates with increasing throughput.

Figure 372 shows the specific surface area of the grit and the specific increment of surface area as functions of the quantity of filling found in the mill [188]. It

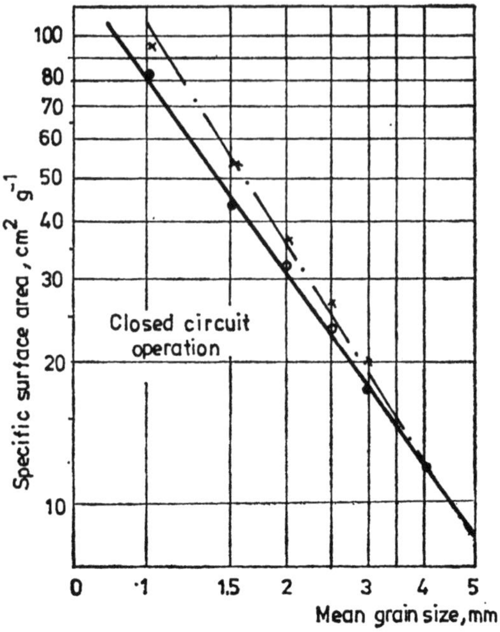

Fig. 373. Dependence of specific surface area on mean grain size

may be seen that the specific surface area decreases with F similarly in every operational state.

With the assumption of a given size distribution, there exists a defined relationship between the specific surface area of the grit and the mean grain diameter. This relationship is shown in Fig. 373 for a D-24 hammer mill both in normal open-circuit operation and in the closed circuit mode [188]. Deviation is observed between the two modes of operation, because closed circuit operation supplies a more homogeneous particle size distribution.

18.3 Size distribution of comminuted products

The grain size in a particle pile obtained during milling is never homogeneous, but shows a certain size distribution. Various functions may be applied to describe the distribution, of which those of importance from the point of view of comminution of agricultural materials are dealt with in the following.

(a) *Rosin–Rammler–Bennet distribution.* This distribution may be written, according to eqn. (9), in the form

$$D(x) = 1 - e^{-(x/x_0)^n}$$

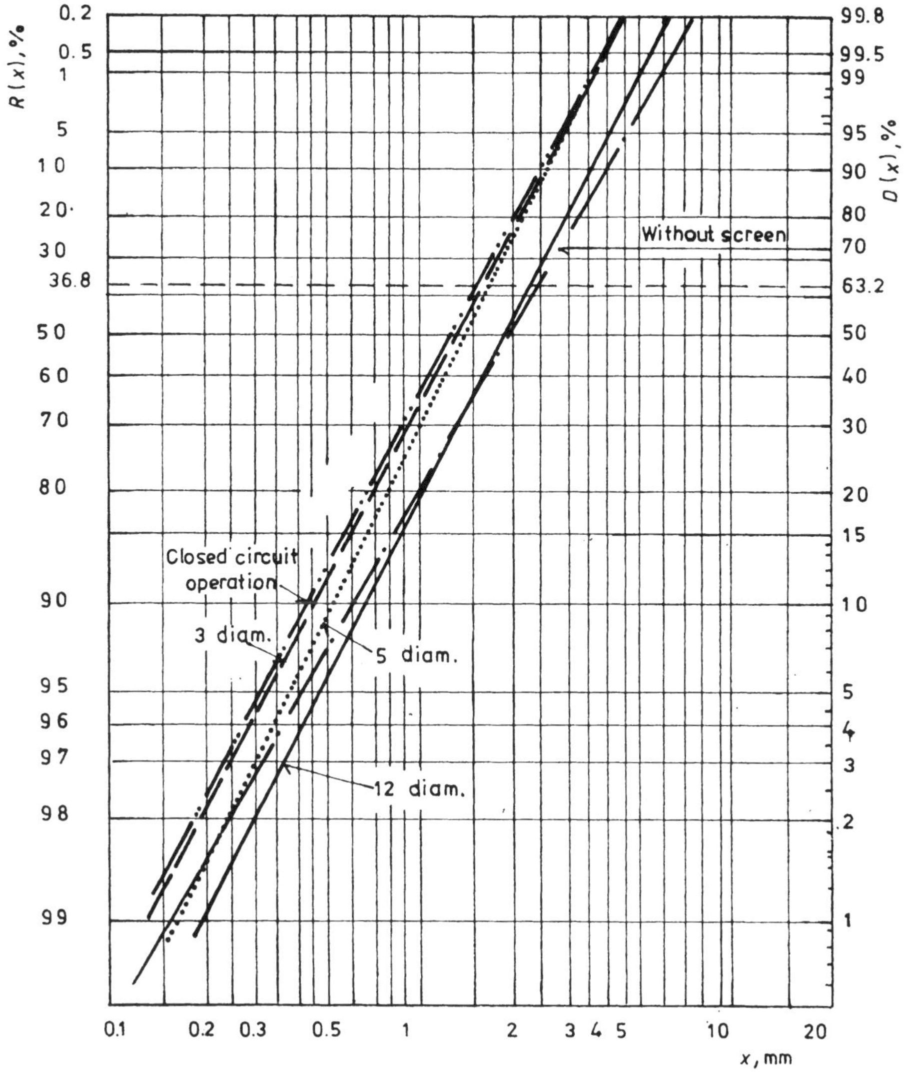

Fig. 374. Screen-pass and screen-residue curves for milling of barley grit (RRB scale)

where x is the grain size, $D(x)$ the characteristic grain size at which 36.8% screen residue is obtained ($1/e = 0.368 = 36.8\%$), and n an exponent characterizing the uniformity of the pile, whose value may vary between 0.5 and 1.3. The Rossin–Rammler–Bennet (RRB) distribution function is well suited for describing the comminution of agricultural products.

Figure 374 shows the distribution curves for the screen residue and the material passed by the screen for barley ground by a D-24 hammer mill with various screens plotted on a special RRB scale [188]. If the actual distribution corresponds to the RRB distribution described by eqn. (9), then straight lines are obtained on this scale. Thus it may rapidly be established from screen analysis results whether or not the actual grain size distribution corresponds to the theoretical distribution.

(b) *Log–normal distribution.* This is an asymmetrical distribution which gives a normal distribution if a logarithmic dimensional coordinate (x coordinate) is applied. This distribution is used frequently in describing comminuted piles, especially in its renormed form (see eqns. (6–8) in Section 2.1). Renorming of the log–normal distribution is necessary because the theoretical distribution is valid between the size limits zero and infinity, while in reality there exist no grains

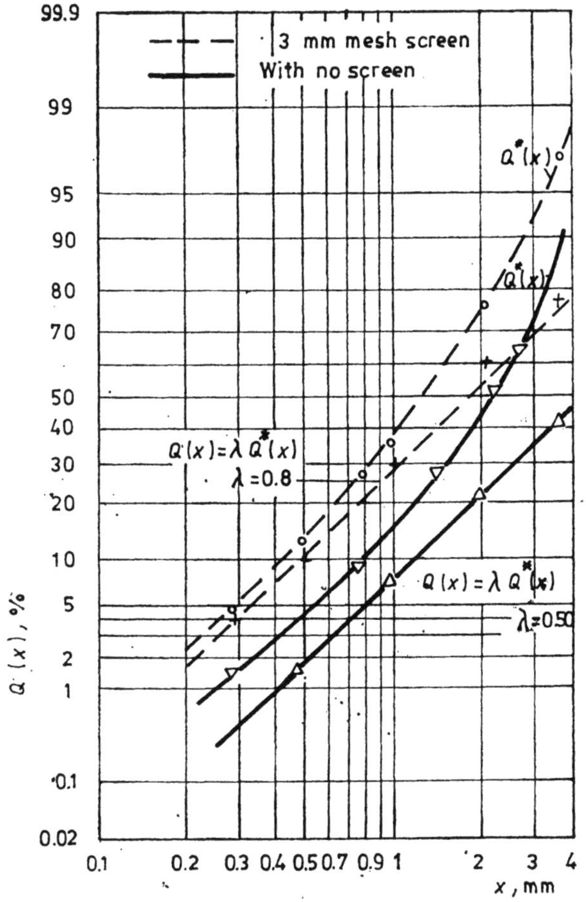

Fig. 375. Screen-pass curves for milling of barley grit and their renorming

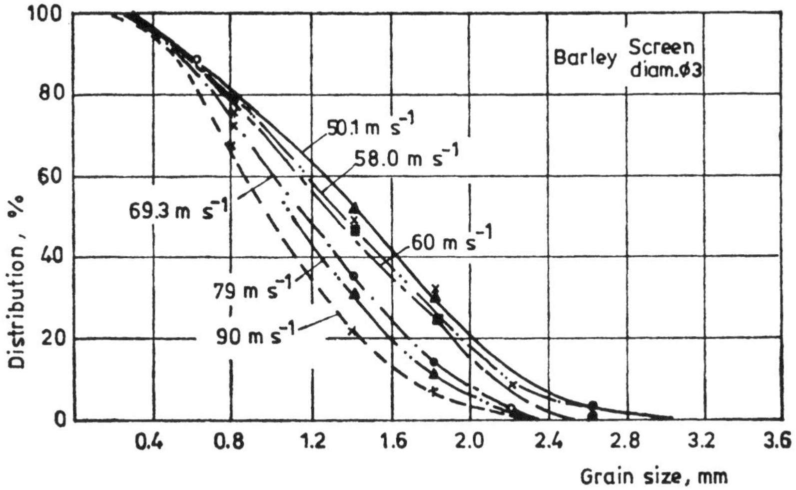

Fig. 376. Distribution curves for milled barley grit (screen hole diameter 3 mm)

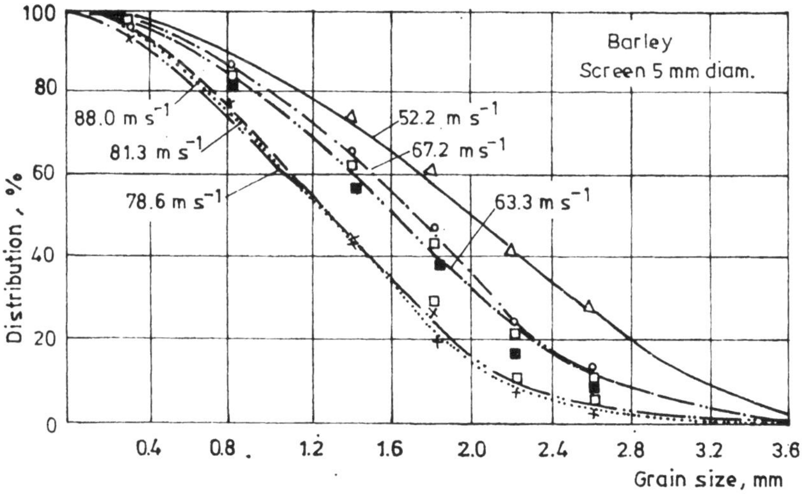

Fig. 377. Distribution curves for milled barley grit (screen hole diameter 5 mm)

larger than the screen perforation or the original grain size. The upper size limit of the theoretical function is changed by renorming from infinity to a maximum value x_{max}. Figure 375 shows distribution curves $Q^*(x)$ for the barley grit passed by a D-24 hammer mill with 3 mm screen mesh and with no screen and the theoretical log–normal curves for the same, straightened by correction factors $\lambda=0.8$ and $\lambda=0.5$ [188].

The fineness of grind is influenced primarily by impact velocity, by the screen hole size and by the mechanical properties of the product. Figures 376 and

377 show as examples the distribution curves for barley grits obtained using two screen types, for various peripheral velocities. On increasing the peripheral velocity the curves are shifted towards smaller diameters, and the milling becomes finer.

18.4 Energy requirements of hammer mills

The energy requirements of hammer mills comprise the idle-run energy and the energy required for grit suction, in addition to the energy required for comminution itself. The idle-run loss may be treated as a friction problem using the methods described in Section 14.9. The grinding energy depends primarily on the mechanical properties of the material, but is not independent of the design of the equipment. The way in which energy is transmitted to the particles and the formation of material streams causing frictional losses in the mill depend to a certain extent on the form of construction, and so the energy required for comminution also depends on the latter. Figure 378 illustrates the energy requirements of a D-24 hammer mill as a function of throughput [188]. The smaller the screen hole size, the higher the energy consumption. The power requirements do not increase linearly with throughput, from which it may be concluded that the grain fineness does not remain constant as a function of throughput. Figure 379 shows the increase of specific surface area as a function of throughput. The above relations are in accordance with the variation of the energy requirements: for a smaller-sized screen the increase of specific surface area is larger, while the value of Δf always decreases slightly with throughput.

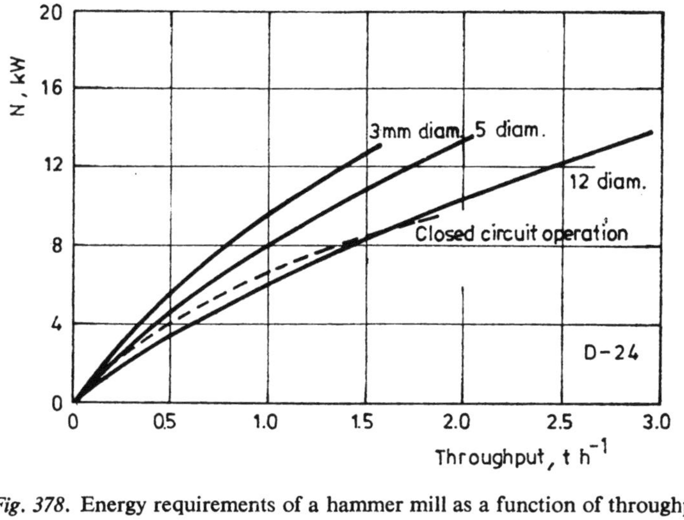

Fig. 378. Energy requirements of a hammer mill as a function of throughput

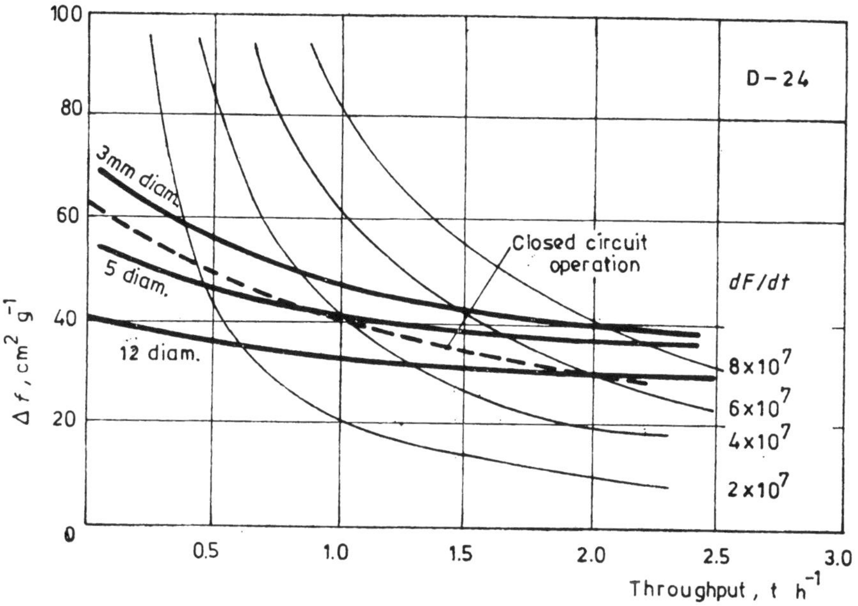

Fig. 379. Relationship between increment of specific surface area and throughput of a hammer mill

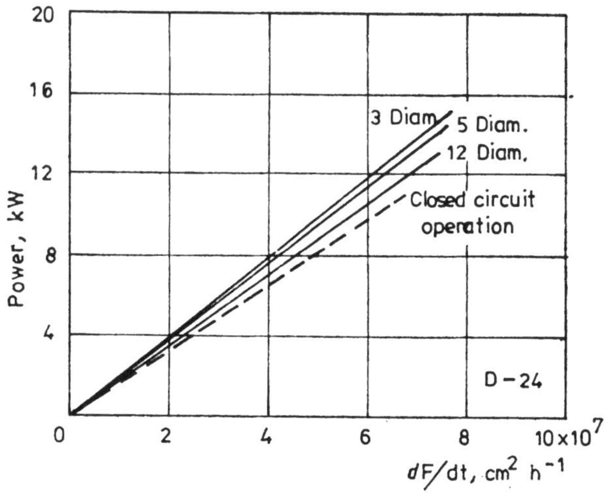

Fig. 380. Relationship between energy requirements and new surface area produced in unit time

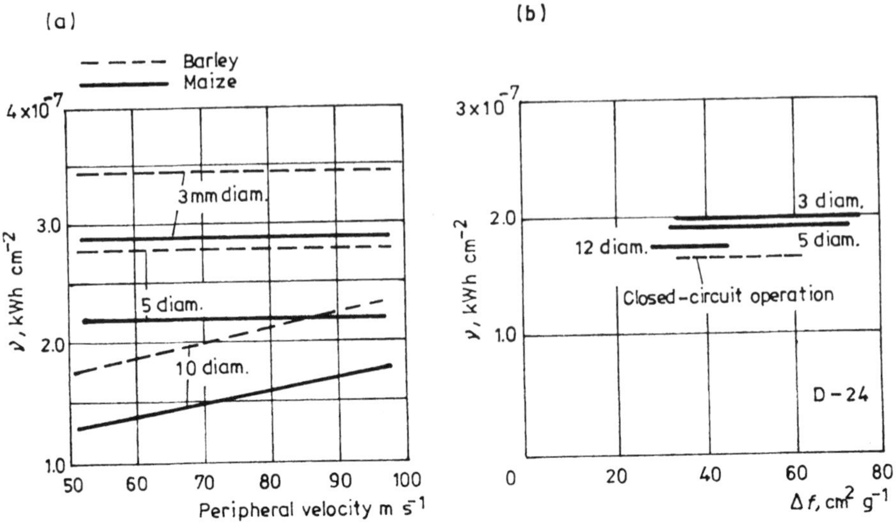

Fig. 381. Specific grinding energy as a function of peripheral velocity and increment of specific surface area

If the energy consumption is related with the help of the two preceding figures to the new surface area obtained in unit time, then straight lines are obtained (Fig. 380). The slopes of the straight lines give the specific energy requirements of comminution, according to eqn. (270).

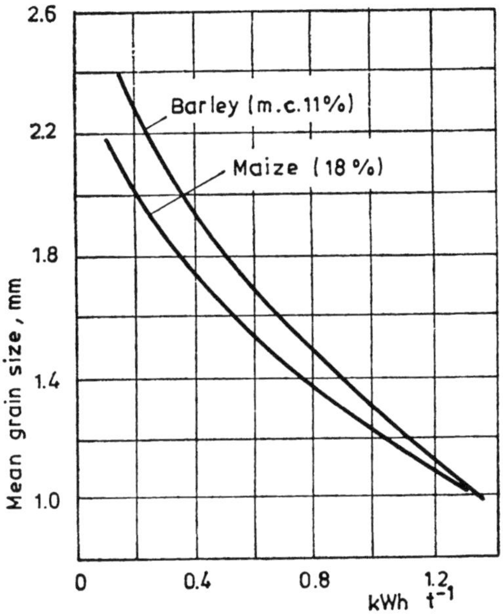

Fig. 382. Relationship between mean grain size and energy consumption in grinding

Figure 381(a) and (b) shows the specific grinding energy as a function of the peripheral velocity and of the specific increase in surface area (for various screens and throughputs). The value of v is generally constant, and variations as a function of peripheral velocity are observed only in the case of large screen perforations. However, the hole size of the screen influences the specific grinding energy, depending on the design of the mill. In practice a direct plot of the energy required to produce grit of a given fineness is also of interest. Such relationships are illustrated in Fig. 382 for barley and maize, comminuted using a hammer mill with 12 hammers.

The energy requirements of comminution are influenced greatly by the moisture content of the product being ground. Wetter products always behave in a more plastic manner and can tolerate a larger deformation without breaking. The friction coefficient also increases with moisture content, and so frictional losses also increase. Generally it may be accepted that the energy requirements of comminution increase linearly with moisture content in the range 10–20%, the extent of the increase is 40–60%.

18.5 Closed-circuit grinding

The essence of closed-circuit grinding is that coarse components are separated from the milling discharged from the mill and recharged. Large-mesh screens are generally used in order to reduce the specific grinding energy. By separating and recirculating the coarse fraction the homogeneity of the final milling is increased compared with that obtained in the open-circuit process.

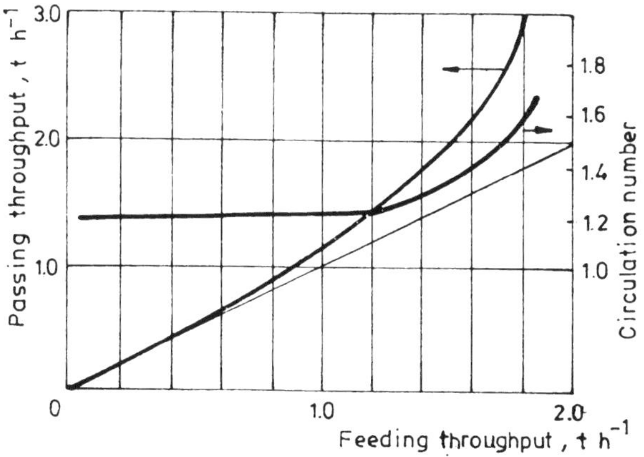

Fig. 383. Relationship between throughput and charged quantities in closed-circuit operation of a hammer mill

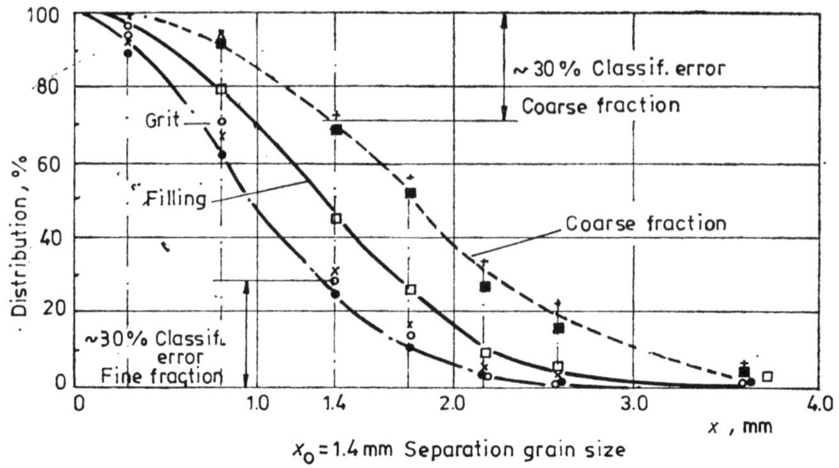

Fig. 384. Distribution curves for closed-circuit grinding

In closed-circuit operation the quantity of material put through the mill is larger than the quantity charged. The difference is the quantity separated by the classifier and returned into the mill. The ratio of the throughput and quantity charged gives the circulation coefficient. Figure 383 shows the characteristics of D-24 hammer mill modified for closed-circuit operation. The circulation coefficient is constant up to a certain throughput, and then increases relatively rapidly.

The filling discharged by the mill, the fine fraction separated and the coarse fraction recirculated show defined size distributions. Figure 384 presents the distribution curves for these functions. Those for the finer and coarse fractions show significant separation, but the classification error is also large. The error may be established simply by finding the grain size x_0 at which the intersections of the fine and coarse fractions at 0 and 100% screen residues are identical. In the present case $x_0 = 1.4$ mm, and the error of classification is 30% [188].

18.6 Grinding of forage materials

Dried forage materials, especially alfalfa, are also comminuted by hammer mills. The material may be charged in the original whole-stem state or in the form of chaff. The latter permits considerably more uniform charging, and so the throughput is greater and the grinding efficiency is improved.

According to practical observations, the throughput of alfalfa chaff 15 mm long is nearly twice as large as that of fibrous material. Figure 385 shows the

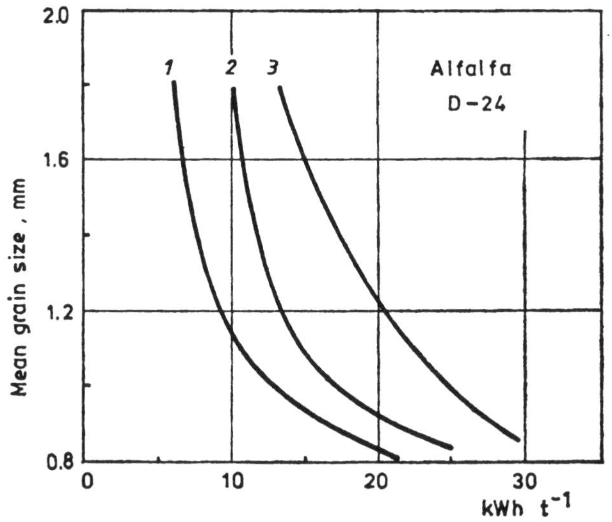

Fig. 385. Energy requirements of alfalfa grinding: *(1)* Alfalfa chaff; *(2)* alfalfa chaff with cutting energy taken into account; *(3)* alfalfa stems

specific energy requirements of alfalfa grinding, as a function of the mean grain size [188]. Curve *(1)* represents the energy requirements for grinding alfalfa chaff; curve *(2)* shows the same with the energy consumption in chaff cutting taken into account. Curve *(3)* illustrates the energy requirements for grinding fibrous alfalfa. Accordingly, it is worthwhile first to cut alfalfa to chaff, since in this way the grinding capacity may be increased and also some energy saved.

REFERENCES

1. Mohsenin, N. *Physical Properties of Plant and Animal Materials.* Pennsylvania State University, 1968.
2. Houston, R. New criterion of size for agricultural products. *Agric. Eng.*, **38** (1957) pp. 856–858.
3. Thompson, R. and Isaacs, G. Porosity determination of grains and seeds. *Trans. Am. Soc. Agric. Eng.*, **10** (1967) pp. 693–696.
4. Jelentés a mezőgazdasági anyagok fizikai-mechanikai tulajdonságainak a kutatásáról. Mezőgépfejlesztési Intézet (Report on research into the physico-mechanical properties of agricultural materials). *Inst. Dev. Agric. Machines,* Budapest (December 1977).
5. Kazarian, E. and Hall, C. Thermal properties of grain. *Trans. Am. Soc. Agric. Eng.*, **8** (1965) pp. 33–38.
6. Voprosy selskokhozyaistvennoi mekhaniki (Problems of agricultural mechanics) Izd. Urozhay, Minsk, 1964, Vol. XII.
7. Jenkins, H. Air flow planimeter for measuring the area of detached leaves. *Plant Physiology,* **34** (1959) pp. 532.
8. Frechette, R. and Zahradnik, J. Thermal properties of McIntosh apple. *Trans. Am. Soc. Agric. Eng.*, **11** (1968) pp. 21–24.
9. Nelson, S. Dielectric properties of grain and seed. *Trans. Am. Soc. Agric. Eng.*, **8** (1965) pp. 38–48.
10. Matthes, R. and Boyd, A. Electrical properties of seed associated with viability and vigor. *Trans. Am. Soc. Agric. Eng.*, **12** (1969) pp. 778–781.
11. Knipper, N. K voprosu primeneniya tokov vysokoi chastoti dlya sushki zerna (On the problem of applying high frequencies for drying seeds). *Tr. VIESH,* (Allunion Inst. El. Selsk. Choz.), **2** (1965) pp. 185–205.
12. Sembery, P. *Mezőgazdasági szemes és szálas anyagok dielektromos jellemzői* (Dielectric properties of granular and forage agricultural materials). Akadémiai Kiadó, Budapest, 1979.
13. Mishina, L. Kolichestvennaya otsenka sveta semyan pri ikh sortirovanii (Quantitative evaluation of the color of grains at sorting). *Mech. El. Soc. Selsk. Choz.*, **2** (1969) pp. 15–17.
14. Birth, G. and Norris, K. Non-destructive measurement of internal color of tomatoes. *Food Technol.*, **11** (1957) pp. 552–557.
15. Palmer, J. Electronic sorting of potatoes and clods by their reflectance. *J. Agric. Eng. Res.*, **6** (1961) pp. 104–111.
16. Norris, K. Measuring light transmittance properties of agricultural commodities. *Agric. Eng.*, **39** (1958) pp. 640–643.

17. HENDERSON, S. A basic concept of equilibrium moisture content. *Agric. Eng.*, **3** 3(1952) pp. 29–32.
18. KUNZE, O. and HALL, C. Moisture adsorption characteristics of brown rice. *Trans. Am. Soc. Agric. Eng.*, **10** (1967) pp. 448–450.
19. CHUNG, D. and PFOST, H. Adsorption and desorption of water vapor by cereal grains and their products. *Trans. Am. Soc. Agric. Eng.*, **10** (1967) pp. 549–557.
20. MURASE, H., MERVA, G., DE BAERDEMAEKER, J. and SEGERLIND, L. Importance of the water potential concept in studying physical properties of plant materials. *Proc. 2nd Int. Conf. on the Physical Properties of Agricultural Materials*, Gödöllő, 1980, Vol. 3.
21. YOUNG, J. and NELSON, G. Research into hysteresis between sorption and desorption isotherm of wheat. *Trans. Am. Soc. Agric. Eng.*, **10** (1967) pp. 756–761.
22. BAKKER-ARKEMA, F., BICKERT, W. and DEXTER, S. Environmental control during storage to prevent cracking of pea beans. *Trans. Am. Soc. Agric. Eng.*, **11** (1968) pp. 380–383.
23. GINZBURG, A., DUBROVSKI, V., KASAKOV, E., OKUNY, G. and RESTSHIKOV, V. Vlaga v zerne (Moisture content of seeds). Izd. Kolos, Moscow, 1969.
24. ZHADAN, V. *Teoreticheskie osnovy konditsionirovaniya vozdukha pri khranenii sochnogo rastitelnogo sirya* (Theoretical bases of air conditioning for the storage of wet plant materials). Izd. Pishchevaya Prom., Moscow, 1972.
25. WHITE, G., ROSS, I. and KLAIBER, J. Moisture equilibrium in mixing of shelled corn. *Trans Am. Soc. Agric. Eng.*, **15** (1972) pp. 508–509.
26. BAKKER-ARKEMA, F., LEREW, L. and ROTH, M. Heating and cooling of potato piles. *Am. Soc. Agric. Eng.*, Pap., 74–6513 (1974).
27. VILLA, L. and BAKKER-ARKEMA, F. Moisture losses from potatoes during storage. *Am. Soc. Agric. Eng.*, Pap. 74–6510 (1974).
28. GALLAHER, G. A method of determining the latent heat of agricultural crops. *Agric. Eng.* **32** (1951) pp. 34 and 38.
29. LIKOV, A. Teoriya sushki (The theory of drying). Mashgiz, Moscow, 1968.
30. ISAACS, G. and SCHEUERMANN, A. Die Berechnung von landwirtschaftlichen Trocknungsanlagen mit dickem Schüttgut. *Landtech. Forsch.*, **14** (1964) pp. 111–120.
31. MÜHLBAUER, W. Untersuchungen über die Trocknung von Körnermais. *Schrif. Arbeitskr. Forsch. Lehre der Max Eyth Ges.*, **1** (1974).
32. STEELE, J, SAUL, R. and HUKILL, W. Deterioration of shelled corn as measured by carbon dioxide production. *Trans. Am. Soc. Agric. Eng.*, **12** (1969) pp. 685–689.
33. WIENEKE, F. Verfahrenstechnik der Halmfutterproduktion. Own Ed. Göttingen, 1972.
34. NELLIST, M. Exposed layer drying of ryegrass seeds. *J. Agric. Eng. Res.*, **21** (1976) pp. 49—66.
35. HUANG, T. and GUNKEL, W. Theoretical and experimental studies of the heating front in a deep bed hygroscopic product. *Trans. Am. Soc. Agric. Eng.*, **17** (1974) pp. 346–354.
36. PFOST, H. Summarizing and reporting equilibrium moisture data for grain. *Am. Soc. Agric. Eng.*, Pap. 76–3520 (1976).
37. NGODDY, P. and BAKKER-ARKEMA, F. A generalized theory of sorption phenomena in biological materials. *Trans. Am. Soc. Agric. Eng.*, **13** (1970) pp. 612–617.
38. CHUNG, D. and CONVERSE, H. Effect of moisturè content on some physical properties of grains. *Am. Soc. Agric. Eng.*, Pap. 69–811 (1969).
39. YOUNG, J. Simultaneous heat and mass transfer in a porous hygroscopic solid. *Trans. Am. Soc. Agric. Eng.*, **12** (1969) pp. 720–725.
40. ROSENAU, J. and BAKKER-ARKEMA, F. Heat and mass transfer in onions. *Am. Soc. Agric. Eng.*, Pap. 70–881 (1970).

41. EKSTROM, G., LILJEDAHL, J. and PEART, R. Thermal expansion and tensile properties of corn kernels and their relationship to cracking during drying. *Trans. Am. Soc. Agric. Eng.*, **9** (1966) pp. 556–561.
42. CHUBIK, L. and MASLOW, A. *Spravochnik po teplofizicheskim kharakteristikam pishchevikh produktov* (Manual of the thermophysical characteristics of food products). Izd. Pishchevaya Prom., Moscow, 1970.
43. ZOERB, G. Instrumentation and measurement techniques for determining physical properties of farm products. *Trans. Am. Soc. Agric. Ang.*, **10** (1967) pp. 100–109.
44. FINDLEY, W., LAI, J. and ONARAN, K. *Creep and Relaxation of Nonlinear Viscoelastic Materials.* North-Holland, Amsterdam, 1976.
45. TIMOSHENKO, S. and GOODIER, J. *Theory of Elasticity.* Mc-Graw-Hill, New York, 1951.
46. CLARK, R., FOX, W. and WELCH, G. Representation of mechanical properties of nonlinear viscoelastic materials by constitutive equations. *Am. Soc. Agric. Eng.*, Pap. 68-811 (1968).
47. DOLGOV, I. and VASILEV, G. *Matematicheskie metody v zemledelcheskoi mekhanike* (Mathematical methods used in agricultural mechanics). Izd. Mashinostroenie, Moscow, 1967.
48. HAMAN, D. Analysis of stress during impact of viscoelastic bodies. *Trans. Am. Soc. Agric. Eng.*, **13** (1970) pp. 893–899.
49. LEE, E. and RADOK, S. The contact problem for viscoelastic bodies. *J. Appl. Mech.*, **27** (1960) pp. 438–444.
50. YANG, W. The contact problem for viscoelastic bodies. *J. Appl. Mech.*, **33** (June 1966) pp. 395—401.
51. MAKHAROBLIDZE, R. *Vliyanie form soprikasayushchichsiya poverkhnostei pri soyudarenii* (The effect of the shape of contact surfaces during impact). Zemled. Mech., **12** (1969) Izd. Mashinostroenie, Moscow.
52. MAKHAROBLIDZE, R. *Isledovanie dinamicheskoi zavisimosti naprazhenii ot deformatsii dlya korneklubneplodov* (Research into the dynamic dependence of stress on deformation for root bulb products). Trudi Cniimesh (Central Res. Inst. Mech. Agr.) **4** (1966) Izd. Urozhai, Minsk.
53. *Voprosy Selskokhozyaistvennoi mekhaniki* (Problems of Agricultural Mechanics) Izd. Urozhai, Minsk, 1967, Vol. XVII.
54. SHARMA, M. and MOHSENIN, N. Mechanics of deformation of fruit subjected to hydrostatic pressure. *J. Agric. Eng., Res.*, **15** (1970) pp. 65–74.
55. HAMANN, D. Dynamic mechanical properties of apple fruit flesh. *Trans. Am. Soc. Agric. Eng.*, **12** (1969) pp. 170–174.
56. FENYVESI, L. and LACK, A. The analysis of mechanical effects on apples. *Proc. 2nd Int. Conf. on the Physical Properties of Agricultural Materials.* Gödöllő, 1980, Vol. 1.
57. FRIDLEY, R. and ADRIAN, P. Mechanical properties of peaches, pears, apricots and apples. *Trans. Am. Soc. Agric. Eng.*, **9** (1966) pp. 135–138.
58. ZOERB, G. and HALL, C. Some mechanical and rheological properties of grains. *J. Agric. Eng. Res.*, **5** (1960) pp. 83–93.
59. PRINCE, R. and BRADWAY, D. Shear stress and modulus of elasticity of selected forages. *Am. Soc. Agric. Eng.*, Pap. 68-816 (1968).
60. FINNEY, E. and HALL, C. Elastic properties of potatoes. *Trans. Am. Soc. Agric. Eng.*, **10** (1967) pp. 4–8.
61. HAMMERLE, J., LUNSFORD, M. and MCCLURE, W. The determination of Poisson's ratio by compression tests of cylindrical specimens. *Am. Soc. Agric. Eng.*, Pap. 69-817 (1969).
62. FENYVESI, L. and LACK, A. Analysis of the mechanical properties influencing the leaf loss of alfalfa. *Rep. Nat. Inst. Agric. Eng.*, Gödöllő (Hungary), (1981).

63. Clevenger, J. and Hamann, D. The behavior of apple skin under tensile loading. *Trans. Am. Soc. Agric. Eng.*, **11** (1968) pp. 34–37.

64. Bright, R. and Kleis, R. Mass shear strength of haylage. *Trans. Am. Soc. Agric. Eng.*, **7** (1964) pp. 100–101.

65. Pickett, L., Liljedahl, J., Haugh, C. and Ullstrup, A. Rheological properties of corn stalks subjected to transverse loading. *Trans. Am. Soc. Agric. Eng.*, **12** (1969) pp. 392–396.

66. Reznik, N. *Silosouborochnye kombainy* (Combines for silo crops). Izd. Mashgiz, Moscow, 1964.

67. Sitkei, Gy. Mezőgazdasági anyagok összenyomhatósága (The compressibility of agricultural materials). *Járművek*, **17** (1970) pp. 414–417.

68. Baganz, K. Untersuchung über den Temperatureinfluss auf verschiedene Festigkeitswerte der Kartoffel, *Thaer Arch.* **12** (1968) pp. 219–226.

69. Nyborg, E. and Coulthard, T. Design parameters for mechanical raspberry harvesters. *Trans. Am. Soc. Agric. Eng.*, **12** (1969) pp. 573–576.

70. Chappel, T. and Hamann, D. Poisson's ratio and Young's modulus for apple flesh under compressive loading. *Trans. Am. Soc. Agric. Eng.*, **11** (1968) pp. 608–610.

71. Bajsz, I. *et al. A szárítás hatása a kukorica mechanikai jellemzőire* (The effect of drying on the mechanical characteristics of maize) Yearb. Agric. College Körmend (Hungary), (1975).

72. Bajsz, I. *et al. Alma és burgonya sérülése dinamikus erők hatására* (Damage to apples and potatoes under the effect of dynamic forces). Rep. Agric. College Körmend (Hungary), (1977).

73. Sitkei, Gy. Alma sérülése ismételt dinamikus erők hatására (Damage to apples under the effect of repeated dynamic forces). *Járművek*, **25** (1978) pp. 131–134.

74. Sitkei, Gy. Effect of drying temperature and cooling rate on the physical properties of shelled corn: a new dynamic measuring method. *Proc. 1st Int. Conf. of Drying*, Montreal, 1978. pp. 236–238.

75. Sitkei, Gy. Reduction of apple fruit damage using cushioning materials. *Acta Tech. Hung.*, **89** (1979) pp. 353–362.

76. Zienkiewich, O. *The Finite-Element Method in Structural and Continuum Mechanics*, McGraw-Hill, London, 1967.

77. Berg, H. *A Finite-Element Method for the Calculation of Transient and Stationary Temperature Fields*. Technical University of Trondheim, Rep. IF/R6 (1971).

78. Rumsey, T. and Friedley, R. Analysis of viscoelastic contact stresses in agricultural products using a finite element method. *Am. Soc. Agric. Eng.*, Pap. 74–3513 (1974).

79. Hermann, L. and Peterson, F. *A Numerical Procedure for Viscoelastic Stress Analysis*. Propulsion Div. Rep. Aerojet General Corporation, Sacramento, Ca.

80. Segerlind, L., Singh, R., DeBaerdemaeker, J. and Gustafson, R. Theoretical aspects of the finite element method. *Am. Soc. Agric. Eng.*, Pap. 74–5501 (1974).

81. Segerlind, L. *Applied Finite Element Analysis*. Wiley, New York, 1976.

82. Mahmoud, M. and Bishara, A. Using finite elements to analyze silo pressure. *Agric. Eng.*, **57** (June 1976) pp. 12–15.

83. Rumsey, T. and Friedley, R. A method for determining the shear relaxation function of agricultural materials. *Trans. Am. Soc. Agric. Eng.*, **20** (1977) pp. 386–389.

84. Misra, R. and Young, J. The finite element approach for solution of transient heat transfer in a sphere. *Trans. Am. Soc. Agric. Eng.*, **22** (1979) pp. 944–949.

85. MISRA, R. and YOUNG, H. Numerical solution of simultaneous moisture diffusion and shrinkage during soybean drying. *Trans. Am. Soc. Agric. Eng.*, **23** (1980) pp. 1277–1282.
86. MISRA, R., YOUNG, J. and HAMANN, D. Finite element procedures for estimating shrinkage stresses during soybean drying. *Trans. Am. Soc. Agric. Eng.*, **24** (1981) pp. 751–755.
87. KUNZE, O. and CHOUDHURY, M. Moisture adsorption related to the tensile strength of rice. *Cereal Chem.*, **49** (1972) pp. 684–696.
88. PRASAD, S., GUPTA, C. and OJHA, T. Mechanics of stress cracking in rice grains. *Proc. 2nd Int. Conf. on Physical Properties of Agricultural Materials*, Gödöllő, 1980. Vol. 4.
89. O'BRIEN, M., GENTRY, J. and GIBSON, R. Vibrating characteristics of fruits as related to in-transit injury. *Trans. Am. Soc. Agric. Eng.*, **8** (1965) pp. 241–243.
90. CHESSON, J. and O'BRIEN, M. Analysis of mechanical vibrations of fruit during transportation. *Am. Soc. Agric. Eng.*, Pap. 69–628 (1969).
91. HOAG, D. and FRIEDLEY, R. Experimental measurement of internal and external damping properties of tree limbs. *Am. Soc. Agric. Eng.*, Pap. 69–818 (1969).
92. FRIEDLEY, R. and O'BRIEN, M. A gyümölcsbetakarítás gépesített módszereinek továbbfejlesztése Kaliforniában (Development of mechanical fruit harvesting in California). *Congr. on Mechanization of Gardening*, Budapest, 1964.
93. O'BRIEN, M., PEARL, R., VILAS, E. and DREISBACH, R. The magnitude and effect of in-transit vibration damage of fruits and vegetables on processing quality and yield. *Trans. Am. Soc. Agric. Eng.*, **12** (1969) pp. 452–455.
94. NELSON, C. and MOHSENIN, N. Maximum allowable static and dynamic loads and effect of temperature for mechanical injury in apples. *J. Agric. Eng. Res.*, **13** (1968) pp. 305–317.
95. JINDAL, V. and MOHSENIN, N. Selection of cushioning materials based on mechanical properties of the cushion and the fruit. *Proc. Workshop at Pennsylvania State University*, 1975.
96. KECK, H. and GOSS, J. Determining aerodynamic drag and terminal velocities of seeds in free fall. *Trans. Am. Soc. Agric. Eng.*, **8** (1965) pp. 553–554.
97. MATTHIES, H. Der Strömungswiderstand beim Belüften landwirtschaftlicher Erntegüter. *VDI Forschungsh.*, 454 (1956).
98. SCHEUERMANN, A. Der Strömungswiderstand bei der Belüftungstrocknung von blattreichem Heu. *Grundlagen Landtech.*, **16** (1966) pp. 140–146.
99. DAY, C. Effect of conditioning and other factors on resistance of hay to air flow. *Trans. Am. Soc. Agric. Eng.*, **6** (1963) pp. 199–201.
100. HASSEBRAUCK, B. Das Trennen von Korn-Häcksel-Gemischen in Sichtern. *Landtech. Forsch.*, **14** (1964) pp. 16–20.
101. HASSEBRAUCK, B. Untersuchungen über die Eignung von Zyklonen zum Trennen von Korn-Strohhäcksel-Gemischen. *Landtech. Forsch.*, **12** (1962) pp. 108–112.
102. GILFILLAND, G. and CROWTHER, A. The behavior of potatoes, stones and clods in a vertical air stream. *J. Agric. Eng. Res.*, **4** (1959) pp. 9–15.
103. LEWALLEN, M. and BROWN, R. Oxygen-permeability of concrete silo wall sections. *Trans. Am. Soc. Agric. Eng.*, **10** (1967) pp. 114–115.
104. LAMPMAN, W. and HUKILL, W. Air resistance of perforated metal supporting grain. *Trans. Am. Soc. Agric. Eng.*, **11** (1968) pp. 837–839.
105. MARCHANT, J. Prediction of fan pressure requirements in the drying of large hay bales. *J. Agric. Eng. Res.*, **21** (1976) pp. 333–346.
106. MARCHANT, J. The prediction of air flows in crop drying systems by the finite element method. *J. Agric. Eng. Res.*, **21** (1976) pp. 417–429.

107. SEGLER, G. *Pneumatic Grain Conveying*. Braunschweig, 1951.
108. WELSCHOF, G. Pneumatische Förderung bei grossen Fördergutkonzentrationen. *VDI Forschungsh.*, **492** (1962).
109. PÁPAI, L. Pneumatikus gabonaszállítás (Pneumatic transport of cereals). *Magy. Tud. Akad. VI. Oszt. Közl.*, **12** (1954) pp. 319–363.
110. KOVÁCS, L. Berechnung des Druckabfalls in 90° horizontal eingebauten Krümmern pneumatischer Getreideförderleitungen. *Period. Politech.* **8** (1964) pp. 447–467.
111. SEGLER, G. and KUENEKE, K. Untersuchungen an Fliessbettförderrinnen. *Landtech. Forsch.*, **15** (1965) pp. 130–136.
112. PÁPAI, L. and SEBESTYÉN, G. Fluidizációs cementszállítás (Fluidized transport of cement). *Gép*, **17** (1965) pp. 475–482.
113. PÁPAI, L. Geschwindigkeits- und Druckverhältnisse bei waagerechter pneumatischer Förderung. *Period. Politech.*, **10** (1966) pp. 397–415.
114. PÁPAI, L. Geschwindigkeits- und Druckverhältnisse bei lotrechter pneumatischer Förderung. *Acta Tech. Hung.*, **69** (1970) pp. 83–104.
115. SIEGEL, W. Experimentelle Untersuchungen zur pneumatischen Förderung körniger Stoffe in waagerechten Rohren. *VDI Forschungsh.*, **538** (1970).
116. FLATOW, J. and SIEGEL, W. Pneumatische Förderung von Körnermais in waagerechten Rohren. *Grundlagen Landtech.*, **19** (1969) pp. 125–128.
117. HARRIS, W. FELTON, K. and BURKHARDT, G. Design data for pneumatic conveying of chopped forage. *Trans. Am. Soc. Agric. Eng.*, **8** (1965) pp. 194–195.
118. NOVIKOV, G. and GOLUSHKO, A. Raschet pnevmaticheskikh ustanovok dlya transportirovaniya svinogo navoza (Calculation of pneumatic swine manure conveying equipment). *Mech. Selsk. Khoz.*, **5** (1970) pp. 29–31.
119. SHOVE, G. and HUKILL, W. Predicting pressure gradients in perforated grain ventilation ducts. *Trans. Am. Soc. Agric. Eng.*, **6** (1963) pp. 115–116.
120. SHOVE, G. Analog computer solution of air distribution in perforated ducts. *Trans. Am. Soc. Agric. Eng.*, **9** (1966) pp. 599–601.
121. LEPEHIN, N. Soprotivlenie kroni plodovykh dereviev vozdushnoi struye (Resistance of fruit-tree crowns against air flow). *Trakt. Selkhozmashiny*, **1** (1969) pp. 34–36.
122. BROOKER, D. Computing air pressure and velocity distribution when air flow through a porous medium and nonlinear velocity-pressure relationships exist. *Trans. Am. Soc. Agric. Eng.*, **12** (1969) pp. 118–121.
123. ALJEKHIN, E. Isledovanie gidrotransporta na otkormochnykh fermakh krupnogo rogatogo skota (Investigation of hydrotransport in cattle fattening farms). *Mech. El. Selsk. Khoz.*, **5** (1969) pp. 31—33.
124. MKRTUMIAN, V. and PEREVEDENTSEV, V. Reologicheskie isledovaniya poluzhidkikh kormovykh smesei (Rheological study of half-liquid fodder mixtures). *Mech. El. Selsk. Khoz.*, **1** (1971) pp. 39–42.
125. MARCHENKO, N. and LIKHMAN, G. Dvizhenie zhidkogo navoza po trubam. (The movement of liquid manure in tubes). *Mech. El. Selsk. Khoz.*, **3** (1972) pp. 23–25.
126. HERUM, F., ISAACS, G. and PEART, R. Flow properties of highly viscous organic pastes and slurries. *Trans. Am. Soc. Agric. Eng.*, **9** (1966) pp. 45–47.
127. ZUEV, V. and TEKUCHEVA, M. Samotekhnie peremeshchenie ekskrementov v kanale (The gravitational movement of excrement in a channel). *Mech. El. Selsk. Khoz.*, **7** (1971) pp. 26–29.
128. KAMP, G. Theoretische und experimentelle Untersuchungen an Wurfgebläsen. *VDI Forschungsh.*, **466** (1958).

129. KROMER, K. H. Untersuchungen an Trommel-feld-häckslern. *KTL Ber. Landtech.*, **114** (1967).
130. WIENECKE, F. Wickel- und Reibungsuntersuchungen an Wellen. *VDI Forschungsh.*, **463** (1957).
131. *Am. Soc. Agric. Eng. Yearb.*, St. Joseph, 1970.
132. NOVIKOV, G. Issledovanie protsessa rezaniya korneplodov (The study of cutting root bulb products). *Trudy VIM Allunion Res. Inst. Agr.*, **16** (1952).
133. BURMISTROVA, M., KOMOLKOVA, T., KLEMM, N., PANINA, M., POLUNOTSHEV, I. *Fiziko-mekhanicheskie svoistva selkokhozyaistvennikh rastenii* (Physico-mechanical properties of agricultural plants). Selkhozgiz, Moscow, 1956.
134. SNYDER, L., ROLLER, W. and HALL, G. Coefficients of kinetic friction of wheat. *Trans. Am. Soc. Agric. Eng.*, **10** (1967) pp. 411–413.
135. STEWART, B. Effect of moisture content and specific weight on the internal friction properties of sorghum grain. *Am. Soc. Agric. Eng.*, Pap. 64–804 (1964).
136. JENIKE, A., ELSEY, P. and WOOLEY, R. Flow properties of bulk solids. *Eng. Exp. Station*, University of Utah, Bull. 95 (1959).
137. JENIKE, A. Storage and flow of solids. *Eng. Exp. Station.* University of Utah, Bull. 123 (1964).
138. HOFFMAN, O. and HESSE, T. Funktionsgerechte Gestaltung von Silos für schwerfliessenden Güter nach dem Jenike-Verfahren. *Grundlagen Landtech.*, **25** (1975) pp. 65–70.
139. ZENKOV, R. *Mechanika nasipnikh gruzov* (Mechanics of bulk loads). Mashgiz., Moscow, 1964.
140. MESTER, L. Kohézió nélküli szemcsés anyagok fizikai-mechanikai alaptörvényei (Basic physico-mechanical laws for cohesionless granular materials). *Járművek*, **24** (1977) pp. 109–114.
141. STROPPEL, A. Spannungszustände in lagernden Haufwerken in der Nähe einer ebenen Wand. *VDI Forschungsh.*, **525** (1968).
142. MESTER, L. Feszültségek a kohéziós szemcsés anyagokban (Stresses in cohesive granular materials). *Járművek*, **25** (1978) pp. 56–60.
143. KVAPIL, R. *Theorie der Schüttgutbewegung.* Springer, Berlin, 1959.
144. HERNE, H. *Theory of Solids Handling.* Nat. Coal Board, London, 1964.
145. WELSCHOF, G. Beitrag zur Messung der Ausflussmengen körniger Güter mit Blenden und Düsen. *Landtech. Forsch.*, **10** (1960) pp. 138–141.
146. JANNSEN, H. Versuche über Getreidedruck in Silozellen. *VDI Z.* **39** (1895) pp. 1045–1049.
147. GUTJAR, E. Raspredelenie davlenii v stenke silozovykh bunkerov (Pressure distribution over silo compartment walls) *Trudy MIMESH*, Moscow Inst. Mech. Agr., 2 (1935).
148. PIEPER, K. and WENZEL, F. *Druckverhältnisse in Silozellen.* Wilhelm Ernst & Sohn, Berlin, 1964.
149. FÜRRL, C. Berechnungsverfahren zur Bestimmung der Lagerungsdichte von Mischfutter. *Dtsch. Agrartech.*, **22** (1972) pp. 188–190.
150. NEUBAUER, L. Simplified equation for silage pressures with moisture variation. *Trans. Am. Soc. Agric. Eng.*, **9** (1966) pp. 295–296.
151. ROBERTS, A. and ARNOLD, P. Discharge chute design for free flowing granular materials. *Am. Soc. Agric. Eng.*, Pap. 69–839 (1969).
152. BÖLÖNI, I. Der Leistungsbedarf von scheiben- und trommelartigen rotierenden Teilen an Landmaschinen. *Landtech. Forsch.*, **14** (1964) pp. 140–145.
153. DOBLER, K. and FLATOW, J. Berechnung der Wurfvorgänge beim Schleuderdüngerstreuer. *Grundlagen Landtech.*, **18** (1968).

154. Klapp, E. Theorie der Verteilung von Feststoffteilchen mittels Schleuderscheiben. *Forsch. Ingenieurwes.*, **31** (1965) pp. 83–86.

155. Mewes, R. Verdichtungsgesetzmässigkeiten nach Presstopfversuchen. *Landtech. Forsch.*, **9** (1959) pp. 68–75.

156. Matthies, H., Entwicklung und Forschung auf dem Gebiete des Verdichtens von Halmgut. *Landtech. Forsch.*, **13** (1963) pp. 157–162.

157. Busse, W. Die Theorie auf dem Gebiete des Verdichtens landwirtschaftlicher Halmgüter. *Landtech. Forsch.*, **14** (1964) pp. 6–15.

158. Sacht, H. Über den Verdichtungsvorgang bei landwirtschaftlichen Halmgütern. *Grundlagen Landtech.*, **17** (1967) pp. 47–52.

159. Busse, W. Das Verdichten von Halmgütern mit hohen Normaldrücken. *Fortschr. Ber. VDI Z.*, **1** (1966).

160. Sacht, H. Das Verdichten von Halmgütern in Strangpressen. *Fortschr. Ber. VDI Z.*, **4** (1966).

161. Voss, H. *Ermittlung von Stoffgesetzen für Halmgut.* Dissertation, Technical University Braunschweig, 1970.

162. Terskov, G. *Raschet zernouborochnykh mashin* (The calculation of cereal harvesting machines). Mashgiz, Moscow, 1961.

163. Fedorov, M. Issledovanie protsessa skhatia solomi (Study of the compression process for straw). *Trakt. Sel'khozmashiny*, **5** (1972) pp. 21–24.

164. Dolgov, I. Zakonomernosti skhatia seno-solomistyk materialov (Laws of the compression of hay–straw type materials). *Mech. El. Selsk. Khoz.*, **5** (1970) pp. 8–11.

165. Vinogradov, V. and Dmitriev, G. Modelirovanie protsessa pressovaniya solomistikh materialov (Modelling the compression process for straw materials). *Zemled. Mech.*, **12** (1969).

166. Kutzbach, H. Die Grundlagen der Halmgutverdichtung. *Fortschr. Ber. VDI Z.*, **16** (1973).

167. Dolgov, I. and Vasilev, G. *Matematicheskie metody v zemledelcheskoi mekhanike.* (Mathematical methods applied in agricultural mechanics). Izd. Mashinostroenie, Moscow, 1967.

168. Schwanghart, H. Messung und Berechnung von Druckverhältnissen und Durchsatz in einer Ringkoller-Strangpresse. *Aufbereit. Tech.*, **10** (1969) pp. 713–722.

169. Schwanghart, H. Festigkeiten von Futtermittelpresslingen. *Aufbereit. Tech.*, **11** (1970) pp. 192–199.

170. Friedrich, W. Das Pelletieren von Mischfutter. *Aufbereit. Tech.*, **19** (1978) pp. 401–406.

171. Hall, G. and Hall, C. Heated die wafer formation of alfalfa and bermudagrass. *Trans. Am. Soc. Agric. Eng.*, **11** (1968) pp. 578–581.

172. Orth, H. and Löwe, R. Influence of temperature on wafering in a continuous extrusion process. *J. Agric. Eng. Res.*, **22** (1977) pp. 283–289.

173. Mohsenin, N. and Zaske, J. Effect of stress relaxation of wafer density, durability and energy requirement on compaction of unconsolidated materials. *Proc. Workshop at Pennsylvania State University*, 1975.

174. Koegel, R. and Bruhn, H. Pressure fractionation characteristics of alfalfa. *Trans. Am. Soc. Agric. Eng.*, **15** (1972) pp. 856–860.

175. Fomin, V. K voprosy mekhanicheskogo obezvozhivaniya travianistykh rastenii (On the problems of mechanical dewatering of forage plants). *Trakt. Sel'khozmasiny.*, **6** (1972) pp. 27–28.

176. Straub, R. and Bruhn, H. Mechanical dewatering of alfalfa protein concentrate. *Trans. Am. Soc. Agric. Eng.*, **21** (1978) pp. 414–418.

177. ANTAL, G. Horizontális mechanikus szőlősajtók műszaki vizsgálata (Technical investigation of horizontal mechanical wine-presses). *Nat. Inst. Agric. Eng.*, Gödöllő (Hungary), Rep., 7, 1971.

178. CHANCELLOR, W. Energy requirements for cutting forage. *Agric. Eng.*, **39** (October 1958) pp. 633–636.

179. BAADER, W. Der Einfluss der Messerbewegung auf das Schnittmoment bei einem Scheibenradschneidwerk mit geraden Messern. *Grundlagen Landtech.*, **16** (1966) pp. 101–105.

180. DOBLER, K. Der freie Schnitt beim Mähen von Halmgut. *Hohenheimer Arbeit.*, **62** (1972).

181. SITKEI, GY. and DUGOVICH, P. Kombinierte Häcksel-Quetsch-Methode in der Grünfutterernte. *VDI Session*, München, 1976.

182. RITTINGER, P. *Lehrbuch der Aufbereitungskunde*. Springer, Berlin, 1867.

183. KÖRMENDY, I. New apparatus to study the pressing process: experimental and evaluation methods. *Acta Aliment. Acad. Sci. Hung.*, **1** (1972) pp. 315–340.

184. BÖLÖNI, I. The required power input of hammer mills. *Acta Tech. Acad. Sci. Hung.*, **45** (1964) pp. 327–344.

185. BÖLÖNI, I. Some regularities of grain distribution and fineness variation as observed in hammer mill products. *Acta Tech. Acad. Sci. Hung.*, **45** (1964) pp. 45–64.

186. BÖLÖNI, I. Some functional regularities of the comminution process in hammer mills. *Acta Tech. Acad. Sci. Hung.*, **41** (1962) pp. 381–398.

187. HENDERSON, S. and BÖLÖNI, I. Closed-circuit grinding of agricultural products. *J. Agric. Eng. Res.*, **11** (1966) pp. 248–254.

188. BÖLÖNI, I. A kalapácsos darálók aprítási mechanizmusa és gépüzemtani összefüggései. (Comminution mechanism of the hammer mill and its mechanical operational relationships). *Doctoral Dissertation*, 1973.

189. SITKEI, GY. General relationships in the breakup of selected agricultural materials for cyclic dynamic loading. *Proc. 2nd Int. Conf. on the Physical Properties of Agricultural Materials*, Gödöllő, 1980, Vol. 1.

190. SITKEI, GY. and BAJSZ, I. Similarity relationships of apple damage for cyclic dynamic loading. *Proc. 2nd Int. Conf. on the Physical Properties of Agricultural Materials*, Gödöllő, 1980. Vol. 1.

191. GUSTAFSON, R., THOMPSON, D. and SOKHANSANJ, S. Temperature and stress analysis for corn kernels: finite-element analysis. *Trans. Am. Soc. Agric. Eng.*, **22** (1979) pp. 955–960.

192. KÖRMENDY, I. Experiments for the determination of the specific resistance of comminuted and pressed apple against its own juice. *Acta Aliment. Acad. Sci. Hung.*, **8** (4) (1979) pp. 321–342.

193. KUTZBACH, H. and SCHERER, R. Thermal conductivity and diffusivity of shelled corn. *Proc. 2nd Int. Conf. on the Physical Properties of Agricultural Materials*, Gödöllő, 1980, Vol. 3.

194. MOSER, E. and SINN, H. Physical, biological and economical research into different transport and storage systems for soft deciduous-tree fruits. *Proc. 2nd Int. Conf. on the Physical Properties of Agricultural Materials*, Gödöllő, 1980. Vol. 3.

SUBJECT INDEX

Printed in Dunstable, United Kingdom

91336870R10275